Agriculture Issues and Policies

nova
science publishers
New York

Agriculture Issues and Policies

Underutilized Crops and Their Value Addition
Romesh Kumar Salgotra, PhD (Author)
Monika Sood, PhD (Author)
Surbhi Jasrotia (Author)
2021. ISBN: 978-1-68507-443-2 (Hardcover)
2021. ISBN: 978-1-68507-481-4 (eBook)

Solanum melongena: Production, Cultivation and Nutrition
Abdul Majid Ansari, PhD (Editor)
Wajid Hasan, PhD (Editor)
M. Prakash, PhD (Editor)
2021. ISBN: 978-1-68507-311-4 (Hardcover)
2021. ISBN: 978-1-68507-323-7 (eBook)

Glyphosate: Agricultural Uses, Ecological Impacts and Potential Carcinogenic Effects
Kassio Ferreira Mendes, PhD (Editor)
2021. ISBN: 978-1-68507-076-2 (Hardcover)
2021. ISBN: 978-1-68507-110-3 (eBook)

The Potato Crop: Management, Production, and Food Security
Pedro Manuel Villa, PhD (Editor)
2021. ISBN: 978-1-68507-096-0 (Hardcover)
2021. ISBN: 978-1-68507-203-2 (eBook)

Oilseeds: Production, Varieties and Management Practices
Lindsey Harvey (Editor)
2019. ISBN: 978-1-53615-635-5 (Softcover)
2019. ISBN: 978-1-53615-636-2 (eBook)

More information about this series can be found at
https://novapublishers.com/product-category/series/agriculture-issues-and-policies/

Hanuman Singh Jatav and Vishnu D. Rajput
Editors

Ecosystem Services

Types, Management and Benefits

nova
science publishers
New York

https://doi.org/10.52305/PFZA6988

NOTICE TO THE READER

Library of Congress Cataloging-in-Publication Data

ISBN: 978-1-68507-614-6

Published by Nova Science Publishers, Inc. † New York

Contents

Preface

The present book, *Ecosystem Services: Types, Management and Benefits,* is written mainly to give systematic information about the ecosystem and its management. Soil ecosystem structure is constituted by dynamic interactive abiotic and biotic compartments. As per FAOUN, ecosystem services are the multitude of benefits that nature provides to society. Biodiversity is the diversity among living organisms, which is essential to ecosystem function and services delivery. The book consists of 21 chapters which mainly give a brief account of comprehensive knowledge to manage the soil ecosystem.

We humbly welcome critical suggestions of the present book readers. The book has covered all relevant aspects and the editors believe that the present book will be helpful to researchers, the scientific community, academicians, business farmers and policymakers.

Hanuman S. Jatav
Vishnu D. Rajput

Acknowledgment

The tenure of humans in this world is supported by many others. It gives me pleasure to remember and express our feelings to those who helped me in the form of motivation to compile this outcome. I, H.S. Jatav, as an editor of this book, "Ecosystem Services: Types, Management and Benefits," thank Professor Jeet Singh Sandhu (FNAAS, HFISPGR, FISPRD) Hon'ble Vice-Chancellor, Sri Karan Narendra Agriculture University, Jobner-Jaipur, for his motivation to compile this work for the scientific community. I am deeply indebted to the renowned Soil Scientist Prof. Saroj Kumar Sanyal (FISSS, FNAAS, FAScT), Former Vice-Chancellor, Bidhan Chandra Krishi Viswavidyalaya, Nadia-West Bengal for his guidance and encouragement to compile this work. I would also like to extend my deepest gratitude to Dr. N.S. Rathore, Hon'ble Vice-Chancellor, Maharana Pratap University of Agriculture and Technology, Udaipur, Rajasthan, India for his motivation and encouragement to compile this work. I am extremely grateful to Professor Satish Kumar Singh (FISSS, FISAC), Banaras Hindu University, Varanasi for his lifelong motivation to compile the work for the students.

I also extend my thanks to Prof. S.R. Dhaka (Dean-COA Fatehpur- Sikar), Dr. Sanjay K. Attar, Dr. Muddser A. Khan, Dr. Kailash Chandra, Dr. Mujhaid Khan, Dr. Champa L. Khatik, Dr. Ramu Meena and Dr. Ajeet Singh Dr. Vishnu D. Rajput for the direct and indirect support to compile this outcome. I am thankful to the Dr. Amitava Rakshit (BHU) FSES, FTWAS.Nxt (Italy), Prof. A.K. Ghosh, Prof. Nirmal De, Prof. Surendra Singh (FISSS), Prof. Priyankar Raha, Prof. P.K. Sharma, Prof. Y.V.Singh and Dr. Ashish Latare for their support and guidance during my studies. I don't have words for the encouragement led by beloved Nikki (OTA Sahab) who has been always with me in my ups and down to compile this work. I am grateful to my school teacher Shri Surendra Sarthak Ji, Shri Ghanshyam Ji and Shri Harkesh Ji, Shri Hanuman Ji. I am also delighted to offer thanks to my friends Dr. Sunil Kumar, Dr. Dinesh Jinger, Dr. Sonu Kumar, Surendra Singh Jatav, Dharamveer Verma, Rajat and last but not least to my family members. I am also thankful to Dr. Eetela Sathyanarayana for his motivation to compile this book.

Chapter 1

Role of Legumes in Cropping Systems for Soil Ecosystem Improvement

Jagadish Jena[1], Sagar Maitra[2,*], Akbar Hossain[3], Biswajit Pramanick[4], Harun I. Gitari[5], Subhashisa Praharaj[2], Tanmoy Shankar[2], Jnana Bharati Palai[2], Akshit Rathore[6], Tanuj Kumar Mandal[7] and Hanuman Singh Jatav[8]

[1]Indira Gandhi Krishi Viswavidyalaya, Raipur, Chhattisgarh, India
[2]Centurion University of Technology and Management, Odisha, Paralakhemundi, India
[3]Bangladesh Wheat and Maize Research Institute, Dinajpur, Bangladesh
[4]Dr. Rajendra Prasad Central Agricultural University, Pusa, Samastipur, Bihar, India
[5]School of Agriculture and Enterprise Development, Kenyatta University, Nairobi, Kenya
[6]Chaudhary Charan Singh Haryana Agricultural University, Hisar, Haryana, India
[7]The Neotia University, Jhinga, Sarisha, West Bengal, India
[8]Sri Karan Narendra Agriculture University, Jobaner-Jaipur, Rajasthan, India

Abstract

Soil health management is a prime concern to ensure agricultural sustainability. Dwindling soil fertility as well as declining soil health is one of the great challenges of the present time to feed the world population. Injudicious use of chemical fertilizers may negatively affect soil health. Further, chemical fertilizer production is associated with release of greenhouse gases to atmosphere and thus, pollutes the environment. In the present context of climate change and global warming, legumes can be considered for mitigation of ill effects of climate change and improving soil health. Legumes are versatile crops, having a huge potential to produce protein-rich grains, fix nitrogen biologically and enrich beneficial microbes in soil. Some legumes are capable to solubilize the unavailable phosphate by exudating organic acids from their roots. Legumes can facilitate rebuilding of soil organic matter and restrict pests and pathogens when incorporated in crop rotation. Besides, legumes are known for weed suppression and check erosion as cover crops. As green and brown manures, legumes add a sizeable quantity of biomass and nutrients to the soil. In cropping system, legumes fit well because of wider diversity and adaptability. The multifaceted benefits of legumes are prominent either in multiple cropping or intercropping, and thus, can be considered as an inevitable choice for soil health management in the journey towards sustainable agriculture.

---

* Corresponding Author's Email: sagar.maitra@cutm.ac.in.

In: Ecosystem Services
Editor: Hanuman Singh Jatav
ISBN: 978-1-68507-614-6

Keywords: legumes, cropping system, green and brown manuring, soil health, agricultural sustainability

1. Introduction

Agriculture is the backbone of the developing nations with a prime focus on optimum productivity of crops and efficient use of natural resources in a suitable manner so that it can provide food and nutritional security in an efficient manner. Crop husbandry is a major component of agriculture and it is influenced by both natural as well as anthropic activities. Human, since the starting of crop husbandry, always seeks to use more and more resources to produce more from the limiting land resource. Cropping system, in its broader sense, is the cropping pattern followed in a farm and its interaction with other farm resources which determine the make-up. After the inception of green-revolution, wheat and rice dominated the cropping systems due to promotional activities and institutional support (Maitra, 2020a). As the high yielding fertilizer and resource responsive cultivars were released, agriculture shifted from "field to mouth" to "field to market." Due to productive as well as marketing advantages along with specialization of crop husbandry, farmers kept on cultivating rice and wheat years after years. This practice of sole cropping led to nutrient mining, weed and pest infestation, declining soil fertility and factor productivity etc. along with yield stagnation (Swarup et al., 2000; Lal, 2006, 2010, 2014, 2016).

Rapid increase in these anthropogenic disturbances always seeking to exploit the natural resources have questioned the agricultural sustainability (Verma et al., 2015a, Lampkin et al., 2015; FAO, 2017). Therefore, it is of prime importance to look into the overall fertility and productivity of the soil as well as agro-ecosystem for the well-being of the future generations (Chen et al., 2011, Stronge et al., 2020). In this direction, growing of legumes may be a suitable option as the crop offers multifaceted benefits in agriculture inclusive of soil health improvement for sustainable agriculture.

2. Legumes, a Brief Description

Global population is going to hit 8.6 billion by 2030 and 9.6 billion by 2050 and along with this population explosion, biotic and abiotic influences of climate change on global food production can't be neglected (Yadav et al., 2015). Therefore, agriculture is under huge pressure to maintain the food as well as the nutritional security, mitigate adverse effect of climatic aberrations and to recover soil health and improvement of other resource bases (Lal 2015). Legume inclusion in cropping systems replacing the existing cereal-cereal cropping systems can register a vital role in sustaining the agricultural production system because of their multirole services, i.e., source of nutritious food for both human and animals (Tharanathan and Mahadevamma, 2003), atmospheric N fixation, mitigation of greenhouse gas emission (Lemke et al., 2007), breaking of weed and pest cycle, etc. Hence, legume inclusion in cropping systems is the need of the hour with high priority to sustain the agricultural activity and to maintain the food as well as nutritional security (Nees et al., 2010).

Pulses, because of their quality protein profile, considered as the "poor man's meet" and occupy a special place in Indian diet. In India, pulses are generally cultivated in resource constraint areas and more emphasis is being given to the cereals. A significant population in India are malnourished. India constitutes nearly 70% of world vegetarian population, so to fulfil this nutritional security there is a scope to increase the area as well as the productivity of pulses using different agro-techniques. Along with nutritional advantages, pulses have ability to enhance soil nutrient status, as they can fix atmospheric nitrogen (N), can sequester carbon in soil and indirectly enrich micro- and macro- nutrient status of soil on their incorporation.

Being the third largest family among flowering plants after the families of *Asteraceae* and *Orchidaceae* in the plant kingdom, the *Leguminosae* comprises of 800 genera and 20,000 species (Lewis et al., 2005). Nutritionally, legumes are far superior than many other food sources containing 126-660 g carbohydrates, 180-430 g protein, 7-400 g fat, 0.9-2.4 g calcium, 28-115 mg iron, 13-39 mg riboflavin, 3-99 mg thiamine and 7-64 g dietic fibers in a kg of dry matter (Berihun and Molla 2017). Grain legumes are broadly grouped into warm-season legumes (i.e., soybean, groundnut, pigeon pea, etc.) and temperate legumes (i.e., pea, lupin, French bean, lentil, vetch, etc.), among which soybean is most widely cultivated legume crop in the world. But, due to more intensive and specialized production system, cultivation and expansion of major cereals are way more extensive than pulses (Preissel et al., 2015). Among global dietary protein need, pulses contribute the highest, i.e., 34% (Schuster 2013). Legumes are also the source of different essential amino acids (Marielle Margier et al., 2018). Legumes along with their multi-role positive effects, i.e., nutritious food for human and cattle, improvement in soil nitrogen pool, soil organic carbon and capable of increasing the productivity of succeeding crop, needed to be focused for their alternate use in cropping systems for achieving the sustainability (Dhakal et al., 2016).

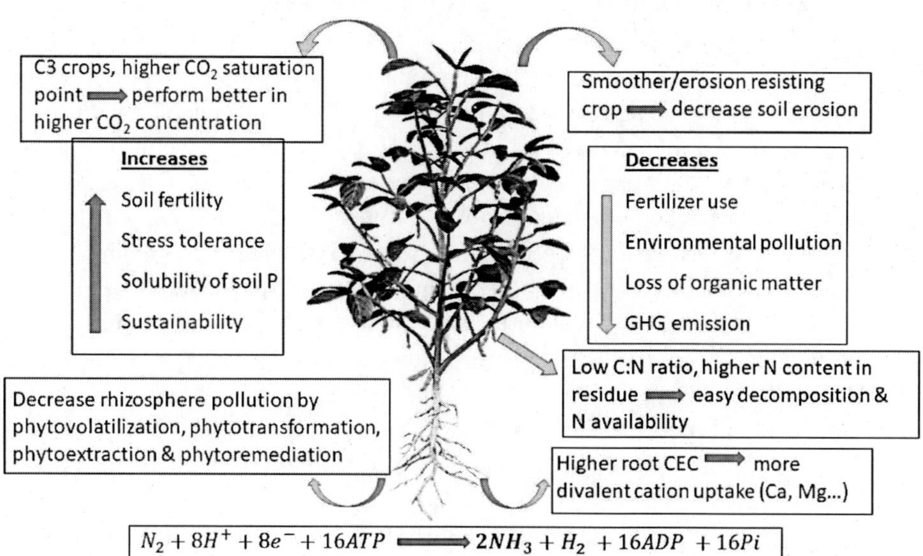

Figure 1. Important characteristics of legume crops.

Agriculture is basically a biological process where crops and livestock are raised for fulfilling day to day human needs, i.e., food, feed, fibre, fuel, medicines, meat, milk, etc. Thus, agriculture is an essential activity for the development and sustenance of world economic system (FAO, 2017). Inclusion of legume in cropping system has gained global attention because of their multiple benefits (Dhakal et al., 2016). Legumes along with dietary needs, act as medicine for animal health (Wattiaux and Howard 2001), enhance soil fertility (Meena et al., 2015a) by nitrogen fixation and solubilization of fixed phosphorous (Jensen 1996) and can also be used as biofuel (Jensen et al., 2012). Besides these positive impacts on soil and animals, inclusion of grain legumes also reduces different biotic and abiotic stresses, i.e., pest, weed, disease etc. The production of grain legume keeps on increasing now a days due to developed agro-techniques, i.e., organic farming, reduced tillage, use of biofertilizers, scheduled input use, etc. (Meena et al., 2016). These yield enhancing practices along with suitable crop rotation and cropping sequence improve agricultural as well as environmental sustainability. 15-25% yield advantages in cereals have also been reported when legume crops are incorporated in cropping systems (Kirkegaard et al., 2008). Legume inclusion also reduces the agrochemical dependence and can lead to sustainable production system in an organic farming system (Verma et al., 2015a, b). The rise in agricultural production costs especially due to rise in fertilizer price can be reduced as addition of legumes in cropping system, increases the availability of soil N, P, K, Ca and Mg significantly over monoculture (Huang, 2007; Stagnari et al., 2017). The reduced fertilizer and tillage need, enhancement in soil C sequestration (Reckling et al., 2014), lower greenhouse gas emission, reduced risk of eutrophication of water bodies and soil acidification (Clark et al., 2013) are some of the potential benefits derived from the inclusion of legumes in cropping system. In this way, legume has important role in enhancing the resource use efficiency along with increase in total system productivity (Lazali et al., 2021; Salama et al., 2021; Varma et al., 2015a).

3. Legumes and Biological Nitrogen Fixation (BNF)

Nitrogen is a primary essential nutrient for crop production and is the vital factor for crops after solar radiation and water. Except legumes, almost all crop plants depend on soil N for their requirement, even the Earth's atmosphere is rich in N gas. Legumes have the capability to fix the atmospheric nitrogen through legume-rhizobia symbiotic association. This biological nitrogen fixation (BNF) not only help the legume to fulfil its nitrogen need, but also enrich the soil nitrogen status to improve the succeeding crop yield (Carranca 2013; Nygren and Leblanc 2015; Chapagain and Riseman 2014; 2015). Plants can get nitrogen from soil either through decomposition of legume residue on their incorporation or from atmospheric nitrogen fixation by leguminous plants (Yong et al., 2015; Varma et al., 2015b). Those leguminous plants which fix and add atmospheric nitrogen to soil called "N-donor" plants and those which receives this soil nitrogen called "N-receiving" plants (Moyer-Henry et al., 2006). This phenomenon of transfer of nitrogen from N-donor to N-receiver varies from 10-85% of the N demand of N-receiving plants (Paynel et al., 2008) and by adoption of suitable legume in crop rotation practices, the nitrogen demand of crops can be fulfilled through BNF (Rahman et al., 2014). The extent of this biological nitrogen fixation varies from zero to several hundred-kilogram nitrogen per hectare (Soumare et al., 2020).

4. Inclusion of Legumes in Cropping System

4.1. Sequential Cropping

Crop rotation is considered to be incomplete if legumes are not included in a cropping system. The amount of nitrogen addition to soil through legume inclusion depends on the legume crop taken for the system (Squire et al., 2019). Crop rotation through suitable legume inclusion not only improves biomass production but also enhance soil carbon and nitrogen status (Lal 2010). This increased C and N status in soil not only makes soil microbes active and supply N in the current season but also benefits the succeeding crops (Akinnifesi et al., 2007; Lithourgidis et al., 2011). Significant yield level in succeeding maize crop even without chemical source of fertilization has been reported when cultivated after legume crops, i.e., *Tephrosia*, *Mucuna*, *Crotalaria*, etc. (Lopez and Mundt 2000).

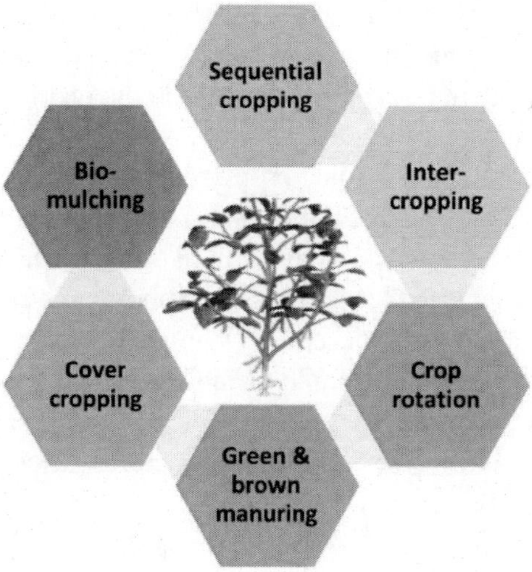

Figure 2. Options for legume crop inclusion.

4.2. Intercropping

Intercropping is the cultivation of two or more than two crops on the same land as they coexist for some period of their life cycle (Andrews and Kassam 1976; Sanchez 1976; Gitari et al., 2020; Maitra, 2020b) which is practiced to achieve some ecological goals, i.e., reducing risks of farmers from complete crop failure against climate abnormalities, increasing resource use efficiency, increasing diversity of the farm promoting species interaction and cybernetics (Lithourgidis et al., 2011). Generally, the main crop or base crop of intercropping should be the dominant crop of the locality and the second or component crop is grown for additional output using the left-out resources. But, in case of legume inclusion in intercropping, the main aim is improving the soil health as well as enhancing the resource efficiency in comparison to pure stands (Inal et al., 2007). Intercropping is broadly grouped into four types i.e., row intercropping, strip intercropping, mixed intercropping and relay intercropping. The higher crop density under intercropping also increases soil microbial diversity, which stabilizes the agroecosystem (Maitra and Ray, 2019; Maitra et al., 2021). Legume inclusion in intercropping system enhances soil moisture and nutrient use efficiency, efficient weed, pest and pathogen management, and maximizes net return of farming increasing the crop yield. In rhizobia-legume symbiosis association, rhizobia receive food from legume and legume crop uses fixed nitrogen in biosynthesis of amino acids and nucleotides. Except legume-rhizobia symbiosis (Fustec et al., 2010) for BNF (Chalka and Nepalia 2006), inclusion of legume also helps in maintaining soil fertility (Hauggaard-Nielsen et al., 2009), enhance soil conservation (Lithourgidis et al., 2011), suppresses weeds population in the crop field (Liang et al., 2010), etc. Thus, it is a low-input agro-technique which follow environment-friendly crop cultivation (Ashoka et al., 2017) to enrich soil nitrogen status and to limit chemical nitrogen fertilization (Dusa 2009, Maitra et al., 2021).

The legumes commonly used as intercrops include chickpea (*Cicer* sp.), pea (*Pisum* sp.), pigeon pea (*Cajanus* sp.), groundnut (*Arachis* sp.), cowpea (*Vigna* sp.), lentil (*Lens* sp.), faba bean (*Vicia* sp.), berseem (*Trifolium* sp.), dry bean (*Phaseolus* sp.), etc. (Blaser et al., 2006), improving the yield of the main crop as well as the system (Sarkar et al., 2000; Ghaley et al., 2005). Numerous principles and aspects are being considered for successful intercropping system (Maitra et al., 2019), i.e., crop species, light interception, soil nutrient status and requirement, soil moisture availability and land requirement (Maitra et al., 2020). The non-legume crop in association with the legume crop in intercropping system may benefited from the biological nitrogen fixation (Sullivan 2003; Garg 2007) or through nitrogen transfer through root exudates, leaf leachates, etc. (Addo-Quaye et al., 2011). Biologically fixed nitrogen by legumes can directly be available to the current season non-legume crop grown together, known as direct N availability (Adeniyan et al., 2007; Dahmardeh et al., 2010) or added to the crop field to supplement the succeeding crop called residual N availability (Raji 2007; Barbosae et al., 2008). Moreover, in the combination of deep-rooted legumes and shallow rooted cereals in an intercropping system, legumes create the facility of bioirrigation under limited soil moisture conditions (Maitra et al., 2021).

Compared to sole cropping, when crops are grown along with legume crops, even nutritionally poor soil can provide better yield. Legume in intercropping can restore the soil nitrogen status by BNF (Fujita et al., 1992; Meena et al., 2017a). Intercropping is one of the principles followed in organic farming and it was reported that, due to its slow growth, legumes are poor competitor of weeds, but when intercropped with cereals act as smoother crops and can effectively control weeds (Maitra et al., 2021). Intercropping of legume and their association with non-legume crops can minimize diseases (Boudreau and Mundt 1992; Fininsa 1996).

4.3. Crop Rotation

Crop rotation is also an intensive strategy with recurrent succession of crops to enhance the output of the system in terms of crop productivity through inclusion of suitable crops. Inclusion of legume in the system is mostly encouraged knowing their multifaceted advantages, i.e., BNF, nutrient recycling, increase soil carbon and nitrogen stock, etc. (Keeler et al., 2009). Legume based crop rotation not only enhance soil quality and system productivity but also breaks weed and pathogen cycle, reduce agrochemical inputs, increase biodiversity of the agricultural ecosystem (Ratnadass et al., 2013). Leguminous crops produce higher biomass and improve soil organic carbon, that further increase the soil microbial population and maintain soil health (Hauggaard-Nielsen et al., 2007). The inclusion of legumes in cropping system produces more biomass using limited resource base, improves soil carbon and nitrogen stock and can be adopted suitably in any cropping system as they can complete their life cycle even in a very short time period and can be used in sustainable land development programs (Lithourgidis et al., 2011). As the legumes are involved in BNF and supply it to both current season as well as succeeding crops in the cropping system, these can also be considered for N-economy (Mahmud et al., 2020; Praharaj and Maitra 2020). The quantity of nitrogen fixed and supplied to current and succeeding crops depends mainly on the legume species, cross inoculation group associated, soil type, climate, duration of the crops, etc. (Chu et al., 2004; Spehn et al., 2002).

4.4. Green and Brown Manuring

Green manuring is the incorporation of green crop residues, preferably legume *in-situ* or *ex-situ* to make the crop residue decompose and supply nutrient to the succeeding crops (Maitra et al., 2018). Legume are preferred for green manuring as they fix atmospheric nitrogen in soil, produce more biomass within short time period, rich in nutrients and low in C:N ratio. Green manuring can partially or completely fulfil the N need of succeeding crop (Tiwari et al., 2004; Meena et al., 2015b). Green manures play a vital role by bringing positive effects on soil physio-chemical and biological properties and thus improves soil health. Further, it replaces fallowness and restricts soil erosion and nutrient loss from the top soil. Amelioration of degraded soils can easily be done by addition of green foliage into the soil (Maitra et al., 2018). Green manuring is beneficial for plant protection aspects also as it manages weeds, diseases and insect pests (Blackshaw et al., 2001; Larkin and Griffin, 2007; Tillman et al., 1999; Agbenin, 2011) and thus, provides a healthy agro-ecosystem for crops. The no-till version of green manuring is brown manuring which can be adopted in cereals cultivation to obtain necessary benefits of green manuring. In rice and maize, brown manuring can be adopted suitable for nutrient saving, weed management and soil health improvement (Maitra and Zaman, 2017) in tropical and sub-tropical climates.

4.5. Cover Crop

Legumes are close growing crops and hence serves as cover crop. Also, the dense foliage of most legumes reduces the erosive action of rainfall to a large extent. Legumes release many root exudates such as organic acids to the soil which acts as a binding agent and reduces soil erodibility by improving aggregate stability (Sanchez-Navarro et al., 2019). As legumes are capable of being grown in relatively resource poor areas, they can act as both a fertility restorer and a measure of erosion control in such areas. Legumes can also be grown in alternate strips along with some erosion susceptible crops to keep the soil loss below acceptable threshold. The benefits of using legume as cover crop is due to the fact that it ensures food and nutritional security while protecting the soil from erosive agents and improving the soil health (Blanchart et al., 2006; Doane et al., 2009).

4.6. Bio-Mulching

After the pods are picked, the residues of legume can be added to soil, which can improve the organic matter content and nutrient status of soil. As the residues of legume have low C:N ratio, they can easily decompose and supply nutrients more quickly as compared to residues of cereals (Tiwari et al., 2004; Chu et al., 2004). Legume residues are comparatively more succulent and hence easy to manage. This can be a prospect that should be taken into consideration while planning a conservation agriculture based cropping system. Maintaining mulch on the soil surface is beneficial in reduction of erosion, enhancement of soil organic matter, addition of nutrient to soil, improvement in soil microbial status and maintenance of proper soil temperature. Moreover, bio-mulching creates as opportunity time for water infiltration, soil moisture conservation through evaporation control and weed management etc. (Kumar et al., 2010; Sumiahadi et al., 2018).

5. Synergy between Legumes and Soil Health Improvement

Modern, intensive agriculture system is mainly agrochemical dependent directing towards consistent degradation of soil health, i.e., physical, chemical and biological quality of soil. Due to intensive cultivation and faulty soil management practices, severe constraints like increase in soil compaction and erosion, reduction in soil productive potential and reduction in soil microbial activity have been well recognized (Unger and Kaspar 1994). Legume inclusion in the soil improves the physicochemical and biological properties by totalling organic substances and root exudates to the soil. It was reported that legume can accumulate about 2.6 kg N ha^{-1} day^{-1}, and their incorporation can be equivalent to 50-100 kg N ha^{-1} application of chemical nitrogenous fertilizers (Ladha et al., 1988). However, Dhaincha (*Sesbania* sp.) at 45-60DAS can accumulate 5.5 kg N ha^{-1} day^{-1}) and can fix about 300 kg N ha^{-1} (Ladha et al., 1988).

Legume crops play a vital role in the world nitrogen cycle fixing atmospheric nitrogen biologically and can supply the available nitrogen to current season crop as well as succeeding crops. The BNF can enhance 9.7-20.5% residual nitrogen content in rice field (Yu et al., 2014). Almost half of the total above ground biomass nitrogen partitioned into below ground biomass i.e., root and nodules (Carranca et al., 2015). Incorporation of legume residue adds about 50–60 kg ha^{-1} of N to the soil that can be used by the succeeding crop and the loss of N in this case is significantly lower than chemical fertilizer application (Singh 2000).

Legume inclusion in soil prevents leaching of essential nutrients, reduce the negative effect of harmful agro-chemicals and also the effect of soil borne pathogens (Dhakal et al., 2016). Soil microflora diversity is enhanced by legume inclusion in soil either as green manure crop or through several cropping system by increasing the soil organic matter status (Kumar et al., 2016). Soil health keeps on decreasing by adopting continuous cereal based cropping system, and as a consequence, there is evident of multi-nutrient deficiencies. To overcome this problem, inclusion of suitable legume in the cropping system i.e., soybean, moong, etc. can be a sustainable option to maintain the soil fertility as well as productivity (Dhaliwal et al., 2011; Singh et al., 2011).

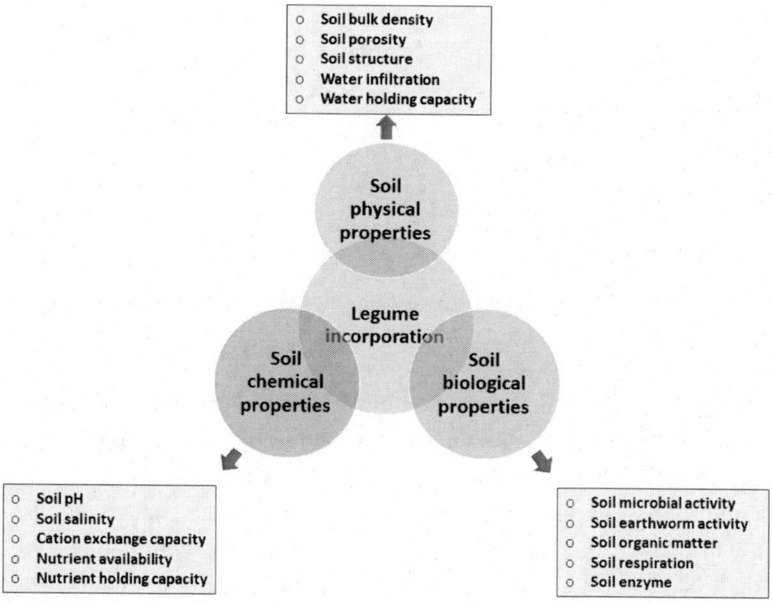

Figure 3. Effect of legume inclusion on soil physical, chemical and biological properties.

Legume crops act as soil amendment with their beneficial effect on soil quality (Binder and Wiek 2007) and can be an essential part of different cropping systems (Hauggaard-Nielsen et al., 2007). The inclusion and incorporation of legumes in cropping systems have several positive impacts on soil health management, i.e., improvement in agricultural productivity, agroecosystem sustainability, soil conservation, increasing soil organic carbon and nitrogen stocks, biological nitrogen fixation, reducing fertilizer use and hence reduction in nitrous oxide emission and nitrate leaching from soil profile, etc. (Nees et al., 2010). Thus, legume crop inclusion in cropping system has a potential for enriching the soil health, productivity and agricultural sustainability (Dhakal et al., 2016).

The residual supply of nitrogen fixed by legume through symbiotic biological nitrogen fixation depends mainly on legume species, crop management practices and climate (Meena et al., 2015a). Adding legumes in cropping system also enhances soil organic carbon (SOC) concentration (Varvel 1994), decline erosion and improve soil water status (Blanchart et al., 2006). In a legume-based cropping system, the residual nitrogen availability to the succeeding crop depends on BNF ability of the legumes, quantity of legume biomass incorporated in the soil and the rate of decomposition of the incorporated legume biomass (Heichel 1987). It was also documented that there is significant enhancement in carbon and nitrogen retention under legume inclusion in cropping systems due to lower C:N ratio of organic matters (Drinkwater et al., 1998) and legume inclusion can maintain higher organic matter status than continuous cropping of non-legume crops (Campbell and Zentner 1993; Stevenson 1982). Legume inclusion in cropping systems has their natural advantages of atmospheric nitrogen fixation and minimal competition with other non-legume crops because of their difference in growth habits (Hauggaard-Nielsen et al., 2007). Nitrogen rich legume residues also contribute to increase in soil bio-pores directly through residue decomposition and indirectly through increasing the population of earthworm in rhizosphere zone, optimizing the balance between air and water in crop root zone (Meena et al., 2015b; Jensen et al., 2012).

The higher C content in legume residue enhances growth of microbial population on their surface and higher N and protein content facilitates the growth of those microbes and easy decomposition in soil and there is report of enhancement of both soil organic carbon as well as humus after legume residue addition in the soil (Lemke et al., 2007). Again, cultivation of legume crop doesn't require intensive tillage, which also help in building up of SOC. The build-up of SOC is more in case of mixed legume-grass cultivation than sole cropping (Yadav et al., 2019). Legume crops through the process of biological N fixation can save 150-200 kg N ha^{-1} $year^{-1}$ (Peyraud et al., 2009) and can fulfil 90% of their own nitrogen requirement on proper rhizobium inoculation. Over the decomposition of these legume residues on incorporation, some N recycled in soil (Meena et al., 2015b) and this N cycling regulated by quality of the residue, soil microbial activity and soil environment, i.e., pH, aeration, etc.

In response to global scale mechanization, increase in oil price and soil erosion, to maintain the soil health, several conservation practices being followed globally and legumes are included in rotation and as cover cropping (Mundt 2002; Srinivasarao et al., 2012). The residues of legumes can improve soil physical, chemical and biological health and quality on their incorporation and increase both nutrient and water holding and supplying ability of the soil (Grandy et al., 2002; Mousavi et al., 2009). It has also been reported that, taking crops having different rooting depths and requiring minimal soil disturbance, i.e., legumes optimize micro- and macro- pores in soil that increases infiltration of water to deeper root zone depth (Kumar and Goh, 2000). Legume crops along with biological nitrogen fixation can add high quality soil organic matter because of their extensive root system and low C:N ratio, helping the growth of soil micro-flora and nutrient cycling (Siddique et al., 2008; Dhakal et al., 2016).

5.1. Improvement of Soil Physical Properties

Physical properties of soil, i.e., structure, texture, density, stability, porosity, etc. are fairly constant towards crop husbandry practices, but are important criteria associated with aeration, erosion, runoff, infiltration rate, nutrient and moisture holding capacity of soil (Dexter 2004). Therefore, proper soil physical condition is essential for optimum tillage, root growth, ground water recharge, prolonged soil moisture availability and deprived soil physical condition may lead to difficulty in farm activities (Schoenholtz et al., 2000). Soil texture is the relative proportion of soil separates, i.e., sand, silt and clays and fairly resistance to change and remain stable for long time, while soil structure can be manipulated using cultural activities i.e., tillage, residue retention, minimum tillage, etc. Bulk density and soil porosity affects root growth significantly and crop management practices, i.e., tillage, residue incorporation, etc. affects these edaphic factors (Beutler et al., 2002). Soil porosity as influenced by soil texture and structure broadly categorised into micro and macro pores, depending on pore size diameter (Dexter 2004), which directly influences on the soil moisture holding capacity and availability to crop plant (Meena et al., 2015a). Inclusion of legumes in soil acts as soil conditioner and improves soil physical properties significantly (Srinivasarao et al., 2012). Leguminous crops used as cover crops as they produce huge biomass and on their incorporation in soil, significantly influence soil physical property and also improves soil microbial population, increasing soil organic matter (Lal 2015). They also protect the soil from different erosive agents and prevent from soil erosion. Furthermore, legume residue inclusion improves soil water stable aggregates and improves soil physical condition preventing soil erosion

(Lithourgidis et al., 2011; Mousavi et al., 2009). Legume crops also improves water status and soil infiltration by increasing soil aggregate stability (Mousavi et al., 2009). Use of legume crop as cover crop also maintain the soil organic carbon status by maintaining optimum C/N ratio of soil (Dexter 2004; Schädler et al., 2011).

5.2. Improvement of Soil Chemical Properties

Soil chemical properties i.e., pH, CEC, SOC, base saturation, buffering capacity and nutrient concentration has direct relationship with the nutrient dynamics and availability in soil to support higher crop yields (Meena et al., 2015a). Legume crop has the properties to release organic root exudates and their incorporation in soil has significant influence on soil chemical properties through addition of organic matter and BNF, sustaining the soil fertility and optimizing the system productivity (Kelly et al., 2009). Legume has well known beneficial effect on available nutrients, soil pH and soil organic carbon stock (Mugwe et al., 2004). Further, legumes included in the cropping system changes the soil pH by releasing organic acids which may enhance soil phosphorous availability (Buerkert et al., 2000; Bado et al., 2004; Meena et al., 2017b) and growth and activity of soil microbes (Juma and Averbeke 2005; Lopez and Mundt 2000), which significantly influence the diseases and nutrient dynamics. Legume soil inclusion not only adds N to soil but also augments huge quantity of essential nutrients, organic matter and sequester atmospheric carbon dioxide (Turnbull and Bowman 2002; Pikul et al., 2008; Sharma and Behera 2009; Ogunwole et al., 2010; Lal 2015). The legume biomass can either be incorporated in-situ or transferred from other field and incorporated i.e., ex-situ. The quantity of available nutrient depends on the nutrient content, decomposition rate, soil type, climate, plant density, crop management, etc. (Adeboye et al., 2005). The biomass decomposition rate and nutrient release varies with pH, aeration, moisture, temperature and other factors (Liang et al., 2010). The incorporation of legume plants of early age makes the added biomass to decompose easily (Bodo et al., 2004) and the decomposed biomass positively influences the soil chemical properties and nutrient dynamics (Melero et al., 2007).

Indiscriminate use of agrochemicals in agriculture lead to pesticide contamination which causes pesticide poisoning and adversely affects soil biodiversity (Prell and Poole 2006; Khan et al., 2009; Jin et al., 2013). Adopting best management practices in agriculture can help recovering from agrochemical contamination and building a holistic agroecosystem (Deakin and Broughton, 2009; Lal, 2011). The remedial action however should take into account the economical, ecological and social strata of sustainability (Schädler et al., 2011). Legume inclusion in soil hold pesticide and never let them carry forward to contaminate soil as well as water bodies preventing erosion and runoff to a greater extent (Fester et al., 2014).

Gram negative bacteria, Rhizobiales, of Alphaproteobacteria class, have some species capable of symbiotic nitrogen fixation with legumes (Sato et al., 2005; Dadhich et al., 2015) by infecting the root of legume plants and fixing the atmospheric nitrogen. This helps the host plant to survive without external supply of nitrogen (Bardos 2003; Bardos et al., 2011). Hydrogen is one of the biproduct of symbiotic nitrogen fixation which is having bioactive property improving the abiotic stress tolerance of plants (Cui et al., 2013). Along with nitrogen fixation Rhizobia has certain advantages i.e., elimination of soil pollutants directly (Jin et al., 2013), or indirectly enhance other degrading microbes and enzymes (Kaiya et al., 2012), acts in metal bioremediation (Fester et al., 2014).

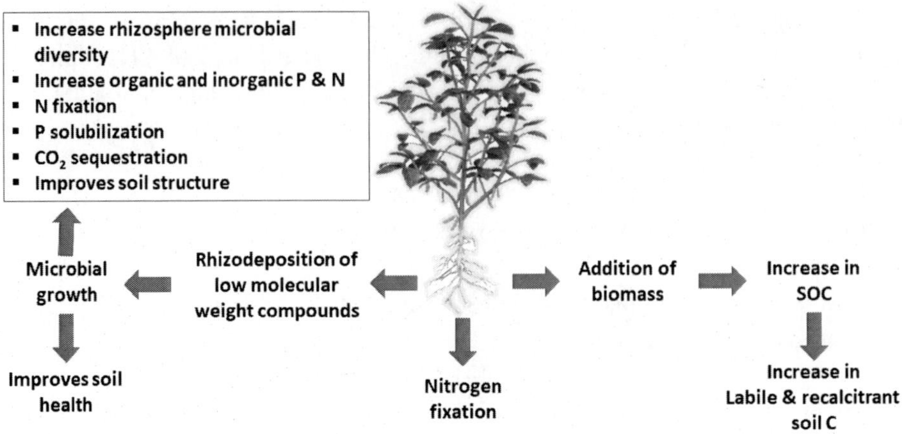

Figure 4. Biological effect of legume inclusion.

5.3. Improvement of Soil Biological Properties

Soil biological properties are most sensitive towards soil management and can be considered as an early indicator of change in soil quality on different management practices (Brookes 1995; Trasar-Cepeda et al., 1998; Suman et al., 2006). Nitrogen is the limiting macro-nutrient in most of the agricultural soil and the requirement of nitrogen in plant is also higher than other mineral nutrients. Rhizobia in association with legume synthesize nitrogenase enzyme which help in atmospheric nitrogen fixation. Nitrogen fixed in biological nitrogen fixation assimilates as protein and glycoproteins in plant biomass (Klauer and Francesch 1997; Lansing and Franceschi 2000). Phosphorous is one of the essential mineral elements for plant growth, but its availability in soil is limited by soil reaction and complexation with Fe, Al, Ca and Mg (Sinclair and Vadez 2002; Meena et al., 2017a). Legume inclusion in cropping system help in releasing soil fixed phosphorous by releasing several acids in the form of root exudates (Shen et al., 2002; Nwoke et al., 2008; Nuruzzaman et al., 2006) and enhancing phosphatase enzyme activity (Gilbert et al., 1999; Helal 1990). Hydrogen gas is released as biproduct during biological nitrogen fixation which encourage microbial activity, microbial carbon and microbial nitrogen in root zone. This microbial carbon and microbial nitrogen comprise of 1-7% of total soil carbon and nearly 5% of total soil nitrogen respectively (McGill et al., 1986; Sørensen 1987; Anderson and Domsch 1989; Insam et al., 1989; Sparling 1992) and this contribute most to the labile carbon and nitrogen fraction in soil (Jenkinson and Ladd 1981). The additions of legumes in cropping system enhances the soil microbial activity as well as increases microbial diversity (Alvey et al., 2003) and small change in the concentration of nitrogen, cellulose, lignin content or C:N ratio of residue trigger the divergence of soil microbial status (Meena et al., 2014). Legume-rhizobium association also influence plant growth promoting rhizobacteria (Schelud'ko et al., 2009) and there is also report of tripartite symbiotic association of mycorrhiza-legume-rhizobium which further enhance the plant nutrition (Hayman 1986; Scheublin et al., 2004). Soil phosphorous has limited mobility and legume generally require more phosphorous for their growth and specially for the nodule formation. The mycorrhizal association of legume help the root to explore more phosphorous and moisture along with other nutrients increasing the effective root zone of the crop (Zahran 1999).

Conclusion

Legume being a short duration crop, is grown world-wide in resource limited conditions. Inclusion of legume in cropping systems has several advantages such as food as well as nutritional security, ecological soundness and creation of an efficient agroecosystem, reduction of soil erosion, enhancement of water and nutrient use efficiency, sustained soil function, biological nitrogen fixation and improvement of soil health, increase in soil organic carbon and nitrogen stock, soil carbon sequestration, and thus, increase in agricultural productivity. In this chapter, an attempt has been made to collect different research outcome and summarize them in such a way to present the importance of legumes and their inclusion in cropping systems to facilitate qualitative improvement of soil and agroecosystem, and their subsequent benefit on the sustainability of the agricultural system. The multifaceted mechanism of legume residue incorporation to enhance soil physical, chemical and biological properties to achieve the long-term sustainability goal has also been addressed. An urgent need to study and understand the whole complex mechanism of developing best management practices taking suitable legume crop in cropping systems is the need of the hour and future researches in the regard can be focused into this direction.

References

Addo-Quaye, A. A., Darkwa, A. A., Ocloo, G. K. (2011). Yield and productivity of component crops in a maize-soybean intercropping system as affected by time of planting and spatial arrangement. *J. Agric. Biol. Sci.* 6(9), 50–57.

Adeboye, M. K. A., Iwuafor, E. N. O., Agbenin, J. O. (2005). Rotation effects of grain and herbaceous legumes on maize yield and chemical properties of an Alfisol in the northern Guinea Savanna, Nigeria. *Niger. J. Soil Res.* 6, 22–31.

Adeniyan, O. N., Akande, S. R., Balogun, M. O., Saka, J. O. (2007). Evaluation of crop yield of African yam bean, maize and kenaf under intercropping systems. *Am. Eurasian J. Agric. Environ. Sci.* 2(1), 99–102.

Agbenin N. O. (2011). Biological control of plant parasitic nematodes: prospects and challenges for the poor Africa farmer. *Plant Prot. Sci.* 47:62-67.

Akinnifesi, F. K., Makumba, W., Sileshi, G., Ajayi, O. C., Mweta, D. (2007). Synergistic effect of inorganic N and P fertilizers and organic inputs from *Gliricidia sepium* on productivity of intercropped maize in Southern Malawi. *Plant Soil.* 294, 203–217.

Alvey, S., Yang, C. H., Buerkert, A., Crowley, D. E. (2003). Cereal/legume rotation effects on rhizosphere bacterial community structure in West African soils. *Biol. Fertil. Soils.* 37, 73–82.

Anderson, T. H., Domsch, K. H. (1989). Ratios of microbial biomass carbon to total organic carbon in arable soils. *Soil Biol. Biochem.* 21, 471–479.

Andrews, D. J., Kassam, A. H. (1976). The importance of multiple cropping in increasing world food supplies, ASA Special Publication, *Am. Soc. Agron.* 27, 1–10.

Ashoka, P., Meena, R. S., Kumar, S., Yadav, G. S., Layek, J. (2017). Green nanotechnology is a key for ecofriendly agriculture. *J. Clean Prod.* 142:4440.

Bado, B. V., Sedogo, M. P., Lompo, F. (2004). Long term effects of mineral fertilizers, phosphate rock, dolomite and manure on the characteristics of an ultisol and maize yield in Burkina Faso. In: Bationo A (ed) *Managing nutrient cycles to sustain soil fertility in sub-Saharan Africa*. Academy Science Publishers, Nairobi, pp. 608.

Barbosae, S., Lima, P. I., Silva, P. S., de Oliveira, O. F., de Sousa, R. P. (2008). Planting times of cowpea intercropped with corn in the weed control. *Revista Caatinga (Mossoró, Brasil)*. 21(1), 113–119.

Bardos, P. (2003). A review of the contaminated land rehabilitation network for environmental technologies in Europe (CLARINET). Part 2: working group findings. *Land Cont. Reclam.* 11(1), 15–30.

Bardos, P., Bone, B., Boyle, R., Ellis, D., Evans, F., Harries, N. D., Smith, J. W. N. (2011). Applying sustainable development principles to contaminated land management using the SuRF-UK framework. *Remed*. 21(2), 1–138.

Berihun, T., Molla, E. (2017). Study on the diversity and use of wild edible plants in Bullen District Northwest Ethiopia. *Aust. J. Bot*. 2017:1–10. https://doi.org/10.1155/2017/ 8383468.

Beutler, A. N., Centurion, J. F., Souza, Z. M., Andrioli, I., Roque, C. G. (2002). Water retention in two oxisols under different uses. *Revista Brasileira de Ciênciado Solo*. 26:829–834.

Binder, C. R., Wiek, A. (2007). The role of transdisciplinary processes in sustainability assessment of agricultural systems. In: Hani, J., Pinte'r, L., Herren, H. R. (eds) *From common principles to common practice. Proceedings and outputs of the first symposium of the international forum on assessing sustainability in agriculture (INFASA)*. International Institute of Sustainable Development and Swiss College of Agriculture, Bern, pp. 33–48.

Blackshaw, R. E., Moyer, J. R., Doram, R. C., Boswell, A. L. (2001) Yellow sweet clover, green manure, and its residues effectively suppress weeds during fallow. *Weed Sci*. 49:406-413.

Blanchart, E., Villenave, C., Viallatoux, A., Barthès, B., Girardin, C., Azontonde, A., & Feller, C. (2006). Long-term effect of a legume cover crop (*Mucuna pruriens* var. utilis) on the communities of soil macrofauna and nematofauna, under maize cultivation, in southern Benin. *Euro. J. Soil Biol. 42*, 136-144.

Blaser, B. C., Gibson, L. R., Singer, J. W., Jannink, J. L. (2006). Optimizing seeding rates for winter cereal grains and frost-seeded red clover intercrops. *Agron. J*. 98, 1041–1049.

Boudreau, M. A., Mundt, C. C. (1992). Mechanisms of alterations in bean rust epidemiology due to intercropping with maize. *Phytopathol*. 82, 1051–1060.

Brookes, P. C. (1995). The use of microbial parameters in monitoring soil pollution by heavy metals. *Biol. Fertil. Soils*. 19, 269–279.

Buerkert, A., Bationo, A., Dossa, K. (2000). Mechanisms of residue mulch induced cereal growth increases in West Africa. *Soil Sci. Soc. Am. J.* 64, 346–358.

Campbell, C. A., Bowren, K. E., Schnitzer, M., Zentner, R. P., Townley-Smith, L. (1991). Effect of crop rotations and fertilization on soil organic matter and some biochemical properties of a thick Black Chernozem. *Can. J. Soil. Sci*. 71, 377–387.

Campbell, C. A., Zentner, R. P. (1993). Soil organic matter as influenced by crop rotation and fertilization. *Soil. Sci. Soc. Am. J.* 57, 1034–1040.

Carranca, C. (2013). Legumes: properties and symbiosis. In: Camisão, A. H., Pedroso, C. C. (eds) *Symbiosis: evolution, biology and ecological effects. Animal science, issues and professions*. Nova Science Publishers, New York, pp. 67-94.

Carranca, C., Torres, M. O., Madeira, M. (2015). Underestimated role of legume roots for soil N fertility. *Agron. Sustain. Dev*. 35:1095-1102.

Chalka, M. K., Nepalia, V. (2006). Nutrient uptake appraisal of maize intercropped with legumes and associated weeds under the influence of weed control. *Indian. J. Agric. Res*. 40, 86–91.

Chapagain, T., Riseman, A. (2014). Barley–pea intercropping: effects on land productivity, carbon and nitrogen transformations. *Field Crop Res*. 166, 18–25.

Chapagain, T., Riseman, A. (2015). Nitrogen and carbon transformations, water use efficiency and ecosystem productivity in monocultures and wheat-bean intercropping systems. *Nutr. Cycl. Agroecosyst*. 101,107–121. https://doi.org/10.1007/s10705-014-9647-4.

Chen, X. P., Cui, Z. L., Vitousek, P. M., Cassman, K. G., Matson, P. A., Bai, J. S., Meng, Q. F., Hou, P., Yue, S. C., Römheld, V., Zhang, F. S. (2011). Integrated soil–crop system management for food security. *Proc. Nat. Acad. Sci.* 108(16), 6399-6404.

Chu, G. X., Shen, Q. R., Cao, J. L. (2004). Nitrogen fixation and N transfer from peanut to rice cultivated in aerobic soil in intercropping system and its effect on soil N-fertility. *Plant Soil* 263:17–27.

Clark, C. M., Yongfei, B., Bowman William, D., Cowles Jane, M., Fenn Mark, E., Gilliam Frank, S., Phoenix Gareth, K., Ilyas, S., Stevens Carly, J., Sverdrup Harald, U., Throop Heather, L. (2013). Nitrogen deposition and terrestrial biodiversity. In: Levin, S. A. (ed) *Encyclopedia of biodiversity*, vol 5, 2nd edn. Academic, Waltham, pp. 519–536.

Cui, W. T., Gao, C. Y., Fang, P., Lin, G. Q., Shen, W. B. (2013). Alleviation of cadmium toxicity in Medicago sativa by hydrogen-rich water. *J. Hazard. Mater*. 260, 715–724.

Dadhich, R. K., Meena, R. S., Reager, M. L., Kansotia, B. C. (2015). Response of bio-regulators to yield and quality of Indian mustard (*Brassica juncea* L. Czernj. and Cosson) under different irrigation environments. *J. App. and Nat. Sci*. 7(1), 52–57.

Dahmardeh, M., Ghanbari, A., Syahsar, B. A., Ramrodi, M. (2010). The role of intercropping maize (*Zea mays* L.) and cowpea (*Vigna unguiculata* L.) on yield and soil chemical properties. *Afr. J. Agric. Res*. 5(8), 631–636.

Deakin, W. J., Broughton, W. J. (2009). Symbiotic use of pathogenic strategies: Rhizobial protein secretion systems. *Appl. Soil Ecol*. 7, 312–320.

Dexter, A. R. (2004). Soil physical quality part I. Theory, effects of soil texture, density, and organic matter, and effects on root growth. *Geoderma*. 120, 201–214.

Dhakal, Y., Meena, R. S., Kumar, S. (2016). Effect of INM on nodulation, yield, quality and available nutrient status in soil after harvest of green gram. *Legume Res*. 39(4), 590–594.

Dhaliwal, S. S., Sadana, U. S., Sidhu, S. S., Singh, G. (2011). Role of cropping systems in amelioration of aggravated micronutrients deficiency in alluvial soils of Punjab. *J. Plant Sci. Res*. 27(1), 21-29.

Doane, T. A., Horwath, W. R., Mitchell, J. P., Jackson, J., Miyao, G., & Brittan, K. (2009). Nitrogen supply from fertilizer and legume cover crop in the transition to no-tillage for irrigated row crops. *Nutr. Cycl. Agroecosys*. 85(3), 253-262.

Drinkwater, L. E., Wagoner, P., Sarrantonio, M. (1998). Legume-based cropping systems have reduced carbon and nitrogen losses. *Nat*. 396, 262–264.

Dusa, E. M. (2009). Researches regarding the productivity of oat-lentil intercropping in the organic agriculture system. *Res. J. Agric. Sci*. 41(1), 22–26.

FAO. (2017). *The future of food and agriculture – Trends and challenges*. Rome, Italy, p. 163.

Fester, T., Giebler, J., Wick, L. Y., Schlosser, D., Kästner, M. (2014). Plant-microbe interactions as drivers of ecosystem functions relevant for the biodegradation of organic contaminants. *Curr. Opin. Biotechnol*. 27, 168–175.

Fininsa, C. (1996). Effect of intercropping bean with maize on bean common bacterial blight and rust diseases. *Int. J. Pest Manag*. 42, 51–54.

Fujita, K., Ofosu-Budu, K. G., Ogata, S. (1992). Biological nitrogen fixation in mixed legume-cereal cropping systems. *Plant Soil*. 141, 155–176.

Fustec. J., Lesuffleur, F., Mahieu, S., Cliquet, J. B. (2010). Nitrogen rhizodeposition of legumes: a review. *Agron. Sustain. Dev*. 30, 57–66.

Garg, N. (2007). Symbiotic nitrogen fixation in legume nodules: process and signaling. A review. *Agron. Sustain. Dev*. 27, 59–68.

Ghaley, B. B., Hauggaard-Nielsen, H., Hogh-Jensen, H., Jensen, E. S. (2005). Intercropping of wheat and pea as influenced by nitrogen fertilization. *Nut. Cycl. Agroecosyst*. 73, 201–212.

Gilbert, G. A., Knight, J. D., Vance, C. P., Allan, D. L. (1999). Acid phosphatase activity in phosphorus deficient white lupin roots. *Plant Cell Environ*. 22, 80.

Grandy, A. S., Porter, G. A., Erich, M. S. (2002). Organic amendment and rotation crop effects on the recovery of soil organic matter and aggregation in potato cropping systems. *Soil Sci. Soc. Am. J*. 66, 1311–1319.

Hauggaard-Nielsen, H., Ambus, P., Jensen, E. S. (2001). Interspecific competition, N use and interference with weeds in pea-barley intercropping. *Field Crop Res*. 70, 101–110.

Hauggaard-Nielsen, H., Gooding, M., Ambus, P., Corre-Hellou, G., Crozat, Y., Dahlmann, C., Dibet, A., von Fragstein, P., Pristeri, A., Monti, M., Jensen, E. S. (2009). Pea–barley intercropping for efficient symbiotic N_2–fixation, soil N acquisition and use of other nutrients in European organic cropping systems. *Field Crop Res*. 113, 64–71.

Hauggaard-Nielsen, H., Jornsgaard, B., Kinane, J., Jensen, E. S. (2007). Grain legume–cereal intercropping: the practical application of diversity, competition and facilitation in arable and organic cropping systems. Renew. *Agric. Food Syst.* 23, 3–12.

Hayman, D. S. (1986). Mycorrhizae of nitrogen fixing legumes. *World J. Microbiol. Biotechnol.* 2(1), 121–145.

Heichel, G. H. (1987). Legume nitrogen: symbiotic fixation and recovery by subsequent crops. In: Heisel, Z. (ed) Energy in plant nutrition and pest control. *Energy World Agric.* 2, 63-80.

Helal, H. M. (1990). Varietal difference in root phosphatase activity as related to the utilization of organic phosphorus. In: Bassam, N. (ed) *Genetic aspects of plant mineral nutrition.* Kluwer Academic Publishers, Dordrecht, pp. 103-105.

Huang, W. Y. (2007). Impact of rising natural gas prices on U.S. ammonia supply, Outlook *Report No. (WRS-0702). USDA Economic Research Service*, Washington, DC, p. 18.

Inal, A., Gunes, A., Zhang, F., Cacmak, I. (2007). Peanut/maize inter-cropping induced changes in rhizosphere and nutrient concentrations in shoots. *Plant Physiol. Biochem.* 45,350-356.

Insam, H., Parkinson, D., Domsch, K. H. (1989). The influence of macroclimate on soil microbial biomass levels. *Soil Biol. Biochem.* 21, 211-221.

Jenkinson, D. S., Ladd, J. N. (1981). Microbial biomass in soil measurement and turnover. In: Paul E. A., Ladd, J. N. (eds) *Soil biochemistry*, Marcel Dekker, New York, 5, pp. 415–471.

Jensen, E. S. (1996). Grain yield, symbiotic N2 fixation and interspecific competition for inorganic N in pea-barley intercrops. *Plant Soil.* 182, 25–38.

Jensen, E. S., Peoples, M. B., Boddey, R. M., Gresshoff, P. M., Hauggaard-Nielsen, H., Alves, B. J. R., Morrison, M. J. (2012). Legumes for mitigation of climate change and the provision of feedstock for biofuels and biorefineries-a review. *Agron. Sustain. Dev.* 32, 329–364.

Jin, Q. J., Zhu, K. K., Cui, W. T., Xie, Y. J., Han, B., Shen, W. B. (2013). Hydrogen gas acts as a novel bioactive molecule in enhancing plant tolerance to paraquat-induced oxidative stress via the modulation of heme oxygenase-1 signalling system. *Plant Cell Environ.* 36,956–969.

Juma, K. A., Averbeke, V. (2005). Response of Muxe to N and P availability in pots. In: Tenywa J. S., Adipala, E., Nampala, P., Tussiime, G., Okori, P., Kyanmuhangire, W. (eds). *Afr Crop Sci Proc.*, Kampala, Uganda, 7, pp 1179–1182.

Kaiya, S., Utsunomiya, S., Suzuki, S., Yoshida, N., Futamata, H., Yamada, T., Hiraishi, A. (2012). Isolation and functional gene analyses of aromatic hydrocarbon-degrading bacteria from a polychlorinated dioxin-dechlorinating process. *Microbes. Environ.* 27, 1112010337-1112010337.

Keeler, B. L., Hobbie, S. E., Kellogg, L. E. (2009). Effects of long-term nitrogen addition on microbial enzyme activity in eight forested and grassland sites: implications for litter and soil organic matter decomposition. *Ecosyst.* 12:1–15.

Kelly, B., Allan, C., Wilson, B. P. (2009). Soil indicators and their use by farmers in the Billabong Catchment, southern New South Wales. *Aust. J. Soil. Res.* 47, 234–242.

Khan, M. S., Zaidi, A., Wani, P. A., Oves, M. (2009). Role of plant growth promoting rhizobacteria in the remediation of metal contaminated soils. *Environ. Chem. Lett.* 7, 1–19.

Kirkegaard, J., Christen, O., Krupinsky, J., Layzell, D. (2008). Break crop benefits in temperate wheat production. *Field Crop Res.* 107(3), 185–195.

Klauer, S. F., Francesch, V. R. (1997). Mechanism of transport of vegetative storage proteins to the vacuole of the paraveinal mesophyll of soybean leaf. *Protoplasma* 200(3), 174–185.

Kumar, K., Goh, K. M. (2000) Crop residues and management practices: effects on soil quality, soil nitrogen dynamics, crop yield and nitrogen recovery. *Adv. Agron.* 68:198–279.

Kumar, R., Reddy, R. M., Sinha, P. S., Tirkey, J., Singh, M. K., & Prasad, B. C. (2010). Impact of Leguminous Biomulching on Soil Properties, Leaf Yield and Cocoon Productivity of Tropical Tasarculture. *J. Entomol.* 7(4), 219-226.

Kumar, S., Sheoran, S., Kumar, S. K., Kumar, P., Meena, R. S. (2016) Drought: a challenge for Indian farmers in context to climate change and variability. *Prog. Res. Int. J.* 11, 6243-6246.

Ladha, J. K., Watanabe, I., Saono, S. (1988). Nitrogen fixation by leguminous green manure and practices for its enhancement in tropical lowland rice. In: *The International Rice Research Institute* (ed.) Sustainable agriculture: green manure in rice farming. IRRI, Los Banos, pp. 165-183.

Lal, R. (2006). Enhancing crop yields in developing countries through restoration of soil organic carbon pool in agricultural lands. *Land Degrad. Dev.* 17:197–206.

Lal, R. (2010). Soil quality and ethics: the human dimension. In: Lal R, Stewart BA (eds) *Food security and soil quality, Advances in soil science*. Taylor & Francis (CRC Press), Boca Raton, pp. 301–308.

Lal, R. (2011). Sequestering carbon in soils of agro-ecosystems. *Food Policy* 36:S33–S39.

Lal, R. (2014) Societal value of soil carbon. *J. Soil Water Conserv.* 69:186A–192A.

Lal, R. (2015). Restoring soil quality to mitigate soil degradation. *Sustain.* 7, 5875–5895. https://doi.org/10.3390/su7055875.

Lal, R. (2016). Soil health and carbon management. *Food. Energy Secur.* 5(4), 212–222. https://doi.org/10.1002/fes3.96.

Lampkin, N. H., Pearce, B. D., Leake, A. R., Creissen, H., Gerrard, C. L., Girling, R., Lloyd, S., Padel, S., Smith, J., Smith, L. G., Vieweger, A., Wolfe, M. S. (2015). The role of agroecology in sustainable intensification. *Report for the Land Use Policy Group. Organic Research Centre, Elm Farm and Game & Wildlife Conservation Trust,* https://www.nature.scot/sites/default/files/2017-06/A1652615.pdf.

Lansing, A. J., Franceschi, V. R. (2000). The paraveinal mesophyll: a specialized path for intermediary transfer of assimilates in legume leaves. *Aust. J. Plant Physiol*. 27, 757–767.

Larkin, R. P., Griffin, T. S. (2007). Control of soilborne potato diseases using Brassica green manures. *Crop Prot*. 26(7), 1067-1077.

Lazali, M., Boudsocq, S., Taschen, E., Farissi, M., Hamdi, W., Ralli, P., Sentenac, H. (2021). CROSYMED Project: Enhancing Nutrient Use Efficiency through Legumes in Agroecosystems of the Mediterranean Basin. *Sustain.* 13, 4695. https://doi.org/10.3390/ su13094695.

Lemke, R. L., Zhonga, Z., Campbell, C. A., Zentner, R. (2007). Can pulse crops play a role in mitigating greenhouse gases from north American agriculture. *Agron. J.* 99(6), 1719–1725.

Lewis, G., Schrire, B., Mackinder, B., Lock, M. (2005). Legumes of the World. Kew: Royal Botanic Gardens, p. 577.

Liang, B., Lehmann, J., Sohi, S. P., Thies, J. E., O'Neill, B., Trujillo, L., Gaunt, J., Solomon, D., Grossman, J., Neves, E. G., FJ, L˜a. (2010). Black carbon affects the cycling of non-black carbon in soil. *Org Geochem.* 41. 206–213.

Lithourgidis, A. S., Dordas, C .A., Damalas, C. A., Vlachostergios, D. N. (2011). Annual intercrops: an alternative pathway for sustainable agriculture. *Aust. J. Crop Sci.* 5, 396–410.

Lopez, C. G., Mundt, C. C. (2000). Using mixing ability analysis from two-way cultivar mixtures to predict the performance of cultivars in complex mixtures. *Field Crop Res.* 68(2), 121–132.

Lopez-Bellido. R. J., Fontána, J. M., López-Bellidob, F. J., López-Bellidoa, L. (2009). Carbon sequestration by tillage, rotation, and nitrogen fertilization in a Mediterranean vertisol. *Agron. J.* 102(1), 310–318.

Mahmud, K., Makaju, S., Ibrahim, R., Missaoui, A. (2020). Current progress in nitrogen fixing plants and microbiome. *Res. Plants* 9, 97; doi:10.3390/plants9010097.

Maitra S., Ray, D. P. (2019). Enrichment of Biodiversity, Influence in Microbial Population Dynamics of Soil and Nutrient Utilization in Cereal-Legume Intercropping Systems: A Review. *Int. J. Biores. Sci.* 6(1), 11-19. doi: 10.30954/2347-9655.01.2019.3.

Maitra S., Zaman, A., Mandal, T. K., Palai, J. B. (2018). Green manures in agriculture: A review, *J. Pharmacog. Phytochem.* 7 (5), 1319-1327.

Maitra, S. (2020a). Potential horizon of brown-top millet cultivation in drylands: A review. *Crop Res.* 55(1-2), 57-63, doi: 10.31830/2454-1761.2020.012.

Maitra, S. (2020b). Intercropping of small millets for agricultural sustainability in drylands: A review. *Crop Res.* 55 (3-4), 162-171.

Maitra, S., Shankar T, Banerjee P (2020). *Potential and Advantages of Maize-Legume Intercropping System, Maize - Production and Use* (Ed. Akbar Hossain), Intechopen, London, UK, pp. 1-14.

Maitra, S., Zaman, A. (2017). Bio-drainage in waterlogged areas for agricultural sustainability. *Int. J. Biores. Sci.* 4 (2), 79-84.

Maitra, S.; Hossain, A.; Brestic, M.; Skalicky, M.; Ondrisik, P.; Gitari, H.; Brahmachari, K.; Shankar, T., Bhadra, P.; Palai, J. B.; Jana, J., Bhattacharya, U., Duvvada, S. K., Lalichetti, S., Sairam, M. (2021). Intercropping–A Low Input Agricultural Strategy for Food and Environmental Security. *Agron.* 11,343. https://doi.org/10.3390/agronomy1102 0343.

Maitra, Sagar, Palai J. B., Manasa, P., Prasanna Kumar D. (2019). Potential of Intercropping System in Sustaining Crop Productivity, *Int. J. Agric. Environ. Biotechnol.* 12 (1), 39-45.

Marielle Margier, M., Georgé, S., Hafnaoui, N., Remond, D., Nowicki, M., Chaffaut, L. D., Amiot, M. J., Reboul, E. (2018). Nutritional Composition and Bioactive Content of Legumes: Characterization of Pulses Frequently Consumed in France and Effect of the Cooking Method. *Nutr.*10, 1668. doi:10.3390/nu10111668

McGill, W. B., Cannon, K. R., Robertson, J. A., Cook, F. D. (1986). Dynamics of soil microbial biomass and water-soluble organic C in Breton L. after 50 years of cropping to two rotations. *Can. J. Soil Sci.* 66:1–19.

Meena, R. S., Bohra, J. S., Singh, S. P., Meena, V. S., Verma, J. P., Verma, S. K., Shiiag, S. K. (2016). Towards the prime response of manure to enhance nutrient use efficiency and soil sustainability a current need: a book review. *J. Clean Prod.* 112, 1258–1260.

Meena, R. S., Gogai, N., Kumar, S. (2017a). Alarming issues on agricultural crop production and environmental stresses. *J. Clean Prod.* 142, 3357–3359.

Meena, R. S., Meena, P. D., Yadav, G. S., Yadav, S. S. (2017b). Phosphate solubilizing microorganisms, principles and application of microphos technology. *J, Clean Prod.* 145,157–158.

Meena, R. S., Meena, V. S., Meena, S. K., Verma, J. P. (2015a). Towards the plant stress mitigate the agricultural productivity: a book review. *J, Clean Prod*, 102, 552–555.

Meena, V. S., Maurya, B. R., Meena, R. S. (2015b). Residual impact of well-grow formulation and NPK on growth and yield of wheat (*Triticum aestivum* L.). *Bangladesh J Bot* 44(1), 143–146.

Meena, V. S., Maurya, B. R., Meena, R. S., Meena, S. K., Singh, N. P., Malik, V. K. (2014). Microbial dynamics as influenced by concentrate manure and inorganic fertilizer in alluvium soil of Varanasi, India. *Afr. J. Microbiol. Res.* 8, 257-263.

Melero, S., Madejon, E., Ruiz, J. C., Herencia, J. F. (2007). Chemical and biochemical properties of a clay soil under dryland agriculture system as affected by organic fertilization. *Eur. J. Agron.* 26, 327–334.

Mousavi, S., Yousefi-Moghadam, S., Mostafazadeh-Fard, B., Hemmat, A., Yazdani, M. R. (2009). Effect of puddling intensity on physical properties of a silty clay soil under laboratory and field conditions. *Paddy Water Environ.* 7(1), 45–5.

Moyer-Henry, K. A., Burton, J. W., Israel, D. W., Rufty, T. W. (2006). Nitrogen transfer between plants: a 15N natural abundance study with crop and weed species. *Plant Soil.* 282, 7–20.

Mugwe, J., Mugendi, D., Okoba, B., Tuwei, P., O'Neill, M. (2004). Soil conservation and fertility improvement using leguminous Shrubs in central highlands of Kenya: NARFP Case study. In: Bationo A (ed) *Managing nutrient cycles to sustain soil fertility in sub-Saharan Africa.* Academy Science Publishers, Nairobi, pp. 608.

Mundt, (2002). Use of multiline cultivars and cultivar mixtures for disease management. *Annu. Rev. Phytopathol.* 40,381–410. https://doi.org/10.1146/annurev.phyto.40.011402. 113723.

Nees, B., Anderberg, S., Olsson, L. (2010). Structuring problems in sustainability science: the multi-level DPSIR framework. *Geoforum* 41(3), 479–488.

Nuruzzaman, M., Lambers, H., Bolland, M. D. A, Veneklaas, E. J. (2006). Distribution of carboxylates and acid phosphatase and depletion of different phosphorus fractions in the rhizosphere of a cereal and three grain legumes. *Plant Soil* 281(1), 109–120.

Nwoke, O. C., Diels, J., Abaidoo, R., Nziguheba, G., Merckx, R. (2008.) Organic acids in the rhizosphere and root characteristics of soybean (*Glycine max*) and cowpea (*Vigna unguiculata*) in relation to phosphorus uptake in poor savanna soils. *Afr. J. Biotech.* 7(20):3620–3627.

Nygren, P., Leblanc, H. A. (2015). Dinitrogen fixation by legume shade trees and direct transfer of fixed N to associated cacao in a tropical agroforestry system. *Tree Physiol.* 35(2), 134-147.

Ogunwole, J. O., Iwuafor, E. N. O., Eche, N. M., Diels, J. (2010). Effect of organic and inorganic soil amendments on soil physical and chemical properties in a West Africa Savanna agroecosystem. *Trop. Subtrop. Agroecosyst.* 12, 247–255.

Paynel, F., Lesuffleur, F., Bigot, J., Diquélou, S., Cliquet, J. B. (2008). A study of 15N transfer between legumes and grasses. *Agron. Sustain. Dev.* 28, 281–290.

Peyraud, J. L., Gall, A. L., Lüscher, A. (2009). Potential food production from forage legume-basedsystems in Europe: An overview. *Irish. J. Agric. Food Res.* 48(2), 115-135.

Pikul, J. L., Johnson, J. M. F. Schumacher, J. T. E., Vigil, M., Riedell, W. E. (2008). Change in surface soil carbon under rotated corn in eastern South Dakota. *Soil. Sci. Soc. Am. J.* 72, 1738–1744.

Praharaj, S., Maitra, S. (2020). Importance of legumes in agricultural production system: An overview. *Agro Econ*. 7(2): 69-71.

Preissel, S., Reckling, M., Schläfke, N., Zander, P. (2015). Magnitude and farm economic value of grain legume pre-crop benefits in Europe: a review. *Field Crop Res*. 175:64–79.

Prell, J., Poole, P. (2006). Metabolic changes of rhizobia in legume nodules. *Trend Biotechnol.* 14:161–168.

Rahman, M. M., Islam, A. M. Azirun, S. M. and Boyce, A. N. (2014). Tropical Legume Crop Rotation and Nitrogen Fertilizer Effects on Agronomic and Nitrogen Efficiency of Rice. *Sci. World J.* Article ID 490841. http://dx.doi.org/10.1155/2014/490841.

Raji, J. A. (2007). Intercropping soybean and maize in a derived savanna ecology. *Afr. J. Biotechnol.* 6(16):1885–1887.

Ratnadass, A., Blanchart, É., Lecompte, P. (2013). Ecological interactions within the biodiversity of cultivated systems. In: *Cultivating biodiversity to transform agriculture*. Hainzelain ed., Springer, Dordrecht, pp. 141–179.

Reckling, M., Preissel, S., Zander, P., Topp, K., Watson, C., Murphy-Bokern, D., Stoddard, F. L. (2014). *Legume futures report 1.6-effects of legume cropping on farming and food systems, legumesupported cropping systems for Europe*. www.legumefutures.de. Accessed on 14 Apr 2019.

Salama, H. S. A., Nawar, A. I., Khalil, H. E., Shaalan, A. M. (2021). Improvement of Maize Productivity and N Use Efficiency in a No-Tillage Irrigated Farming System: Effect of Cropping Sequence and Fertilization Management. *Plants.* 10, 1459. https://doi.org/10. 3390/plants10071459.

Sanchez, P. A. (1976). *Properties and management of soils in the tropics*. John Wiley and Sons, New York, p. 618.

Sánchez-Navarro, V., Zornoza, R., Faz, Á., Fernández. J. A. (2019). Comparing legumes for use in multiple cropping to enhance soil organic carbon, soil fertility, aggregates stability and vegetables yields under semi-arid conditions. *Sci. Horti.* 246, 835-41.

Sarkar, R. K., Shit, D. and Maitra, S. (2000). Competition functions, productivity and economics of chickpea (*Cicer arietinum*)-based intercropping system. *Ind. J. Agron.* 45 (4): 681-86.

Sato, Y., Monincová, M., Chaloupková, R., Prokop, Z., Ohtsubo, Y., Minamisawa, K., Tsuda, M., Damborský, J., Nagata, Y. (2005). Two rhizobial strains, Mesorhizobium loti MAFF303099 and Bradyrhizobium japonicum USDA110, encode haloalkane dehalogenases with novel structures and substrate specificities. *Appl. Environ. Microbiol.* 71, 4372–4379.

Schädler, S., Morio, M., Bartke, S., Rohr-Zänker, R., Finkel, M. (2011). Designing sustainable and economically attractive brownfield revitalization options using an integrated assessment model. *J. Environ. Manag.* 92, 827–837.

Schelud'ko, A. V., Makrushin, K. V., Tugarova, A. V., Krestinenko, V. A., Panasenko, V. I., Antonyuk, L. P., Katsy, E. I. (2009). Changes in motility of the rhizobacterium Azospirillum brasilense in the presence of plant lectins. *Microbiol. Res.* 164, 149–156.

Scheublin, T. R., Ridgway, K. P., Young, J. P. W., van der Heijden, M. G. A. (2004). Nonlegumes, legumes, and root nodules harbor different arbuscular mycorrhizal fungal communities. *Appl. Environ. Microbiol.* 70, 6240–6246.

Schoenholtz, S. H., VamMiegroet, H., Burger, J. A. (2000). A review of chemical and physical properties as indicators of forest soil quality: challenges and opportunities. *Forest Ecol. Manag.* 138, 335–356.

Schuster, G. I. (2013). Cultivated, primarily as food source. In: Nutritional Aspects of Legumes, vol. I. *Encyclopedia of Life Support System* (EOLSS). www.eolss.net/ samplechapters/c10/e5-02- 02.pdf.

Sharma, A. R., Behera, U. K. (2009). Nitrogen contribution through Sesbania green manure and dual – purpose legumes in maize–wheat cropping systems: agronomic and economic considerations. *Plant Soil.* 325, 289–304.

Shen, H., Yan, X., Zhao, M., Zheng, S., Wang, X. (2002). Exudation of organic acids in common bean as related to mobilization of aluminium- and iron-bound phosphates. *Environ. Exp. Bot.* 48(1), 1–9.

Siddique, K. H. M., Johansen, C., Kumarrao, J. V. D. K., Ali, M. (2008). Legumes in sustainable cropping systems. In: Kharkwal MC (ed) Food legumes for nutritional security and sustainable agriculture.

Proceedings of the fourth international food legumes research conference (IFLRC-IV), October 18–22, 2005, New Delhi, India, vol 1. *Indian Society of Genetics and Plant Breeding*, New Delhi, pp. 787–819.

Sinclair, T. R., Vadez, V. (2002). Physiological traits for crop yield improvement in low N and P environments. *Plant Soil*. 245, 1–15.

Singh, G., Dhaliwal, S. S., Sadana, U. S., Walia, S. S. (2011). Surface and subsurface distribution of Zn, Cu, Fe and Mn as influenced by different cropping systems in Typic Ustocrepts soils of Punjab, India. *J. Plant Sci. Res.* 27(2), 175-188.

Singh, V. P. (2000). Planting geometry in maize (*Zea mays*) and blackgram (*Phaseolus mungo*) intercropping system under rainfed low hill valley of Kumaon. *Ind. J. Agron*. 45(2), 274–278.

Sørensen, L. H. (1987). Organic matter and microbial biomass in a soil incubated in the field for 20 years with 14C-labelled barley straw. *Soil Biol. Biochem.* 19, 39–42.

Soumare, A., Diedhiou, A. G., Thuita, M., Hafidi, M., Ouhdouch, Y., Gopalakrishnan, S., Kouisni, L. (2020). Exploiting Biological Nitrogen Fixation: A Route Towards a Sustainable Agriculture. *Plants*. 9, 1011. doi:10.3390/plants9081011.

Sparling, G.P. (1992). Ratio of microbial biomass carbon to soil organic carbon as a sensitive indicator of changes in soil organic matter. *Aust. J. Soil Res.* 30, 195-207.

Spehn, E., Scherer Lorenzen, M., Schmid, B., Hector, A., Caldeira, M., Dimitrakopoulos, P., Finn, J., Jumpponen, A., O'donnovan, G., Pereira, J. (2002). The role of legumes as a component of biodiversity in a cross-European study of grassland biomass nitrogen. *Oikos* 98, 205–218.

Squire, G. R., Quesada, N., Begg, G. S., Iannetta, P. P. M. (2019). Transitions to greater legume inclusion in cropland: Defining opportunities and estimating benefits for the nitrogen economy. *Food Energy Secur.* 8,e00175. https://doi.org/10.1002/fes3.175.

Srinivasarao, C., Venkateswarlu, B., Lal, R. (2012). Long-term effects of soil fertility management on carbon sequestration in a rice-lentil cropping system of the Indo-Gangetic plains. *Soil Sci. Soc. Am. J.* 76(1), 167–178.

Stagnari, F., Maggio, A., Galieni, A., Pisante, M. (2017). Multiple benefits of legumes for agriculture sustainability: an overview. *Chem. Biol. Technol. Agric.* 4, 2. https://doi.org/ 10.1186/s40538-016-0085-1.

Stevenson, F. J. (1982). Origin and distribution of nitrogen in soil. *Nitrogen Agric, Soils*, 22, 67-122.

Stronge, C., Stevenson, B. A., Harmsworth, G. R., Kannemeyer, R. L. (2020). A Well-Being Approach to Soil Health—Insights from Aotearoa New Zealand Dean, *Sustain*. 12, 7719. doi:10.3390/su12187719.

Sullivan, P. (2003). Intercropping principles and production practices. *Appropriate Technology Transfer for Rural Areas Publication.* http://www.attra.ncat.org.

Suman, A., Lal, M., Singh, A. K., Gaur, A. (2006). Microbial biomass turnover in Indian subtropical soils under different sugarcane intercropping systems. *Agron. J.* 98(3), 698–704.

Sumiahadi, A., Chozin, M. A., Guntoro, D. (2018). Effectiveness of *Arachis pintoi* Karp. & Greg. as biomulch to reduce soil erosion and increase soil fertility on maize cultivation. *Proc. Book, 5th Int Conf Sustain Agric Environ*. (ICSAE 5), Hammamet City, Tunisia, pp. 62-69.

Swarup, A., Wanjari, R.H. (2000). *Three Decades of All India Coordinated Research Project on Long-Term Fertilizer Experiments to Study Changes in Soil Quality, Crop Productivity and Sustainability*; IISS: Bhopal, India, http://www.iiss.nic.in/Rti%20Appl/ RTI%2080.pdf, accessed 29 August, 2021.

Tharanathan, R. N., Mahadevamma, S. (2003). Grain legumes—a boon to human nutrition. *Trend Food Sci. Tech.* 14, 507–18.

Tillman, G., Schomberg, H., Phatak, S., Mullinix, B., Lachnicht, S., Timper, P. and Olson, D. (2004). Influence of cover crops on insect pests and predators in conservation-tillage cotton. *J. Econ. Entomol*. 97:1217-1232.

Tiwari, R. C., Sharma, P. K., Khandelwal, S. K. (2004). Effect of green manuring through Sesbaniacannabina and Sesbaniarostrata and nitrogen application through maize (*Zea mays*) in maize– wheat (*Triticum aestivum*) cropping system. *Ind. J. Agron.* 49(1), 15–17.

Trasar-Cepeda, C., Leiros, C., Gil-Sotres, F., Seoane, S. (1998). Towards biochemical quality index for soils: an expression relating several biological and biochemical properties. *Biol. Fertil. Soil*. 26, 100–106.

Turnbull, J., Bowman, W. D. (2002). Variable effects of nitrogen additions on the stability and turnover of soil carbon. *Nat.* 419, 915–917.

Unger, P. W., Kaspar, T. C. (1994). Soil compaction and root growth: a review. *Agron J.* 86:759-766.

Vandenkoornhuyse, P., Husband, R., Daniell, T. J., Watson, I. J., Duck, J. M., Fitter, A. H., Young, J. P. W. (2002). Arbuscular mycorrhizal community composition associated with two plant species in a grassland ecosystem. *Mol. Ecol.* 11, 1555–156.

Varvel, G. E. (1994). Rotation and nitrogen fertilization effects on changes in soil carbon and nitrogen. *Agron. J.* 86,319–325.

Verma, J. P., Jaiswal, D. K., Meena, V. S., Meena, R. S. (2015a). Current need of organic farming for enhancing sustainable agriculture. *J. Clean Prod.* 102, 545–547.

Verma, S. K., Singh, S. B., Prasad, S. K., Meena, R. N., Meena, R. S. (2015b). Influence of irrigation regimes and weed management practices on water use and nutrient uptake in wheat (*Triticum aestivum* L. Emend. Fiori and Paol.). *Bangladesh J. Bot.* 44(3), 437–442.

Wattiaux, M. A., Howard, T. M. (2001). *Technical dairy guide: nutrition and feeding*. University of Wisconsin. http://babcock.cals.wisc.edu/de/html/ch6/nutrition_eng_ ch6.html. Accessed on 14 Apr 2019.

Yadav, G. S., Lal, R., Meena, R. S., Babu, S., Das, A., Bhowmik, S. N., Datta, M., Layak, J., Saha, P. (2019). Conservation tillage and nutrient management effects on productivity and soil carbon sequestration under double cropping of rice in north eastern region of India. *Ecol. Indicators* 105, 303-315.

Yadav, S. S., Hunter, D., Redden, B., Nang, M., Yadava, D. K., Habibi, A. B. (2015). Impact of climate change on agriculture production, food, and nutritional security. In: Redden, R., Yadav, S. S., Maxted, N., Dulloo, M. S., Guarino, L., Smith, P. editors. *Crop wild relatives and climate change.* New Jersey, USA, John Wiley, pp. 1–23.

Yong, T., Liu, X., Yang, F., Song, C., Wang, X., Liu, W., Su, B., Zhou, L., Yang, W. (2015). Characteristics of nitrogen uptake, use and transfer in a wheat-maize-soybean relay intercropping system. *Plant Prod. Sci.* 18, 388–397.

Yu, Y., Xue, L., Yang, L. (2014). A winter legume in rice crop rotations reduce nitrogen loss, and improve rice yield and soil nitrogen supply. *Agron. Sustain. Dev.* 34, 633-640.

Zahran, H. H. (1999). Rhizobium-legume symbiosis and nitrogen fixation under severe conditions and in an arid climate. *Microbiol. Mol. Biol. Rev.* 63, 968–989.

Chapter 2

Novel Silver Nanoparticles as a Modern Defense Tool in Agro-Ecosystems

Anamika Verma* **and Mohita Srivastava**
School of Agriculture, Lovely Professional University,
Phagwara, Punjab, India

Abstract

Silver nanoparticles are the most fascinating metallic nanomaterials involved in a wide range of agricultural applications. The employment of silver nanoparticles (AgNPs) as antimicrobial agents and antimicrobial coatings for crop protection is trending these days. Silver nanoparticles ranging from 1nm to 100nm in size constantly release a low concentration of silver ions that confer defense against serious pests and pathogens. Ample research work in the concerned area implicating the antimicrobial property of AgNPs has been conducted over the last decade. Wherein, AgNPs synthesis has been reported from various methods however its implication and the safety issues are yet to be addressed by the scientific community. Hereby, we discuss the properties, synthesis methods, and characterization of AgNPs. We extensively discuss the advancements and achievements of AgNPs concerning anti-bacterial, anti-fungal, and anti-pathogenic activities along with environmental nanoremediation, highlighting the key processes involved. Henceforth, we will attempt to accentuate the potential of AgNPs towards revolutionizing defense technology for sustained agro-ecosystem shortly.

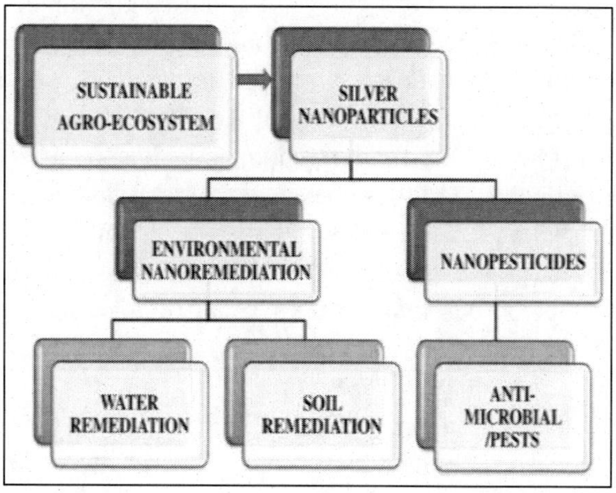

* Corresponding Author's Email: verma.anamika02@gmail.com.

In: Ecosystem Services
Editor: Hanuman Singh Jatav
ISBN: 978-1-68507-614-6

Keywords: green synthesis, nano-sensors, nanoremediation, nanopesticides, silver nanoparticles

1. Introduction

Progressive urbanization and expanding population accompanied by subsequent degraded agricultural lands have grave repercussions for sustainable food production as well as the ecosystem altogether. Henceforth researchers ought to pursue a sustainable solution towards agroecosystem management and monitoring. In the quest to achieve crop protection and production during green revolution, an indiscriminate administration of chemical pesticides triggered pesticide resistance in the pests. Consequently, contaminating the environment and shrinking natural resources (aquatic, air, and soil) for future generations leaving the sustainable agriculture concept far behind. High productivity with better quality accompanied by soil and water conservation etc. are the prime determinants of sustainable agriculture which mean environmental health is of utmost importance. Hence adopting immediate measures employing innovations is desired to further stop this exploitation. In this regard, nano-technological applications or innovations sound promising. Within this framework, silver nanoparticles specially provide an alternative method for plant protection and bioremediation (Mukhopadhyay 2014, Kar et al. 2016). Today silver nanoparticles (AgNPs) are picking up prevalence owing to their physic-chemical properties exhibiting excellent antibacterial (bacteriostatic and bactericidal) activity (Abdeen et al. 2014). Its biosynthesis can be achieved from metals by chemical and physical synthesis, where electrical, optical, and chemical properties are critical factors. However, recently many researchers are taking into hand green synthesis of silver nanoparticles (AgNPs) using biological systems providing a safer alternative to conventional methods (Dawadi et al. 2021). Green synthesis is becoming increasingly common these days, using biological agents such as plants, algae, fungi, bacteria, viruses, and plants for silver nanoparticles (AgNPs) development (Ahmed et al. 2016, Verma and Khanam 2020). Silver nano-particle (material that ranges between 10-100 nm dimensions) can alone or as a nanocarrier or nano-encapsulated pesticide (Verma and Khanam 2020; Verma, 2016) release a minute amount of chemical dosage with enhanced solubility, specificity, permeability, and stability (Bhattacharyya et al. 2016; Worrall et al. 2018). Thus, prompting the least chemical exposure to humans and the environment making it an obvious eco-friendly approach promoting sustainable agriculture (Nuruzzaman et al. 2016). Blending nano pesticides and nano sensor technologies can upgrade the agricultural system without impeding the soil microbes and the least chemical leaching, subsequently maintaining soil fertility and water purity. Nanoremediation is an emerging industry employing nanoparticles to treat groundwater (Bardos et al. 2014), wastewater (Theron et al. 2008), soil, soil particles, sediments (Gomes et al. 2013), wastelands and barren lands. However, evaluation of the current environmental risk assessment approach of silver nanoparticles and framing of their experimental protocols altogether is still pending. The alarming question of the environmental risk that nano products can pose owing to the probability of their accumulation in the food chain, resulting into some unpredictable behavior in the nature altogether necessitates biosafety measures formulation. Investigations concerning the toxicity level of nanoparticles, their residual impact on the food chain as well as their behavior studies related to the environment are the foremost research aspects to be addressed prior their marketing (Verma and Khanam 2020). In the lieu of the above, we hereby abridge the vast application of silver nanoparticles as pesticides themselves and their potential employment in agriculture and environmental system. We also provide to the reader, an insight into the future of the novel silver nanoparticles in agro-ecosystem altogether.

2. Silver Nanoparticle and Green Synthesis

The prime challenge of nanoparticle research is the development of a highly controllable nanoparticle synthesis approach, efficient characterization tools with better models, and theories analyzing research observations. Owing to certain physic-chemical properties and widespread applications in agriculture as fungicides, bactericides, and insecticides offering uniform distribution in a controlled manner with reduced plant damage has triggered its biosynthesis (Golinska et al. 2014; Iravani et al. 2014; John et al. 2020). The efficacy of AgNPs could be anticipated by having killed approximately 650 pathogenic types of microbes within 30 minutes (Shahrokh et al. 2014). Many researchers have found various physical and chemical methods of synthesis of AgNPs at the room as well as high temperature using microwave irradiation or sonication or light-induced synthesis (Iravani et al. 2014). Conventional AgNPs synthesis method is divided into two approaches viz. top-down and bottom-up. Recently green synthesis of AgNPs green synthesis has been picking up stream owing to its significant environment-friendly approach (Sreelakshmy and Deepa 2016). This method involves microbes, plant extracts, or natural reagents (sugars, biodegradable polymers like chitosan) as reducing or capping agents to synthesize AgNPs (Ahmed and Ikram 2015). These biosynthesized AgNPs are further characterized using transmission electron microscopy (TEM), scanning electron microscopy (SEM), energy-dispersive X-ray spectroscopy (EDX), X-ray diffraction (XRD), and Fourier transform infrared spectroscopy (FTIR) (Rao and Tang, 2017). The physical and chemical synthesis is considered more hazardous and labor-intensive as compared to biological/ green synthesis (Lee and Jun, 2019).

2.1. Synthesis *via* Micro-Organisms

Microorganisms led biosynthesis of AgNPs includes certain fungi, bacteria, and alga which are considered effective in absorbing and extracting metals via secreting significant quantities of enzymes hydrolyzing metals leading to enzymatic metal ion reduction (Anand and Bhagat 2019). In this context, an array of plant pathogenic fungus, *Fusarium oxysporum, Trichoderma viride, Aspergillus flavus, Aspergillus fumigates, Verticillium sp., Fusarium oxysporum, Cladosporium cladosporioides, Chrysosporium tropicum, Phoma glomerata,* etc. can generate Ag NPs extracellularly rapidly (Li et al. 2012). Similarly, Ag NPs have been isolated exclusively from various bacteria (*Bacillus licheniformis, Escherichia coli, Corynebacterium glutamicum, Bacillus cereus, Pseudomonas stutzeri, Ureibacillus thermosphaericus, Pseudomonas aeruginosa*) (Juibari et al. 2011; Pandian et al. 2014; Huq, 2020). Anti-microbial properties of AgNPs are also highly recognized against food spoiling microbes in the processing industries. In this aspect, Huq (2020) in his study synthesized AgNPs biologically *via* bacteria *Pseudoduganella eburnae* (which was isolated first from lagoon sediments as studied and isolated by Jin et al. (2017)) and studied their antibacterial mechanism. Various studies have recorded the potent activity of biosynthesized AgNPs against *Bacillus subtilis, Staphylococcus aureus* (Loretta et al. 2016), *Pseudomonas aeruginosa, Escherichia coli, Klebsiella pneumonia, shigella sp.,* etc. To say the least, recently John et al. (2020) reported an efficient and rapid synthesis of stable AgNPs from a new *Pseudomonas* strain isolated from protozoa *Euplotes focardii* with high antimicrobial activity against *Escherichia coli, Staphylococcus aureus,* and *Candida albicans*, prospected its further exploitation for bioremediation and in antibiotics production.

2.2. Synthesis *via* Plant Extracts

The AgNPs biosynthesis exploiting micro-organisms is reliable but it usually deals with isolation and incubation of living cells as well as reduction of silver nitrate which in turn is time-consuming and a complex process that takes at least 24 hours for AgNPs production (Rao and Tang, 2017). Hence, the synthesis of AgNPs using various plant extracts is advantageous concerning convenience, reaction rate, feasibility, stability, etc. In green synthesis via plants, the aqueous plant extract or fruit extracts are mixed with silver nitrate which acts as reducing or capping agents and kept at high temperature for 30 to 90 minutes, ultimately reducing silver ions into AgNPs (indication being the change in color to brown) (Figure 1). Ahmed et al. (2016) in his review paper have enlisted various sources for the synthesis of Ag NPs as reported by

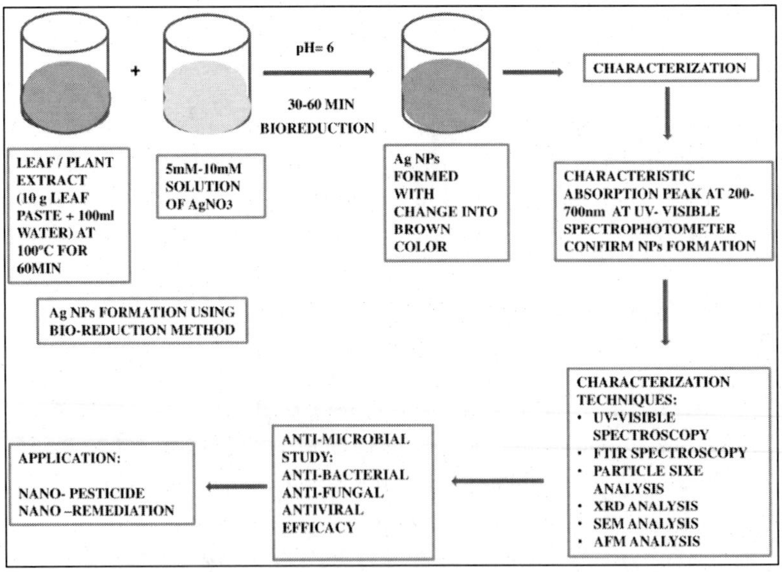

Figure 1. Silver nanoparticles green synthesis via plant extracts.

various researchers utilizing an array of the plant (powdered and aqueous) extracts viz. beetroot, *Terminalia arjuna* (Ahmed and Ikram, 2015), *Ocimum sanctum* (Bindhani and Panigrahi, 2015), *Aloe vera* (Dinesh et al. 2015), *Piper nigrum* (Aparna Mani et al. 2015), *Elettaria cardamom* (Sharada and Omprakash 2015), Papaya peel (Prasad and Srinivasulu, 2015), etc. to name a few and further went on to test their anti-microbial activity. Similarly, Verma and

Khanam (2020) have also mentioned many plant-derived sources for AgNPs biological synthesis. Recently, some AgNPs has been reported to be synthesized from leaf extracts of *Cucumis prophetarum* (Hemlata et al. 2020), *Tragopogon collinus* (Seifipour et al. 2020), *Cinnamomum tamala* (Nahar et al. 2020), *Impatiens balsamina* and *Lantana camara* fresh leaves (Aritonang et al. 2019), *Solanum mammosum* fruit extracts (Pilaquinga et al. 2019), *Eriobotrya japonica* (Rao and Tang 2017), etc. maximum of which conferred anti-microbial activity. The potential of higher plants as a source for this AgNPs formation and in terms of the extent of resistance towards an entire class of pathogen and pests is still largely unexplored. Toxicity studies of AgNPs on wide spectra of plant pathogens as well as its immediate and long-term effect on the environment opens up the door for a new range of plant antimicrobial agents demanding comprehensive study (Jain, 2014).

3. Silver Nanoparticles as Plant Protectant

According to the literature, AgNPs alone can impart higher insecticidal or anti-pathogenic activities owing to high stability and longevity (Athanassiou et al. 2018) (Figure 2). Ashraf et al. (2020) biologically synthesized AgNPs through *Melia Azadirachta* and found them effective against *Fusarium* wilt in tomato crops. Similarly, Kumari et al. (2017) reported AgNPs conferring resistance against early blight disease of tomato caused by pathogen *Alternaria solani*. In one instance, microbial spores incubated with AgNPs in the laboratory recorded significantly reduced *Fusarium culmorum* mycelial growth (Kasprowicz et al. 2010) and inhibition of fungal hyphae conidia germination in other (Lamsal et al. 2011). Besides systematic trials performed by several scientists presented documentation of AgNP's broader applicability to regulate a range of plant pathogens namely *Setosphaeria turcica* (Huang et al. 2020), *Colletotrichum* sp. (Lamsal et al. 2011), sclerotium-forming fungus (Min et al. 2009), cucumber powdery mildew (Lamsal et al. 2011), *Pseudomonas syringae* pv. Tomato (Kim et al. 2011), gray mold in strawberry (Moussa et al. 2013), etc. Application of Ag NPs against various pathogenic micro-organisms such as *Alternaria* spp., *Sclerotinia* spp, *Rhizoctonia*, *Botrytis* spp. as well as yellow bean mosaic virus has been enlisted by various researchers in their review (Bardos et al. 2014; Mishra and Singh, 2015). According to some researchers application of Ag NPs reportedly increase the host plant resistance, increase chlorophyll and anti-oxidants content, improved enzyme activity especially in the phenyl propenoid pathway which associates with cell-wall strengthening and production of bioactive compounds like lignin, suberin, quinones, etc. which destroy pathogen's proteolytic enzymes (Kumari et al. 2017; Ashraf et al. 2020). Likewise, AgNPs have proved their potential in pest control coupled with or without chemical pesticides (Table 1.). *Bacillus thuringiensis kurstaki* led low cost- and ecofriendly synthesis of AgNPs is reported to act as a nanobiopesticide for insect pest control (Sayed et al. 2017). Several synthesized NPs have been reported as highly efficient against arthropod pests like moths, beetles, ticks, etc. The cotton bollworm, *Helicoverpa armigera* was found to be very susceptible to AgNPs synthesized by leaf aqueous extract of *Euphorbia hirta* since all larval instars and pupae exhibited high mortality levels (C80%) after only 4 days exposure in cotton, *Gossypium hirsutum* L., leaves that had been treated with the NPs at 10 ppm (Devi et al. 2014).

When green synthesized AgNPs are applied as a bioweapon on microbial cells they cause chemical transformation via cytotoxicity which consists of various steps culminating into cell death or change in the structure of the cell membrane as illustrated in Figure 3. Just after application of AgNPs as nano pesticides to the crops interaction between the microbial cells and AgNPs in the environment and affected plant area or diseased areas commences. Firstly AgNPs adheres to the microbial cell membrane causing ruptures or permeability and modifying the cell membrane bi-lipid layer. It leaks the cellular content and impairs transport activity. Secondly, the permeable cell membrane now allows the AgNPs to penetrate the cell resulting in Ag ions interacting with numerous cellular organelles and biomolecules affecting the corresponding cellular function preferably destabilizing, or denaturing the proteins and DNA. Thirdly, ROS (Reactive Oxygen Species) and free radicals generation start causing cellular toxicity *via* proteins and lipids oxidation, mitochondrial dysfunction, ribosome destabilization, and endoplasmic reticulum stress. Finally, genotoxicity due to DNA damage via strands breakage, transcription and translation inhibition, and cellular mutations occur ultimately leading to cell death or apoptosis. As AgNPs have a wide surface area they get oxidized and release Ag^{+} ions in presence of water as in the cases with microbial cells. However, the rate of Ag^{+} ions release also depends on the size and shape of the nanoparticles as well as the microbial enzymes and proteins as well as the ROS formed due to AgNPs also determine the silver ion release and cytotoxicity. Other factors include the type of microbes (bacteria, fungus, and virus) and thickness, architecture, the composition of the concerned microbe. According to some studies, gram-negative bacteria are more susceptible to gram-positive cells owing to the differences between their cell membranes as reviewed by Lee and Jun (2019).

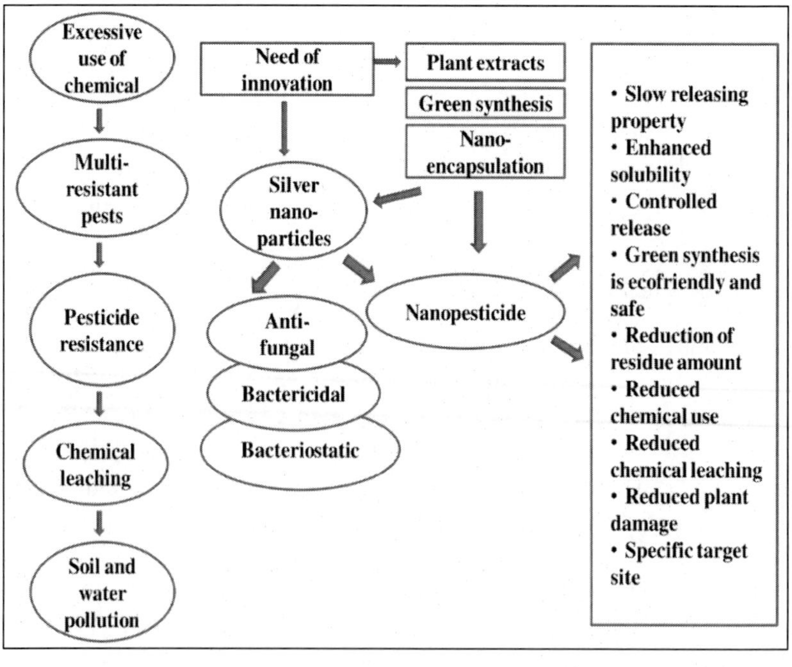

Figure 2. Anti-pathogenic and pesticidal activity of silver nanoparticles (adopted from Verma and Khanam, 2020).

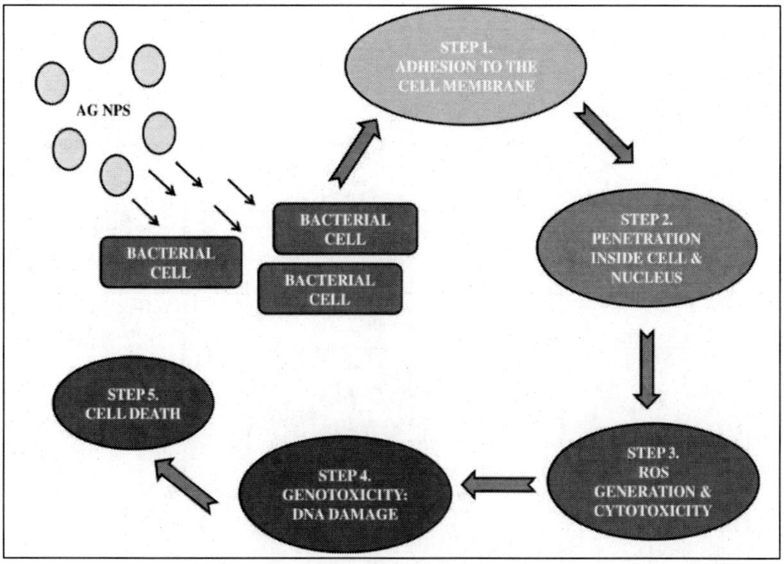

Figure 3. Mechanism of action of silver nanoparticles against microbial cells.

Table 1. Silver nanoparticle, their stabilizing agents and anti-pathogenic activities

AgNPs (protectants)	Stabilizing agent/Surfactants	Target microbe/ pests	Category	Function/effects	Ref.
AgNPs	Microalgae led biosynthesis	Various agricultural pathogens	Fungus and bacteria	Anti-microbial property	Terra et al. (2019)
AgNPs	Fructose stabilized silver nanoparticles	*Erwinia amylovora, Clavibacter michiganensis, Ralstonia solanacearum, Xanthomonas campestris pv. campestris* and *Dickeya solani*	Bacteria	Inhibitory properties	Dzimitrowicz et al. (2018)
AgNPs	Citrate-coated silver nanoparticles	*Aspergillus parasiticus*	Fungus (Mycotoxin producing filamentous)	Reduction in aflatoxin biosynthesis	Mitra et al. (2017)
AgNPs	*Stenotrophomonas sp.* mediated	*Sclerotium rolfsii*	Fungi	Reduced sclerotia germinatio	Mishra et al. (2017)
AgNPs	SDS, SDBS, TX-100, and Tween 80	*Ralstonia solanacearum*	Bacteria	Enhancement of antibacterial activity	Chen et al. (2016)
AgNPs	Graphene-oxide-silver nanocomposite	*Fusarium graminearum*	Fungus	Restricts spore formation	Chen et al. (2016)
AgNPs	Biogenic nanoparticles	*Candida albicans, Fusarium oxysporum,* and *Aspergillus flavus*	Fungi	Anti-fungal properties	Aziz et al. (2016)
AgNPs	*Bacillus pumilus, B. persicus*, and *Bacillus licheniformis mediated*	Bean Yellow Mosaic Virus	Virus	Anti-viral activity	Elbeshehy et al. (2015)

Table 1. (Continued)

AgNPs (protectants)	Stabilizing agent/Surfactants	Target microbe/ pests	Category	Function/effects	Ref.
AgNPs	30-150mg/ml	*Meloidogyne* sp.	Nematode	Reduced number of nematodes by 99%	Cromwell et al. (2014)
AgNPs	DNA directed Silver nanocomposite	*Xanthomonas perforans*	Bacteria	Cell death or decreased viability	Oscoy et al. (2013)
AgNPs	*Carisa carandus* berry water extracts	Pathogenic fungi	Fungi	Anti-pathogenic effects	Kim et al., 2012
AgNPs	Acalypha indica leaf extracts	*Alternaria alternata, Sclerotinia sclerotiorum, Macrophomina phaseolina, Rhizoctonia solani, Botrytis cinerea* and *Curvularia lunata*	Fungi	Inhibitory effect at 15 mg concentration	Krishnaraj et al. (2012)
AgNPs	Colloidal silver nanoparticles	*Colletotricum* sp. (Pepper anthracnose)	Fungus	High efficiency in controlling spore producing fungi	Lamsal et al. (2011)
AgNPs	Silver ions and nanoparticles	*Bipolaris sorokiniana* and *Magnaporthe grisea*	Fungi	Anti-fungal property	Jo et al. (2009)

3.1. Silver Nanoparticles Encapsulated Pesticides

Environmentally sustainable nano-encapsulated pesticides with target pest specificity and improved stability (Anand and Bhagat, 2019; John et al. 2020), show improved solubility and stabilities of active ingredients overcoming pesticides low water solubility issues, decreased volatilization, improved stability with a slow sustained release (Venugopal and Sainadh, 2016; Worrall et al. 2018). Rouhani et al. (2012) synthesized Ag NPs *via* the solvo-thermal method and mixed them with different concentrations of insecticides and found an effective control over *Aphis nerii*, an oleander aphid a common pest to ornamental plants. A bioassay of AgNPs and malathion was performed against red flour beetle (*Tribolium castaneum*) which provided the conclusion that the mixture of both showed high insecticidal efficacy (Alisha and Thangapandiyan, 2019). Malandrakis et al. (2020) coated benzimidazoles (BEN-R) thiophanare methyl (TM) and carbendazim with AgNPs and found increased resistance via fungi-toxic effects against *Monilia fructicola* major pest of apple. Similarly, Ihegwuagu and coworkers (2015) encapsulated dichlorvos and chlorpyriphos with starch-AgNPs and recorded high encapsulation efficiency and enhanced insecticidal delivery. Though AgNPs, can be used against a wide spectrum of noxious pest species, very little research has been conducted concerning AgNPs encapsulated pesticides, most of the experiments have been conducted utilizing other metal nanomaterials. Past surveys have shown that AgNPs can alleviate the deleterious effects of synthetic pesticides and provide target-specific management of crop pests. Therefore, could help establish a smart nanosystem to mitigate environmental imbalances ensuring food safety and food productivity (Nuruzzaman et al. 2016). Nevertheless, the techniques for substantial contributions to agricultural systems ought to be developed, as some facets enlisted by Fraceto et al. (2016) in their review. Adequate work and information are expected to better comprehend the nanopesticidal-related effects and environmental fate.

4. Silver Nanoparticle Led Environmental Remediation

Nanoremediation is an emerging industry using nanoparticles to treat groundwater (Bardos et al. 2014), wastewater (Theron et al. 2008), soil, soil particles, sediments (Gomes et al. 2013), wastelands, barren lands, etc. Wherein, a nanoparticle is brought into direct contact with the varied contaminant which either immobilizes the contaminant or detoxifies it (Table 2).

Table 2. Environmental nanoremediation of various agro-chemicals accumulated in water

Target Chemical	Form of AgNPs	Technique	Function/ property	Action	Ref.
Chromium ions and carbaryl	Reduced graphene-oxide based AgNPs	Localized surface plasmon resonance (LSPR) colorimetric sensor	Insecticide	Calorimetric probes in detection of pesticides in water	Minh et al. (2020)
Chlorpyriphos	Hibiscus leaf extract meadiated AgNPs coated and fused into polyurethane foam and glass beads	Mineralization	Pesticide	Removal of pesticide from water	Varghese et al. (2019)
Fipronil	*Brassica*-AgNps, *Ipomoea-AgNps*, *Camellia*-AgNps, and *Plantago*-AgNps	Pesticide degradation	Insecticide	Remediation of contaminated water and flooded soils	Romeh et al. (2018)
Chlorpyriphos and malathion	Cellulose-acetate membrane incorporated AgNPs	Pesticide degradation (nanoremediation)	Organophosph orothioate pesticides	Time-dependent mineralization of pesticicides	Manimegalai et al. (2014)
Permethrin	Chitosan-salt- AgNPs Composite	Bio-nanoremediation studies	Organo-chlorine pesticide	High pesticide removal from water	Rahmanifer and Dehagi et al. (2014)
Chlorpyriphos and Malthion	polyurethane foam (PUF) based silver nanoparticles	Bio-nanoremediation	Organophosph orothioate/ chlorine based pesticides	Mineralisation of pesticides from water	Manimegalai et al. (2012)
Chlorpyriphos	Ag@citrate NPs on neutral alumina	Degradation	Organophosph orothioate pesticide	Drinking water purification	Bootharaju and Pradeep (2012)
Atrazine	Chitosan-silver Nanoparticles Composite	Bio-nanoremediation	Pre-emergence broad-leaf herbicide	Drinking water filtration system; reduction in pesticide content	Saifuddin et al. (2011)
Chlorpyriphos and Malthion	Alumina-power loaded with AgNPs	Pesticide degradation/ mineralisation (nanoremediation)	Organophosph orothioate/ chlorine based pesticides	Elimination of pesticides from drinking water	Nair and Pradeep (2007)
Endosulphan	AgNPs	Pesticide degradation/ mineralisation (nanoremediation)	Cancerous property	Removal of pesticide from water	Nair et al. (2003)

Drainage of agro-industrial wastage directly into water bodies and leaching of fertilizers in dissolved form i.e., nitrate from the soil due to nitrification of nitrogen by soil microorganisms causes water contamination. In ideal conditions, 50% of nitrogen fertilizers applied are used by plants, 20% evaporates, 15-25% remain in the soil and 10% of it interferes with the groundwater (Kar et al. 2016). Now, these nitrates continuously accumulate toxic compounds in water bodies creating an oxygen-less environment resulting in eutrophication in water bodies (Bardos et al. 2014). In this context, silver nanoparticles have enormous potential as wastewater treatment catalysts and as redox-active media and are used as antimicrobial compounds for coliform found in wastewater, active biocides against Gram-positive and Gram-negative bacteria including *Escherichia coli*, *Staphylococcus aureus*, *Klebsiella pneumonia,* and *Pseudomonas aeruginosa*. The application begins with the injection of silver nanoparticles into a contaminated water medium which reaches the contaminants places through water flow. Upon contact the nanoparticle sequester contaminants *via* adsorption, complexation, or immobilization and in advanced cases by detoxifying them completely or converting them to the least harmful forms. Because of its wide surface area by volume ratio and the inclusion of a higher number of reactive sites, nanoparticles can be extremely reactive. This enables maximum pollutant interaction, resulting in a rapid significant decrease in potential contaminants. Since they are so small, Brownian or random motions govern their physical movement in the water, rather than wander wall effects. In terms of size and high density, the motion of micrometer-sized particles, especially micro-scale metal particles, is entirely controlled by gravity-induced sedimentation. Nano-sized particles can be conveniently suspended in water during the development and production stages in the absence of adequate electrostatic surface forces, thereby providing a flexible remediation device. This allows automatic penetration further into the subsurface whereby chemicals are contained as a liquid. In addition to nanoparticles, a nanosensor can be added to search and reach the contaminants site (discussed later).

Common pollutants like heavy metals, agro-based pesticides, and fertilizers are responsible for soil pollution leaving them barren (Awasthi et al. 2016). Primarily, soil condition (physical and chemical) acts as indicators of contamination such as a decrease in soil microbial community, physical appearance, soil humus, soil basal respiration and soil testing of degraded lands (Azarbad et al. 2016; El-Ramady et al. 2017). Hence, monitoring the extent of soil pollution and types of toxic contaminants and remediate it by using various advanced techniques like nanoremediation is a must (Lu et al. 2015). Soil remediation methods are of three types i.e., physical, chemical, and biological. Physic-chemical engineering methods demand excessive digging up of soils and then removal of the wastes and contaminants from that area which are ultimately heaped elsewhere (Kenessov et al. 2016). This is a very expensive and labor-intensive or machine-oriented method along the handling of pollutants or hazardous material becomes a tedious job (Adki et al. 2014). On the other hand, biological nanoremediation involving microbes (microbial remediation) and plant (phytoremediation) led nanoparticles to eliminate contaminants provides a more promising technology to curb this menace (Mustafa and Komatsu 2016). There are many phytoremediation studies of soil using nanoparticles like silver, copper, zinc, aluminium, etc. that have given positive results (Shah et al. 2017). Nanoparticles promote pollutant degradation via numerous means including certain redox reactions, surface processes, adsorption, and exchange of ions, surface complexion, and electrostatic interaction. Given these accounts for nanoparticle's potential, nevertheless, they appear to have self-aggregation and nanoparticles may interact with suspended solids or sediments, whereby they can bioaccumulate and join the food chain or water table. So these fate processes must depend on both particle and the ecological system characteristics (Askari et al. 2012).

4.1. Nanoremediation: Understanding the Mechanism

A well-planned nanoremediation project starts from an in-depth site investigation along with designing a conceptual site model and overall study of contamination pathways that directly contact living organisms. Selection of the most suitable nanoparticle or single remediation technology is needed to be followed. A pre-screening is done to assure whether the nanoparticle application is the only option to remediate the site and if so what particular property of a nanoparticle would offer a better result. This also depends on the contaminant size, spread, depth, amount, nature, physical and chemical properties, general site conditions, hydrochemical site conditions or hydrogeological conditions, etc., and hence a chemical or physical remediation technique is selected. After pre-screening, depending upon favorable factors and risk factors particles are selected and then are investigated or tested for the probable reactions and their ultimate fate in the environment after remediation (Figure 4). Now, the application of nanoparticle selected depends further on numerous conditions like subsurface, water table depth, hydraulic conductivity, extent and amount of contaminant (general condition), groundwater velocity, stoichiometry, redox potential, sulfate and nitrate content in water or soil (hydro-chemical condition) and type of particle injection, the pressure needed, expected gas production and contaminant distribution (hydro-geological condition) (NanoRem, 2017). Pre-consideration of the basic functioning of a particle in a contaminated site containing different categories of contaminants like non-reducible pollutants [hydrophobic (pollutants with aromatically bound halogen, pesticides, pharmaceuticals, etc.) and hydrophilic (acids, alcohols, fuel oxygenates etc.)] and reducible pollutants (halogenated hydrocarbons, metal/metalloids, pesticides, pharmaceuticals with aliphatically bound halogens, nitro compounds, etc.) is necessary. So far there are three basic modes of action of nanoparticles on pollutants *viz.* reduction (most common), oxidation (rare), and adsorption (mostly iron nanoparticles). On-site parameters for site-specific tests are also required for a better understanding of mobility and reactivity of nanoparticles which can be performed in the lab before or directly in the field if all the parameters lie in an optimum range. After injection, site-specific mobility tests are conducted and a distinction between the radius of influence (total volume of porous media affected by the injection or volume reached by the transport fluid) and the radius of transport (distance from point of injection within which 80% of the injected nanoparticles are emplaced) is needed to be formed. Mostly radius of influence is greater than the radius of transport and to increase the latter some cascading column tests in conjunction with a numerical model are executed. Site-specific mobility tests and numerical modeling has been explained in guidelines given by NanoRem (2017). Later if nanoparticles transportation and distribution is achieved as per expectations, characterization of nanoparticles, numerical, and experimental up-scaling are done. Characterization of the nanoparticle's interactions with contaminants is executed by Atomic Force Microscopy (AFM), Transmission Electron Microscopy (TEM), Laser Confocal Microscopy, etc. Thus, raw data collected consists of site condition, reactivity test, and characterization of particles, the product formed, bi-product and intermediates formed along with life-time of reactive particles or their ultimate environmental fate. These raw data coupled with monitoring tools can determine the reaction products to demonstrate successful remediation (Figure 5). During chemical, remediation monitoring is essential to analyze the products and by-products formed as a result of reactions. There are two approaches of monitoring one is integrative type [(access parameters like contaminant mass, emissions, etc.) and the other is point type (provide details of chemical processes (intermediates, biproducts, etc.)] approach combination of both ensures successful remediation. Hence, long-term monitoring ensures whether the nanoremediation goal is completely achieved or not. The failure is usually due to incomplete reduction and underestimation of the longevity vs process time. In such a case, the researcher is left with two options either re-injection of particles and their monitoring or search alternate probably more appropriate nano-remediation technology keeping in mind the regulatory issues.

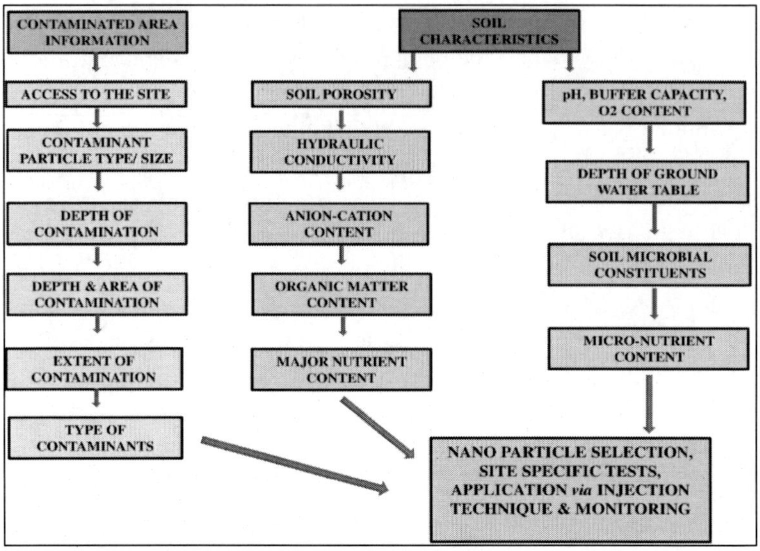

Figure 4. Pre-screening techniques for selection of nanoparticles for site-specific remediation.

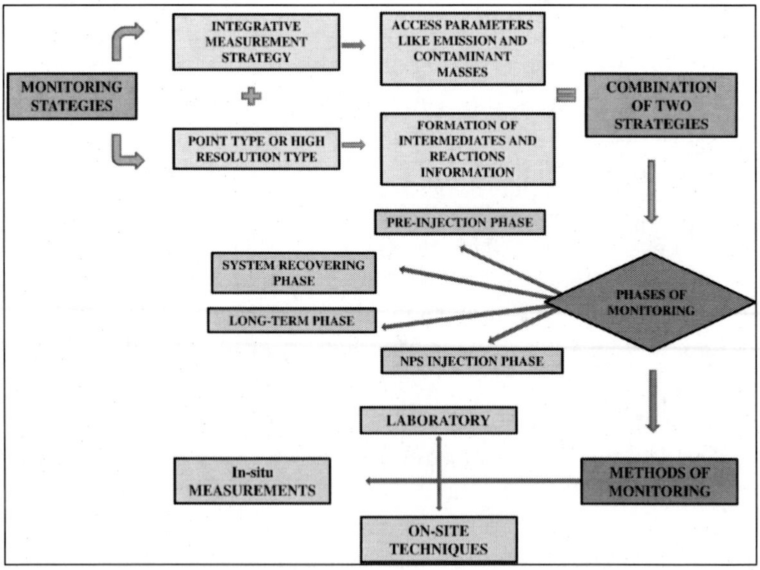

Figure 5. Monitoring techniques for successful environmental nanoremediation.

4.2. Nano-Biosensors Led Environmental Remediation

Decontamination and purification of drinking water and other water bodies need to be addressed in developing countries on an urgent basis. This can easily be achieved via nano-bio sensors with biosensing devices offering ease in production, in situ solicitation, and a broad range of coverage as compared to the traditional analytical procedures (Grawe et al. 2015). Typically, a sensor is composed of two parts: the receptor and the transducer (Wilson, 2005). Wherein, the receptor coheres to a distinct analyte and is developed via biological or green process adding up to the affinity, explicitness, reaction flow, and lifespan to the biosensors in a particular contaminated environment (Turner, 2013). The second component i.e., transducer is independent, which operates on optical, electrochemical, and varying monitoring techniques responsible of converting one form of energy to another (Turner, 2013). The mechanism follows as initially the bioreceptor identifies and coheres to the contaminant and their biological processes are then transformed into various signal displays in accordance to the presence of the number of contaminants in a particular site (Touhami, 2014). The sensory signals produced are of different types viz. florescence, electric pulse, amplitude, photo-luminance, etc. (Xu and Ying, 2011). According to this, the biosensors belongs to two major categories such as optical and electrochemical biosensors (Khanam et al. 2020). The optical sensing mainly works on colorimetric, fluorescent, and surf surface plasmon resonance principle whereas, electrochemical techniques can be employed *via*. voltammetry, amperometry, potentiometry, and impedemetric. Among them, voltammetry is widely studied for analysis of trace metals and pesticides, involving electrochemical accumulation of analyte at the surface of the electrode, followed by its oxidation during a linear sweep or pulsed reverse scan, providing detection limits much smaller than those of expensive spectroscopy/chromatography methods(Touhami, 2014; Balakrishnan et al. 2019). Reportedly, biosensors have made a breakthrough in monitoring pesticide and heavy metal contamination in the aqueous systems (Justino et al. 2017; Khanam et al. 2020).

5. Research and Development

Limited advancement in R&D industry because of confidentiality issues and unclear applications has been observed. Moreover, the regulatory burdens in some countries (e.g., The EU) have reduced the pace of its development and researches. Big agrochemical and pesticide companies that were actively involved in nano-based product formation a decade ago, now have ceased their work as a result of scarce researches owing to nanoparticle toxicity linkage with environmental and health issues. The uncertainity is the sole reason why we don't see a single registered pesticide in the market that contains nanomaterials to date (Kah et al. 2015). If we analyze the current scenario, broadly, there are five prime objectives of research and development affecting nanoparticle-led technology incorporation into agriculture and the environment. First, demand for solid research related to the understanding of the interaction of nanopesticides with plants. Second, methods or implications related to the better understanding of the fate of nanoparticles and nanopesticides in soil and the environment. Third, nano-enabled agrochemical and soil microbiome interaction study. Fourth, use of nanotechnology in search of the least toxic/ harmful nano-formulation. Fifth, ratify regulatory frameworks to deal with nanoformulations-environmental safety aspects. Taking into account the above points, some latest researches/progress done on a small scale (or laboratory) includes size-dependent uptake of nanomaterials where minute size of AgNPs (3.5 to 20nm) enough to perform an antimicrobial activity is advocated rather than bigger size (<50nm) with greater chances to enter into the plants through their stomatal opening and cuticle layer (Cai et al. 2020; Li et al. 2020). AgNPs uptake via plants is also surface-charge dependent, per se, the plant cell wall is negatively charged and so it favors the uptake of cationic nanomaterials, where on the other hand, anionic nanomaterials possess the high rate of translocation once it gets entry inside plants (Liu et al. 2019). Selection of nanoparticles can be done based on anatomy of the plants which can also hasten or reduce the absorption rate (Spielman-Sun et al. 2019). Studies related to abiotic stress resistance are also a new dimension of nanomaterials which was explored recently (Hojjat and Kamyab, 2017; Iqbal et al. 2019; Nguyen et al. 2020). Uptake, quantification, assimilation, translocation, and interaction have been studied by various researchers wherein, various characterization techniques come into play an important role (Mozhayeva and Engelhard, 2019). Various scientific works related to the physiochemical response of plants towards nanomaterial provides an insight to the interactions and the ultimate fate. Nano-based seed priming or coating techniques lead to higher absorption of water and minerals in the plants, there, nanoparticles may get easily translocated into the plant and start functioning at the molecular level by activating various genes at DNA as well as mRNA levels governing cell division, photosynthesis, growth, yield and stress resistance hence making it a promising technology for future agriculture (Boykov et al. 2019). Various researches have been conducted successfully using nano-based biosensors for pesticide residue detection as well as pathogen detection capable of remediating the contamination in the environment (Zhang and Fang, 2010; Fang et al. 2014). Every aspect of nano-based sustainable agriculture and its future in the modern-day agriculture industry has been compiled and portrayed adequately by Mittal and coworkers (2020) in their review work. Finally, very few countries (EU and Switzerland) are examining the appropriateness of the regulatory framework for handling nanoformulations. These countries have presented the regulatory measures for Agri/food/feed, including legislation and guidance for safety assessment in EU and non-EU countries (Amenta et al. 2015).

Constraint and Policy

Some constraints urging steadfast framing and enactment of certain policies pertaining to release of nanoparticles formulations in the agro-ecosystem are enlisted hereby:

a) Green synthesis of nanoparticles leads to bio-nanopesticides genesis, however, its subsequent formation may draw numerous complications demanding proper acknowledgment.
b) The challenges associated are the size regulation and distribution of the nanoparticles, high reproducibility of biological processes, and its upgraded synthesis.
c) Felicitous research paving the way to the inter-relationship studies between nanoparticles and the target species are prime requisite.
d) More stable silver nanoparticle synthesis procedures from the green route need to be prepared and standardized.

e) Research and incorporation of more suitable and effective stabilizer agents and optimization of synthesis parameters especially silver/stabilizer ratio is a top priority.
f) Biocidal properties of AgNPs need to be fine-tuned using appropriate reducing agents leading to broad-spectrum insect and microbe inhibition.
g) Natural polymers as stabilizers with anti-microbial properties required for AgNPs synthesis need to work out.
h) Toxicological repercussions owing to the interactions with the target and non-target species accompanying the environment heeds proper attention.
i) Minute particle size, high surface area with efficient absorbing properties may lead to its inadvertent penetration into human bodies via food and water resources leading towards toxicity at the cellular level or cytotoxicity along with prolonged accumulation in the tissues.
j) Henceforth, demanding striking research work oriented towards different possible absorption paths of nanomaterial to enter into the human bodies and its accumulation therein.
k) Meager knowledge regarding the fate and behavior process of such formulations on an account of the poor regulation, demands exacting standards before approval.
l) Regulatory frameworks demanding risk assessment, approval, pre and post-marketing surveillance is imperative.

Future Prospects

Green synthesis of nanoparticles must be encouraged to narrow down the chemical accumulation in soil and their selection criteria must rely on their risk related to the environment and humans. Majorly future projects must focus on the optimization of synthetic protocols, in vitro analysis of simulated silver nano-constructs and their rate of leaching in the soil, surface analysis of nano-constructs and their cellular interaction with plants, the toxicity of AgNPs to non-target pests, synthesis of a library of various sizes of AgNPs for their specific targets and their translocation studies, etc. The exact mechanism of action of these nanoparticles on living and non-living entities and the fate of their bi-products formed after remediation in the environment needed to be explored. New approaches like nano pesticides, nanoremediation, and nanosensors must be incorporated more and more for a sustained food production along with environmental safety after standardizing the safety protocols. The future thrust is to develop nanoproducts that are environment friendly and can compete with existing chemical formulations. The latest emerging techniques, biosensors utilizing endophytes seem promising approaches for sustained agro-ecosystem. As recently endophytes have received much appreciation for conferring abiotic stress resistance to crops and helping in resilience agriculture essential to fulfill our present needs and future demands. Now, endophytes-led nanosensors have been enlisted in various reviews as their research is being put forward with immense interest by the researchers (Khanam et al. 2020). Hence, the collaboration of all these components may result in a promising alternative to better address the agricultural and environmental problems altogether. And lastly, agricultural scientists, researchers, and farmers need to take into consideration the agro-ecosystem as a single component or as a whole rather than focusing only on the agricultural (high productivity) components all alone setting the environment at risk.

Conclusion

Silver nanoparticles indeed have enough potential to be considered as a promising candidate for modern defence tools in agro-ecosystem. The anti-microbial capacity of silver nanoparticles adds up to its potential where it can act as a protectant itself as well as act like a nanocarrier for chemical pesticides. Large-scale experiments for AgNPs application must be organized to know their fate and behaviour in a better way. No doubt if handled properly and given permission to be applied on farms after passing through trials and proper regulatory policies, AgNPs have enormous potential to revolutionize agriculture and conserve natural resources as a whole.

Conflict of Interest

Author declares no conflict of interest.

Acknowledgment

We authors are thankful to Dr. Abhishek Raj, Assistant Professor, Lovely Professional University, Punjab, India, and Editor, Frontiers in forests and global change for his suggestions which helped us in preceding our work into right direction.

Funding Agency

This research did not receive any specific grant from funding agencies in the public, commercial, or not-for-profit sectors.

References

Abdeen, S., Geo, S., Sukanya, S., Praseetha, P. K., Dhanya, R. P. (2014). Biosynthesis of Silver nanoparticles from Actinomycetes for therapeutic applications. *Int. J. Nano Dimension* 5 (2), 155–162. https://doi.org/10.7508/IJND.2014.02.008.

Adki V. S., Jadhav, J. P., Bapat, V. A. (2014). At the cross roads of environmental pollutants and phytoremediation: a promising bio remedial approach. *J. Plant Biochem. Biotechnol.* 23(2), 125–140. https://doi.org/10.1007/s13562-013-0250-6.

Ahmed, S., Saifullah, Ahmed, M., Swami, B. L., Ikram, S. (2016). Green synthesis of silver nanoparticles using *Azadirachta indica* aqueous leaf extract. *J. Rad. Res. App. Sci.* 9, 1-7. https://doi.org/10.1016/j.jrras.2015.06.006.

Ahmed, S., Ikram, S. (2015). Silver nanoparticles: one pot green synthesis using *Terminalia arjuna* extract for biological application. *J. Nanomed. Nanotechnol.* 6,309. https://doi.org/10.4172/2157-7439.1000309.

Alisha, A., Thangapandiyan S. (2019). Comparative bioassay of silver nanoparticles and malathion on infestation of red flour beetle, *Tribolium castaneum*. *J. Basic Appl. Zool.* 80(55). https://doi.org/10.1186/s41936-019-0124-0.

Amenta, V., Aschberger, K., Arena, M., Bouwmeester, H., Moniz, F. B., Brandhoff, P. (2015). Regulatory aspects of nanotechnology in the agri/feed/food sector in EU and non-EU countries. *Regul. Toxicol. Pharmacol.* 73, 463–476. https://doi.org/10.1016/j.yrtph.2015.06.016.

Anand, R., Bhagat, M. (2019). Silver nanoparticles (AgNPs): as nanopesticides and nanofertilizers. *MOJ Biol. Med.* 4(1):19–20. https://doi.org/10.15406/mojbm.2019.04.00107.

Aparna Mani, K. M. A., Seethalakshmi, S., Gopal, V. (2015). Evaluation of In-vitro Anti-Inflammatory Activity of Silver Nanoparticles Synthesised using *Piper Nigrum* Extract. *J. Nanomed. Nanotechnol.* 6, 268. https://doi.org/10.4172/2157-7439.1000268.

Aritonang, HF., Koleangan, H., Wuntu, A. (2019). Synthesis of silver nanoparticles using aqueous extract of medicinal plants' (*Impatiens balsamina* and *Lantana camara*) fresh leaves and analysis of antimicrobial activity. https://doi.org/10.1155/2019/8642303.

Ashraf, H., Anjum, T., Naseem, S. (2020). Microwave assisted green synthesis and characterization of silver nanoparticles using *Melia azadirachta* for the management of *Fusarium* wilt in tomato. *Front. Microbio.* 11, Article 238. https://doi.org/10.3389/fmicb.2020.00238.

Askari/Asgari, M., Hossein, K., Payam, N., Kamran, S., Marjan, P. (2012). Investigation of water remediation by nanoparticles. *World Acad. Sci. Engg. Tech.* 61. https://doi.org/10.13140/2.1.3110.2082.

Athanassiou, C., Kavallieratos, N., Benelli, G., Losic, D., Rani, P., Desneux, N. (2017). Nanoparticles for pest control: current status and future perspectives. *J. Pest Sci.* 1-15. https://doi.org/10.1007/s10340-017-0898-0.

Awasthi, A.K. Zeng, X., Li, J. (2016). Environmental pollution of electronic waste recycling in India: A critical review. *Env. Pollution* 211, 259-270. https://doi.org/10.1016/j.envpol.2015.11.027.

Azarbad, H., Van Straalen, N. M., Laskowski, R., Nikiel, K., Röling, W. F. M., Niklinska, M. (2016). Susceptibility to additional stressors in metal-tolerant soil microbial communities from two pollution gradients. *Appl. Soil Ecol.* 98, 233–242. http://dx.doi.org/10.1016/j. apsoil.2015.10.020.

Aziz, N., Faraz, M., Pandey, R., Sakir, M., et al. (2015). Facile algae-derived route to biogenic silver nanoparticles: synthesis, antibacterial and photocatalytic properties. *Langmuir* 31:11605–11612.

Balakrishnan, R. M., Uddandarao, P., Raval, K., Raval, R. (2019). A perspective of advanced biosensors for environmental monitoring. In tools, techniques and protocols for monitoring *Env. Contam.* pp. 19–51. https://doi.org/10.1016/B978-0-12-814679-8.00002-9.

Bardos, P., Bone, B., Daly, P., Elliott, D., Jones, S., Lowry, G., Merly, C. (2014). A risk/benefit appraisal for the application of Nano-Scale Zero Valent Iron (nZVI) for the remediation of contaminated sites. *NanoRem* WP9. https://www.researchgate.net/publication/264788812.

Bhattacharyya, A., Duraisamy, P., Govindarajan, M., Buhroo, A. A., Prasad, R. (2016). Nano-biofungicides: emerging trend in insect pest control, in *Advances and Applications through Fungal Nanobiotechnology,* ed. R. Prasad (Cham: Springer International Publishing), 307–319. https://doi.org/10.1007/978-3-319-42990-8_15.

Bindhani, B. K., Panigrahi, A. K. (2015). Biosynthesis and characterization of silver nanoparticles by using leaf extracts of *Ocimum sanctum* L. (Tulsi) and study of its antibacterial activities. *J. Nanomed. Nanotechnol.* S6, 008. https://doi.org/10.4172/2157-7439.S6-008.

Bootharaju, M. S., Pradeep, T. (2012). Understanding the degradation pathway of the pesticide, chlorpyrifos by noble metal nanoparticles. *Langmuir* 28(5), 2671-2679. https://doi.org/10.1021/la2050515.

Boykov, I. N., Shuford, E., Zhang, B. (2019). Nanoparticle titanium dioxide affects the growth and microRNA expression of switchgrass (*Panicum virgatum*). *Genomics* 111, 450–456. https://doi.org/10.1016/j.ygeno.2018.03.002.

Cai, L., Cai, L., Jia, H., Liu, C., et al. (2020). Foliar exposure of Fe3O4 nanoparticles on *Nicotiana benthamiana*: evidence for nanoparticles uptake, plant growth promoter and defense response elicitor against plant virus. *J. Hazardous Matter* 393, 122415. https://doi.org/10.1016/j.jhazmat.2020.122415.

Chen, J., Sun, L., Cheng, Y., Lu, Z., et al. (2016). Graphene oxide-silver nanocomposite: novel agricultural antifungal agent against *Fusarium graminearum* for crop disease prevention. *ACS Appl. Mat. Interfaces* 8(36), 24057-24070. https://doi.org/10.1021/acsami.6b05730.

Cromwell, W. A., Yang, J., Starr, J. L., Jo, Y. K. (2014). Nematicidal effects of silver nanoparticles on root-knot nematode in bermudagrass. *J. Nemat.* 46, 261–266. https://www.ncbi.nlm.nih.gov/pmc/articles/PMC4176408/.

Dawadi, S., Katuwal, S., Gupta, A., Lamichhane, U., et al. (2021). Current Research on Silver Nanoparticles: Synthesis, Characterization, and Applications. *J. Nanomat.* ID 6687290. https://doi.org/10.1155/2021/6687290.

Devi, D. G., Murugan, K., Panneer, S. C. (2014). Green synthesis of silver nanoparticles using *Euphorbia hirta* (Euphorbiaceae) leaf extract against crop pest of cotton bollworm, *Helicoverpa armigera* (Lepidoptera: Noctuidae). *J. Biopestic.* 7, 54–66. https://www.jbiopest.com/users/LW8/efiles/vol_7_0_54-66.pdf.

Dzimitrowicz, A., Motyka-Pomagruk, A., Cyganowski, P., Babinska, W., et al. (2018). Antibacterial activity of fructose-stabilized silver nanoparticles produced by direct current atmospheric pressure glow discharge towards quarantine pests. *Nanomat.* 8(10), 751. https://dx.doi.org/10.3390%2Fnano8100751.

Elbeshehy, E. K., Elazzazy, A. M., Aggelis, G. (2015). Silver nanoparticles synthesis mediated by new isolates of Bacillus spp., nanoparticle characterization and their activity against Bean Yellow Mosaic Virus and human pathogens. *Front. Microbio.* 6,453. https://doi.org/10.3389/fmicb.2015.00453.

El-Ramady, H., Alshaal, T., El-Henawy, A., Abdalla, N., Taha, H., Elmahrouk, M., Shalaby, T., et al. (2017). Environmental nanoremediation under changing climate. *Env. Biodiv. Soil Security* 1, 109-128. https://doi.org/10.21608/jenvbs.2017.1550.1009.

Fang, Y., Umasankar, Y., Ramasamy, R. P. (2014). Electrochemical detection of p-ethylguaiacol, a fungi infected fruit volatile using metal oxide nanoparticles. *Analyst.* 139, 3804–3810. https://doi.org/10.1039/c4an00384e.

Golinska, P., Wypij, M., Ingle, A. P., Gupta, I., Dahm, H., Rai, M. (2014). Biogenic synthesis of metal nanoparticles from actinomycetes: biomedical applications and cytotoxicity. *Appl. Microbiol. Biotechnol.* 98(19), 8083–8097. https://doi.org/10.1007/s00253-014-5953-7.

Gomes, H. I., Ferreira, C. D., Ribeiro, A. B. (2013). Overview of in situ and ex situ remediation technologies for PCB-contaminated soils and sediments and obstacles for full-scale application. *Sci. Total Env.* 237: 445–446. https://doi.org/10.1016/j.scitotenv.2012.11.098.

Grawe, G. F., de Oliveira, T. R., de Andrade, Narciso, E., Moccelini, S. K., Terezo, A. J., Soares, M. A., Castilho, M. (2015). Electrochemical biosensor for carbofuran pesticide based on esterases from *Eupenicillium shearii* FREI-39 endophytic fungus. *Biosens. Bioelect.* 63, 407–413. https://doi.org/10.1016/j.bios.2014.07.069.

Hemlata, Meena P. R., Singh A. P., Tejavath K. K. (2020). Biosynthesis of silver nanoparticles using *Cucumis prophetarum* aqueous leaf extract and their antibacterial and antiproliferative activity against cancer cell lines. *ACS Omega 5* 10, 5520–5528. https://doi.org/10.1021/acsomega.0c00155.

Hojjat, S.S., Kamyab, M. (2017). The effect of silver nanoparticle on Fenugreek seed germination under salinity levels. *Russian Agric. Sci.* 43, 61–65. https://doi.org/10.3103/S1068367417010189.

Huang, W., Yan, M., Duan, H., Bi, Y., Cheng, X., Yu, H. (2020). Synergistic antifungal activity of green synthesized silver nanoparticles and Epoxiconazole against *Setosphaeria turcica. J. Nanomat.* Article ID 9535432. https://doi.org/10.1155/2020/9535432.

Huq, M. A. (2020). Green synthesis of silver nanoparticles using *Pseudoduganella eburnea* MAHUQ-39 and their antimicrobial mechanisms investigation against drug resistant human pathogens. *Int. J. Mol. Sci.* 21, 1510. https://doi.org/10.3390/ijms21041510.

Ihegwuagu, N., Rufus, S., Tor-Anyiin, T. et al. (2015). Facile formulation of starch–silver-nanoparticle encapsulated dichlorvos and chlorpyrifos for enhanced insecticide delivery. *NZ. J. Chem.* 40. https://doi.org/10.1039/C5NJ01831E.

Iqbal, M., Raja, N. I., Hussain, M., Ejaz, M., et al. (2019). Effect of silver nanoparticles on growth of wheat under heat stress. *Iranian J. Sci. Technol. Transactions A, Sci.* 43, 387–395.

Iravani, S., Korbekandi, H., Mirmohammadi, S. V., Zolfaghari, B. (2014). Synthesis of silver nanoparticles: chemical, physical and biological methods. *Res. Pharm. Sci.* 9(6), 385–406. https://www.ncbi.nlm.nih.gov/pmc/articles/PMC4326978/.

Jain, D., Kothari, S. (2014). Green synthesis of silver nanoparticles and their application in plant virus inhibition. *J. Mycol. Plant Pathol.* 44, 21–24. https://www.researchgate.net/publication/260981626.

Jo, Y. K., Kim, B. H., Jung, G. (2009). Antifungal activity of silver ions and nanoparticles on phytopathogenic fungi. *Plant disease* 93(10), 1037–1043. https://doi.org/10.1094/PDIS-93-10-1037.

John, M. S., Nagoth, J. A., Ramasamy, K. P., Mancini, A., Giuli, G., Natalello, A., et al. (2020). Synthesis of bioactive silver nanoparticles by a *Pseudomonas* Strain associated with the Antarctic psychrophilic protozoan *Euplotes focardii*. *Mar. Drugs*, 18, 38. https://doi.org/10.3390/md18010038.

Juibari, M. M., Abbasalizadeh, A., Jouzani, G. H. S., Noruzi, M. (2011). Intensified biosynthesis of silver nanoparticles using a native extremophilic *Ureibacillus thermosphaericus* strain. *Mater. Lett.* 65, 1014–1017. http://dx.doi.org/10.1016%2Fj.matlet.2010.12.056.

Justino, C. I., Duarte, A. C., Rocha-Santos, T. A. (2017). Recent progress in biosensors for environmental monitoring: A review. *Sensors* 17(12), 2918. .https://doi.org/10.3390/s17122918.

Kar, S., Rathore, V. S., Champati, P. K., Sharma, R., Swain, S. K. (2016). Classification of river water pollution using Hyperian data. *J. Hydro.* 537, 221-233. https://doi.org/10.1016/j.jhydrol.2016.03.047.

Kalimuthu, K., Panneerselvam, C., Chou, C., Tseng, L. C., et al. (2017). Control of dengue and Zika virus vector *Aedes aegypti* using the predatory copepod *Megacyclops formosanus*: Synergy with *Hedychium coronarium*- synthesized silver nanoparticles and related histological changes in targeted mosquitoes. *Process Safety Env. Prot.* 109, 82-96. https://doi.org/0.1016/j.psep.2017.03.027.

Kasprowicz, M. J., Kozioł, M., Gorczyca, A. (2010). The effect of silver nanoparticles on phytopathogenic spores of *Fusarium culmorum*. *Can. J. Microbiol.* 56(3), 247–253. https://doi.org/10.1139/w10-012.

Kenessov, B., Koziel, J. A., Bakaikina, N. V., Orazbayeva, D. (2016). Perspectives and challenges of on-site quantification of organic pollutants in soils using solid-phase microextraction. *Trends Anal. Chem.* http://dx.doi.org/10.1016/j.trac.2016.04.007.

Khanam, Z., Gupta, S., Verma, A. (2020). Endophytic fungi-based biosensors for environmental contaminants- A perspective. *S. Afr. J. Bot.* 134, 401-406. http://doi.org/10.1016/j.sajb.2020.08.007.

Kim, H. J., Park, H. J., Choi, S. H. (2011). Antimicrobial action effect and stability of nanosized silica hybrid Ag complex. *J. Nanosci. Nanotechnol.* 11(7), 5781-5787.

Kim, S. W., Jung, J. H., Lamsal, K., Kim, Y. S., et al. (2012). Antifungal effects of silver nanoparticles (AgNPs) against various plant pathogenic fungi. *Mycobiol.* 40(1), 53-58. https://dx.doi.org/10.5941%2FMYCO.2012.40.1.053.

Krishnaraj, C., Rajan, R., Mohan, K., Kalaichelvan, P. (2012). Optimization for rapid synthesis of silver nanoparticles and its effect on phytopathogenic fungi. *Spectrochimica acta. Part A, Mol. Biomol. Spect.* 93, 95-9. https://doi.org/10.1016/j.saa.2012.03.002.

Kumari, M., Pandey, S., Bhattacharya, A., Mishra, A., Nautiyal, C. S. (2017). Protective role of biosynthesized silver nanoparticles against early blight disease in *Solanum lycopersicum*. *Plant Physiol. Biochem.* 121, 216–225. https://doi.org/10.1016/j.plaphy.2017.11.004.

Lamsal, K., Sang-Woo, K., Jung J. H., Kim, Y. S., Kim, K. U., Lee, Y. S. (2010). Inhibition effects of silver nanoparticles (AgNPs) against powdery mildew on cucumber and pumpkin. *Microbio.* 39(1), 26-32. https://doi.org/10.4489/MYCO.2011.39.1.026.

Lamsal, K., Kim, S. W., Jung, J. H., Kim, Y. S., Kim, K. U., Lee, Y. S. (2011). Application of silver nanoparticles for the control of *Colletotrichum sp. in vitro* and pepper anthracnose disease in field. *Mycobiol.* 39(3), 194–199. https://doi.org/10.5941/myco.2011.39.3.194.

Lee, S., Jun, B. H. (2019). Silver nanoparticles: synthesis and application for nanomedicine. *Int. J. Mol. Sci.* 20, 865. https://doi.org/10.3390/ijms20040865.

Li, G., He, D., Qian, Y., Guan, B., et al. (2012). Fungus-mediated green synthesis of silver nanoparticles using *Aspergillus terreus*. *Int. J. Mol. Sci.* 13(1), 466-476. https://dx.doi.org/10.3390%2Fijms13010466.

Li, W.Q., Qing, T., Li, C.C., Li, F., et al. (2020). Integration of subcellular partitioning and chemical forms to understand silver nanoparticles toxicity to lettuce (*Lactuca sativa* L.) under different exposure pathways. *Chemosphere* 258, 127349. https://doi.org/10.1016/j.chemosphere.2020.127349.

Liu, M., Feng, S., Ma, Y., Xie, C., He, X., et al. (2019). Influence of surface charge on the phytotoxicity, transformation, and translocation of CeO2 nanoparticles in cucumber plants. *ACS Appl. Mater Interf.* 11, 16905–16913. https://doi.org/10.1021/acsami.9b01627.

Loretta, O. O., Stephen, E., Ezeata, A., Usman, E. (2016). In vitro biodegradation of palm oil mill effluent (POME) by *Bacillus subtilis*, *Pseudomonas aeruginosa* and *Aspergillus niger*. *J. Bioremed. Biodeg.* 7, 361. https://doi.org/10.4172/2155-6199.1000361.

Lu, Y., Song, S., Wang, R., Liu, Z., Meng, J., Sweetman, A. J., Jenkins, A., Ferrier, R. C., Li, H., Luo, W., Wang, T. (2015). Impacts of soil and water pollution on food safety and health risks in China. *Environ. Int.* 77, 5–15. .https://doi.org/10.1016/j.envint.2014.12.010.

Manimegalai, G., Kumar, S., Sharma, C. (2012). Pesticide mineralization in water using silver nanoparticles incorporated on polyurethane foam. *Int. J. Sci. Res.* (IJSR). 1(3), 91–94. https://www.ijsr.net/search_index_results_paperid.php?id=IJSR13010110.

Manimegalai, G., Shanthakumar, S., Sharma, C. (2014). Silver nanoparticles: synthesis and application in mineralization of pesticides using membrane support. *Int. Nano Letter* 4,105. https://ui.adsabs.harvard.edu/link_gateway/2014INL.....4..105M/doi:10.1007/s40089-014-0105-8.

Min, J. S., Kim, K. S., Kim, S. W., Jung, J. H., Lamsal, K., Kim, S. B., Jung, M. Y., Lee, Y. S. (2009). Effects of colloidal silver nanoparticles on sclerotium-forming phytopathogenic fungi. *Plant Pathol J.* 25, 376-380. https://doi.org/10.5423/PPJ.2009.25.4.376.

Minh, P. N., Hoang, V. T., Dinh, N. X., Hoang, O. V., et al. (2020). Reduced graphene oxide-wrapped silver nanoparticles for applications in ultrasensitive colorimetric detection of Cr(vi) ions and the carbaryl pesticide. *NZ. J. Chem.* 44, 7611-7620. https://doi.org/10.1039/D0NJ00947D.

Mishra, S., Singh, H. (2015). Biosynthesized silver nanoparticles as a nanoweapon against phytopathogens: exploring their scope and potential in agriculture. *Appl. Microbiol. Biotechnol.* 99, 1097–1107. https://doi.org/10.1007/s00253-014-6296-0.

Mishra, S., Singh, B., Naqvi, A., et al. (2017). Potential of biosynthesized silver nanoparticles using Stenotrophomonas sp. BHU-S7 (MTCC 5978) for management of soil-borne and foliar phytopathogens. *Science Replica* **7**, 451-54. https://doi.org/10.1038/srep45154.

Mitra, C., Gummadidala, P. M., Afshinnia, K., Merrifield, R. C., et al. (2017). Citrate-Coated Silver Nanoparticles Growth-Independently Inhibit Aflatoxin Synthesis in *Aspergillus parasiticus*. *Env. Sci. Technol.* 51(14), 8085-8093. https://doi.org/10.1021/acs.est.7b01230.

Mittal, D., Kaur, G., Singh, P., Yadav, K., Azmal, A. S. (2020). Nanoparticle-Based Sustainable Agriculture and Food Science: Recent Advances and Future Outlook. *Front. Nanotech.* 2. https://doi.org/10.3389/fnano.2020.579954.

Moussa, S., Tayel, A., Alsohim, A., Abdullah, R. (2013). Botryticidal activity of nanosized silver-chitosan composite and its application for the control of gray mold in strawberry. *J. food Sci.* 78. https://doi.org/10.1111/1750-3841.12247.

Mozhayeva, D., Engelhard, C. (2020). A critical review of single particle inductively coupled plasma mass spectrometry–a step towards an ideal method for nanomaterial characterization. *J. Analyt. Atomic Spect.* 43, 5290–5294. https://doi.org/10.1039/C9JA00206E.

Mukhopadhyay, S. S. (2014). Nanotechnology in agriculture: prospects and constraints. *Nanotechnol. Sci. Appl.* 7, 63–71. https://doi.org/10.2147/NSA.S39409.

Mustafa, G., Komatsu, S. (2016). Toxicity of heavy metals and metal-containing nanoparticles on plants. *Biochimica et Biophysica Acta (BBA)-Proteins and Proteomics* 1864(8), 932–944. https://doi.org/10.1016/j.bbapap.2016.02.020.

Nahar, K., Aziz, S., Bashar, M., Haque, M., Al-Reza, S. (2020). Synthesis and characterization of Silver nanoparticles from *Cinnamomum tamala* leaf extract and its antibacterial potential. *Int. J. Nano Dimen.*, 11(1), 88-98. http://www.ijnd.ir/article_667484.html.

Nair, S., Pradeep, T. (2007). Extraction of chlorpyrifos and malathion from water by metal nanoparticles. *J. Nanosci. Nanotech.* 7,1-7. https://doi.org/10.1166/jnn.2007.733.

Nair, A.S., Tom, R.T., Pradeep, T. (2003). Detection and extraction of endosulfan by metal nanoparticles. *J. Environ. Monitoring* 2. https://doi.org/10.1039/B300107E.

NanoRem, 2017. Generalized guideline for application of nanoremediation (Edited by Juergen Braun). www.nanorem.eu/toolbox.

Nuruzzaman, M., Rahman, M. M., Liu, Y., Naidu, R. (2016). Nanoencapsulation, nano-guard for pesticides: a new window for safe application. *J. Agric. Food Chem.* 64, 1447–1483. https://doi.org/10.1021/acs.jafc.5b05214.

Nguyen, D. V., Nguyen, H. M., Le, N. T., Nguyen, K. H., et al. (2020). Copper nanoparticle application enhances plant growth and grain yield in maize under drought stress conditions. *J. Plant Growth Reg.* http://dx.doi.org/10.1007/s00344-021-10301-w.

Ocsoy, I., Paret, M. L., Ocsoy, M. A., Kunwar, S., et al. (2013). Nanotechnology in plant disease management: DNA-directed silver nanoparticles on graphene oxide as an antibacterial against *Xanthomonas perforans*. *Acs Nano* 7, 8972–8980. https://doi.org/10.1021/nn4034794.

Pandian, S. S. G., Mariaamalraj, S., Achary, A. (2014). Influence of biosurfactant in biosynthesis of silver nanoparticles by *Pseudomonas Aeruginosa* AMB AS7. *IJIRSET*. 3, 1111-1115. http://www.ijirset.com/upload/2014/iciet/mech/13_788.pdf.

Pilaquinga, F., Morejón, B., Ganchala, D., Morey, J., Piña, N., Debut, A., et al. (2019). Green synthesis of silver nanoparticles using *Solanum mammosum* L. (Solanaceae) fruit extract and their larvicidal activity against *Aedes aegypti* L. (Diptera: Culicidae). *PLoS ONE* 14(10), e0224109. https://doi.org/10.1371/journal.pone.0224109.

Rao, B., Tang, R. C. (2017). Green synthesis of silver nanoparticles with antibacterial activities using aqueous *Eriobotrya japonica* leaf extract. *Adv. Nat. Sci. Nanosci. Nanotechnol.* .https://doi.org/10.1088/2043-6254/aa5983.

Prasad, C. H., Srinivasulu, K. (2015). Catalytic reduction of 4-nitrophenol using biogenic silver nanoparticles derived from papaya (*Carica papaya*) peel extract. *Indust. Chem.* 01(01). https://doi.org/10.4172/2469-9764.1000104.

Rouhani, M., Mohammad, A. S., Kalantari, S., 2012. Insecticide effect of silver and zinc nanoparticles against *Aphis nerii* Boyer de fonscolombe (Hemiptera: aphididae). Chilean *J. Agric. Res.* 72(4) october-december 2012. https://www.bioline.org.br/pdf?cj12089.

Rahmanifar, B., Dehaghi, M. S. (2014). Removal of organochlorine pesticides by chitosan loaded with silver oxide nanoparticles from water. *Clean Tech. Env. Policy* 16, 1781–1786.

Rao, B., Tang, R. C. (2017). Green synthesis of silver-nanoparticles with antibacterial activities using aqueous eriobotrya japonica leaf extract. *Adv. Nat. Sci: Nanosci. Nanotechnol.* https://doi.org/10.1007/s10098-013-0692-5.

Romeh, A. A. A. (2018). Green silver nanoparticles for enhancing the phytoremediation of soil and water contaminated by fipronil and degradation products. *Water Air Soil Poll.* 229, 147. https://doi.org/10.1007/s11270-018-3792-3.

Saifuddin, N., Nian, C. Y., Zhan, L. W., Ning, K. X. (2011). Chitosan-silver nanoparticles composite as point-of-use drinking water filtration system for household to remove pesticides in water. *Asian J. Biochem.* 6, 142-159. https://dx.doi.org/10.3923/ajb.2011.142.159.

Sayed, A., Kim, S., Behle, R. (2017). Characterisation of silver nanoparticles synthesised by Bacillus thuringiensis as a nanobiopesticide for insect pest control. *Biocontrol Sci. Tech.* https://doi.org/10.1080/09583157.2017.1397597.

Seifipour, R., Nozari, M., Pishkar, L. (2020). Green synthesis of silver nanoparticles using *Tragopogon Collinus* leaf extract and study of their antibacterial effects. *J. Inorg. Organomet. Polym.* https://doi.org/10.1007/s10904-020-01441-9.

Shah, G., Jan, M., Afreen, M., Anees, M., Rehman, S., Daud, M. K., Malook, I., Jamil, M. (2016). Halophilic bacteria mediated phytoremediation of salt-affected soils cultivated with rice. *J. Geochem. Explor.* https://doi.org/10.1016/j.gexplo.2016.03.011.

Shahrokh, S., Hosseinkhani, B., Emtiazi, G. (2014). The impact of nano-silver on bacterial aerobic nitrate reductase. *J. Bioproc. Biotechnol.* 4, 162. https://doi.org/10.4172/2155-9821.1000162.

Sharada, S., Omprakash, V. (2015). Green synthesis and characterization of silver nanoparticles and evaluation of their antibacterial activity using *Elettaria cardamom* seeds. *J. Nanomed. Nanotech.* 6(02). https://doi.org/10.4172/2157-7439.1000266.

Spielman-Sun, E., A. Avellan, G. D. Bland, R. V. Tappero, et al. (2019). Nanoparticle surface charge influences translocation and leaf distribution in vascular plants with contrasting anatomy. *Environ. Sci. Nano J.* 6, 2508–2519. https://doi.org/10.1039/C9EN00626E.

Sreelakshmy, V., Deepa, M. K., Mridula, P. (2016). Green Synthesis of Silver Nanoparticles from Glycyrrhiza glabra Root Extract for the Treatment of Gastric Ulcer. *J. Develop. Drugs*, 5,152. https://doi.org/10.4172/2329-6631.1000152.

Suresh, U., Murugan, K., Benelli, G., et al. (2015). Tackling the growing threat of dengue: *Phyllanthus niruri*-mediated synthesis of silver nanoparticles and their mosquitocidal properties against the dengue vector

Aedes aegypti (Diptera: Culicidae). *Parasitol. Res.* 114, 1551–1562. https://doi.org/10.1007/s00436-015-4339-9.

Terra, A. L. M., Kosinski, R. C., Moreira, J. B., Costa, J. A. V., Morais, M. G. (2019). Microalgae biosynthesis of silver nanoparticles for application in the control of agricultural pathogens. *J. Environ. Sci. Health, Part B*, 54(8), 709-716.

Theron, J., Walker, J. A., Cloete, T. E. (2008). Nanotechnology and Water Treatment: Applications and Emerging Opportunities. *Crit. Rev. Microbio.* 34(1), 43–69. https://doi.org/10.1080/10408410701710442.

Touhami, A. (2014). Biosensors and nanobiosensors: design and applications. *Nanomedicine* 15, 374-403. ISBN: 978-1-910086-01-8. http://www.onecentralpress.com/wp-content/uploads/2014/11/CHAPTER-15-NM-08-LATEST.pdf.

Turner, A. P. (2013). Biosensors: sense and sensibility. *Chem. Soc. Rev.* 42, 3184–3196. https://doi.org/10.1039/c3cs35528d.

Varghese, J., Chandan, M. R., Shanthakumar, S. (2020). Fixed bed column study for pesticide removal using silver nanoparticles-embedded polyurethane foam and glass beads. *Chem. Eng. Comm.* 207(10), 1337-1346. https://doi.org/10.1080/00986445.2019.1647181.

Venugopal, N. V., Sainadh, N. V. (2016). Novel polymeric nanoformulation of mancozeb-an eco-friendly nanomaterial. *Int. J. Nanosci.*, 15(4), 1650016. https://doi.org/10.1142/S0219581X16500162.

Verma, A. (2016). Nanotechnology in agriculture: an emerging vista. *Everyman's Sci.* LI (3), 168-171. ISSN: 0531-495X. http://sciencecongress.nic.in/everymans_science.php.

Verma, A., Khanam, Z. (2020). Phytonanotechnology and agriculture-chapter-14. Phytonanotechnology: Challenges and prospects [Eds: Thajuddin, N. and Matthew, S], Elsevier. ISBN: 978-0-12-822348-2. https://doi.org/10.1016/B978-0-12-822348-2.00014-0.

Wilson, J. S. (2005). *Sensor Technology Handbook.* Elsevier, Amsterdam/Boston. https://www.kelm.ftn.uns.ac.rs/literatura/si/pdf/Sensor%20Technology%20Handbook.pdf.

Worrall, E.A., Hamid, A., Mody, K. T., Mitter N., Pappu H. R. (2018). Nanotechnology for plant disease management. *Agron.* 8, 285. https://doi.org/10.3390/agronomy8120285.

Xu, X., Ying, Y. (2011). Microbial biosensors for environmental monitoring and food analysis. *Food Rev. Int.* 27(3), 300–329. https://doi.org/10.1080/87559129.2011.563393.

Zhang, L., Fang, M. (2010). Nanomaterials in pollution trace detection and environmental improvement. *Nano Today* 5, 128–142. http://dx.doi.org/10.1016/j.nantod.2010.03.002.

Chapter 3

Nano Techniques for Efficient Fertilizer Management

Pratibha Singh[1,*], Mahendra Singh[2], Shilpi Gupta[3], A. P. Singh[4], Ajeet Singh[1], Shweta Gupta[1] and Seema Sharma[5]

[1]SKNAU, Jobner, Rajasthan, India
[2]BAU, Sabour, Bihar, India
[3]College of Agriculture, GBPUAT, Pantnagar
[4]IFFCO, Jaipur, Rajasthan, India
[5]SKNAU, Jobner, India

Abstract

The indiscriminate use of agrochemicals for increasing agricultural production is resulting in soil and groundwater pollution due to their low efficiency and high environmental risk. For sustaining the agricultural productivity and conserving the soil health it is necessary to identify environment friendly approach with high nutrient use efficiency. Nanotechnology is becoming important for the agricultural sector day by day. Different studies reported promising results and applications in the areas of pesticides delivery, biopesticides, fertilizers, and genetic material for plant transformation. Use of nanomaterials is expected to reduce the dosage and controlled release of fertilizers. Nanotechnology may also provide solutions for degrading persistent chemicals into harmless useful components. This approach has promising impact in the use of natural resources such as chitosan and diatomite for production of coating material which are useful for many agricultural applications and in mitigation of drought stress tolerance of crops. The present chapter gives extensive information about applications of nanoparticles in agriculture, types of nanoparticles, characterization of nanoparticles and nano fertilisers etc.

1. Introduction

Nanotechnology is the integration of agriculture, food sciences, earth science and engineering. It has great significance in agricultural research, resulting in solutions for various agricultural problems. Crop productivity is declining day by day the main reason behind this is deterioration in soil health and fertility status of soil due to various kinds of erosion, nutrient depletion, scarcity of water, change in soil pH, reduction in organic matter and poor drainage. However, population is increasing globally and to satisfy the requirement of this ever-increasing

* Corresponding Author's Email: pratibha.soils@sknau.ac.in.

In: Ecosystem Services
Editor: Hanuman Singh Jatav
ISBN: 978-1-68507-614-6

population, all the resources should be the utilized very efficiently in the very precise manner. Keeping in view all these challenges, it is the need of hour to identify and utilize the advance knowledge and technological tools for enhancing the use efficiency of agricultural resources and inputs. Materials that are nanoscale in three dimensions are nanoparticles, for example, precipitates, dendrimers, fullerenes, colloids, and tiny particles of semiconductor materials (Sekhon 2014). Nanoparticles (NPs) may be defined as particulate matter with at least one dimension less than 100 nm (Auffan et al., 2009). Most of the fertilizers become partially available to plants caused by several inhibiting factors like leaching, photolytic degradation, hydrolysis, and decomposition. Nutrient losses may be minimized through the exploitation of new applications with the utilization of nanomaterials present in soil which may also result in yield enhancement (Siddiqui et al., 2015). Nutrients derived from nanofertilzers or nanoencapsulation might have properties that are effective to crops demand and controlled release of chemical fertilizers that regulate plantgrowth and enhance target activity (De Rosa et al., 2010; Nair et al., 2010).

Both macro nutrients such as nitrogen, phosphorous, and potassium compoundsand micronutrients may effectively be delivered by controlled-release (CR) fertilizer system. Apart from the reduction of costs by increasing the efficiency of fertilizer use, CR also reduces the negative impact caused by fertilizer application (Gumbo et al., 2008; Anderson 2009; Davidson et al., 2013). Utilization of nanoscale carriers may be successfully employed for the efficient delivery of fertilizers, pesticides, herbicides, plant growth regulators, etc. (Prasadet al., 2012). The three important mechanisms responsible for efficient delivery, better storage, and controlled release are (i) encapsulation and entrapment (ii) polymers and dendrimers (iii) surface ionic and weak bond attachments (Sawant et al., 2006; Johnston 2010). These mechanisms enhance stability against degradation in the environment by reducing the amount of application and runoff losses (Ditta et al., 2015). These carriers should be designed in such a way that they can anchor the plant roots to the surrounding rhizosphere environment. An understanding of molecular and conformational mechanisms between the nanoscale delivery and targeted structures and the soil fraction is required to reveal the anchoring mechanism (NAAS 2013). These may elucidate slower uptake of active ingredients, thereby reducing the consumption of inputs and also reducing the waste material production.

2. Applications of Nanoparticles in Agriculture

Nanotechnology has the prospective to improve the agriculture and food industry with novel nanotools for the controlling of rapid disease diagnostic, enhancing the capacity of plants to absorb nutrients among others (Prasad et al., 2017). This technology has gained good drive in the last decade with a profusion of public funding, but the stage of expansion is good, even though many methods became under the sunshade of agriculture. This might be attributed to a unique nature of farm production, which functions as an open system whereby energy and matter are exchanged freely. The scale of demand of input materials is always being large in contrast with industrial nanoproducts with the absence of control over the input of the nanomaterials in contrast with industrial nanoproducts (Mukhopadhyay, 2014). The nanotechnology can play an important role in the productivity through control of nutrients (Gruere, 2012; Mukhopadhyay, 2014) as well as it can also contribute in the monitoring of water quality and pesticides for sustainable improvement of agriculture (Prasad et al., 2014).

Nanomaterials have such diverse assets and activities that are impossible to deliver a general assessment of their health and environmental risks (Prasad et al., 2014). Properties (other than size) of NPs have the influence on toxicity include chemical composition, shape, surface structure, surface charge, behavior, extent of particle aggregation (clumping) or disaggregation, etc. may associate with engineered NPs (Ion et al., 2010). For this reason, even nanomaterials of the same chemical composition that have different sizes or shapes can exhibit their different toxicity level. The implication of the nanotechnology research in the agricultural sector is become to be necessary even key factor for the sustainable developments of agriculture. In the agri-food areas pertinent applications of nanotubes, fullerenes, biosensors, controlled delivery systems, nano filtration, etc. were observed (Ion et al., 2010; Sabir et al., 2014). The potential of nanotechnology in agriculture is immense, but a few issues are still to be addressed as the risk assessment. In this respect, some nanoparticles attractants are derived from biopolymers such as proteins and carbohydrates with low effect on human health and the environment (Ghidan and Antary, 2019). Nanotechnology has many uses in all stages of production, processing, storing, packaging and transport of agricultural products. Applications of nano biosensors are nursing soil conditions for the monitoring of herbicides, pesticides and fertilizers, crop growth, detecting the food borne contaminants and heavy metals and residue analysis (Singh et al., 2021). Nanotechnology will transform agriculture and food industry such as in case of farming techniques, enhancing the ability of plants to absorb nutrients, disease detection and control pests.

2.1. Nanotechnology in Fertilizer and Pesticides

Now days, sustainable crop production is needed. It may be implicit to present a good move towards of ecosystem for long run. Agricultural practices that can cause long term damage to soil include excessive tilling of the soil which leads to erosion and loss of nutrients. This will lead to salinization. This is to satisfy human being food, animal feed and fibre needs.

In the recent decade nanofertilizers are freely available in the market, but particularly the agricultural fertilizers are still not shaped by the major chemical companies (Table 1). Nanofertilizers may contain nano zinc, silica, iron and titanium dioxide, ZnCdSe/ZnS core shell QDs, InP/ZnS core shell QDs, Mn/ZnSe QDs, gold nanorods, core shell QDs, etc. as well as should endorse control release and improve the its quality (Prasad et al., 2017). Studies of the uptake, biological fate and toxicity of several metal oxide NPs, viz. Al_2O_3, TiO_2, CeO_2, FeO, and ZnONPs were carried out intensively in the present decade for agricultural production (Dimkpa, 2014; Zhang et al., 2016). The deficiency of zinc has been documented as one of the main problems in limiting agricultural productivity in the alkaline nature of soils (Sadeghzadeh, 2013).

Nanomaterials act as fertilizers might have the properties such as crop improvement and with less eco-toxicity. Plants can give an important way for their bioaccumulation into the food chain. The recent developments in agriculture cover the applications of NPs for more effective and safe use of chemicals for plants. Efficient, effective and drought-tolerant microorganisms may have vital role in production of various crops (Singh et al., 2021). The effects of different NPs on plant growth and phytotoxicity were reported by several workers including magnetite (Fe3O4) nanoparticles and plant growth (Shankramma et., 2016), alumina, zinc, and zinc oxide on seed germination and root growth of five higher plant species; radish, rape, lettuce, corn, and cucumber, silver nanoparticles and seedling growth in wheat (Lin et al., 2007), sulfur nanoparticles on tomato (Salem et al., 2016), zinc oxide in mungbean, nanoparticles of AlO, CuO, FeO, MnO, NiO, and ZnO (Ghidan et al., 2018a, 2018b). Silver nanoparticles can stimulate wheat growth and yield. Soil applied 25 ppm SNPs had highly favourable growth promoting effects on wheat growth and yield.

Table 1. A number of commercial product of nanofertilizers

Commercial product	Content	Company
Nano-GroTM	Plant growth regulator and immunity enhancer	Agro Nanotechnology Corp., FL, United States
Nano Green	Extracts of corn, grain, soybeans, potatoes, coconut, and palm	Nano Green Sciences, Inc., India
Nano-Ag Answer R	Microorganism, sea kelp, and mineral electrolyte	Urth Agriculture, CA, United States
Biozar Nano-Fertilizer	Combination of organic materials, micronutrients, and macromolecules	Fanavar Nano-Pazhoohesh Markazi Company, Iran
Nano Max NPK Fertilizer	Multiple organic acids chelated with major nutrients, amino acids, organic carbon, organic micro nutrients/trace elements, vitamins, and probiotic	JU Agri Sciences Pvt. Ltd, Janakpuri, New Delhi, India
Master Nano Chitosan Organic Fertilizer	Water soluble liquid chitosan, organic acid and salicylic acids, phenolic compounds	Pannaraj Intertrade, Thailand
TAG NANO (NPK, PhoS, Zinc, Cal, etc.) fertilizers	Proteino-lacto-gluconate chelated with micronutrients, vitamins, probiotics, seaweed extracts, humic acid	Tropical Agrosystem India (P) Ltd, India

Source: Prasad et al., 2017.

Zinc has been considered as an essential micronutrient for metabolic activities in plants although it is required in trace amounts in plants. It was found that zinc has an important role in management of reactive oxygen species and protection of plant cells against oxidative stresses. Zinc has important functions in the synthesis of auxin or indoleacetic acid (IAA) from tryptophan as well as in biochemical reactions required for formation of chlorophyll and carbohydrates. The crop yield and quality of produce can be affected by deficiency of Zn. The development of insecticide resistance in pest insects has been an increasing problem for agriculture and public health (Ghidan and Antary, 2019). Magnesium oxide (MgO) is important inorganic materials with many uses such as adsorbents, fire retardants, advanced ceramics, toxic waste remediation, and photo electronic materials. Therefore, various techniques and routes for synthesis of MgONPs have been reported (Ghidan et al., 2017a). MgOH was synthesized by green methods using nontoxic neem leaves extract (Moorthy et al., 2015), Citrus limon leaves extract, acacia gum (Awwad and Ahmad, 2014).

2.2. Nanotechnology in Disease Control

Fusarium wilt is a destructive disease of tomato, brinjal and pulses in several countries due to its severe production loss, long-lasting survival of fungus in the soil and generation of resistant strain. The disease can be reduced up to some extent with the use of resistant cultivars and pesticides. However, the occurrence and development of new pathogenic strains is a regular problem, and the use of chemicals is expensive and not always effective and ecofriendly. In recent years, the uses of nanoparticles have been considered as an alternative solution to control plant pathogens.

3. Types of Nanoparticles

Nanoparticles (NPs) are broadly divided into various categories depending on their morphology, size and chemical properties. Based on physical and chemical characteristics, some of the well known classes of NPs are given as below.

3.1. Carbon-Based Nanoparticles

Fullerenes and carbon nanotubes (CNTs) represent two major classes of carbon-based NPs. Fullerenes contain nanomaterials that are made of globular hollow cage such as allotropic forms of carbon (Khan et al., 2019). They have created noteworthy commercial interest due to their electrical conductivity, high strength, structure, electron affinity, and versatility (Astefanei et al., 2015). With the advent of nanoscience, the use of carbon-based nanoparticles has now been considered to evaluate the reproductive performance of plants (Zaytseva and Neumann, 2016).

3.2. Metal Nanoparticles

Metal nano particles are purely made of the metals precursors. Due to well-known localized surface plasmon resonance (LSPR) characteristics, these NPs possess unique optoelectrical properties. NPs of the alkali and noble metals i.e., Cu, Ag and Au have a broad absorption band in the visible zone of the electromagnetic solar spectrum. The facet, size and shape-controlled synthesis of metal NPs is important in present day cutting-edge materials (Dreaden et al., 2012). Due to their advanced optical properties, metal NPs find applications in many research areas.

3.3. Ceramics Nanoparticles

These ceramics nanoparticles are inorganic non-metallic solids, synthesized via heat and consecutive cooling. They can be found in amorphous, polycrystalline, dense, porous or hollow forms (Sigmund et al., 2006). Therefore, these NPs are getting great attention of researchers due to their use in applications such as catalysis, photocatalysis, photodegradation of dyes, and imaging applications. (Thomas et al., 2015).

3.4. Semiconductor Nanoparticles

These materials have properties between metals and non-metals and therefore they found various applications in the literature due to this property (Ali et al., 2017; Khan et al., 2017a).

Semiconductor NPs possess wide bandgaps and therefore showed significant alteration in their properties with bandgap tuning. As an example, variety of semiconductor NPs are found exceptionally efficient in water splitting applications, due to their suitable bandgap and bandedge positions (Hisatomi et al., 2014).

3.5. Polymeric Nanoparticles

Polymeric nanoparticles are mostly nanospheres or nanocapsular shaped (Mansha et al., 2017). The former are matrix particles whose overall mass is generally solid and the other molecules are adsorbed at the outer boundary of the spherical surface. In the latter case the solid mass is encapsulated within the particle completely (Rao and Geckeler, 2011). The PNPs are readily functionalize and thus find bundles of applications in the literature (Abd Ellah and Abouelmagd, 2016; Abouelmagd et al., 2016).

3.6. Lipid-Based Nanoparticles

Lipid nanotechnology (Mashaghi et al., 2013) is a special field, which focus the designing and synthesis of lipid NPs for various applications such as drug carriers and delivery (Puri et al., 2009) and RNA release in cancer therapy (Gujrati et al., 2014). These NPs contain lipid moieties and effectively using in many biomedical applications. Generally, a lipid NP is characteristically spherical with diameter ranging from 10 to 1000 nm. Like polymeric NPs, lipid NPs possess a solid core made of lipid and a matrix contains soluble lipophilic molecules. Surfactants or emulsifiers stabilized the external core of these NPs (Rawat et al., 2011).

4. Characterization of Nanoparticles

Different characterization techniques have been practiced for the analysis of various physicochemical properties of NPs. These techniques include such as X-ray diffraction (XRD), X-ray photoelectron spectroscopy (XPS), infrared (IR), SEM, TEM, Brunauer-Emmet-Teller (BET) and particle size analysis (Khan et al., 2019). The morphological features of NPs always attain great interest since morphology always influences most of the properties of the NPs. There are different characterization techniques for morphological studies but microscopic techniques such as polarized optical microscopy (POM), SEM and TEM are the most important.

4.1. Particle Size

Characterizations of nanoparticles are primarily evaluated by the particle size distribution and their morphology. With the aid of electron microscopy it's now possible to ascertain the morphology as well as the size of nanoparticles. Smaller the size of nanoparticles larger surface area, which results in to fast drug release. Loaded drug when exposed to the particle surface area causes significant drug release. In contrast, inside the nanoparticles drugs slow diffusion of larger particles occurs. Consequently, smaller particles tend to aggregate during storage and transportation of nanoparticles dispersion (Bhatia, 2016).

4.2. Morphological Characterizations

The morphological features of nanoparticles always attain great interest since morphology always influences most of the properties of the NPs. There are various characterization techniques for morphological studies of nanoparticles, but microscopic techniques such as polarized optical microscopy (POM), SEM and TEM are the most important among these.

SEM method is based on electron scanning principle, and it provides all available information about the NPs at desired nanoscale level. The dispersion of SWNTs in the polymer matrix poly (butylene) terephthalate (PBT) and nylon-6 revealed through this technique (Saeed and Khan, 2016). The same group also provides POM study of their materials, which showed star-like spherulites of the formed materials, whose size was decreased with the incremental filling of SWNTs. The morphological features of ZnO modified metal organic frameworks (MOFs) were studied through SEM technique, which indicates the ZnO NPs dispersion and morphologies of MOFs at different reaction conditions (Figure 1) (Mirzadeh and Akhbari, 2016).

Figure 1. SEM images of ZnO modified MOFs at different temperatures (Mirzadeh and Akhbari, 2016).

Similarly, TEM is based on electron transmittance principle, so it can provide information of the bulk material from very low to higher magnification. The different morphologies of gold

NPs are studied via this technique (Khan et al., 2019). Figure 2 provides some TEM micrographs showing various morphologies of gold NPs, prepared via different methods (Khlebtsov and Dykman, 2011, 2010a, 2010b). TEM also provides essential information about two or more layer materials, such as the quadrupolar hollow shell structure of Co3O4 NPs observed through TEM. Porous multicell structure induces shorter Li+ diffusion path length with adequate annulled space to buffer the volume expansion, good cycling performance, greater rate capacity, and specific capacity as well (Wang et al., 2013).

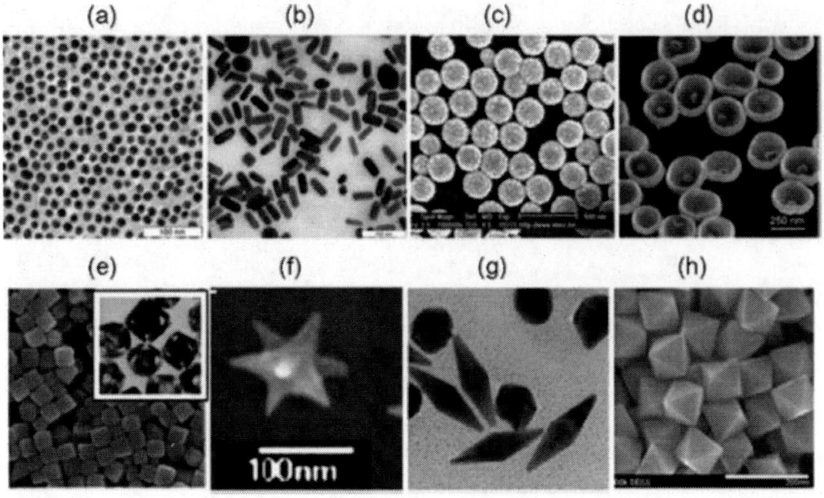

Figure 2. TEM images of different form of gold NPs, synthesized by different techniques (Khlebtsov and Dykman, 2011, 2010a, 2010b).

4.3. Structural Characterizations

X-Ray Differection (XRD) is one of the most significant characterization techniques to reveal the structural properties of nano particles (Khan et al., 2019). It gives enough information about the crystallinity and phase of NPs. It also provides rough idea about the particle size through Debye Scherer formula (Khan et al., 2017b, 2017c; Ullah et al., 2017). This technique worked well in both single and multiphase NPs identification (Emery et al., 2016). Nevertheless, in the case of smaller NPs having size less than hundreds of atoms, the acquisition and correct measurement of structural and other parameters may be difficult. Moreover, NPs having more amorphous characteristics with varied inter atomic lengths can influence the XRD diffractogram. In that case, proper comparison of the diffractograms of bimetallic NPs with those of the corresponding monometallic NPs and their physical mixtures is required to obtain accurate information. Comparison of computer simulated structural model of bimetallic NPs with observed XRD spectra is the best way to get good contrast (Ingham, 2015).

5. Nano Fertilisers

Degradation in Soil health is due to indiscriminate use of agricultural inputs viz. fertilizer, other agricultural chemicals, degradation of water resources and uneven distribution of weather parameters resulting in low input use efficiency. Nutritional deficiency is the third most important in limiting factor for plant growth and productivity next to drought and salinity (Singh and Singh, 2021). Continuous use of NPK fertilizers in cropping system has remarkedly increased food production but simultaneously revealed the widespread deficiency of micronutrients in soil (Sahai et al., 2006). To overcome these problems and sustaining the agricultural production and maintaining soil health and environment, there is urgent need to focus on advanced knowledge, tools and techniques for improving the resource use efficiency.

Development of smart materials may be an effective solution that can systematically release chemicals to specific targeted sites in plants which may be beneficial in controlling nutrition deficiency in agriculture. "Smart delivery system" refers the integration of specifically targeted, highly controlled, remotely regulated and multifunctional characteristic for avoiding biological barriers for successful targeting (Nair et al., 2010). Large scale production of nanoparticles of physiologically important metals due to technological advancement has improved fertilizer formulation by minimizing nutrient loss and increased uptake in plant cell (Naderi and Danesh-Shahraki 2013). The properties like high surface area, sorption capacity, and controlled-release kinetics of nano-fertilizers (NFs) towards targeted sites attributed them as smart delivery system. The NFs are the NMs that act as macro or micronutrients essential for crop plants (macro- or micronutrient fertilizers) or act as carriers of the conventional chemical fertilizers (nanocarriers) for efficient utilization of the nutrients.

Nanotechnology has provided the feasibility of exploring nanoscale or nanostructured materials as fertilizer carrier or controlled-release vectors for building of the so-called smart fertilizers as new facilities to enhance the nutrient use efficiency and reduce the cost of environmental pollution (Chinnamuthu and Boopati 2009). Surface coatings of nanomaterials on fertilizer particles hold the material more strongly due to higher surface tension than the conventional surfaces and thus help in controlled release (Brady and Weil 1999). Tarafdar and Adhikari (2015) described the role of various nanomaterials in soil and plant system (Table 2).

As mentioned in Table 3 nano-fertilizers show controlled release of agrochemicals, site targeted delivery, reduction in toxicity, and enhanced nutrient utilization of delivered fertilizers (Cui et al., 2010) due to their high surface area, increased solubility, and specific targeting due to small size, enhanced mobility and low toxicity (Sasson et al., 2007).

Nano fertilizers are modified form of conventional fertilizers or extracted from different vegetative or reproductive parts of the plant by different chemical, physical, mechanical or biological methods with the help of nanotechnological intervention for improving soil fertility, productivity and quality of agricultural produces. Tarafdar and Adhikari, 2015 reported the properties of nano-fertilizers having greater surface area due to very less size of particles led to more site of facilitation of different metabolic process in the plant system resulting in higher production of metabolites.

Table 2. Nanomaterials and their functions

Nanomaterials	**Functions**
Carbon nanotubes	Seed germination
Nanonutrients	Plant/animal/human nutrition
Nanopesticides	Plant protection
Nanaoscale carriers	Efficient delivery of fertilizer & pesticides
Nanosensor	To detect nutrients and contaminants

Table 3. Comparison of nanotechnology-based formulations and conventional fertilizers applications (Cui et al., 2010)

S. N.	Properties	Nano-fertilizers-enabled technologies	Conventional technology
1	Solubility and dispersion of mineral micronutrients	Nano-sized formulation of mineral micronutrients may improve solubility and dispersion of insoluble nutrients in soil, reduce soil absorption and fixation, and increase the bioavailability	Less bioavailability to plants due to large particle size and less solubility
2	Nutrient uptake efficiency	Nanostructure formulation might increase fertilizer efficiency and uptake ratio of the soil nutrients in crop production and save fertilizer resource	Bulk composite is not available for roots and decrease efficiency
3	Controlled release modes	Both release rate and release pattern of nutrients for water soluble fertilizers might be precisely controlled through encapsulation in envelope forms of semi permeable membranes coated by resin-polymer, waxes, and sulfur	Excess release of fertilizers may produce toxicity and destroy ecological balance of soil
4	Effective duration of nutrient release	Nanostructure formulation can extend effective duration of nutrient supply of fertilizers into soil	Used by the plants at the time of delivery, the rest is converted into insoluble salts in the soil
5	Loss rate of fertilizer nutrients	Nanostructure formulation can reduce loss rate of fertilizer nutrients into soil by leaching and/or leaking	High loss rate by leaching, rain off, and drift

6. Smart Delivery System of Nano Fertilizers

Nano-fertilizers are also termed as "Smart Fertilizers." Nano-fertilizers are synthesized or modified form of traditional fertilizers, bulk materials or extracted from different vegetative or reproductive parts of the plant with different chemical, physical, mechanical or biological methods along with the help of nanotechnology used to improve soil fertility, productivity and quality of agricultural produce (Tarafdar et al., 2012).

Nanofertilizers are nutrient carriers of nanodimensions ranging from 30 to 40 nm (10−9m or one-billionth of a meter) and capable of holding bountiful of nutrient ions due to their high surface area and release it slowly and steadily that commensurate with crop demand (Subramanian et al., 2015). Nanofertilizers have a profound influence on crop production. There are many issues with the use of traditional chemical fertilizers; however, low use efficiency is theprominent one, which not only increases the cost of production but also causesenvironmental pollution (Wilson et al., 2008). Nanomaterials with the large surfacearea could solve this problem due to their nanosize. These could be utilized asnanocoatings, e.g., sulfur nanocoating (≤100-nm layer), ensuring their controlledrelease, surface protection, and ultimately boosting up their use efficiency (Bradyand Weil 1999; Santoso et al., 1995). Nanofertilizers may be looked upon as analternative and more efficient to the ordinary fertilizers. These nanoformulations of nitrogenous fertilizer minimize nitrogen losses by leaching, emissions, and soilmicrobial immobilization (Liu et al., 2006). Subramanian et al., (2008) reported that nanofertilizers and nanocomposites can be used to control the release of nutrientsfrom the fertilizer granules so as to improve the nutrient use efficiency whilepreventing the nutrient ions either get fixed or lost to the environment. Nanofertilizers can be delivered in a timely manner to a rhizospheric target due to their high use efficiency.

Controlled-release fertilizers may also improve the soil by decreasing toxic effects associated with over-application of traditional chemical fertilizers (Suman et al., 2010).

7. Controlled-Release of Nanofertilizers

The most successful use of nanoparticles is as slow-release fertilizers. Because ofthe high surface tension, they will hold material more strongly from the plant than conventional surfaces. Moreover, nanocoatings can also provide surface protection for larger particles. Direct application of large amounts of fertilizer, in the form of ammonium salts, urea, nitrate, or phosphate compounds, may produce extremely high local concentrations which are harmful. Much of the fertilizer may be dissolved in runoff water and cause adverse effects such as pollution and will not be available to the plants of interest (Wilson et al., 2008).

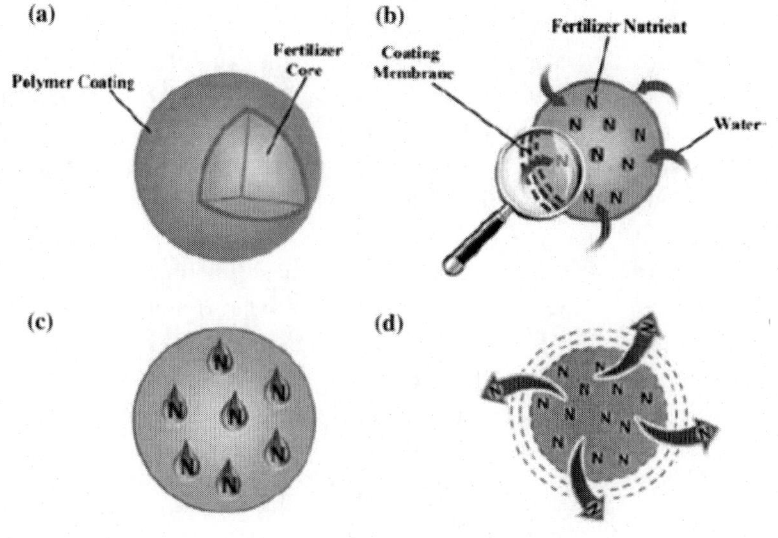

Figure 3. Diffusion mechanism of controlled release; (a) fertilizer core with polymer coating; (b) water penetrates into the coating and core granule; (c) fertilizer dissolution and osmotic pressure development; (d) controlled release of nutrient through swollen coating membrane. (Adapted from Azeem et al., 2014).

Generally, the controlled-release mechanism depends on numerous factors like the nature of the coating material, the type of fertilizer, agronomic conditions, and soil pH. Liu et al., (2008) and Shaviv (2005) postulated a possible release mechanism for coated fertilizers known as the multistage diffusion model. This model described that when applying the coated fertilizer, irrigation water penetrates the solid fertilizer core leading to partial nutrient dissolution (Figure 3). Subsequently, as osmotic pressure builds within the containment, the granule consequently swells and causes two processes. In the first, when osmotic pressure surpasses threshold membrane resistance, the coating bursts and the entire core released spontaneously. This is known as the "failure mechanism" or "catastrophic release" as coined by Goertz (1993). Secondly, if the membrane withstands the developing pressure, core fertilizer is thought to be released slowly via diffusion for which the driving force may be a concentration or pressure gradient, or combination thereof called the "diffusion mechanism." The failure mechanism is generally observed in frail coatings (e.g., sulphur or modified sulphur), while polymer coatings (e.g., polyolefin) are expected to exhibit the diffusion release mechanism. A pictographic representation of both mechanisms is given in Figure 3.

The sequential process for the release mechanism is diffusion/swelling of the fertilizer followed by degradation of the polymer coating, and subsequent fracture or dissolution. Similar release mechanism was reported by several researchers (Guo et al., 2005; Liang et al., 2007; Liu et al., 2007; Wu and Liu, 2008).

Conclusion

The extensive use of agrochemical leads to disturbance of soil health. So, nanoformulations could give the feasible option to minimise the use of chemical-based fertilizers. At present we have a need of nitrogenous and phosphatic fertilizers in large quantities as these are primary nutrients and play very important role in growth and development of plants. These fertilizers also have problem of low nutrient use efficiency and losses are more. Thus, more efforts might be focused on N and P-nano fertilizers. Comparison studies should also be conducted between Ca-NMs from CaCO3 with other Ca sources like CaCl2 or CaSO4. Research should also be focused on multinutrient comparison like that of Ca- and N-NMs using soluble Ca (NO3)2 as control. Intensive research might be conducted on micronutrients to reduce the effect of factors affecting their availability under field conditions. Biofortification studies may also be conducted to investigate the role of micronutrients in enhancing the quality of food in comparison to conventional fertilizers. Application of nano micronutrients through fertigation should also be tested to compare the efficacy of these nanoformulations over commercially available micronutrient fertilizers. Toxicity of nanomaterials is the main constraint with these formulations therefore efficient and effective dose should be calculated for each and every crop without any toxic effects may also be investigated with the help of research studies. Hence, nanotechnology may be a powerful tool for addressing the current issues of environmental challenges and benefits of human being or mankind.

References

Abd Ellah, N. H., Abouelmagd, S. A. (2016). Surface functionalization of polymeric nanoparticles for tumor drug delivery: approaches and challenges. *Expert Opin. Drug Deliv.* 1–14. http://dx.doi.org/10.1080/17425247.2016.1213238.

Abouelmagd, S. A., Meng, F., Kim, B.-K., Hyun, H., Yeo, Y. (2016). Tannic acid-mediated surface functionalization of polymeric nanoparticles. *ACS Biomater. Sci. Eng.*, 6b00497 http://dx.doi.org/10.1021/acsbiomaterials.6b004.

Ali, S., Khan, I., Khan, S. A., Sohail, M., Ahmed, R., Rehman, A., Ur Ansari, M. S., Morsy, M. A. (2017). Electrocatalytic performance of Ni@Pt core–shell nanoparticles supported on carbon nanotubes for methanol oxidation reaction. *J. Electroanal. Chem.* 795, 17–25. http://dx.doi.org/10.1016/ j.jelechem.2017.04.040.

Anderson DM (2009) Approaches to monitoring, control and management of harmful algal blooms (HABs). *Ocean Coast Manag* 52:342–347.

Astefanei, A., Nunez, O., Galceran, M. T., (2015). Characterisation and determination of fullerenes: a critical review. *Anal. Chim. Acta* 882, 1–21. http://dx.doi.org/10.1016/ j.aca.2015.03.025.

Auffan M, Rose J, Bottero JY, Lowry GV, Jolivet JP, Wiesner MR (2009) Towards a definition of inorganic nanoparticles from an environmental, health and safety perspective. *Nat Nanotechnol* 4:634–641.

Awwad A, Ahmad A. Biosynthesis, characterization, and optical properties of magnesium hydroxide and oxide nanoflakes using Citrus limon leaf extract. *Arab Journal of Physical Chemistry*. 2014; 1:65-70.

Azeem B, Kushaari K, Man ZB, Basit A, Thanh TH (2014) Review on materials & methods to produce controlled release coated urea fertilizer. *J Control Release* 181:11–21.

Bhatia, S. (2016). Nanoparticles Types, Classifi cation, Characterization, Fabrication Methods and Drug Delivery Applications Springer International Publishing Switzerland. *Natural Polymer Drug Delivery Systems*, doi 10.1007/978-3-319-41129-3_2.

Brady NR, Weil RR (1999) In: Brady NR, Weil RR (eds) *The nature and properties of soils*. Prentice Hall, New Jersey, pp 415–473.

Chinnamuthu CR, Boopati PM (2009) Nanotechnology and agroecosystem. *Madras Agric J* 96:17–31.

Cui HX, Sun CJ, Liu Q, Jiang J, Gu W (2010) Applications of nanotechnology in agrochemical formulation, perspectives, challenges and strategies. In: *International conference on Nanoagri*, Sao Pedro, Brazil, pp 28–33.

Cui HX, Sun CJ, Liu Q, Jiang J, Gu W (2010) Applications of nanotechnology in agrochemical formulation, perspectives, challenges and strategies. In: *International conference on Nanoagri*, Sao Pedro, Brazil, 20–25 June 2010, pp 28–33.

Davidson DW, Verma MS, Gu FX (2013) *Controlled root targeted delivery of fertilizer using an ionically crosslinked carboxymethyl cellulose hydrogel matrix*. Springer Plus 2:318. doi:10. 1186/2193-1801-2-318.

De Rosa MC, Monreal C, Schnitzer M, Walsh R, Sultan Y (2010) Nanotechnology in fertilizers. *Nat Nanotechnol* 5:91–94.

Dimkpa, C. O. (2014). Can nanotechnology deliver the promised benefits without negatively impacting soil microbial life? *J. Basic Microbiol.* 54, 889–904. doi: 10.1002/jobm. 201400298.

Ditta A, Arshad M, Ibrahim M (2015) Nanoparticles in sustainable agricultural crop production: applications and erspectives. In: Siddiqui MH, Al-Whaibi MH, Mohammad F (eds) *Nanotechnology and plant sciences*. Springer, Switzerland, pp 55–75.

Dreaden, E. C., Alkilany, A. M., Huang, X., Murphy, C. J., El-Sayed, M. A. (2012). The golden age: gold nanoparticles for biomedicine. *Chem. Soc. Rev.* 41, 2740–2779. http://dx.doi.org/10.1039/C1CS15237H.

Emery, A. A., Saal, J. E., Kirklin, S., Hegde, V. I., Wolverton, C., 2016. High-throughput computational screening of perovskites for thermochemical water splitting applications. *Chem. Mater.* 28. http://dx.doi.org/10.1021/acs.chemmater.6b01182.

Ghidan AY, Al-Antary TM, Awwad AM, Ghidan OY, Al Araj SE, Ateyyat MA. Comparison of different green synthesized nanomaterials on green peach aphid as aphicidal potential. *Fresenius Environmental Bulletin*. 2018a, 2018a;27(10):7009-7016.

Ghidan AY, Al-Antary TM, Salem NM, Awwad AM. Facile green synthetic route to the zinc oxide (ZnONPs) nanoparticles: Effect on green peach aphid and antibacterial activity. *Journal of Agricultural Science*. 2017a;9(2):131-138.

Ghidan, A. Y., and Antary, T, M, Al. (2019). Applications of Nanotechnology in Agriculture. *IntechOpen*. doi: 10.5772/intechopen.88390.

Gruere, G. P. (2012). Implications of nanotechnology growth in food and agriculture in ECD countries. *Food Policy* 37, 191–198. doi: 10.1016/j.jhazmat.2014.05.079.

Gujrati, M., Malamas, A., Shin, T., Jin, E., Sun, Y., Lu, Z.-R. (2014). Multifunctional cationic lipid-based nanoparticles facilitate endosomal escape and reduction-triggered cytosolic siRNA release. *Mol. Pharm.* 11, 2734–2744. http://dx.doi.org/10.1021/mp400787s.

Gumbo RJ, Ross G, Cloete ET (2008) Biological control of microcystis dominated harmful algal blooms. *Afr J Biotechnol* 7:4765–4773.

Guo M, Zhu M, FaluWuL (2005) Preparation and properties of a slow-release membrane-encapsulated urea fertilizer with superabsorbent and moisture preservation. *Ind Eng Chem Res* 44(12):4206–4211.

Hisatomi, T., Kubota, J., Domen, K. (2014). Recent advances in semiconductors for photocatalytic and photoelectrochemical water splitting. *Chem. Soc. Rev.* 43, 7520–7535. http://dx.doi.org/10.1039/C3CS60378D.

Ingham, B., 2015. X-ray scattering characterisation of nanoparticles. *Crystallogr. Rev.* 21, 229–303. http://dx.doi.org/10.1080/0889311X.2015.1024114.

Ion, A. C., Ion, I., and Culetu, A. (2010). Carbon-based nanomaterials: Environmental applications. *Univ. Politehn. Bucharest* 38, 129–132.

Johnston CT (2010) Probing the nanoscale architecture of clay minerals. *Clay Miner* 45:245–279.

Khan, I., Abdalla, A., Qurashi, A. (2017a). Synthesis of hierarchical WO3 and Bi2O3/WO3 nanocomposite for solar-driven water splitting applications. *Int. J. Hydrogen Energy* 42, 3431–3439. http://dx.doi.org/10.1016/j.ijhydene.2016.11.105.

Khan, I., Abdalla, A., Qurashi, A., 2017a. Synthesis of hierarchical WO3 and Bi2O3/WO3 nanocomposite for solar-driven water splitting applications. *Int. J. Hydrogen Energy* 42, 3431–343. http://dx.doi.org/10.1016/j.ijhydene.2016.11.105.

Khan, I., Ali, S., Mansha, M., Qurashi, A., 2017b. Sonochemical assisted hydrothermal synthesis of pseudo-flower shaped Bismuth vanadate (BiVO4) and their solar-driven water splitting application. *Ultrason. Sonochem.* 36, 386–392. http://dx.doi.org/10.1016/j.ultsonch.2016.12.014.

Khan, I., Ibrahim, A.A.M., Sohail, M., Qurashi, A., 2017c. Sonochemicalassisted synthesis of RGO/ZnO nanowire arrays for photoelectrochemical water splitting. *Ultrason. Sonochem.* 37, 669–675. http://dx.doi.org/10.1016/j.ultsonch.2017.02.029.

Khan, I., Saeed, K and Khan, I. (2019). Nanoparticles: Properties, applications and toxicities. *Arabian Journal of Chemistry*. 12:908-931. https://doi.org/10.1016/j.arabjc.2017.05.011.

Khlebtsov, N., Dykman, L., 2010. *Plasmonic nanoparticles*. pp. 37–85. http://dx.doi.org/10.1201/9781439806296-c2.

Khlebtsov, N., Dykman, L., 2011. Biodistribution and toxicity of engineered gold nanoparticles: a review of in vitro and in vivo studies. *Chem. Soc. Rev*. 40, 1647–1671. http://dx.doi.org/10.1039/C0CS00018C.

Khlebtsov, N. G., Dykman, L. A., 2010b. Optical properties and biomedical applications of plasmonic nanoparticles. *J. Quant. Spectrosc. Radiat. Transf.* 111, 1–35. http://dx.doi. org/10.1016/j.jqsrt.2009.07.012.

Liang R, Liu M, Wu L (2007) Controlled release NPK compound fertilizer with the function of water retention. *React FunctPolym* 67(9):769–779.

Lin D, Xing B. Phytotoxicity of nanoparticles: Inhibition of seed germination and root growth. *Environmental Pollution*. 2007;150(2):243-250. doi: 10.1016/j. envpol.2007.01. 016.

Liu LS, Kost J, Fishman M, Hicks KB (2008) A review: controlled release systems for agricultural and food applications. In: Parris N, Liu LS, Song C, Shastri VP (eds) New delivery systems for controlled drug release from naturally occurring materials, *ACS Symposium series*, vol 992. pp 265–281.

Liu M, Liang R, Zhan F, Liu Z, Niu A (2007) Preparation of superabsorbent slow-release nitrogen fertilizer by inverse suspension polymerization. *Polym Int* 56(6):729–737.

Liu X, Feng Z, Zhang F, Zhang S, He X (2006) Preparation and testing of cementing and coating nano-subnanocomposites of slow/controlled-release fertilizer. *Agric Sci China* 5:700–706.

Mansha, M., Khan, I., Ullah, N., Qurashi, A. (2017). Synthesis, characterization and visible-light-driven photoelectrochemical hydrogen evolution reaction of carbazole-containing conjugated polymers. *Int. J. Hydrogen Energy*. http://dx.doi.org/10.1016/j.ijhydene. 2017.02.053.

Mashaghi, S., Jadidi, T., Koenderink, G., Mashaghi, A. (2013). Lipid nanotechnology. *Int. J. Mol. Sci.* 14, 4242–4282. http://dx.doi.org/10.3390/ijms14024242.

Mirzadeh, E., Akhbari, K. (2016). Synthesis of nanomaterials with desirable morphologies from metal–organic frameworks for various applications. *CrystEngComm* 18, 7410–7424. http://dx.doi.org/10.1039/C6CE01076H.

Moorthy SK, Ashok CH, Rao KV, Viswanathana C. Synthesis and characterization of MgO nanoparticles by neem leaves through green method. *Materials Today: Proceedings*. 2015; 2:4360-4368.

Mukhopadhyay SS. (2014). Nanotechnology in agriculture: Prospects and constraints. *Nanotechnology, Science and Applications*. 7:63-71.

NAAS (2013) *Nanotechnology in agriculture: scope and current relevance. Policy Paper No 63*, National Academy of Agricultural Sciences, New Delhi, India, pp 1–20.

Nair R, Varghese SH, Nair BG, Maekawa T, Yoshida Y, Kumar DS (2010) Nanoparticulate material delivery to plants. *Plant Sci* 179:154–163.

Prasad R, Bagde US, Varma A (2012) Intellectual property rights and agricultural biotechnology: an overview. *Afr J Biotechnol* 11(73):13746–13752.

Prasad, R., Kumar, V., and Prasad, K. S. (2014). Nanotechnology in sustainable agriculture: present concerns and future aspects. *Afr. J. Biotechnol.* 13, 705–713. doi: 10.5897/ AJBX2013.13554.

Prasad, Ram., Bhattacharyya, A., and Nguyen, Q. D. (2017). Nanotechnology in Sustainable Agriculture: Recent Developments, Challenges, and Perspectives. *Frontiers in Microbiology.* doi: 10.3389/fmicb.2017.01014.

Puri, A., Loomis, K., Smith, B., Lee, J.-H., Yavlovich, A., Heldman, E., Blumenthal, R. (2009). Lipid-based nanoparticles as pharmaceutical drug carriers: from concepts to clinic. *Crit. Rev. Ther. Drug Carrier Syst.* 26, 523–580.

Rao, J. P., Geckeler, K. E. (2011). Polymer nanoparticles: preparation techniques and size-control parameters. *Prog. Polym. Sci.* 36, 887– 913. http://dx.doi.org/10.1016/j.progpolymsci.2011.01.001.

Rawat, M. K., Jain, A., Singh, S., Mehnert, W., Thunemann, A. F., Souto, E. B., Mehta, A., Vyas, S. P. (2011). Studies on binary lipid matrix based solid lipid nanoparticles of repaglinide: in vitro and in vivo evaluation. *J. Pharm. Sci.* 100, 2366–2378. http://dx. doi.org/10.1002/jps.22435.

Sabir, S., Arshad, M., and Chaudhari, S. K. (2014). Zinc oxide nanoparticles for revolutionizing agriculture: synthesis and applications. *Sci. World J.* 2014:8. doi: 10.1155/2014/925494.

Sadeghzadeh, B. (2013). A review of zinc nutrition and plant breeding. *J. Soil Sci. Plant Nutr.* 13, 905–927. doi: 10.4067/S0718-95162013005000072.

Saeed, K., Khan, I., 2014. Preparation and properties of single-walled carbon nanotubes/poly (butylene terephthalate) nanocomposites Iran. *Polym. J.* 23, 53–58. http://dx.doi. org/10.1007/s13726-013-0199-2.

Saeed, K., Khan, I., 2016. Preparation and characterization of singlewalled carbon nanotube/nylon 6,6 nanocomposites. *Instrum Sci. Technol.* 44, 435–444. http://dx.doi.org/10.1080/10739149.2015.1127256.

Sahai P., Srivastava PC, Singh SK and Singh AP (2006) Evaluation of organics incubated with zinc sulphate as zinc source for rice-wheat rotation. *J. Eco-friendly Agriculture* 1(2): 120-125.

Salem NM, Albanna LS, Abdeen A, Ibrahim OQ, Awwad AI. Sulfur nanoparticles improves root and shoot growth of tomato. *Journal of Agricultural Science*. 2016;8(4):179-185. doi: 10.5539/jas.v8n4p179.

Santoso D, Lefroy RDB, Blair GJ (1995) Sulfur and phosphorus dynamics in an acid soil/crop system. *Aust J Soil Res* 33:113–124.

Sasson Y, Levy-Ruso G, Toledano O, Ishaaya I (2007) Nanosuspensions: emerging novel agrochemical formulations. In: Ishaaya I, Horowitz AR, Nauen R (eds) *Insecticides design using advanced technologies.* Springer, Berlin, pp 1–39.

Sawant R, Hurley J, Salmaso S, Kale A, Tolcheva E, Levchenko T, Torchilin V (2006) Smart drug delivery systems: double-targeted pH-responsive pharmaceutical nanocarriers. *Bioconj Chem* 17(4):943–949.

Sekhon BS (2014) Nanotechnology in agri-food production: an overview. *Nanotechnol Sci Appl* 7:31–53.

Shankramma K, Yallappa S, Shivanna MB, Manjanna J. Fe2O3 magnetic nanoparticles to enhance S. lycopersicum (tomato) plant growth and their biomineralization. *Applied Nanoscience*. 2016;6:983-990. doi: 10.1007/s13204-015-0510-y.

Shaviv A (2005) Controlled release fertilizers, *IFA International Workshop on Enhanced-efficiency Fertilizers,* Frankfurt, International Fertilizer Industry Association, Paris, France.

Shaviv A (2005) *Controlled release fertilizers, IFA International Workshop on Enhanced-efficiency Fertilizers,* Frankfurt, International Fertilizer Industry Association, Paris, France.

Siddiqui MH, Al-Whaibi MH, Firoz M, Al-Khaishany MY (2015) Role of nanoparticles in plants. In: Siddiqui MH, Al-Whaibi MH, Mohammad F, (eds) *Nanotechnology and plant sciences*, Springer, Switzerland, pp 19–35.

Sigmund, W., Yuh, J., Park, H., Maneeratana, V., Pyrgiotakis, G., Daga, A., Taylor, J., Nino, J. C., 2006. Processing and structure relationships in electrospinning of ceramic fiber systems. *J. Am. Ceram. Soc.* 89, 395–407. http://dx.doi.org/10.1111/j.1551- 2916.2005. 00807.x.

Singh P., Singh A. P. (2021) Nanomaterials in Soil Health Management and Crop Production: Potentials and Limitations. In: Kharissova O. V., Martínez L. M. T., Kharisov B. I. (eds) *Handbook of Nanomaterials and Nanocomposites for Energy and Environmental Applications*. Springer, Cham. https://doi.org/10.1007/978-3-030-11155-7_35-1.

Singh, Ajeet, Rajput P, Chopra R, Ankush, Singh S, Singh A and Singh Pratibha (2021) Application of Nanomaterials in Agriculture and Environment Management. *Microbial Synthesis of Nanomaterials.* Nova Science Publishers pp163-178.

Singh, P, Sammauria R, Singh M, Singh SK, Jatav HS, Mehjabeen, Yadav MR (2021) Rhizobia: A Potent Tool for Amelioration of Drought Stress in Legumes. In: Tariq Aftab, Khalid Rehman Hakeem (eds) *Plant Growth Regulators*. Springer, pp 421-439 https://doi.org/10.1007/978-3-030-61153-8.

Subramanian KS, Manikandan A, Thirunavukkarasu M, Rahale CS (2015) Nano-fertilizers for balanced crop nutrition. In: Rai M, Ribeiro C, Mattoso L, Duran N (eds) *Nanotechnologies in food and agriculture*. Springer, Switzerland, pp 69–80.

Subramanian KS, Paulraj C, Natarajan S (2008) Nanotechnological approaches in nutrient management. In: Chinnamuthu CR, Chandrasekaran B, Ramasamy C (eds) Nanotechnology applications in agriculture. *TNAU technical bulletin*, Coimbatore, India, pp 37–42.

Suman PR, Jain VK, Varma A (2010) Role of nanomaterials in symbiotic fungus growth enhancement. *Curr Sci* 99:1189–1191.

Suman PR, Jain VK, Varma A (2010) Role of nanomaterials in symbiotic fungus growth enhancement. *Curr Sci* 99:1189–1191.

Sun, S. (2000). Monodisperse Fe Pt nanoparticles and ferromagnetic FePt nanocrystal superlattices. *Science* 80 (287), 1989–1992. http://dx.doi.org/10.1126/science.287. 5460.1989.

Tarafdar J. C. and Adhikari Tapan (2015) *Soil Science: An Introduction, Chapter: Nanotechnology in Soil Science.*, Editors: Eds. Rattan, R. K. et al., 775-807.

Tarafdar JC, Raliya R, and Rathore I (2012) Microbial synthesis of phosphorous nanoparticle from tri- calcium phosphate using Aspergillus tubingensis TFR-5. *J. Bionanosci*. 6: 84–89.

Thomas, S., Harshita, B. S. P., Mishra, P., Talegaonkar, S. (2015). Ceramic nanoparticles: fabrication methods and applications in drug delivery. *Curr. Pharm. Des*. 21, 6165–6188. http://dx.doi.org/10.2174/1381612821666151027153246.

Ullah, H., Khan, I., Yamani, Z. H., Qurashi, A., 2017. Sonochemicaldriven ultrafast facile synthesis of SnO2 nanoparticles: growth mechanism structural electrical and hydrogen gas sensing properties. *Ultrason. Sonochem*. 34, 484–490. http://dx.doi.org/10.1016/j.ultsonch.2016.06.025.

Vyas, S. P. and Khar, R. K. (2002). *Targeted and controlled drug delivery*. CBS publishers and distributers. New Delhi. 1:331–43.

Wang, J., Yang, N., Tang, H., Dong, Z., Jin, Q., Yang, M., Kisailus, D., Zhao, H., Tang, Z., Wang, D., 2013. Accurate control of multishelled Co3O4 hollow microspheres as high-performance anode materials in lithium-ion batteries. *Angew. Chemie Int*. Ed. 52, 6417–6420. http://dx.doi.org/10.1002/anie.201301622.

Wilson MA, Tran NH, Milev AS, Kannangara G, Volk H, Lu G (2008) Nanomaterials in soils. *Geoderma* 146(1):291–302.

Zaytseva, O., Neumann, G. (2016). Carbon nanomaterials: production, impact on plant development, agricultural and environmental applications, *Chem. Biol. Technol. Agric*. 3: 1–26.

Zhang, Q., Han, L., Jing, H., Blom, D. A., Lin, Y., Xin, H. L., et al., (2016). Facet control of gold nanorods. *ACS Nano* 10, 2960–2974. doi: 10.1021/acsnano. 6b00258.

Chapter 4

Nitrogen Management in Field Crops and Strategies for Enhancing Its Efficiency

Asik Dutta[1,*], Ankita Trivedi[2], Narendra Kumar[1], C. P. Nath[1], K. K. Hazra[1], C. S. Praharaj[1], Hanuman Singh Jatav[3] and Eetela Sathyanarayana[4]

[1]ICAR-Indian Institute of Pulses Research, Kanpur, Uttar Pradesh, India
[2]ICAR-Indian Agricultural Research Institute, New Delhi, Delhi, India
[3]S.K.N. Agriculture University-Jobner, Jaipur, Rajasthan, India
[4]Agriculture College (PJTSAU) -Palem, Nagarkurnool, Telengana, India

1. Introduction

Soil is the critical component of the Earth ecosystem, and plays pivotal role in agricultural production system. Healthy and productive soils sustain life on the Earth. However, poor fertility of Indian soils is reported long back during 1928 by the Royal Commission on Agriculture in India. The commission reported about widespread nitrogen (N) deficiency in Indian soils. At present also, N is universally deficient in Indian soils with more than 99% of agricultural soils in India responding to the application of N-fertilizer application. Being the main constituent of proteins (also nucleic proteins), N holds the key role to the life on the Earth. Nitrogen (N) shares paramount position in group "V" with other non-metals and metalloids like Phosphorus (P), Arsenic (As), Antimony (Sb) and so on popularly known as 'Pnicogens'. Although, the lithosphere i.e., soil, rock and silicate minerals shares majority of N (98%), the atmosphere is the undeniable store house of N in our environment. Aside, structural elements N rank first in the list of essential elements and having an indispensable role on plant phenology and physiology. N is highly mobile inside the plant and plant tissue in general contains 1.5% N but, the content may be as high as 5% in the cell walls. Before coming to the main field, let's have a glance on the role of N on crop growth:

1. Crucial part of genetic materials like deoxyribonucleic acid (DNA) and ribonucleic acid (RNA) and conveying genetic code to the next generation.
2. Vital component of all biochemical compounds like amino acids, proteins, porphyrins, flavins, both purine and pyrimidine nucleotides, enzymes, co-enzymes and alkaloids.

[*] Corresponding Author's Email: asikdutta975@gmail.com.

In: Ecosystem Services
Editor: Hanuman Singh Jatav
ISBN: 978-1-68507-614-6

3. In the chlorophyll four N-molecules forms the base by encircling the central magnesium ion thus helps the plants to capture the solar energy and produce food.
4. In food grains N fertilisation increases the amounts of essential amino acids like glutamic acid, proline, cysteine, and methionine.
5. Moreover, N improves the quality of economical plant parts and improves the vigour of crop plants.
6. N promotes better utilisation of other major elements like Phosphorus (P) and potassium (K).

As it is highly mobile the deficiency symptoms of N mostly appears in the lower and older leaves and proteolysis i.e., yellow colouration is the primary indication of N deprivation. Under severe condition the entire leaves became yellow and plants eventually dies. Whereas, excessive application of N-fertilisers does exorbitant vegetative growth promoting succulent, taller and heavier plants leads to lodging. Over application of N increases the risk of more pest and disease infestation. The uptake and utilisation efficiency of other nutrients deviates a lot due to unnecessary fertilisation.

2. Status of Nitrogen in Indian Soil

Soil fertility is the inherent ability of the soil to supply nutrient elements to the plants. Most of the Indian soils are low in N content i.e., 0.03 to 0.07% (Dey and Sekhon, 2016). The low status of N in soil is due to low soil organic matter (SOM) content as the turn-over rate of SOM is very high in tropical and sub-tropical conditions. But Pathak (2010) reported that N fertility status in majority of the states were medium except in the North East India, where soils are rich in available N (Figure 1). Many reports showed positive N balance in Indian soils (Pathak et al., 2010; Murgan and Dhadwal, 2007) whereas, reports from FAI (2018) discloses the negative N-balance. The controversy has been justified as exclusion of N through irrigation, rain and crop residues in the study. However, on the contrary, world agriculture has witnessed N driven higher production and so is the case in India. But, like many other countries across the globe including India clearly mentioned the poor status of available N and showcased the need of balanced N-fertilisation (Dey and Sekhon, 2016). Increased food demand, subsidized nitrogenous fertilizer, intensive agricultural practices has caused irrational use of nitrogenous fertilizer and has aggravated the problem of environmental pollution. From Figure 2, it is clear that over the decades, consumption of nitrogenous fertilizer has been increased substantially. On present day, the consumption ratio of N, P and K fertilizer is 6.7:2.4:1, whereas, the recommended usage ratio is 4:2:1. Those numerical figures clearly show that fertilizer application in Indian agriculture is highly skewed towards N fertilizer application. Further, the use efficiency of N- fertilizer is very low which means the voluminous N fertilizer applied goes mostly unutilized by the crop projecting money drainage scenario. As a result, the excessive reactive N leaves the agro-ecosystems through leaching as nitrate or dissolved form of organic N, or through gaseous emissions to the atmosphere in the form of NH_3, NO, N_2O or N_2. All these avenues of loss, with the important exception of N_2, are linked to one or more local, regional or global environmental hazard (Singh et al., 2007). It gets lost to the water bodies, atmosphere and soil thereby causing environmental hazard. So, judicious use of nitrogenous fertilizer is the call of the hour, needing immediate global attention. Policy makers need to think beyond politics when it comes to environmental sustainability.

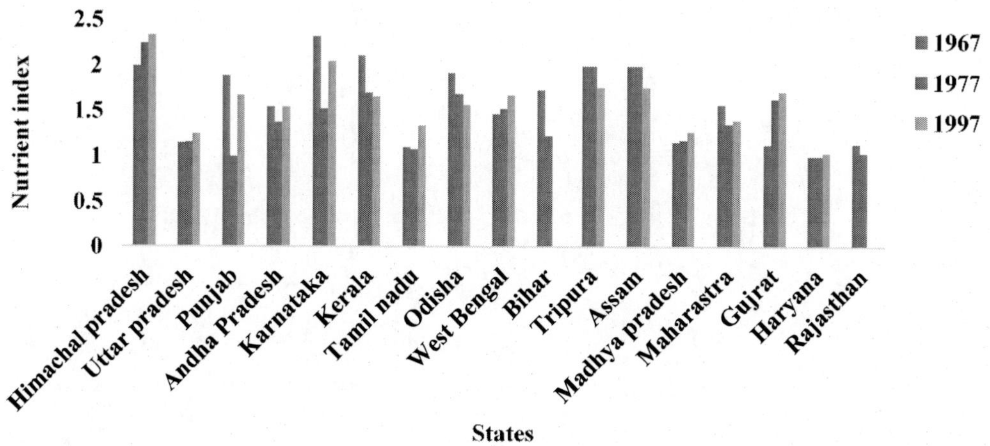

Figure 1. Nutrient index of N agricultural soil in different states of the country over the years (Source: Pathak, 2010).

The fertility status of N in soil can be improved by improving nitrogen use efficiency (NUE) or by increasing the storage of N in the Indian soil through crop rotation, checking the loss of N, integrated nutrient management, improved good agricultural practices, real time N application, site specific nutrient management, slow release fertilizer application and better policy formulations. Nutrient mining in agricultural practices for improved agricultural production needs to be refilled immediately before the damage comes to a stage which cannot be further repaired.

3. Production and Consumption of N-Fertilisers in Indian Agriculture

Due to population outburst the agriculture and allied sectors have huge responsibility of feeding more than 9 billion people across the globe by 2050 (FAO, 2013). Fertiliser responsive high yielding crop varieties are the preclusive necessities for satisfying the need of hungry stomachs. From the last few decades fertiliser sector helps a lot to achieve the title of food secure country rather than being a dependent one. Trend of nitrogenous fertiliser production and consumption in India is presented in Figure 2. It is clear that over the year the production and consumption of N-fertilisers increased dramatically, whereas the consumption is two steps ahead of production. Production of fertilizer-N in India increased from 0.1 million tons (MT) in 1960-61 to 13.4 MT in 2017-18 which is at presently produced by 30 numbers of urea manufacturing units in the country. Whereas, the consumption of fertilizer-N increased from 0.6 mt in 1960-61 to 17.0 mt during 2017-18. Thus, country emerged as 2nd largest producer and consumer of fertilizer-N in the world. About 70% of total N-fertilizer consumption in India goes to 5 major crops namely rice, wheat, maize, sugarcane and cotton. Similarly, about 50% of total N-fertilizer consumption is in 5 states – Uttar Pradesh, Maharashtra, Punjab, Gujarat and Madhya Pradesh. An estimate shows that demand and supply of fertilizer-N in India by 2030 would be

23.45 MT and production 18.76 MT, respectively, leaving a gap of 4.69 MT (Tewatia and Chanda, 2017).

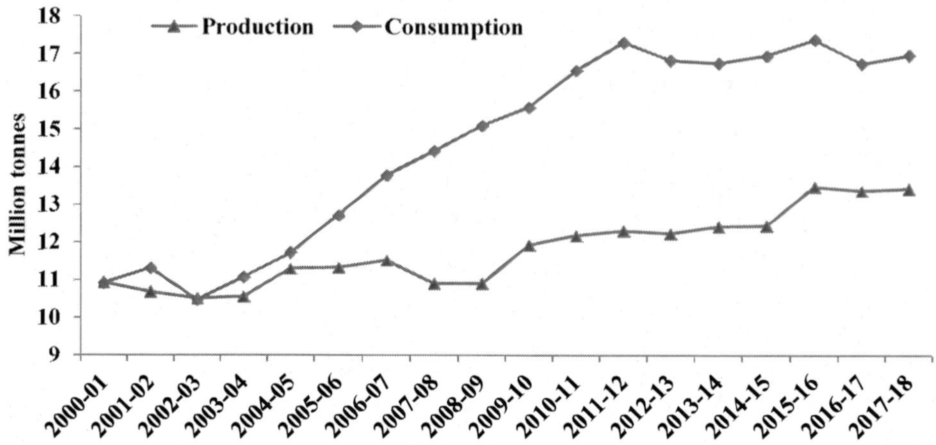

Figure 2. Production ('000 tonnes) and consumption trend of nitrogenous fertilizers in India (FAI, 2018).

Among different parts of the country, Northern part mostly Indo-Gangetic Plain (IGP) and southern zone consume more fertilisers than western and eastern zones (Figure 3). Among different zones, over the year N-fertiliser consumption follows a decreasing trend except the western part of the country.

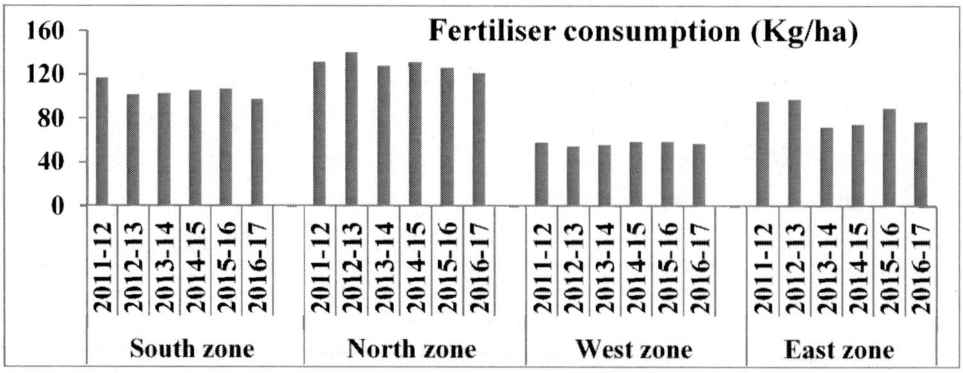

Figure 3. Fertilizers consumption trend in different zones of the country (Source: DAC, 2017).

Over the year there is bumper jump in fertiliser consumption in three major crops such as rice, wheat and sugarcane (Figure 4). The major reason for this surging in fertiliser consumption is cultivation of high yielding fertiliser responsive varieties to meet the demand of supply and drop the status of 'begging bowl'.

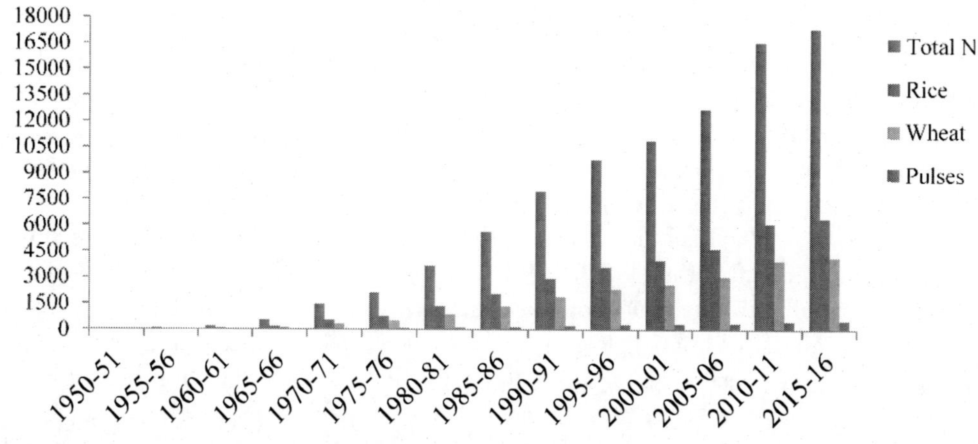

Figure 4. Trend of crop wise fertiliser consumption ('000 tonnes).

4. Strategies for Improving Nitrogen Use Efficiency (NUE)

Balanced fertilisation is one of the prerequisite inputs of healthy plant growth but, due to different loopholes the targeted goal is still untouched and at the same time polluting our mother nature. In case of nitrogenous fertilisers the use efficiency is hardly half of the applied amount and in actual case under rainfed and irrigated condition the proficiency hardly exceeds 25 and 35% irrespective of any crop (Rakshit et al., 2015a). Ladha et al. (2005) revealed that for rice, wheat and maize the NUE is 46, 57 and 65% respectively. So, for getting productive NUE there must be a synchrony between crop demand and supply so that we can achieve our two major objectives:

1. Surging the NUE and uniform nutrient supply throughout the growing period.
2. Tapering the loss, saving and better utilisation for the future crop.

So, to hit the target advanced resource conserving agronomic, microbial and molecular approaches should come forward and work hand to hand. In the following section four main strategies are briefed which could justify the fore mentioned objectives.

4.1. Strategy-1: Soil and Input Management

4.1.1. Selection of Fertiliser

There are two major plant available forms of N in the soil, these are nitrate (NO_3^-) and ammonium (NH_4^+), and the former is the most loving one. Though, few crops like paddy, tomato favour the later one. For this reason, the two ion supplying fertilisers are preferred over the straight one. Ammonical fertilisers are less prone to N loss than NO_3^- as NH_4^+ gets attached with the negatively charged sites of the soil.

4.1.2. 4R of Fertiliser Management

After selecting N-fertiliser suitable dose for specific location and crop, it is ubiquitous to follow the steps of 4R for attaining higher NUE. 4R is basically acronym of right place, right method, right time and right dose. Required fertiliser dose is paramount as over and under application does not satisfy the need besides causing environmental pollution. Soil testing and fertiliser recommendation (STCR) is the soundest approach for calculating the fertiliser dose depending on the crop requirement. Other than this, omission plot technique is also a viable option for N-fertiliser application and getting the targeted yield (Dobermann and Witt, 2002). A well-co-ordinated timing of fertiliser application is the rudiment of higher use efficiency. For example, in sandy soils to check the leaching loss of N fertilisers, it is advisable to split the entire amount in two or three halves depending on the crop requirement. Whereas, in calcareous soils a pre-irrigation before fertiliser application is advocated. To judge the actual time of fertiliser application on-field N management strategies such as, chlorophyll meter, leaf colour charts and off-field technique like tissue testing is recommended. The right method is closely linked with right place because, depending upon the fertiliser these two varies. For instance, N-fertilisers are applied in splits whereas, P fertilisers are mostly band placed in close proximity to the roots as these are fixed by iron or aluminium oxides or calcium hydroxides depending on soil pH. In lieu, with P and K, N-fertilisers are applied even after planting as top dress or subsurface side-dress. As, N-fertilisers are highly soluble and have negative impact on environment age old methodologies are replaced by new agro-techniques. A combination of FYM that to by 25% of recommended nitrogen dose and rest by urea showed greater Agronomic nitrogen use efficiency (ANUE) and apparent nitrogen recovery (ANR) than supplying the whole amount by urea (Singh and Ahlawat, 2012). While, our natural resources are plunging at a blistering rate so, inter-cropping with legumes, green manuring, and organic practices can curtail the dose of fertilisers. As, there is both positive and antagonistic relation among the nutrients for that reason, a well-timed site specific methodology which can satisfy the plant need is the indubitable option for higher use efficiency.

4.1.3. Slow Release N-Fertilisers and Nitrification Inhibitors

Applied urea generally gets dissolved within 3-7 days of application and unable to satisfy plant need. For that, numerous natural and synthetic products are used in practise for quenching the crop need and minimize the pollution. Unlike the normal prilled urea the slow release materials free up the N in a linear fashion correlating with the plant need. Slow release materials are prepared by size variation of the normal product or coating the product with natural or artificial substances like sulphur, polymer or neem oil. The most commonly used slow release materials are urea super granule (USG), sulphur coated urea (SCU), neemcoated urea (NCU) and polymer coated product. Bandaogo et al. (2015) reported that deep placed (5-7 cm.) USG is performing better than surface placed prilled forms (PU) with respect to yield attributes grain yield of rice. Like, the USG, NCU is also highly effective in increasing the use efficiency of N particularly in transplanted rice. Meena et al. (2018) reported that NCU at 125% of recommended amount in 3 splits (50:25:25) significantly increased yield parameters of rice than supplying cent percent of N through prilled forms.

Nitrification is a two-step microbial mediated process of conversion of ammonium (NH_4^+) into nitrate (NO_3^-) and nitrification inhibitors (NIs) are synthetic and natural chemicals which have negative impact on nitrifiers thereby reduce NO_3^- availability by plodding the first step of nitrification (Prasad and Power, 1995). NIs not only saves costly inputs but also, duration of

NH_4^+ in the soil increases. At present, many synthetic, natural and indigenous NIs are commercially available for use.

4.1.4. Resource Conservation Technologies (RCTs)

Resource conservation technologies can be defined as package of practices that can orderly utilise the precious resource thereby increases its efficiency and save it for the future i.e., sustainable management of resources for the current and hoard for the upcoming time. In the upcoming section the importance of reduced tillage, site specific nutrient management and planting methods is being highlighted under RCTs.

4.1.4.1. Land and Tillage Management

Land levelling more specifically GPS based laser land levelling is trailblazer of modern day RCTs as it is highly effective in regularize the land surface up to 2 cm from its normal altitude making a gentle slope throughout crop field concomitantly expanding the net cropped area by 3-6%. Land levelling is very important for efficient water use and boosting the yield by 7.5% and 11% respectively (Jat et al., 2009). Reduced or no till system is unique in the sense as, only one pass is required to disturb the previous crop residue nominally for sowing operation. It is having a momentous impact since it saves time, labour, money and amplifies the yield. Sharma et al. (2005) from his experimental evidences concluded that, with fertiliser dose of 180:90:60 Kg ha^{-1} of N: P_2O_5: K_2O wheat yield increased by 17% under zero tillage. The viable reasons are build-up of nutrient particularly N and organic carbon, limiting the water evaporation, temperature fluctuation and inordinate microbial activity. Hilton et al. (1994) showed that under N-loss was 1.6 times higher in tracked soils than the untracked and he also concluded that residue which is left on the soil surface keeps the surface moist and acts as a food source for the microbial community. A two year (2013-15) experiment with a combination of tillage, N and residue management came up with significantly higher rice yield under 75% recommended done of N in zero tillage with residue retention and with Green Seeker™ (GS) than 100 recommended N fertilization under conventional tillage without residue (Nath et al., 2017) (Table 1). They also reported that Green seeker ™ (GS) based N-management saves 8 and 9.3% of N in first and 10.7 and 12.7% N in the second year without waiving the yield. Similar finding in a seven years field trial on zero tillage and residue retention was also reported by Kumar et al. (2019).

Table 1. Effect of zero tillage (ZT) and residue management on wheat yield

Treatment	Grain yield (t ha^{-1})	Straw Yield (t ha^{-1})
CT - R + 100 N	4.48c	7.21c
CT + R + 100 N	4.63bc	7.42bc
CT + R+ 75 N + GS	4.91ab	7.79ab
ZT - R + 100 N	4.51c	7.33c
ZT + R+ 100 N	5.00ab	8.00a
ZT+ R+ 75 N + GS	5.11a	8.20a

* Note- CT: Conventional tillage, R: Residue, ZT: Zero tillage, N: Nitrogen, GS: Green Seeker (Source: Nath et al., 2017).

Development of modernised farm implements such as, happy turbo seeder, rotary till drill can do multiple farm operations like sowing, incorporation of crop residue and green manure, weeding as a result shares dual interest of farming community and environment.

4.1.4.2. Site Specific Nutrient Management (SSNM)

It is a combined soil and plant based approach of precision nutrient management. SSNM serves the dual property i.e., corrective and prescriptive. In the prescriptive approach nutrient supplying capacity of soil, fertiliser calculation based on targeted yield, environmental damages and other biotic, abiotic factors are taking into account whereas, in the corrective method nutrient status of the standing crop is judged like for N- leaf colour chart (LCC), Soil plant analysis development (SPAD) and green seeker ™ are mostly used.

(i) Leaf Colour Chart (LCC)

Leaf colour chart or LCC is a plastic strip where graded shades of green colours are embedded. It was first developed in Japan and later on many improvisations came making it more precise. International Rice Research Institute (IRRI) in collaboration with agriculture departments of several nations made a six panel. A team of scientists at Zheijiang Agricultural University, China went one step ahead and developed an eight colour shaded (3, 4, 5, 5.5, 6, 6.5, 7 and 8) LCC (ZAU-LCC) which is suitable for all the three eco-types of rice (Indica, Japonica and Hybrid) (Yang et al., 2003). To make the LCC more inch-perfect scientists at IRRI came together and brought a 4 (green colour shade from 2 to 5) panel IRRI-LCC (Fairhurst et al., 2007). Fully grown young 'Y-leaf' is highly pertinent characteristics leaf for SPAD and LCC in rice and wheat. LCC is a cost effective, handy and user friendly tool for measurement of N status in green leaves based on the wavelengths emitted from the plant leaves. Generally, at 7-10 days interval starting form 15 DAT or 21 DAS till flowering and whenever, the leaf colour falls below LCC the fertiliser should be applied. In case of hybrid rice and basmati rice cultivars the LCC values are 4 and 3 respectively. Several studies reported that N-recommendation using LCC can save up to 20-25% fertiliser without any yield suspension.

(ii) Optical Sensors Like SPAD and Chlorophyll Meter

Generally, two major approaches are there for fertiliser calculation under SPAD. These are described below-

- Fixed threshold value approach: In this approach N fertiliser are applied below a critical SPAD value to avoid yield penalty. Peng et al. (1996) started work and found SPAD value 35 is most suitable for IR-72 dominating variety in south-Asia. SPAD-35 actually indicates 1.4 g N m^2 of leaf area and giving satisfactory result in all season. This value is well suited to also other part of world like India though; it is not universal in true sense as due to overcast condition the value should be between 30-32 in the wet season (Balasubramanian et al., 1998) whereas, under north Indian and Pakistan condition 37.5 SPAD value is best suited (Bijay-Singh et al., 2002; Hussain et al., 2003). Kyaw (2003) obtained significantly higher yield and 7.5% saving of N-fertiliser dose of rice cultivar BRRI-dhan 28 than the normal broadcasting method in Bangladesh using SPAD value of 35. SPAD vale not only have a spatial variation but also, the critical value changes accordingly with the varieties. For instance SPAD-37 well correlated with rice variety IET-4094 in West Bengal saving 35 kg ha^{-1} without hampering the production (Maiti et al., 2004).
- Sufficiency index value approach: In this approach N-fertilisers are applied whenever, sufficiency index moves below a prefixed threshold value. The merit of this approach

is it can self-modify according to the soil, season and cultivar (Singh et al., 2010). To understand the applicability of sufficiency index value approach Bijay-Singh et al. (2006) from a study concluded that N-fertilizer application with SPAD<90% increased NUE and also saved N-fertilizers.

So, at last it can be concluded that. LCC may not be effective as SPAD as the single difference between two colour points is 4-5 SPAD. But, a combination of both the application can play the role of effective N-management.

4.1.5. Planting Method

4.1.5.1. Furrow Irrigated Raised Bed (FIRB)

Raised beds of 45 cm are prepared and irrigation done through 30 cm wide furrows. Generally, in each bed 3 rows of wheat, 2 rows of mung, soybean, chickpea, and single row of maize and pigeonpea is maintained. It can save 30-40% of irrigation water and 25% of fertilisers (Kumar et al., 2015).

4.1.5.2. Surface Seeding

In the low lying areas where there is problem of water stagnation surface seeding is the easiest way for sowing followed in rice growing areas of India, Bangladesh and Nepal. These areas are termed as rice-fallow and lentil or lathyrus is grown under para/relay cropping. Surface seeding saves precious time and does not need any costly machine to do. System is so designed that second crop is grown on residual soil moisture under rainfed without any fertilizer application (Kumar et al., 2016).

4.2. Stratregy-2: Microbial Approaches

A colossal microbial diversity is existing in our natural ecosystem substantially in the rhizosphere and phyllosphere of plants having a gigantic potential for increasing use efficiency of nutrients. Among the microbes the noteworthy are N fixers, P-solubilizers, K-solubilizer and mobilizer, plant growth promoting rhizobacteria (PGPR) and hormone producers meeting crop nutrient requirement vis-à-vis safeguarding the plants from biotic and abiotic stresses. So, the three fold complex interlinkage between crop- microbial community- nutrients must be examined to understand this beneficiary process.

4.2.1. Cyanobacteria

Cyanobacteria or blue- green algae (BGA) have an acute impact on revamping soil health and nourishing crop growth especially in rice. The role of cyanobacteria is not only supplying mineral nutrients or build up the organic matter status in the soil but also, aiding the crop growth by improving soil aggregation, water holding capacity (WHC) by virtue of their filamentous structure and secreting different mucilaginous and bio-chemical compounds in the soil such as, plant-hormones, vitamins, amino acids and release of secondary metabolites like phenols, flavonoids etc. Out of the above mentioned bunch of activities, cyanobacteria are blessed with inherent ability of fixing nitrogen gas by a specially designed cell known as heterocyst. The uniqueness about heterocyst is that it mostly present in the terminal part and quite a big than normal vegetative cells. Lang (1968) examined the ultrastructure of heterocyst and reported that it is a special tri-layer cellular structure, where, except the inner layer the other layers are composed of non-cellulolytic polysaccharides which don't allow the atmospheric oxygen to diffuse in, because enzymes associated with BNF principally nitrogenase is highly oxygen sensitive. Sah (2008) also reported that surprisingly if oxygen moves inside it reduced by the polar plugs in the heterocyst and react with evolving hydrogen to make water. Cyanobacteria mediated N fixation is governed by hydrogen uptake (hup) genes (Margheri et al., 1991). Apart from N fixation, hydrogen (H_2) gas generate at the same time of which more than 1/3rd being reused by hup- genes and rest part is used as a green fuel (Dutta et al., 2005). Mostly, heterocyst's lacks free oxygen generating photosystem II (PS-II) of light reaction and owns ATP inducing PS-I though there are countless number of N-fixing mechanism for example, *Trichodesmium* sp. fix N- aerobically followed by enzyme expression in the light period of photocycle, (Capone et al., 1997). Azolla- water fern developed a quirky symbiotic relationship with a cyanobacterium Anabaena sustaining the productivity of rice based eco-system fixing up to 120-150 kg N ha^{-1} (Wagner, 2011). It is well suited to the coastal tracts of Odisha, Andhra Pradesh, Tamil Nadu and West Bengal areas having temperature range in between 20-30° C throughout the year with sufficient supply of P. Cyanobacteria based bio-fertilisers are good options for curtailing the amount of inorganic fertilisers and detrimental impact on eco system. National Rice Research Institute, Cuttack from an experiment concluded that application of *Aulosira* sp. in the soil at the rate of 60 Kg ha^{-1} can upturn the soil-N content by 14% and progressive release of half of soil-N in NH_4^+form even up to 50 days (Syiem, 2005). Prasanna et al. (2012) from a pot experiment in rice cultivar 'Pusa 1460' verified the efficacy of bacterial (*Providencia* sp., *Brevundimonas* sp. and *Ochrobactrum* sp.) and cyanobacterial (2 species of *Anabaena* sp. and 1 species of *Calothrix* sp.) consortia against recommended fertiliser practise as control and reported the pragmatic impact on soil N- status and rice productivity of consortia over inorganics. The constructive impact is due to build up in soil organic matter content and further increase in enzymatic activity (El-Gaml, 2006). Beside as a nutrient provider in rice, cyanobacterial culture is highly effective in other crops like wheat, lettuce and maize as it increases the nitrogenase activity, photosynthetic pigments (Chlorophyll A, B) and the effect is more noticeable with vermicompost or inoculation with other beneficial microorganism like Azotobacter (Abd-Alla et al., 1994; Sukor, 2013). In conjunction with the chemical fertilisers cyanobacterial cultures foster the activity of fertilisers in wheat variety (HD 2687). Prasanna et al. (2009) reported that soil ecosystem under wheat based cropping system harbours algal species like *Calothrix, Anabaena, Oscillatoria* and *Phormidium*. Though, in rice ecosystem the importance of cyanobacteria for succouring the fertility and productivity is beyond words as these cultures scale down the dose of fertilisers in input intensive rice based cropping system and making it more eco-friendly and profitable. Other than, NUE or fertiliser saving cyanobacteria secretes indole-3 acetic acid stimulates the microbial folk in soil particularly in rhizosphere aiding the roots to forage N form the soil (Karthikeyan et al., 2007; Spaepen et al., 2007).

4.2.2. Trichoderma

Trichoderma sp. is mostly popular for their plant protective role and highly useful against numerous biotic stresses thus favoured by our farmer friends. Multiple research reports came out with the fact that, beside a plant protectant the potential of Trichoderma is immense as it can promote below ground growth and development, safe guarding the plant from abiotic stresses and increasing the use efficiency also uptake of essential nutrients ultimately plant yield. Many researchers proclaimed the role of multiple Trichoderma sp. on few profit-oriented crops such as cotton, lettuce, cabbage, pea and Chinese radish. Like Cyanobacteria, Trichoderma sp. can promote colonization of microbes in the root zone by secreting IAA (Zhang et al., 2013). Nitrogen is the foremost nutrient element for crop life cycle and its role in plant life cycle is ineffable. Pre-sowing seed treatment alone can reduce the N- dose by 1/3rd to ½th of the recommended dose (Harman, 2011; Shoresh et al., 2010). These bio-based formulations are antecedent of future research particularly for the developing countries and marginal farmers. *T. harzianum* (Th 37) based treatment @ 20 kg ha^{-1} on previous crop stubbles at the preliminary stage of ratoon sugarcane intensify N, P and K availability 27, 65, and 44%, respectively (Singh et al., 2010). Shukla et al. (2008) showed that application of Trichoderma and Gluconacetobacter enriched FYM increased available-N content by 27.45% over nitrogenous fertilisers alone. Similar, positive out-turns are also delineated by other researchers (Srivastava et al., 2006; Yadav et al., 2008).

4.2.3. Bio-Priming

Bio-priming may be defined as an amalgamation of seed hydration with priming agents and inoculation with biological agents to protect the seed physiologically and biologically from numerous biotic stresses. Bio-priming or more pointedly seed-priming not protect form soil or seed borne diseases but also, enhance the utilisation efficacy of major and minor nutrients. Bio-priming based seed or seedling treatment has broken the age old plant protection measures and started new horizon of crop protection. Mostly endophytic bacterial or fungal culture and crop specific plant growth promoting rhizo-bacteria (PGPR) are in use (Rakshit et al., 2015). Analogous with the other bio-priming methods, much emphasis is given on this course of plant protection means as along with better plant growth seed vigour and quality improved a lot under this. Generally, the method of application is coating the seeds with microbial culture or culturing the seeds in microbes for a definite period before sowing. Inadequateness of need based or pathogen specific plant protection measures or resistance cultivars seed priming bio-agents is assured method of plant health management. So, in stressful conditions and low-input agriculture more indicatively inorganic excluded organic agriculture system bio-priming is a win-win strategy for the commercial sectors and farmers. The role of bio-agents is highly exclusive to crop-cultivar and genotype (Table 2). Gupta et al. (2000) revealed several mechanisms which can singularly or conjointly helps the plants from different biotic and abiotic stresses. These are detailed in Table 3.

Table 2. Mode of action of bio-primers

Strategy	Mechanism
1	More nutrient mobilisation from the non-labile pools and promotes biological nitrogen fixation
2	Providing immunity to the plants by production of hydrogen cyanide (HCN), producing siderophores and antibiotics
3	Protection from abiotic stresses like salinity, acidity and heavy metal stress
4	Production of indole-3-acetic acid (IAA) favours microbial colonization and lowering the ethylene production in seedlings protecting them from senescence

Table 3. Bio-primers used in different crops for increasing NUE

Bio-agent	**Crop**	**% increase in NUE**	**Country**	**Reference**
Azospirillum amazonense	Rice	3.5 to 18.5%	Brazil	Rodrigues et al., 2008
T. harzianum *T. virens* *T. atroviride*	Soybean	15.8% 5.2% 11%	Iran	Entesari et al., 2013
T. viride	Cucumber	5.9%	Egypt	Moharam and Negim, 2012
T. harzianum	Tomato	2.5%	Bangladesh	Molla et al., 2012
T. harzianum	Tea	65%	India	Thomas et al., 2010
Glomus intraradices	Melon	7.9%	Spain	Martínez- Medina et al., 2009
P. fluorescens	Broccoli	235.4%	India	Tanwar et al., 2013

4.3. Startegy-3: Molecular and Physiological Approach

As the time progresses we need more efficient crop cultivars because of ever shrinking natural resources and environmental perturbation, satisfying the hungry stomach of billions and sustainably increase the crop yield. Nutrient efficient cultivars can be defined as those can sustain one self and maintains higher production even in nutrient poor soils. Crops are highly specific in their mechanism of nutrient uptake and having differential response too for example, most of the legumes secrete Phosphorus cycling enzymes and root exudates form the roots and making the non-labile P available (Graham, 1984). These idiosyncratic mechanisms are actually a delightful thing to study and divulge the survival strategy even in poor fertile soils. Particularly for the breeders and molecular biologists these properties are important for selecting and developing new cultivars.

4.3.1. Physiological Traits Governing NUE

4.3.1.1. Root Length and System Structure

The major form of nitrogenous fertiliser loss in leaching and one of the major ways to reduce this by revamping the root traits like root length and its density. For maximum NO_3^- uptake explicitly in the lower depths the critical root length density is 1 cm cm^{-3} (Reynolds et al., 2006). Among the cereals, profound genetic disparity is found in wheat but, soil characteristics and nutrient availability governs root traits a lot (Hoad et al., 2001; Ehdaie and Waines, 2003). These discrepancy in the root system is tool for breeding material selection in different crops, whereas, the other subsidiary selection parameters are lower canopy temperature and higher stomatal conductance. Development of wheat cultivars with superior root system i.e., more vigorous and deeper can be developed by subsuming D genomes from *Triticum tauschii* commonly known as Tausch's goat grass (2n = 2x = 14) (Reynolds et al., 2009). Carvalho et al. (2014) reported that no extra carbon (C) is needed to modify the root length density. Plants have their own survival strategy to endure the livelihood by maintain a higher root to shoot ratio or higher root to plant biomass i.e., root dry weight ratio (RDWR) in the N deficient soils, although the data pertaining to the response of RDWR to N supply is still missing. Generally, root traits are controlled by multiple genes therefore, it is immensely difficult to decide on (Hall and Richards, 2013). For example, root hairs which aid the plants by absorbing more N are controlled by RTH1 and RTH3 in maize (Hochholdinger and Tuberosa, 2009). Hence, breeding programme for developing varieties for specific or developed root system is highly restricted. Now days, molecular breeding programme using marker-assisted selection is been in use to amalgam numerous root traits such as root length, density, and RDWR conspicuously in major cereals like wheat for better nutrient efficient crop cultivar development.

Microbial Importance of Nitrogen Uptake

Plant roots secrete oodles of organic substrate like secretions, lysates, excretions and rhizo-deposits which can biochemically alter the composition in the rhizosphere hence plant-microbe interaction and nutrient availability. Rhizosphere is basically a biological hot-spot containing more than 10^8-10^9 no of bacteria gram^{-1} and 1-1.5 m fungal filament cm^{-2} of soil respectively (Moënne-Loccoz et al., 2015) and not only in terms of quantity, rhizosphere in unique in terms of microbial diversity too. Microorganisms mediated change in the N-dynamics is highly distinguishable in the rhizosphere as both plant and microbes compete for the same source of N. Not only that, microbes mediated N-fixation (biological nitrogen fixation) and N-transformation (Denitrification) reorient the entire N-dynamics (Herold et al., 2012). Bardon et al. (2014) reported that many plants can reshape the denitrification process by releasing allelopathic chemicals but, concrete evidence is still to come. Another, N-use strategy is accelerating the microbe umpired mineralization or soil organic nitrogen (SON) resulting boosting in the population of beneficial microbe population such as, protozoan predators, vesicular-arbuscular mycorrhizae and bacteria (Atul-Nayyar et al., 2009). Legumes have novel property of fixing the atmospheric N- by having a symbiotic relationship with rhizobium species but, non-legumes too can utilise the atmospheric N- by having a synergistic relationship with microbial species such as, Azolla-anabaena symbiosis with rice; Frankia mediated N-fixation in tree species. Several reports are there saying increase in wheat growth and yield by application of N-fixing bacteria yet, it is highly cultivar specific. Disregarding, with the direct N- utilisation mechanisms the indirect measures influencing the microbial population and plant-microbe association is production of phyto-hormones like auxins, gibberellic acids (GA) and cytokinins. For e.g., rhizosperic bacterial and fungal species of several wheat cultivars secretes GAs stimulating better root hair growth (Upadhyay et al., 2009). Inocluation with PGPR can act as a shield from several biotic stresses for instance, Pseudobactin-a siderophore produced by *Pseudomonas sp.*control the soft-rot disease in potato by withdrawing the iron present in the rhizosphere (Hetrick et al., 1995). Hence, it is the need of an hour to carry out breeding strategies serving dual purpose of higher N- efficiency and phyto-protection from abiotic and biotic stresses.

4.3.2. Physiological Traits Superintending the N-Utilisation Efficiency

4.3.1.2. Nitrate (NO_3^-) Assimilation

The first step of nitrate (NO_3^-) assimilation is reduction of nitrate reductase (NR) mediated nitrate (NO_3^-) reduction into nitrite (NO_2^-) in the cytosols utilising NAD(P)H as an electron acceptor. The plastid is the next place second step, where NO_2^- is converted into NH_4^+ by the enzyme nitrite reductase (NiR) and here ferrodoxin act as an electron donor (Setif et al., 2009; Sakakibara et al., 2012). After conversion into $NH_4^{+,}$ it is assimilated inside the plant body with the help of an enzyme duo as glutamate. Glutamine synthetase (GS) takes care of the first step of reaction and in association with its counterpart glutamate synthase (GOGAT), glutamate reacts with 2- oxoglutarate (donates the C-skeleton) in the reaction pathway and forms glutamine. After that, the amino acids (glutamine and glutamate) shares their amino (-NH_2) groups with other N-containing bio-chemical compounds most commonly other amino acids for future use, act as a building block for proteins and bare-bones for genetic materials like DNA and RNA (Lea and Miflin, 2011). They also reported glutamate dehydrogenase (GDH) mediated proxy pathway for Glutamate formation as ammonia (NH_3) fused with 2-oxoglutarate. Though, detailed scientific studies with radio-isotope (N^{15}) and mutants lacking GS-GOGAT pathway disapproved the GDH pathway and proclaimed that 95% of the plants us the former for NO_3^- assimilation not the later but, afterwards, coherent results showing GDH mediated pathway is functional if the plant above and below ground parts are C-limited (Labboun et al., 2009; Fontaine et al., 2012). Research taking the help of mutants and genetically modified crops our understanding related to amino acid metabolism improved a lot particularly derived from glutamine or glutamate.

4.3.1.3. Photosynthesis

Nearly 3/4th of leaf N present as chloroplast Rubisco in wheat likely involved in the photosynthetic cycle (Evans, 1983). For that, N deficient plants often suffer from multiple physiological problems like curtailment of tissue N-concentration, total leaf area, stomatal conductance, rate of photosynthesis and most importantly chlorophyll content resulting poor interception of insolation ultimately yield (Foulkes et al. 2009). A way out from this troublesome explicitly under low N content is soaring specific leaf N area (SLN) along with extinction coefficient (Kc) for example, modern bread wheat cultivars having Kc value of 0.55 for optimum photosynthetic light utilisation. Genotypes with higher embryo size, prior canopy closure and upright leaf angle outperform others. Canopy orientation is also a highly influencing characteristic for better photosynthesis like prostrate and upright leaves in the vegetative and later stages of crop, respectively. Thus, in the cellular level there are number of avenues to upsurge the current photosynthetic rate. These are turning down the Rubisco activity as it consumes the pre-fixed C and N by photorespiration, speedy recovery of ribulose-1,5-bisphosphate (RuBP), change in photochemistry in the C4 plants like C3 plants under changing climatic condition and better canopy and stem management for utilisation of intercepted photosynthetically active radiation (PAR) (Murchie et al., 2009; Parry et al., 2010).

4.3.1.4. N-Dynamics in Later Crop Stage

In wheat, N dynamics in pre and post anthesis period changes a lot as 35–42%, 14-20%, 20-31% and 16-23% remains in the leaf lamina, sheath, stem and ear under optimum N supply (Barraclough et al., 2014). But, picture is not the same under sub-optimal dose as; the N accumulates more in the ear (Gaju et al., 2014). Pask et al. (2012) opined that under restricted N-supply there was uniform distribution of N in different plant parts whereas, under ample condition nearly half of the total-N is located in the above ground parts. Recently, phenotypes having 'Stay-green' traits are getting lots of attention from the breeding point of view as leafs remains green during the entire grain filling stage. In wheat ($2n = 6x = 42$) stay-green genotypes with broader genetic variation is dominant (Gaju et al., 2011; Naruoka et al., 2012). While, normally during anthesis most of the wheat genotypes remains green and there was no cut down of photo assimilates under ideal N-fertilisation but, according to some researchers in the post-anthesis period the supply of assimilates hindered under short-fall of N dose (Borras et al.,

2004; Gaju et al., 2011). Apart from genotypic and phenotypic variation an optimum N-aupply throughout the growing season is of high priority to draw a line between supply and demand particularly in grain filling stage irrespective of any crop. Therefore, understanding the pre and post anthesis N-demand, translocation, utilisation vis-à-vis environment factors governing N-availability is vital to develop breeding programme for higher yield and quality produce.

4.3.1.5. Molecular Strategy

Molecular strategies utilises the data present in the genetic level for genotype improvement using a number of pathways at transcriptional level, post-transcriptional level or through allosteric regulation. This fore mentioned pathways relies on multiple factors namely N-application rate, organs, genotypes and its developmental stage, thus reproving the choice of promoters. For example, alanine aminotransferase (AlaAT), is a highly successful strategy to improve the NUE in oil-seeds and rice and it has been tested in wheat (Good et al., 2007, Shrawat et al., 2008). The researchers also concluded that except the enzymes and proteins associated with N uptake and metabolism can be targeted as there is least chance of interruption from the post-transcriptional factors. Undeniably, because of their crucial role in N-assimilation enzymes such as, NR, NiR, GS and GOGAT thought to be prime regulators of NUE in plants. Zhao et al. (2013) showed that, incorporation of NR gene form tobacco into wheat resulted increased seed protein level and that to without any further addition of N. Passively, through correlation study it was proved that leaf GS activity in wheat perhaps used as a marker for NUE (Kichey et al., 2007). Additionally, various quantitative trait loci (QTL) associated with both quantitative and qualitative character present either in cytosolic GS1 (Habash et al., 2007; Gadaleta et al., 2014) or in plastidic GS2 (Bordes et al., 2013) were considered mostly in C4 crops like maize for improving NUE. Nigro et al. (2013) figured out that not cytosolic GS1, it was NADH-GOGAT regulated the grain protein content in durum wheat. So, such interesting findings laid the way for future research in the arena of molecular biology but, it's not a low hanging fruit as NUE govern by polygenic traits and genotype specific. In modern days, several SNP arrays (90-820K) and uniting multiple loci with modest will help to meet the knowledge gap of genotype and environment interaction to develop promising genotypes suitable for higher NUE in different environments.

4.4. Strategy-4: Advanced Techniques for Optimizing NUE

4.4.1. Simulation Modelling Techniques

Simulation modelling is now a day's need of the hour to optimise the N-dose with predicted crop yield. Crop simulation models along with secondary data like soil, field and meteorological used to prescribe N dose in the beginning as well as in the standing crop. Beside N-management these models can highlight other aspects of crop growth like agronomic interventions and biotic and abiotic factors influencing yield, yield attributes and NUE (Ladha et al., 2005).

4.4.2. Remote Sensing and Geographic Information System (GIS) Technologies

Remote sensing and GIS are modern tools to precise N-management particularly in larger areas. A well-coordination existed between N-content in plant parts with different vegetation index like Normalised deviation vegetation index (NDVI) measured through different ground or aerial sensors. Along with, the plant parts different index could determine the N-content in the soil accordingly N-management strategies for the successive crops can be formulated. In association with remote sensing data, GIS can fine-tune the data of fertiliser recommendation over the farmer practise (Ladha et al., 2005).

Conclusion

Nitrogen (N) may come 4th in the list of essentiality but, in true sense the role of N is transcendent. Nitrogenous fertilisers are the most commonly used in agriculture with a share of more than 50% and have important role to sustain the productivity of agricultural production systems. The fate of applied fertiliser is decided by multiple components climate, soil type, cultivar and socio-economical condition of the farmers. In well manged research plots fertilizer N-recovery is 50-60% but, farmers field with more heterogeneity it comes down to 25-35%. So, due to management gap the use efficiency of N-fertilisers are not up to the mark, hence creating environmental issues and monetary loss. To appraise the challenge a combined approach of agronomy (e.g., resource conservation technology, use if nitrification inhibitors), soil science (e.g., Site specific nutrient management, use of microbes), physiology (pre and post-anthesis traits associated with N-uptake and utilisation) and breeding (marker based molecular approach) for development of efficient N-cultivars needs to be embrace for satisfying the hungry stomachs with quality produce. Though, there are number of hurdles like mechanisation, lack of nutrient stewardship strategies among the institutes and farmers, adversity of developing efficient phenotypes and poor extension slow down the progress. Although, future tools like 'omics and associated studies' and collaborative research in private-public model may open the Pandora's Box and develop high-throughput better performing genotypes under various spatio-temporal condition.

References

Abd-Alla, M. H., Mahmoud, A. L. E., and Issa, A. A. (1994). Cyanobacterial biofertilizer improved growth of wheat. *Phyton*, *34*, 11-18.

Atul-Nayyar, A., Hamel, C., Hanson, K., and Germida, J. (2009). The arbuscular mycorrhizal symbiosis links N mineralization to plant demand. *Mycorrhiza*, *19*(4), 239-246.

Balasubramanian, V., Morales, A. C., Cruz, R. T., and Abdulrachman, S. (1998). On-farm adaptation of knowledge-intensive nitrogen management technologies for rice systems. *Nutrient Cycling in Agroecosystems*, *53*(1), 59-69.

Bandaogo, A., Bidjokazo, F., Youl, S., Safo, E., Abaidoo, R., and Andrews, O. (2015). Effect of fertilizer deep placement with urea supergranule on nitrogen use efficiency of irrigated rice in Sourou Valley (Burkina Faso). *Nutrient cycling in agroecosystems*, *102*(1), 79-89.

Bardon, C., Piola, F., Bellvert, F., Haichar, F. E. Z., Comte, G., Meiffren, G., and Poly, F. (2014). Evidence for biological denitrification inhibition (BDI) by plant secondary metabolites. *New Phytologist,* 204(3), 620-630.

Barraclough, P. B., Lopez-Bellido, R., and Hawkesford, M. J. (2014). Genotypic variation in the uptake, partitioning and remobilisation of nitrogen during grain-filling in wheat. *Field Crops Research*, *156*, 242-248.

Bordes, J., Ravel, C., Jaubertie, J. P., Duperrier, B., Gardet, O., Heumez, E., and Balfourier, F. (2013). Genomic regions associated with the nitrogen limitation response revealed in a global wheat core collection. *Theoretical and applied genetics*, *126*(3), 805-822.

Borrás, L., Slafer, G. A., and Otegui, M. E. (2004). Seed dry weight response to source–sink manipulations in wheat, maize and soybean: a quantitative reappraisal. *Field Crops Research*, *86*(2-3), 131-146.

Capone, D. G., Zehr, J. P., Paerl, H. W., Bergman, B., and Carpenter, E. J. (1997). Trichodesmium, a globally significant marine cyanobacterium. *Science*, *276*(5316), 1221-1229.

Carvalho, P., Azam-Ali, S., and Foulkes, M. J. (2014). Quantifying relationships between rooting traits and water uptake under drought in Mediterranean barley and durum wheat. *Journal of integrative plant biology*, *56*(5), 455-469.

DAC (2017). *Annual report.* http://agricoop.nic.in/annual-report.

Dey, P. and Sekhon, B. S. (2016). Nitrogen Fertility Status of the Indian Soils vis-a-vis the World Soils. *Indian Journal of Fertilisers,* Vol. 12 (4), pp.36-43.

Doberman, A., and Witt, C. (2002). The evolution of site specific nutrient management in irrigated systems of Asia. *Field Crops Research*, *87*, 167-178.

Dutta, D., De, D., Chaudhuri, S., and Bhattacharya, S. K. (2005). Hydrogen production by cyanobacteria. *Microbial cell factories*, *4*(1), 36.

Ehdaie, B., and Waines, J. G. (2003). 1RS translocation increases root biomass in Veery-type wheat isogenic lines and associates with grain yield. In *Proceedings of the 10th International wheat genetics symposium, ISC Paestum, Italy, Rome* (pp. 693-695).

El-Gaml, M. (2006). *Studies on cyanobacteria and their effect on some soil properties* (Doctoral dissertation, Thesis, Benha University. Enhanced-Efficiency Fertilizers, Frankfurt, Germany. Digital Commons, University of Nebraska).

Entesari, M., Sharifzadeh, F., Ahmadzadeh, M., and Farhangfar, M. (2013). Seed biopriming with Trichoderma species and Pseudomonas fluorescent on growth parameters, enzymes activity and nutritional status of soybean. *International Journal of Agronomy and Plant Production*, *4*(4), 610-619.

Evans, J. R. (1983). Nitrogen and photosynthesis in the flag leaf of wheat (Triticumaestivum L.). *Plant physiology*, *72*(2), 297-302.

FAI (2018). *Fertiliser statistics.* https://www.faidelhi.org/general/con-npk.pdf

Fairhurst, T., Witt, C., Buresh, R., Dobermann, A., and Fairhurst, T. (2007). *Rice: A practical guide to nutrient management*. Int. Rice Res. Inst.

FAO (2013). *Food and Agricultural Organization.* FAOSTAT. http://faostat.fao.org/Rome

Fontaine, J. X., Tercé-Laforgue, T., Armengaud, P., Clément, G., Renou, J. P., Pelletier, S., and Hirel, B. (2012). Characterization of a NADH-dependent glutamate dehydrogenase mutant of Arabidopsis demonstrates the key role of this enzyme in root carbon and nitrogen metabolism. *The Plant Cell*, *24*(10), 4044-4065.

Foulkes, M. J., Hawkesford, M. J., Barraclough, P. B., Holdsworth, M. J., Kerr, S., Kightley, S., and Shewry, P. R. (2009). Identifying traits to improve the nitrogen economy of wheat: recent advances and future prospects. *Field Crops Research*, *114*(3), 329-342.

Gadaleta, A., Nigro, D., Marcotuli, I., Giancaspro, A., Giove, S. L., and Blanco, A. (2014). Isolation and characterisation of cytosolic glutamine synthetase (GSe) genes and association with grain protein content in durum wheat. *Crop and Pasture Science*, *65*(1), 38-45.

Gaju, O., Allard, V., Martre, P., Le Gouis, J., Moreau, D., Bogard, M., and Foulkes, M. J. (2014). Nitrogen partitioning and remobilization in relation to leaf senescence, grain yield and grain nitrogen concentration in wheat cultivars. *Field Crops Research*, *155*, 213-223.

Gaju, O., Allard, V., Martre, P., Snape, J. W., Heumez, E., LeGouis, J., and Hubbart, S. (2011). Identification of traits to improve the nitrogen-use efficiency of wheat genotypes. *Field Crops Research*, *123*(2), 139-152.

Good, A. G., Johnson, S. J., De Pauw, M., Carroll, R. T., Savidov, N., Vidmar, J., and Stroeher, V. (2007). Engineering nitrogen use efficiency with alanine aminotransferase. *Botany*, *85*(3), 252-262.

Graham, R. D. (1984). Breeding for nutritional characterstics in cereals. *Advances in plant nutrition*, *1*, 57-102.

Gupta, A., Gopal, M., and Tilak, K. V. B. (2000). Mechanism of plant growth promotion by rhizobacteria.

Habash, D. Z., Bernard, S., Schondelmaier, J., Weyen, J., and Quarrie, S. A. (2007). The genetics of nitrogen use in hexaploid wheat: N utilisation, development and yield. *Theoretical and Applied Genetics*, *114*(3), 403-419.

Hall, A. J., and Richards, R. A. (2013). Prognosis for genetic improvement of yield potential and water-limited yield of major grain crops. *Field Crops Research*, *143*, 18-33.

Harman, G. E. (2011). Multifunctional fungal plant symbionts: new tools to enhance plant growth and productivity. *New Phytologist*, *189*(3), 647-649.

Herold, M. B., Baggs, E. M., and Daniell, T. J. (2012). Fungal and bacterial denitrification are differently affected by long-term pH amendment and cultivation of arable soil. *Soil Biology and Biochemistry*, *54*, 25-35.

Hetrick, B. A. D., Wilson, G. W. T., Gill, B. S., and Cox, T. S. (1995). Chromosome location of mycorrhizal responsive genes in wheat. *Canadian Journal of Botany*, *73*(6), 891-897.

Hilton, B. R., Fixen, P. E., and Woodard, H. J. (1994). Effects of tillage, nitrogen placement, and wheel compaction on denitrification rates in the corn cycle of a corn-oats rotation. *Journal of plant nutrition*, *17*(8), 1341-1357.

Hoad, S. P., Russell, G., Lucas, M. E., and Bingham, I. J. (2001). *The management of wheat, barley, and oat root systems.*

Hochholdinger, F., and Tuberosa, R. (2009). Genetic and genomic dissection of maize root development and architecture. *Current opinion in plant biology*, *12*(2), 172-177.

Hussain, F., Zia, M. S., Akhtar, M. E., and Yasin, M. (2003). Nitrogen management and use efficiency with chlorophyll meter and leafcolour chart. *Pakistan Journal of Soil Science (Pakistan).*

Jat, M. L., Gupta, R., Ramasundaram, P., Gathala, M. K., Sidhu, H. S., Singh, S., and Ladha, J. K. (2009). Laser assisted precision land leveling: a potential technology for resource conservation in irrigated intensive production systems of Indo-Gangetic plains. *Integrated Crop and Resource Management in the Rice-Wheat System of South Asia, International Rice Research Institute, Los Banos*, 223-238.

Karthikeyan, N., Prasanna, R., Nain, L., and Kaushik, B. D. (2007). Evaluating the potential of plant growth promoting cyanobacteria as inoculants for wheat. *European Journal of Soil Biology*, *43*(1), 23-30.

Kichey, T., Hirel, B., Heumez, E., Dubois, F., and Le Gouis, J. (2007). In winter wheat (*Triticum aestivum* L.), post-anthesis nitrogen uptake and remobilisation to the grain correlates with agronomic traits and nitrogen physiological markers. *Field Crops Research*, *102*(1), 22-32.

Kumar N, C. P. Nath, K. K. Hazra, Krishnashis Das, M. S. Venkatesh, M. K. Singh, S. S. Singh, C. S. Praharaj and N. P. Singh. 2019. Impact of zero-till residue management and crop diversification with legumes on soil aggregation and carbon sequestration. *Soil & Tillage Research* 189: 158–167.

Kumar N, KK Hazra, Saumya Singh and N Nadarajan. 2016. Constraints and prospects of growing pulses in rice fallows of India. *Indian Farming* 66(6): 13-16.

Kumar N, Singh MK, Praharaj CS, Singh U and Singh SS. 2015. Performance of chickpea under different planting method, seed rate and irrigation level in Indo-Gangetic Plains of India. *Journal of Food Legumes* 28(1): 40-44.

Labboun, S., Tercé-Laforgue, T., Roscher, A., Bedu, M., Restivo, F. M., Velanis, C. N., and Hirel, B. (2009). Resolving the role of plant glutamate dehydrogenase. I. In vivo real time nuclear magnetic resonance spectroscopy experiments. *Plant and cell physiology*, *50*(10), 1761-1773.

Ladha, J. K., Pathak, H., Krupnik, T. J., Six, J., and van Kessel, C. (2005). Efficiency of fertilizer nitrogen in cereal production: retrospects and prospects. *Advances in agronomy*, *87*, 85-156.

Lang, N. J. (1968). The fine structure of blue-green algae. *Annual Reviews in Microbiology*, *22*(1), 15-46.

Lea, P. J., and Miflin, B. J. (2018). Nitrogen assimilation and its relevance to crop improvement. *Annual Plant Reviews online*, 1-40.

Maiti, D., Das, D. K., Karak, T., and Banerjee, M. (2004). Management of nitrogen through the use of leaf color chart (LCC) and soil plant analysis development (SPAD) or chlorophyll meter in rice under irrigated ecosystem. *The Scientific World Journal*, *4*, 838-846.

Margheri, M. C., Tredici, M. R., Allotta, G., and Vagnoli, L. (1991). Heterotrophic metabolism and regulation of uptake hydrogenase activity in symbiotic cyanobacteria. In *Nitrogen Fixation* (pp. 481-486). Springer, Dordrecht.

Martínez-Medina, A., Roldán, A., and Pascual, J. A. (2009). Performance of a Trichoderma harzianum bentonite–vermiculite formulation against Fusarium wilt in seedling nursery melon plants. *HortScience, 44*(7), 2025-2027.

Meena, A. K., Singh D. K., Pandey, P. C., and Nanda, G. (2018). Growth, yield, economics, and nitrogen use efficiency of transplanted rice (*Oryza sativa*L.) as influenced by different nitrogen management practices through neem (*Azadirachta indica*) coated urea. *Int. J. Chem. Stud.* 6(3): 1388-1395

Moënne-Loccoz, Y., Mavingui, P., Combes, C., Normand, P., and Steinberg, C. (2015). Microorganisms and biotic interactions. In *Environmental Microbiology: Fundamentals and Applications* (pp. 395-444). Springer, Dordrecht.

Moharam, M. H., and Negim, O. O. (2012). Biocontrol of Fusarium wilt disease in cucumber with improvement of growth and mineral uptake using some antagonistic formulations. *Communications in agricultural and applied biological sciences, 77*(3), 53-63.

Molla, A. H., Haque, M. M., Haque, M. A., and Ilias, G. N. M. (2012). Trichoderma-enriched biofertilizer enhances production and nutritional quality of tomato (Lycopersicones-culentum Mill.) and minimizes NPK fertilizer use. *Agricultural Research, 1*(3), 265-272.

Murchie, E. H., Pinto, M., and Horton, P. (2009). Agriculture and the new challenges for photosynthesis research. *New Phytologist, 181*(3), 532-552.

Murugan A. V. and Dadhwal V. K., 2007. *Indian agriculture and nitrogen cycle. In Agricultural nitrogen use & its environmental implications* (EdsAbrol YP, Raghuram N, Sachdev MS), IK International Publishing House Pvt. Ltd. New Delhi, 2007, pp. 9-28.

Naruoka, Y., Sherman, J. D., Lanning, S. P., Blake, N. K., Martin, J. M., and Talbert, L. E. (2012). Genetic analysis of green leaf duration in spring wheat. *Crop science, 52*(1), 99-109.

Nath, C. P., Das, T. K., Bhattacharyya, R., Pathak, H., Paul, S., Chakraborty, D., and Hazra, K. K. (2019). Nitrogen effects on productivity and soil properties in conventional and zero tilled wheat with different residue management. *Proceedings of the National Academy of Sciences, India Section B: Biological Sciences, 89*(1), 123-135.

Nigro, D., Gu, Y. Q., Huo, N., Marcotuli, I., Blanco, A., Gadaleta, A., and Anderson, O. D. (2013). Structural analysis of the wheat genes encoding NADH-dependent glutamine-2-oxoglutarate amidotransferases and correlation with grain protein content. *PloS one, 8*(9), 73751.

Parry, M. A., Reynolds, M., Salvucci, M. E., Raines, C., Andralojc, P. J., Zhu, X. G., and Furbank, R. T. (2010). Raising yield potential of wheat. II. Increasing photosynthetic capacity and efficiency. *Journal of experimental botany, 62*(2), 453-467.

Pask, A. J. D., Sylvester-Bradley, R., Jamieson, P. D., and Foulkes, M. J. (2012). Quantifying how winter wheat crops accumulate and use nitrogen reserves during growth. *Field Crops Research, 126*, 104-118.

Pathak, H., Mohanty, S., Jain, N., & Bhatia, A. (2010). Nitrogen, phosphorus, and potassium budgets in Indian agriculture. *Nutrient Cycling in Agroecosystems, 86*(3), 287-299.

Peng, S., Garcia, F. V., Laza, R. C., Sanico, A. L., Visperas, R. M., and Cassman, K. G. (1996). Increased N-use efficiency using a chlorophyll meter on high-yielding irrigated rice. *Field Crops Research, 47*(2-3), 243-252.

Prasad, R. and Power J. F. (1995). Nitrification inhibitors for agriculture, health, and the environment. *Adv. Agron., 54*, 233-281.

Prasanna, R., Joshi, M., Rana, A., Shivay, Y. S., and Nain, L. (2012). Influence of co-inoculation of bacteria-cyanobacteria on crop yield and C–N sequestration in soil under rice crop. *World Journal of Microbiology and Biotechnology, 28*(3), 1223-1235.

Prasanna, R., Nain, L., Ancha, R., Srikrishna, J., Joshi, M., and Kaushik, B. D. (2009). Rhizosphere dynamics of inoculated cyanobacteria and their growth-promoting role in rice crop. *Egyptian Journal of Biology, 11*(1): 26-36.

Rakshit, A., Singh, H. B., and Sen, A. (Eds.). (2015a). *Nutrient use efficiency: from basics to advances*. New Delhi: Springer.

Rakshit, A., Sunita, K., Pal, S., Singh, A., and Singh, H. B. (2015). Bio-priming mediated nutrient use efficiency of crop species. In *Nutrient use efficiency: From basics to advances*(pp. 181-191). Springer, New Delhi.

Reynolds, M., Dreccer, F., and Trethowan, R. (2006). Drought-adaptive traits derived from wheat wild relatives and landraces. *Journal of experimental botany*, *58*(2), 177-186.

Reynolds, M., Manes, Y., Izanloo, A., and Langridge, P. (2009). Phenotyping approaches for physiological breeding and gene discovery in wheat. *Annals of Applied Biology*, *155*(3), 309-320.

Rodrigues, E. P., Rodrigues, L. S., de Oliveira, A. L. M., Baldani, V. L. D., dos Santos Teixeira, K. R., Urquiaga, S., and Reis, V. M. (2008). Azospirillumamazonense inoculation: effects on growth, yield and N 2 fixation of rice (Oryzasativa L.). *Plant and Soil*, *302*(1-2), 249-261.

Sah, P. (2008). Understanding the physiology of heterocyst and nitrogen fixation in cyanobacteria or blue-green algae. *Nat Sci*, *6*, 28-33.

Sakakibara, Y., Kimura, H., Iwamura, A., Saitoh, T., Ikegami, T., Kurisu, G., and Hase, T. (2012). A new structural insight into differential interaction of cyanobacterial and plant ferredoxins with nitrite reductase as revealed by NMR and X-ray crystallographic studies. *The journal of biochemistry*, *151*(5), 483-492.

Sétif, P., Hirasawa, M., Cassan, N., Lagoutte, B., Tripathy, J. N., and Knaff, D. B. (2009). New insights into the catalytic cycle of plant nitrite reductase. Electron transfer kinetics and charge storage. *Biochemistry*, *48*(12), 2828-2838.

Sharma, R. K., Tripathi, S. C., Kharub, A. S., Chhoker, R. S., Mongia, A. D., Shoran, J., and Nagrajan, S. (2005). A decade of research on zero tillage and crop establishment. *DWR, Karnal–132001. Res. Bulletin*, *18*, 5-8.

Shoresh, M., Harman, G. E., and Mastouri, F. (2010). Induced systemic resistance and plant responses to fungal biocontrol agents. *Annual review of phytopathology*, *48*, 21-43.

Shrawat, A. K., Carroll, R. T., DePauw, M., Taylor, G. J., and Good, A. G. (2008). Genetic engineering of improved nitrogen use efficiency in rice by the tissue-specific expression of alanine aminotransferase. *Plant biotechnology journal*, *6*(7), 722-732.

Shukla, S. K., Yadav, R. L., Suman, A., and Singh, P. N. (2008). Improving rhizospheric environment and sugarcane ratoon yield through bioagents amended farm yard manure in udic ustochrept soil. *Soil and Tillage Research*, *99*(2), 158-168.

Singh, R. J., and Ahlawat I. P. S. (2012). Dry matter, nitrogen, phosphorous, and potassium partitioning, accumulation and use efficiency in transgenic cotton based cropping systems. *Commun Soil Sci Plant Anal* 43 (20):2633–2650.

Singh, V., Singh B., Singh Y., Thind, H. S., and Gupta, R. K. (2010). Need based nitrogen management using the chlorophyll meter and leaf colour chart in rice and wheat in South Asia: a review. *Nutrient Cycling in Agroecosystems*, *88*(3), 361-380.

Singh, B., Gupta, R. K., Singh, Y., Gupta, S. K., Singh, J., Bains, J. S., and Vashishta, M. (2006). Need-based nitrogen management using leaf color chart in wet direct-seeded rice in northwestern India. *Journal of New Seeds*, *8*(1), 35-47.

Singh, B., Singh, Y., Ladha, J. K., Bronson, K. F., Balasubramanian, V., Singh, J., and Khind, C. S. (2002). Chlorophyll meter–and leaf color chart–based nitrogen management for rice and wheat in Northwestern India. *Agronomy Journal*, *94*(4), 821-829.

Spaepen, S., Vanderleyden, J., and Remans, R. (2007). Indole-3-acetic acid in microbial and microorganism-plant signaling. *FEMS microbiology reviews*, *31*(4), 425-448.

Srivastava, S. N., Singh, V., and Awasthi, S. K. (2006). Trichoderma induced improve-ment in growth, yield and quality of sugarcane. *Sugar Tech*, *8*(2-3), 166-169.

Sukor, A. (2013). *Effects of cyanobacterial fertilizers compared to commonly-used organic fertilizers on nitrogen availability, lettuce growth, and nitrogen use efficiency on different soil textures* (Doctoral dissertation, Colorado State University. Libraries).

Syiem, M. B. (2005). *Entrapped cyanobacteria: Implications for biotechnology.*

Tanwar, A., Aggarwal, A., Kaushish, S., and Chauhan, S. (2013). Interactive effect of AM fungi with Trichoderma viride and Pseudomonas fluorescens on growth and yield of broccoli. *Plant Protection Science*, *49*(3), 137-145.

Tewatia R. K. and Chanda T. K. 2017. Trends in Fertilizer Nitrogen Production and Consumption in India. In Editor(s) Yash P. Abrol, Tapan K. Adhya, Viney P. Aneja, Nandula Raghuram, Himanshu Pathak, Umesh Kulshrestha, Chhemendra Sharma and Bijay Singh. *'The Indian Nitrogen Assessment'.* Elsevier; Pages 45-56.

Thomas, J., Ajay, D., Kumar, R. R., and Mandal, A. K. A. (2010). Influence of beneficial microorganisms during in vivo acclimatization of in vitro-derived tea (Camellia sinensis) plants. *Plant Cell, Tissue and Organ Culture (PCTOC), 101*(3), 365-370.

Upadhyay, S. K., Singh, D. P., and Saikia, R. (2009). Genetic diversity of plant growth promoting rhizobacteria isolated from rhizospheric soil of wheat under saline condition. *Current microbiology, 59*(5), 489-496.

Varinderpal-Singh, D., Yadvinder-Singh, Bijay-Singh, Baldev-Singh, Gupta, R. K., Jagmohan-Singh, and Balasubramanian, V. (2007). Performance of site-specific nitrogen management for irrigated transplanted rice in northwestern India. *Archives of Agronomy and Soil Science, 53*(5), 567-579.

Wagner, S. C. (2011). Biological Nitrogen Fixation. Nature Education Knowledge 3 (10): 15.

Win, K. K. (2003). *Plot Specific N Fertilizer Management for Improved N-use Efficiency in Rice Based Systems of Bangladesh.* Cuvillier.

Yadav, R. L., Singh V., Srivastav S. N., Lal R. J., Sangeeta S., Awasthi S. K., Joshi B. B. (2008). Use of Trichoderma harzianum for the control of red rot disease of sugarcane. *SugarcInt* (UK) 26:28–33.

Yang, W. H., Peng, S., Huang, J., Sanico, A. L., Buresh, R. J., and Witt, C. (2003). Using leaf color charts to estimate leaf nitrogen status of rice. *Agronomy Journal, 95*(1), 212-217.

Zhang, F., Yuan, J., Yang, X., Cui, Y., Chen, L., Ran, W., and Shen, Q. (2013). Putative Trichoderma harzianum mutant promotes cucumber growth by enhanced production of indole acetic acid and plant colonization. *Plant and soil, 368*(1-2), 433-444.

Zhao, X. Q., Nie, X. L., and Xiao, X. G. (2013). Over-expression of a tobacco nitrate reductase gene in wheat (Triticumaestivum L.) increases seed protein content and weight without augmenting nitrogen supplying. *PloS one, 8*(9), e74678.

Chapter 5

Unravelling the Significance of ACC-Deaminase in Sustainable Arid Soil Management

Anandkumar Naorem[1,*], Abhishek Patel[1], Shiva Kumar Udayana[2], M. Ashajyoti[3], P. Helena[4] and Truptimayee Suna[5]

[1]Indian Council of Agricultural Research-Central Arid Zone Research Institute, RRS-Bhuj, Gujarat, India
[2]MS Swaminathan School of Agriculture, Centurion University of Technology and Management, Paralakhemundi, Gajapati, Odisha, India
[3]Plant Protection Lab, ICAR – Central Agroforestry Research Institute, Jhansi, Uttar Pradesh
[4]Institute of Agricultural Sciences, Banaras Hindu University, U.P.
[5]Indian Council of Agricultural Research-Indian Agricultural Research Institute, New Delhi, India

Abstract

Drought stress is one of the most key environmental stresses that limit crop productivity, especially in dryland soils. The stress caused by low soil water availability could lead to huge economic losses. Such issues demand a sustainable approach that could either increase the tolerance of the crops to low water stress or increase water availability, which is of high concern during the era of climate change. Mining the beneficial microbes from soils and plants for their multifaceted activities could stimulate crop growth and enhance drought tolerance. The drought stress often results in an increased accumulation of ethylene in the crop, reducing the crop growth and yield. Soil dwelling beneficial microbes could play an imperative role in this regard. These microbes produce an enzyme, ACC-deaminase (1-aminocyclopropane-1-carboxylate deaminase) that are produced in response to the increased load of ethylene in the plant. The main reason for releasing this remarkable enzyme under a stressed environment is to break the ACC (an immediate precursor of ethylene) to ammonia and α-ketobutyrate, which are further utilised as nitrogen sources by the microbes. The conversion of ACC to these by-products lowers the ethylene levels in the plants and reduces drought stress. Therefore, understanding and unravelling the importance of ACC-deaminase producing microbes will benefit crop production in arid and semi-arid regions.

Keywords: beneficial microbes, climate change, drought stress, soil sustainability, soil microbiome

* Corresponding Author's Email: Anandkumar.naorem@icar.gov.in.

In: Ecosystem Services
Editor: Hanuman Singh Jatav
ISBN: 978-1-68507-614-6

1. Introduction

Several physiological and environmental factors limit plant growth and productivity. This is due to their effects on the soil nutrient availability and alteration of soil physicochemical properties (Singh et al. 2015). Several biotic and abiotic stress factors have negative impacts on plant growth and productivity. The abiotic variables include salinity, extreme temperatures, drought, waterlogging, mechanical injury, heavy metal toxicity, and other contamination of soil with several types of organic and inorganic hazardous chemicals (Gamalero and Glick, 2012). For most of the crops, the loss experienced as a result of these variables is estimated to be greater than 50%. (Bray et al., 2000). As a result, abiotic stressors are one of the most significant factors affecting agricultural productivity around the world. As a result, the global food supply must be doubled to meet the rising food demand of the increasing human population (Tilman et al., 2002). The problems caused by biotic and nutritional variables may be solved with pesticides and biofertilizers, respectively, but resolving abiotic disasters with non-biological techniques is relatively difficult. Plants respond to the aforementioned stressors by altering the levels of different hormones, which then cause the development of stress-related proteins necessary for defense against the negative consequences of stressors. Ethylene is one of the most frequent plant hormones that modulate the stress response. When ethylene production exceeds a certain threshold, it is converted to "stress ethylene," which is unfavourable in terms of root/shoot proliferation and other growth factors, and so inhibits plant growth and development. Certain plant-associated bacteria that produce the enzyme 1-aminocyclopropane-1-carboxylate deaminase (ACCD) can lessen the stress ethylene has on plants (Glick, 2007). ACCD converts ACC, an immediate precursor of ethylene, to α-ketobutyrate and ammonia, lowering the level of ethylene in plants (Figure 1) and allowing root/shoot growth to resume (Glick, 2014). This ability to assign resistance to abiotic stressors through ACCD activity and some other mechanisms of plant growth-promoting bacteria (PGPB) to alleviate stresses in host plants is termed as induced systemic tolerance (Yang et al., 2009).

2. Plant Growth-Promoting Bacteria in Action

Using the beneficial microbiome of plants, also known as plant growth-promoting bacteria (PGPB), is an alternate technique. PGPB are found mostly in the soil around plant roots, a micro-ecosystem known as the rhizosphere, as well as in vegetal tissues (endosphere) and on the aerial surfaces of leaves and stems (phyllosphere). The number of cells and diversity of bacteria in the rhizosphere is known to be substantially higher than in rootless soil (Arshad and Frankenberger, 1997). This is owing to the existence of nutrients emitted by the plant's roots, which include sugars, amino acids, organic acids, and other tiny molecules that can account for up to a third of the carbon fixed by the plant (Glick, 2012). Up to 10^8 or 10^9 microbial cells per gram of soil can be found in unstressed soils, however, the microbial population in stressed soils can be decreased to 10^4 cells per gram of soil (Timmusk et al., 2011). Dodd and Pérez-Alfocea (2012) found that many microorganisms are favourable to plant growth and boost plant tolerance to salt stress. Various bacterial genera from the rhizosphere, phyllosphere and endosphere are among them (Shrivastava and Kumar, 2015). Bacilli (and other sporulating bacteria) stand out in this regard because of their diverse direct and indirect mechanisms for boosting plant growth, as well as their ability to sporulate and withstand a variety of environmental stresses (Santoyo et al., 2012). Bacilli are thus great choices for serving as the primary biological agents in a variety of bio-inoculants, such as biostimulants, biofertilizers, and bio fungicides. Bacilli have been successfully employed in the field for decades, especially to fight conditions like salt stress (Glick and Skof, 1986). New bacilli strains with high saline tolerance are now being discovered and described, demonstrating a broad potential to stimulate the growth of a variety of plant species. Bacilli exhibit a variety of positive mechanisms during interactions with plants (Tahir et al., 2019; Vílchez et al., 2018; Chinnaswamy et al., 2018). As a result, PGPB with ACCD activity is critical in decreasing the negative effects of environmental stresses.

3. Interlinking Ethylene and ACC-Deaminase Activity

Phytohormones are organic substances produced in specific plant organs. They are either transferred to different sites or active in the same tissue where they are synthesized. They are signal molecules that control overall plant growth. The expression of plant intrinsic genes is regulated by them. Phytohormones are available for the plants by two sources. They are either endogenously synthesized by plant tissues themselves or exogenously synthesized by associated PGPR.

Ethylene regulates plant growth and development at the hormonal level, and its increased biosynthesis under stress alters the plant response, resulting in a reduction in biomass output (Taiz and Zeiger, 2010). As a result, regulating ethylene levels in plants under stress is difficult. Plants are always helped by a variety of beneficial bacteria in nature to withstand/tolerate/resist biotic and abiotic stress. Beneficial rhizobacteria that produce 1-aminocyclopropane-1-carboxylic acid deaminase (ACCD) reduce stress ethylene levels by decomposing ACC (an immediate precursor of ethylene) into α-ketobutyrate and ammonium, which are then utilised by microbes as a carbon and nitrogen source (Glick et al. 1998).

PGPR consists of an enzyme ACC-deaminase which is a weapon to stress ethylene and is able to enhance plant growth (Gontia-Mishra et al. 2014). PGPR synthesizing ACC-deaminase minimizes drought stress and induces tolerance against high salt concentration in plants (Zahir et al. 2008, 2009). Its activity has been observed in several rhizobacteria like Acinetobacter, Azospirillum, Agrobacterium, Achromobacter, Ralstonia, Enterobacter, Bacillus, Rhizobium, Burkholderia, Serratia, Pseudomonas, etc. (Kang et al. 2010). Recently, ACC-deaminase synthesizing Pseudomonas stutzeri A1501 is reported to facilitate rice cultivation in salt and heavy metal presence (Han et al. 2015). Apart from this, Gupta and Pandey, (2019) conducted an experiment of isolation of ACC deaminase producing bacteria from rhizospheric soil of Garlic and reported that two potential strains, ACC02 and ACC06, were found to possess other growth-promoting potentials like IAA production, phosphate solubilization, siderophore and ammonia production. Rhizobacteria capable of synthesizing ACC-deaminase also provide stress resistance from radiation, poly-aromatic hydrocarbons, high salt concentration, insect predation, wounding, and high light intensity (Glick 2012). ACC-deaminase synthesizing rhizobacterial inoculation induces the plant shoot growth and root elongation and increases nodulations in rhizobium. It also promotes mycorrhizal colonization and thus nutrient uptake in several crops (Glick 2012).

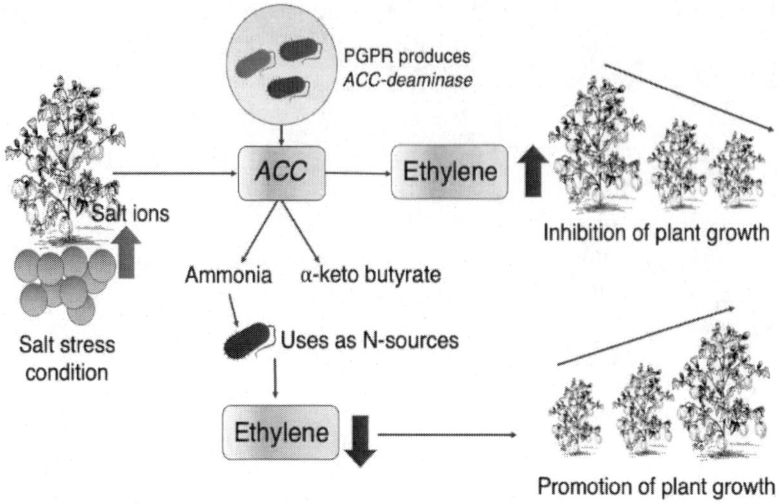

Figure 1. The action of ACC-deaminase on reducing the ethylene levels in plants under abiotic stress (In this case, an example of salt stress is studied).

4. Producers and Diversity of ACC Deaminase

Soil is a place where diverse forms of microbes are being habituated. According to Glick, 2014, the rhizospheric region of soil accounts for a higher population of microbiological flora. The bacteria present in the rhizosphere soil plays an important role in promoting growth and development of plants following different mechanisms *viz.*, organic and inorganic nutrient solubilization, siderophore production, water uptake, soil fertility, segregation of stress-induced ethylene precursor, ACC by ACC deaminase and so on (Aeron *et al.*, 2019). From all the involved mechanisms of plant growth promotion, the enzyme ACC deaminase shows the key mechanism as it can degrade the ACC.

ACC deaminase was first found in soil microorganisms, where it was demonstrated to convert ACC to ammonia and α-ketobutyrate, which were then metabolised by microbes (Honma and Shimomura, 1978). It is a pyridoxal phosphate-dependent enzyme that requires roughly 3 mol of enzyme-bound pyridoxal phosphate per mol of enzyme or 1 mol per trimeric subunit for enzyme activity (Honma, 1985). *Pseudomonas sp.* strain ACP was used to purify ACC deaminase, which was then partially purified from *Pseudomonas chloroaphis* 6G5 (Klee et al., 1991) and *Pseudomonas putida* GR12-2 (Jacobson et al., 1994). The molecular mass and shape of the enzyme isolated from all three sources appear to be identical. The enzyme activity of ACCD was tested in a variety of bacterial species over a wide pH range, with the maximum activity found at pH 8.0–8.5. (Zhao et al., 2003). The ideal temperature for *P. putida* GR12-2 ACC deaminase activity is 30°C (Glick et al., 1998). In the presence of its substrate ACC, ACC deaminase is an inducible enzyme whose synthesis is induced. In *Pseudomonas sp.* strain ACP and *P. putida* GR12-2, the minimal substrate level for induction was determined to be 100 nM. ACCD induction is a lengthy and difficult process. It shows activity in the first several hours after being incubated with the substrate, but it steadily declines after that (Jacobson et al., 1994). In minimum medium supplemented with ammonium sulphate as a nitrogen source, the basal level of enzyme activity is observed. Honma (1983) demonstrated that transferring the bacteria from a nutrient-rich medium to a minimum medium supplied with ACC as the sole nitrogen source resulted in increased activity. It shows that the induction of enzyme activity is proportional to the amount of ACC in the substrate. Other amino acids, including L-alanine, DL-alanine, and D-serine, increase enzyme activity and expression of ACCD to a lesser amount than ACC. Furthermore, both ACC and amino isobutyric acid were shown to produce the same degree of enzyme activity in *Pseudomonas sp.* strain ACP (Honma, 1983). According to Glick et al. (1998), a large percentage of ACC is secreted from plant roots or seeds, taken up by soil microorganisms, and hydrolyzed to ammonia and α-ketobutyrate. The amount of ACC outside the plant roots decreases as a result of ACC absorption and hydrolysis. Furthermore, by exuding more ACC into the rhizosphere, the balance between the internal and external ACC levels is maintained. As a result, a reduction in ACC levels influences the production of the stress hormone ethylene in host plants, stimulating plant growth (Glick et al., 1998). The essential feature of the reaction catalyzed by ACC deaminase is the opening of the ACC cyclopropane ring. Although the chemical mechanism is unknown, the most likely mechanisms for cleaving the cyclopropane bond are nucleophilic addition and elimination (Ortíz-Castro et al., 2009). L-isomers of amino acids such as L-alanine, L-serine, L-homoserine, and L-aminobutyricacid inhibit ACC deaminase competitively, with L-alanine and L-serine showing the highest inhibition. ACC-related compounds such as 2-alkyl-ACC and vinyl-ACC can also be used as substrates for ACC deaminase obtained from *Pseudomonas sp.* strain ACP, although the enzyme has an unusual preference for D-amino acids and is inert with any L-amino acids or derivatives (Walsh et al., 1981; Honma, 1985). Therefore, the present chapter addresses importance of ACC deaminase which reduces the level of stress ethylene in plants and protects them from the negative effects of environmental stressors.

The diversity of ACC deaminase activity is ubiquitous in bacteria including *Alphaproteobacteria*(Ma et al. 2003a), *Betaproteobacteria* (Gontia-Mishra et al. 2017b) and *Gammaproteobacteria*(Gontia-Mishra et al. 2017a), *Actinobacteria* (Siddikee et al. 2010; Gontia et al. 2011; Jha et al. 2012), *Firmicutes* (Siddikee et al. 2010; Timmusk et al. 2011), *Bacteroidetes* and *Flavobacterium* (Marques et al. 2010; Mesa et al. 2015).

For a better understanding of the ACC deaminase producing bacteria, a list is shown in Table 1 from all the studies carried out by different researchers. Rhizobacteria having ACC deaminase activity are effective in promoting plant growth and water use efficiency under drought conditions, by reducing the ethylene or ACC accumulation. Many studies have been conducted so far to check the diversity and as well how effective in the development of plants with the higher secretion of ACC deaminase in the rhizosphere soil.

Table 1. List of ACC deaminase producing bacteria

Bacterial strain producing ACC deaminase	ACC deaminase activity (nMaKB $mg^{-1}h^{-1}$)	Sources
Aneurinibacillusaneurinilyticus ACC02	~ 1500	Gupta and Pandey (2019)
Paenibacillus sp. ACC06	~ 1500	Gupta and Pandey (2019)
Achromobacterxylosoxidans A551	400 ± 4	Belimov et al. (2001, 2005)
A. xylosoxidans Bm1	90 ± 4	Belimov et al. (2001, 2005)
Achromobacter sp. strain *CM1*	130 ± 3	Belimov et al. (2001, 2005)
Acidovorax facilis 4p-6	3,080 ± 120	Belimov et al. (2001, 2005)
Pseudomonas sp. UW4	-	Orozco Mosqueda et al. (2019)
*Cronobactersakazakii*EF059843	-	M.S. Afridi et al. (2019)

Table 1. (Continued)

Bacterial strain producing ACC deaminase	ACC deaminase activity (nMaKB $mg^{-1}h^{-1}$)	Sources
Enterobacter cloacae		Cheng et al. (2007); Shaharoona et al. (2007); Ahmad et al. (2011); Barra et al. (2016)
Enterobacter aerogenes CAL3	16 ± 12	Shah et al. (1998)
Bacillus licheniformis	-	Tiwari et al. (2018)
B. subtilis	-	Tiwari et al. (2018)
Pseudomonas fluorescens	-	Zarei et al. (2020)
Pseudomonas putida UW4	3,030 ± 60	Hontzeas et al. (2006)
P. syringae GR12-2	3,470 ± 30	Belimov et al. (2001, 2005)
P. brassicacearum Am3	5,660 ± 12	Belimov et al. (2001, 2005)
P. putida BM3	3,780 ± 32	Belimov et al. (2001, 2005)
P. marginalis DP3	4,054 ± 27	Belimov et al. (2001, 2005)
Rhizobium leguminosarum 128C53K	5 ± 1	Belimov et al. (2001, 2005)
R. hedysari ATCC 43676	20 ± 0.1	Ma et al. (2003b)
R. leguminosarum 99A1	8 ± 3	Ma et al. (2003b)
Rhodococcus sp. strain Fp2	7,320 ± 400	Belimov et al. (2001, 2005)
Rhodococcus sp. strain 4N-4	12,970 ± 440	Belimov et al. (2001, 2005)
Serratia quinivirans SUD165	12 ± 15	Belimov et al. (2001, 2005)
Variovorax paradoxus 3P -3	3,700 ± 90	Belimov et al. (2001, 2005)
V. paradoxus 2C-1	3,588 ± 26	Belimov et al. (2001, 2005)

5. Role of ACC-Deaminase in Plant Nutrition and Stress Management

The stresses (biotic and abiotic) harmfully impact plant development and commodity yield. However, the soil and plant rhizosphere microbial communities act as saviours through stress management (deploying various stress alleviation and growth promotion mechanisms). In general, when plants are exposed to any of these stressors the ethylene, which is an essential gaseous hormone involved in various normal physiological activities of plants, is identified to be elevated. The increased expression of ethylene is correlated with the transcriptional activation of defense mechanism by synthesizing the Pathogenesis Related proteins (PR) and subsequent induction of acquired resistance to sustain plant growth for a longer period (Van Loon et al. 2006). The elevated expression level of this particular hormone beyond threshold levels affect the normal plant physiology and decrease the root and shoot development, also inducing senescence, chlorosis and leaf abscission. It was proved this type of physiological change can even predispose the plants to potential plant pathogens.

The effect of ethylene levels on various stages of disease process studied in tomato for the vascular wilt disease caused by the *Verticillium* spp. by application of synthetic ethylene synthesis inhibitor aminoethoxy vinyl glycine (AVG) and ethylene biosynthetic precursor 1-aminocyclopropane-1-carboxylate (ACC). The disease severity was found to be statistically reduced in plants that have 1-aminocyclopropane-1-carboxylate (ACC) initially at the time of initiation of infection whereas the same ACC when present in the post-infection stage aggravated disease in tomato to *Verticillium dahlia* infection (Robison et al. 2001). Here the importance of ethylene synthesis inhibitors such as AVG and ACC deaminase in the reduction of losses and enhancing the survival rate of plants can be clearly understood. Many Plant Growth Promoting Rhizobacteria (PGPR) and several other fungi can produce the enzyme 1-

aminocyclopropane-1-carboxylate (ACC) deaminase which can facilitate plant growth and perform stress alleviation limiting the ethylene production by converting ACC into α-ketobuyrate acid and ammonia (Honma and Shimora, 1978; Glick et al. 2007). It is found that the presence of ACC- deaminase synthesizing gene is ubiquitous among soil microorganisms including Rhizobium strains such as *Rhizobium leguminosarum* and *Rhizobium gallicum, Azospirillum, Pseudomonas, Ralstonia, Enterobacter etc.,* (Duan, 2007). Many of these genera are also known to have biological control activity and disease suppressive behaviour. Wang et al. (2000) proved that the transformed *Pseudomonas fluorescens* strain *CHO1* with ACC deaminase gene able to protect cucumber seedlings from damping-off disease (*Pythium* spp.) and potato tubers from soft rot disease (*Erwinia caratovora*). However, the same strain is unable to give protection in the case of Fusarium crown and root rot disease induced in tomatoes. This indicates either the external application or genetic manipulation in plants/microorganisms for ACC deaminase activity will benefit in the fight against pests and diseases also other environmental induced stress stimuli. Also, the transformed *Pseudomonas fluorescens* strain *CHO1* altered plant growth response under biotic stress by increasing the root length which was also anticipated due to the lowered level of ethylene production. Another bacterial species *Enterobacter cloacae* UW4 was used to study the ACC deaminase activity in canola roots by using RNA arbitrarily primed – PCR technique also proved it will confer stress tolerance and root growth elongation in canola (Hontzeas et al. 2004). Now it is well understood that the role of ACC deaminase producing microorganisms in biotic stress alleviation and plant protection is significant.

The abiotic factors like water and salt stress are two major production limiting factors in cropping systems under both irrigated and arid zones and also sole contributors in land degradation globally (Bharti and Barnawal, 2019). The role of ACC deaminase producing PGPR and other fungi need to be considered to address these important problems also. *Achromobacter piechaudii* ARV8 originally isolated from drought and salt-affected areas conferred salt and water stress tolerance to tomato and pepper seedlings under induced stress conditions (Mayak et al. 2004a, b). It even helped both the plant species in getting quick recovery after water stress when the stress condition was resumed. An important fungal species *Trichoderma longibrachiatum* strain T6 (TL-6) known to produce both the IAA and ACC deaminase. Wheat (*Triticum aestivum* L.) treated with and without the *Trichoderma longibrachiatum* T6 gave significant results in terms of plant growth and enhancement in tolerance to NaCl stress. The activity of ACC deaminase increased to 26% at NaCl concentration of 10 mg mL^{-1} and the same elevated up to 31% at 20 mg mL^{-1} salt concentration (Zhang et al. 2019). This study stated that under the influence of TL6 wheat seedlings showed transcriptional level changes in growth promotion, root elongation and salt tolerance significantly in extreme salt concentration levels. Flood is another abiotic problem that occurs in high or erratic rainfall receiving areas. Waterlogging causes an anaerobic condition in roots which trigger excess ACC synthesis, and the translocation of the same into stem causes subsequent leaf epinasty, chlorosis and necrosis symptoms. To inhibit this phenomenon ACC-deaminase producing microorganisms application or the genetic engineering of the plants to express this gene under flood experiencing conditions especially in roots also discussed in some studies (Grichko and Glick, 2001). Reclamation of heavy metal polluted sites and alleviation of the metals and other chemicals induced stress can also be possible by the application of ACC- deaminase producing microorganisms. Li et al. 2006 showed*Pseudomonas putida* UW4 treatment in nickel contaminated soils enhanced the growth of tobacco plants. In another study transgenic canola (*Brassica napus*) expressing *Enterobacter cloacae* UW4 derived 1-aminocyclopropane-1-carboxylate (ACC) deaminase gene showed better stress tolerance level to six different metal ions including arsenate (Nie et al. 2002). These studies indicate the genetic engineering for the expression of (ACC) deaminase gene works for plant pest and diseases management and most importantly for the abiotic factors like drought, salt, temperature etc.,

The soil microorganisms especially present in arid zones, degraded lands, heavy metal polluted areas, complex cropping systems such agroforestry lands can be explored and screened for the identification of organisms with ACC deaminase activity. These microorganisms will work as biosensors for stress indication and help in ecosystem sustainability (Wang et al. 2000).

6. Role of ACC-Deaminase in Alleviating Salt Stress

Salinity, being one of the abiotic stresses, impacts physiological, biochemical and microbial activities of plants through ionic and osmotic stresses (Ashraf 2004; Clergue et al., 2005; Slama et al. 2015). Salt outside the plant immediately hinders plant growth via problematic water uptake and root growth (Munns and Tester 2008). Nevertheless, salt accumulation inside the plant hinders enzymic and photosynthesis activities (Munns 2005; Roy et al. 2014). Several researchers found the salinity as one of the major causes for yield reduction of different crops (Promila and Kumar, 1982; Ghanem et al., 2009; Pushpavalli et al., 2016) due to alteration in natural soil-plant-microorganism interaction (Clergue et al., 2005). Plants interact with soil upon various biochemical and molecular levels to function its growth properly. However, in salt-induced stress conditions, this functionality alters either to combat from or surrender to adversity. In this mechanism, the plant uptakes salt ions, transports them to different parts, store them at basic molecular and whole plant level, biosynthesise the suitable solutes, alter the functional pathways and induce antioxidant and hormones (Roy et al. 2014). This happens at the cellular level when associated corresponding stress perceptions induce signals to a transcription factor, which in turn activate different genes to respond salinity effect on the plant through hormonal activity (Agarwal et al. 2013; Chen et al. 2014; Kurusu et al. 2015). The hormonal activities may include detoxification, the release of phytohormone, ion homeostasis, osmoprotectant, and cell integrity, alone or in combination, as per the gene activation. Phytohormones serve towards plant growth and its development. Under salinity stress, to response salinity, phytohormones such as Abscisic acid, ABA: Abscisic acid, Jasmonic Acid, IAA: Indole Acetic Acid, Ethylene, Salicylic Acid, and Brassinosteroids are released or controlled (Ryu and Cho 2015; Wani et al. 2016).

Ethylene (C_2H_4) is a gaseous phytohormone released by plants with demand-driven concentration as for its growth under stress conditions (Glick 2014). There are several studies has been done for ethylene signalling under salinity (Cao et al. 2008; Ryu and Cho 2015) and found that whenever a plant undergoes salinity stress it releases stress ethylene. The suitable ethylene concentration supports seed germination, root growth, flowering etc. in plants and ripening of the fruits. However uncontrolled excess concentration restricts root growth, defoliation, inhibit photosynthesis and affect chlorophyll (Singh et al. 2015). Hence, regulation on ethylene releases under salinity is a must for the sustainable growth of the plant.

As the plant response (sensitivity or tolerance) is the microbial and biochemical expression of the plant to salinity stress, the use of advantageous bacteria/microbiome/ enzymes called PGPBhas been studied extensively (Lugtenberg and Kamilova, 2009; Glick, 2012; 2014;

Numan et al., 2018) and these assist plant growth under salt-stressed conditions. PGPB is highly concentrated in the rhizosphere (vicinity of plant root), live in the endosphere (plant tissues system) and phyllosphere (steam and leaf surface system). The salt-tolerant PGPR have been isolated from extreme soils (alkaline, saline, and sodic). These PGPRs function for plant growth under salt stress as:

3. Having salt-tolerant mechanisms such as $Na^{+}(K^{+})/H^{+}$ antiporters, production of osmolyte, ACC deaminase, antioxidant etc.
4. Produce phytohormones produced which modulate pant physiology
5. Produce exopolysaccharides (EPS) which enhance water and ionic uptake and stabilize soil structure and aggregation.

Table 2. ACC deaminase-containing PGPB and their positive effects on crops under salt stress

Crop	ACC deaminase containing PGPB	Positive effects on crop	Sources
Wheat	*P. putida*, *P. aeruginosa* and *S. proteamacula*	Increased, grain weight and yield Better ionic (N^{+}/K^{+}) balance	Zahir et al. (2009)
	Bacillus mojavensis	Reduces ethylene production and improves yield	Pourbabaee et al. (2016)
Rice	*P. fluorescens* MSP-393	Osmotolerance mechanisms helped to maintain the root colonization of bacteria	Paul and Nair (2008)
	Pseudomonas stutzeri A1501	Improved rice growth	Han et al. (2015)
Cucumber	*P. putida*	Increased Root-shoot weight and number of leaves	Gamalero et al. (2009)
Okra	*Bacillus megaterium*	Higher root-shoot and biomass	Habib et al. (2015)

The ACC deaminase was initially considered from the Pseudomonas sp. strain by Honma and Shimomura (1978). No doubt different PGPB adopts several mechanisms to felicitate the plant growth and development (Glick 2012), perhaps, ACC deaminase has possession of the bacterial activity required for plant growth (Glick 2014). This enzyme cleavage ACC (ethylene precursor) into ammonia and α-ketobutyrate, hence dropping ethylene concentration in stressed plants (Jacobson et al. 1994; Glick et al. 1999). In the plant root-soil system the rhizosphere bacteria act upon the exudates from roots and release IAA, and the plant takes some of it. This IAA level can activate ACS to synthesize ACC, in this way IAA act as a stimulator for ethylene production in the plant. Some ACC is exudated through roots and consumed by bacteria and then cleaved by ACC deaminase (Penrose et al. 2001). As a net result, bacterium works as the sink for ACC, moreover, lowering ACC level reduces the potential rise of ethylene level in the plant. However, under salt-stressed condition, when the plant keeps on secreting IAA but do not able to synthesize ACC deaminase, the ethylene level increases accordingly and adversely affect plant growth and production. Whereas ACC deaminase-containing PGPB secretes both well in balance and subsequently, decrease ethylene inhibited plant growth in salt stress conditions. Several studies confirmed the ACC deaminase affectability for several crops such as tomato (Tank and Saraf, 2010; Yan et al., 2014), cotton (Yue et al., 2007; Wu et al., 2012) etc. against salt stress (Table 2). This results in the better tolerance ability of plants against salt stress which may be attributed to elongated roots, better enzymic functionality and more resistance towards ethylene-induced growth inhibition.

Conclusion and Future Considerations

To feed the growing human population in a climate-changing scenario, we need to address the constraints in crop production. Therefore, several innovative techniques and approaches are developed to achieve the goal of increasing food production without jeopardizing the environment. One of these widely acceptable technologies is the use of plant growth-promoting bacteria. Research on these beneficial bacteria has revealed several beneficial microbes have multifaceted plant growth-promoting characteristics, including the production of ACC-deaminase enzyme. Although, in general, plants have their protective mechanisms to grow under abiotic stress. However, these mechanisms are not sufficient to improve productivity while mitigating the effects of abiotic stress. The interaction between crop and plant has generated information on some areas where it can be used for improving crop production. ACC-deaminase enzyme helps in reducing the ethylene levels in the plants under several types of biotic stress. The chapter discusses the role of ACC-deaminase producing microorganisms that have the potential to modify the stress response of the crops, which further improves crop productivity. Further research must be conducted to unravel the insights on the microbe-host interaction under several types of abiotic stresses so that it can be exploited for improving crop productivity in marginal or degraded soils. The increasing trend of research in ACC-deaminase inclusion in bio formulations has improved crop production under normal as well as salt-affected soils. However, trials based on field experiments must be executed to evaluate the efficiency of the microbial strains (those producing ACC-deaminase) in field conditions. Both the endophytic as well as rhizobacteria must be evaluated to improve our understanding of how they differ in their ACC-deaminase producing behaviour and its associated effects on plant growth.

References

Aeron, A., Khare, E., Jha, C.K., Meena, V.S., Aziz, S.M.A., Islam, M.T., Kim, K., Meena, S.K., Pattanayak, A., Rajashekara, H. (2019). Revisiting the plant growth-promoting rhizobacteria: lessons from the past and objectives for the future. *Archives of Microbiology*. 1–12.

Afridi, M.S., Amna, Sumaira, Mahmood, T., Salam, A., Mukhtar, T., Mahmood, S., Ali, J., Khatoon, Z., Bibi, M., Javed, M.T., Sultan, T., Chaudhary, H.J. (2019). Induction of tolerance to salinity in wheat genotypes by plant growth-promoting endophytes: Involvement of ACC deaminase and antioxidant enzymes. *Plant Physiology et Biochemistry*, doi: https://doi.org/10.1016/ j.plaphy.2019.03.041.

Agarwal P.K., Shukla PS, Gupta K, Jha B (2013) Bioengineering for salinity tolerance in plants: state of the art. *Mol Biotechnol* 54:102–123.

Ahmad, M., Zahir, Z.A., Asghar, H.N., Asghar, M. (2011). Inducing salt tolerance in mung bean through coinoculation with rhizobia and plant-growth-promoting rhizobacteria containing 1-aminocyclopropane-1-carboxylate deaminase. *Canadian Journal of Microbiology*. 57, 578- 589.

Arshad, M., Frankenberger Jr., W.T., 1997. Plant growth-regulating substances in the rhizosphere: microbial production and functions. *Advances in Agronomy* Vol. 62. Academic Press, pp. 45–151.

Ashraf M. (2004). Some important physiological selection criteria for salt tolerance in plants. *Flora* 199: 361–376.

Barra, P.J., Inostrozab, N.G., Acuñab, J.J., Morab, M.L., Crowley, D.E., Jorquerab, M.A. (2016). Formulation of bacterial consortia from avocado (*Perseaamericana Mill.)* and their effect on growth, biomass and superoxide dismutase activity of wheat seedlings under salt stress. *Applied Soil Ecology.* 102, 80-91.

Belimov, A.A., Hontzeas, N., Safronova, V.I., Demchinskaya, S.V., Piluzza, G., Bullitta, S., Glick, B.R. (2005). Cadmium-tolerant plant growth-promoting bacteria associated with the roots of Indian mustard (Brassica juncea L. Czern.). *Soil Biology and Biochemistry.* 37:241–250.

Belimov, A.A., Safronova, V.I., Sergeyeva, T.A., Egorova, T.N., Matveyeva, V.A., Tsyganov, V.E., Borisov, A.Y., Tikhonovich, I.A., Kluge, C., Preisfeld, A., Dietz, K.J., Stepanok, V.V. (2001). Characterization of plant growth promoting rhizobacteria isolated from polluted soils and containing 1-aminocy- clopropane-1-carboxylate deaminase. *Canadian Journal of Microbiology.* 47:642–652.

Bharti, N. and Barnawal, D., 2019. Amelioration of salinity stress by PGPR: ACC deaminase and ROS scavenging enzymes activity. In *PGPR amelioration in sustainable agriculture* (pp. 85-106). Woodhead Publishing.

Bray, E.A., Bailey-Serres, J., and Weretilnyk, E. (2000). "Responses to abiotic stresses," in *Biochemistry and Molecular Biology of Plants*, eds W. Gruissem, B. Buchannan, and R. Jones (Rockville, MD: American Society of Plant Physiologists), 1158–1249.

Cao Y.-R., Chen S.-Y., Zhang J.-S. (2008). Ethylene signaling regulates salt stress response: an overview. *Plant Signaling & Behavior.* 3:761–763.

Chen M.-X., Lung S.-C., Du Z.-Y., Chye M.-L. (2014). Engineering plants to tolerate abiotic stresses. *Biocatalysis and agricultural biotechnology* 3:81–87.

Cheng, Z., Park, E., Glick, B.R., 2007. 1-Aminocyclopropane-1-carboxylate deaminase from *Pseudomonas putida* UW4 facilitates the growth of canola in the presence of salt. *Canadian Journal of Microbiology.* 53, 912-918.

Chinnaswamy, A., Coba de la Peña, T., Stoll, A., de la Peña Rojo, D., Bravo, J., Rincón, A., Pueyo, J.J., 2018. A nodule endophytic Bacillus megaterium strain isolated from Medicago polymorpha enhances growth, promotes nodulation by Ensifermedicae and alleviates salt stress in alfalfa plants. *Ann. Appl. Biol.* 172 (3), 295–308.

Clergue, B., Amiaud, B., Pervanchon, F., Lasserre-Joulin, F., Plantureux, S., 2005. Biodiversity: function and assessment in agricultural areas. A review. *Agron. Sustain.* Dev. 25 (1), 1–15.

Dodd, I.C., Pérez-Alfocea, F., 2012. Microbial amelioration of crop salinity stress. *J. Exp. Bot.* 63 (9), 3415–3428.

Duan, J., 2007. *1-aminocyclopropane-1-carboxylate (ACC) deaminase genes in Rhizobia: Isolation and characterization* (Master's thesis, University of Waterloo).

Gamalero E., Berta G., Glick B.R. (2009). The use of microorganisms to facilitate the growth of plants in saline soils. In: *Microbial strategies for crop improvement.* Springer, Berlin, pp 1–22.

Gamalero, E., and Glick, B.R. (2012). "Plant growth-promoting bacteria and metal phytoremediation," in *Phytotechnologies: Remediation of Environmental Contaminants*, eds N.A. Anjum, M.E. Pereira, I. Ahmad, A.C. Duarte, S. Umar, and N.A. Khan (Boca Raton, FL: CRC Press), 361–376.

Ghanem, M.E., van Elteren, J., Albacete, A., Quinet, M., Martínez-Andújar, C., Kinet, J.M., et al. (2009). Impact of salinity on early reproductive physiology of tomato (Solanum lycopersicum) in relation to a heterogeneous distribution of toxic ions in flower organs. *Funct. Plant. Biol.* 36, 125–136.

Glick B, Penrose D, Li J (1998) A model for the lowering of plant ethylene concentrations by plant growth-promoting bacteria. *J Theor Biol* 190: 63–68.

Glick, B.R. (2007). Promotion of plant growth by bacterial ACC deaminase. *Crit. Rev. Plant Sci.* 26, 227–242. doi: 10.1080/07352680701572966.

Glick, B.R., 2012. Plant growth-promoting bacteria: mechanisms and applications. *Scientifica.*

Glick, B.R., 2014. Bacteria with ACC deaminase can promote plant growth and help to feed the world. *Microbiol. Res.* 169 (1), 30–39.

Glick, B.R., Patten, C.L., Holguin, G., et al. (1999). *Biochemical and genetic mechanisms used by plant growth promoting bacteria.* Imperial College Press, London.

Glick, B.R., Penrose, D.M., Li, J. (1998). A model for the lowering of plant ethylene concentrations by plant growth promoting bacteria. *Journal of Theoretical Biology.* 190:63–68.

Glick, B.R., Skof, Y.C., 1986. Environmental implications of recombinant DNA tech- nology. *Biotechnol. Adv.* 4, 261–277.

Glick, B.R., Todorovic, B., Czarny, J., Cheng, Z., Duan, J. and McConkey, B., 2007. Promotion of plant growth by bacterial ACC deaminase. *Critical Reviews in Plant Sciences*, *26*(5-6), pp.227-242.

Gontia, I., Kavita, K., Schmid, M., et al. (2011). *Brachybacteriumsaurashtrense*sp. nov., a halotolerant root associated bacterium with plant growth-promoting potential. *International Journal of Systematic and Evolutionary Microbiology.* 61:2799–2804.

Gontia-Mishra I., Sasidharan S., Tiwari S. (2014). Recent developments in use of 1-aminocyclopropane-1-carboxylate (ACC) deaminase for conferring tolerance to biotic and abiotic stress. *Biotechnol Lett.* doi:10.1007/s10529-014-1458-9.

Gontia-Mishra, I., Sapre, S., Tiwari, S. (2017b). Zinc solubilizing bacteria from the rhizosphere of rice as prospective modulator of zinc biofortification in rice. *Rhizosphere*. 3:185–190.

Gontia-Mishra, I., Sapre. S., Kachare, S., et al. (2017a). Molecular diversity of 1-aminocyclopropane- 1-carboxylate (ACC) deaminase producing PGPR from wheat (*Triticum aestivum*L.) rhizosphere. *Plant Soil.* 414:213–227.

Grichko, V.P. and Glick, B.R., 2001. Amelioration of flooding stress by ACC deaminase-containingplant growth-promoting bacteria. *Plant Physiology and Biochemistry*, *39*(1), pp.11-17.

Gupta, S., Pandey, S., 2019. ACC deaminase producing Bacteria with multifarious plant growth promoting traits alleviates salinity stress in french bean (Phaseolus vulgaris) plants. *Frontiers in Microbiol*ogy. 10, 1506. https://doi.org/10.3389/fmicb.2019.01506.

Habib S.H., Kausar H., Saud H.M. (2015). Plant growth-promoting rhizobacteria enhance salinity stress tolerance in okra through ROS-scavenging enzymes. *BioMed Res Int* (Hindawi Publishing Corporation) https://doi.org/10.1155/2016/6284547

Han Y., Wang R., Yang Z., Zhan Y., Ma Y., Ping S., Zhang L., Lin M., Yan Y. (2015) 1-Aminocyclopropane-1-carboxylate deaminase from Pseudomonas stutzeri A1501 facilitates the growth of Rice in the presence of salt or heavy metals. *J Microbiol Biotechnol* 25(7):1119–1128.

Honma M., Shimomura T. (1978) Metabolism of 1-aminocyclopropane-1- carboxylic acid. *Agric Biol Chem* 42: 1825–1831.

Honma, M. (1983). Enzymatic determination of 1-aminocyclopropane- 1-carboxylate deaminase. *Agric. Biol. Chem*. 47, 617–618. doi: 10.1271/bbb1961.47.617.

Honma, M. (1985). Chemically reactive sulfhydryl groups of l-aminocyclopropane-1-carboxylate deaminase, *Agric. Biol. Chem.* 49, 567-571.

Hontzeas, N., Hontzeas, C.E., Glick, B.R. (2006). Reaction mechanisms of bacterial enzyme 1-amino-cyclopropane-1-carboxylate deaminase. *Biotechnology Advances.* 24:420–426.

Hontzeas, N., Saleh, S.S. and Glick, B.R., 2004. Changes in gene expression in canola roots induced by ACC-deaminase-containing plant-growth-promoting bacteria. *Molecular plant-microbe interactions*, *17*(8), pp.865-871.

Jacobson, C.B., Pasternak, J.J., and Glick, B.R. (1994). Partial purification and characterization of 1-aminocyclopropane-1-carboxylate deaminase from the plant growth promoting rhizobacterium Pseudomonas putida GR12-2. *Can. J. Microbiol.* 40, 1019–1025. doi: 10.1139/m94-162.

Jha, B., Gontia, I., Hartmann, A. (2012). The roots of the halophyte *Salicornia brachiata*are a source of new halotolerant diazotrophic bacteria with plant growth-promoting potential. *Plant Soil.* 356:265–277.

Kang B.G., Kim W.T., Yun H.S. (2010). Use of plant growth-promoting rhizobacteria to control stress responses of plant roots. *Plant Biotechnol Rep* 4:179–183.

Klee, H.J., Hayford, M.B., Kretzmer, K.A., Barry, G.F., and Kishore, G.M. (1991). Control of ethylene synthesis by expression of bacterial enzyme in transgenic tomato plants. *Plant Cell* 3, 1187–1193. doi: 10.1105/tpc.3. 11.1187.

Kurusu T., Kuchitsu K. and Tada Y. (2015) Plant signaling networks involving Ca2+ and Rboh/Nox-mediated ROS production under salinity stress. *Front. Plant Sci.*6:427.

Li, Q., Shah, S., Saleh-Lakha, S. and Glick, B.R., 2006. Growth of tobacco in nickel-contaminated soil in the presence of the plant growth-promoting bacterium *Pseudomonas putida* UW4. *Current Microbiology*.

Lugtenberg, B., Kamilova, F., 2009. Plant-growth-promoting rhizobacteria. *Annu. Rev. Microbiol.* 63, 541–556.

Ma, W., Guinel, F.C., Glick, B.R. (2003b). Rhizobium leguminosarum biovar viciae 1-aminocyclopro- pane-1-carboxylate deaminase promotes nodulation of pea plants. *Applied and Environmental Microbiology.* 69:4396–4402.

Ma, W., Sebestianova, S.B., Sebestian, J., et al. (2003a). Prevalence of 1-aminocyclo-propane-1- carboxylate deaminase in *Rhizobium* spp. *Antonie Van Leeuwenhoek.* 83:285–291.

Marques, A.P.G.C., Pires, C., Moreira, H. (2010). Assessment of the plant growth promotion abilities of six bacterial isolates using *Zea mays* as indicator plant. *Soil Biology and Biochemistry.* 42:1229–1235.

Mayak, S., Tirosh, T. and Glick, B.R., 2004a. Plant growth-promoting bacteria that confer resistance to water stress in tomatoes and peppers. *Plant Science*, *166*(2), pp.525-530.

Mayak, S., Tirosh, T. and Glick, B.R., 2004b. Plant growth-promoting bacteria confer resistance in tomato plants to salt stress. *Plant physiology and Biochemistry*, *42*(6), pp.565-572.

Mesa, J., Mateos-Naranjo, E., Caviedes, M.A., et al. (2015). Scouting contaminated estuaries: heavy metal resistant and plant growth promoting rhizobacteria in the native metal rhizoaccumulator*Spartina maritima. Marine Pollution Bulletin.* 90:150–159.

Munns (2005). Genes and salt tolerance: bringing them together. *New Phytologist.* 167: 645– 63.

Munns R and Tester M. (2008). Mechanisms of salinity tolerance. *Annu. Rev Plant Biol* 59:651–681.

Nie, L., Shah, S., Rashid, A., Burd, G.I., Dixon, D.G. and Glick, B.R., 2002. Phytoremediation of arsenate contaminated soil by transgenic canola and the plant growth-promoting bacterium Enterobacter cloacae CAL2. *Plant Physiology and Biochemistry*, *40*(4), pp.355-361.

Numan, M., Bashir, S., Khan, Y., Mumtaz, R., Shinwari, Z.K., Khan, A.L., Ahmed, A.H., 2018. Plant growth promoting bacteria as an alternative strategy for salt tolerance in plants: a review. *Microbiol. Res.* 209, 21–32.

Orozco-Mosqueda, M. del C., Duan, J., DiBernardo, M., Zetter, E., Campos-Garcia, J., Glick, B.R., Santoyo, G. (2019). The Production of ACC Deaminase and Trehalose by the Plant Growth Promoting Bacterium Pseudomonas sp. UW4 Synergistically Protect Tomato Plants Against Salt Stress. *Frontiers in Microbiology*. 10. http://dx.doi.org/10.339/fmicb.2019.01392.

Ortíz-Castro, R., Contreras-Cornejo, H.A., Macías-Rodríguez, L., and López- Bucio, J. (2009). The role of microbial signals in plant growth and development. *Plant Signal. Behav.* 4, 701–712. doi: 10.4161/psb.4.8.9047.

Paul, D., & Nair, S. (2008). Stress adaptations in a plant growth promoting rhizobacterium (PGPR) with increasing salinity in the coastal agricultural soils. *Journal of basic microbiology*, 48(5), 378-384.

Penrose D.M., Moffatt B.A., Glick B.R. Determination of 1-aminocyclopropane-1- carboxylic acid (ACC) to assess the effects of ACC deaminase-containing bacteria on roots of canola seedlings. *Can J Microbiol* 2001;47:77–80.

Pourbabaee A.A., Bahmani E., Alikhani H.A., Emami S. (2016). Promotion of wheat growth under salt stress by halotolerant bacteria containing ACC deaminase. *J Agr Sci Tech* 18:855–864.

Promila, K., and Kumar, S. (1982). Effect of salinity on flowering and yield characters in pigeonpea. *Ind. J. Plant Physiol.* 25, 252–257.

Pushpavalli, R., Quealy, J., Colmer, T., Turner, N., Siddique, K., Rao, M., et al. (2016). Salt stress delayed flowering and reduced reproductive success of chickpea (Cicer arietinum L.), a response associated with Na+ accumulation in leaves. *J. Agron. Crop Sci.* 202, 125–138. doi: 10.1111/jac.12128.

Robison, M.M., Griffith, M., Pauls, K.P. and Glick, B.R., 2001. Dual role for ethylene in susceptibility of tomato to Verticillium wilt. *Journal of Phytopathology*, *149*(7-8), pp.385-388.

Roy S.J., Negrao S., Tester M. (2014). Salt resistant crop plants, Current Opinion in Biotechnology 26:115–124.

Ryu H, Cho Y-G (2015) Plant hormones in salt stress tolerance. *J Plant Biol* 58:147–155.

Santoyo, G., Orozco-Mosqueda, M.D.C., Govindappa, M., 2012. Mechanisms of biocontrol and plant growth-promoting activity in soil bacterial species of Bacillus and Pseudomonas: a review. *Biocont. Sci. Technol.* 22 (8), 855–872.

Shah, S., Li, J., Moffatt, B.A., Glick, B.R. (1998). Isolation and characterization of ACC deaminase genes from two different plant growth-promoting rhizobacteria. *Canadian Journal of Microbiology*. 44:833–843.

Shaharoona, B., Arshad, M., Khalid, A. (2007). Differential response of etiolated pea seedlings to inoculation with rhizobacteria capable of utilizing 1-aminocyclopropane-1- carboxylate or L-methionine. *Journal of Microbiology.* 45, 15-20.

Shrivastava, P., Kumar, R., 2015. Soil salinity: a serious environmental issue and plant growth promoting bacteria as one of the tools for its alleviation. *Saudi J. Biol. Sci.* 22 (2), 123–131.

Siddikee, M.A., Chauan, P.S., Anandham, R., et al. (2010). Isolation, characterization, and use for plant growth promotion under salt stress, of ACC deaminase-producing halotolerant bacteria derived from coastal soil. *Journal of Microbiology and Biotechnology.* 20:1577–1584.

Singh R.P., Shelke G.M., Kumar A., Jha P.N. (2015). Biochemistry and genetics of ACC deaminase: a weapon to "stress ethylene" produced in plants. *Front Microbiol* 6(937):2015. https://doi. org/ 10.3389/fmicb.2015.00937.

Slama I., Abdelly C., Bouchereau A., Flowers T., Savour A. (2015). Diversity, distribution and roles of osmoprotective compounds accumulated in halophytes under abiotic stress. *Ann of Bot* 1–15.

Tahir, M., Ahmad, I., Shahid, M., Shah, G.M., Farooq, A.B.U., Akram, M., Zakir, A., 2019. Regulation of antioxidant production, ion uptake and productivity in potato (Solanum tuberosum L.) plant inoculated with growth promoting salt tolerant Bacillus strains. *Ecotoxicol. Environ. Saf.* 178, 33–42.

Taiz, L. & Zeiger, E. *Plant physiology* Fifth ed. (Sinauer Associates Inc.) 1–778, (Sunderland Massachusetts U.S.A 2010).

Tank N., Saraf M. (2010). Salinity-resistant plant growth promoting rhizobacteria ameliorates sodium chloride stress on tomato plants. *J Plant Interact* 5:51–58.

Tilman, D., Cassman, K.G., Matson, P.A., Naylor, R., and Polasky, S. (2002). Agricultural sustainability and intensive production practices. *Nature* 418, 671–677. doi: 10.1038/nature01014.

Timmusk, S., Paalme, V., Pavlicek, T., Bergquist, J., Vangala, A., Danilas, T., Nevo, E., 2011. Bacterial distribution in the rhizosphere of wild barley under contrasting mi- croclimates. *PLoS One* 6 (3), e17968.

Tiwari, G., Duraivadivel, P., Sharma, S. et al. (2018). 1-Aminocyclopropane-1-carboxylic acid deaminase producing beneficial rhizobacteria ameliorate the biomass characters of *Panicum maximum* Jacq. by mitigating drought and salt stress. *Scientific Report.* 8, 17513. https://doi.org/10.1038/s41598-018-35565-3.

van Loon, L.C., Geraats, B.P. and Linthorst, H.J., 2006. Ethylene as a modulator of disease resistance in plants. *Trends in plant science*, *11*(4), pp.184-191.

Vílchez, J.I., Tang, Q., Kaushal, R., Chen, S., Liu, R., Zhang, H., 2018. Genome sequence of Bacillus cereus strain TG1-6, a plant-beneficial rhizobacterium that is highly salt tol- erant. *Genome Announ* 6 (19), e00351–18.

Walsh, C., Pascal, R.A., Hohnston, M., Raines, R., Dikshit, D., Krantz, A., et al. (1981). Mechanistic studies on the pyridoxal phosphate enzyme 1- aminocyclopropane-1-carboxylate deaminase form Pseudomonas sp. *Biochemistry* 20, 7509–7519. doi: 10.1021/bi00529a028.

Wang, C., Knill, E., Glick, B.R. and Défago, G., 2000. Effect of transferring 1-aminocyclo-propane-1-carboxylic acid (ACC) deaminase genes into Pseudomonas fluorescens strain CHA0 and its gac A derivative CHA96 on their growth-promoting and disease-suppressive capacities. *Canadian Journal of Microbiology*, *46*(10), pp.898-907.

Wani S.H., Kumar V., Shriram V., Sah .SK. (2016). Phytohormones and their metabolic engineering for abiotic stress tolerance in crop plants. *The crop j.* 4: 162–176.

Wu Z., Yue H., Lu J., Li C. (2012). Characterization of rhizobacterial strain Rs-2 with ACC deaminase activity and its performance in promoting cotton growth under salinity stress. *World J Microbiol Biotechnol* 28:2383–2393.

Yan J, Smith MD, Glick BR, Liang Y (2014) Effects of ACC deaminase containing rhizobacteria on plant growth and expression of Toc GTPases in tomato (Solanum lycopersicum) under salt stress. *Botany* 92(11):775–781.

Yang, J., Kloepper, J.W., and Ryu, C.M. (2009). Rhizosphere bacteria help plants tolerate abiotic stress. *Trends Plant Sci.* 14, 1–4. doi: 10.1016/j.tplants.2008.10.004.

Yue H., Mo W., Li C., Zheng Y., Li H. (2007). The salt stress relief and growth promotion effect of RS-5 on cotton. *Plant Soil* 297:139–145.

Zahir Z.A., Ghani U., Naveed M., Nadeem S.M., Asghar H.N. (2009) Comparative effectiveness of Pseudomonas and Serratia sp. containing ACC deaminase for improving growth and yield of wheat (Triticum aestivum L.) under salt-stressed conditions. *Arch Microbiol 191*:415–424.

Zahir Z.A., Munir A., Asghar H.N. (2008). Effectiveness of rhizobacteria containing ACCdeaminase for growth promotion of pea (Pisum sativum) under drought conditions. *J Microbiol Biotechnol* 18:958–963.

Zarei, T., Moradi, A., Kazemeini, S.A. et al. (2020). The role of ACC deaminase producing bacteria in improving sweet corn (*Zea mays* L. var saccharata) productivity under limited availability of irrigation water. *Scientific Report.* 10, 20361. https://doi.org/10.1038/s41598-020-77305-6.

Zhang, S., Gan, Y. and Xu, B., 2019. Mechanisms of the IAA and ACC-deaminase producing strain of Trichoderma longibrachiatum T6 in enhancing wheat seedling tolerance to NaCl stress. *BMC plant biology*, *19*(1), pp.1-18.

Zhao, H., Chen, K., Li, K., Du, W., He, S., and Liu, H.W. (2003). Reaction of 1-amino-2-methylenecyclopropane-1-carboxylate with 1- aminocyclopropane-1-carboxylate deaminase: analysis and mechanistic implications. *Biochemistry* 42, 2089–2103. doi: 10.1021/bi020567n.

Chapter 6

Reclamation of Deteriorated Soils Utilizing Green Approaches

Nafeesa Farooq Khan[1,*] and Summia Rehman
[1]Biological Invasion Lab, University of Kashmir, Jammu and Kashmir, India

Abstract

The realm of green technology alternatives is fast growing and there are many different definitions of what green technology is and the benefits it provides. Simply defined, green technology is a framework for developing new clean technologies that combine one or more components of ecologically friendly solutions. Soil quality has an indisputable impact on global population distribution, necessitating the development of new desirability and sustainability requirements. However, it plays an important, although frequently neglected, role in community structure. It's crucial to remember that soils can't be considered in isolation. Furthermore, technological advancements have the potential to influence the effectiveness of soil conditions. In recent years, the application of new technologies has substantially increased the profitability of agriculture in many places of the world that were previously unsuitable for cultivation. The present chapter has given focus on the possible approach for the reclamation of deteriorated soils utilizing green approaches.

Introduction

It is an established fact that without human intervention, 90 percent of soil will be polluted by 2050. A contaminated/degraded soil lacks the ability to supply basic sustenance to living organisms. High salinity, a decrease in fertility, a decline in organic matter (altered soil structure), soil erodibility, as well as a rise in alkalinity and pH are all indicators of a degraded soil. Over a one-third part of the globe's soil has been depleted and the IPCC predicts that by 2050 that number will have risen to 90%. In addition, man's prevailing agricultural practices has disrupted the soil structure and drainage capacity stressing water, air and nutrients movements, and consequently impacts plant growth (Golchin et al. 1995; Tejada et al. 2008; Sinha et al. 2009). Furthermore, synthetic chemical usage has raised soil salinity or alkalinity. Besides, the existing human population of more than 7 billion people is putting a strain on agriculture's ability to meet present food demands without jeopardizing natural sources' ability to feed subsequent generations (Robertson & Swinton, 2005). Also, it has been predicted that

* Corresponding Author's Email: khnnafisa@gmail.com.

In: Ecosystem Services
Editor: Hanuman Singh Jatav
ISBN: 978-1-68507-614-6

by 2100, the world's population would approximately fall between 8.5 to 12 billion people and this huge population growth would further cause food shortages leading to the evolution of resource-depletion agriculture and culminating into an irreversible environmental harm. It is also reported that marginally degraded soil yields nearly 30% less food and holds half as much water in comparison to good healthy soil. Thus, disputes between humans and the environment, which are mostly the root of severe pandemics, will only get worse if nothing is done. By securing potential food supplies, reducing water stress, and mitigating climate change visa viz soil repair will go a long way toward resolving this dispute. In this regard ongoing food protection and environmental degradation are now major concerns of the present world (Singh et al. 2019). The term "green technology" is broad and mainly concerned with the use of scientific research and technologies to preserve the surroundings. This concept encompasses a wide range of approaches, including the use of green chemistry, pollution management, etc. Much of this has to do with preserving the land, water and air sources. Such technologies are used to restore life to an environment that has been compromised. So altogether such advancements are also known as clean technology or environmental technology. Even though there are options for restoring damaged soils, more information and tools are required to strengthen potential long-term and large-scale impacts. The main goal of this chapter is to discuss green technologies which help preserve nature while still debating how humans can use such unharming technologies. Green technology has received a lot of attention ever since the nineties. It has many advantages for not only ecology but also for a green and clean lifestyle. Such methods safeguard the planet's health for all life to thrive. In order to address global concerns, a renewed emphasis on green tech in soil remediation has attracted a lot of research interest, which has been focusing on strategies that are both sustainable and ecologically friendly for the past several years. Using these approaches that use green amendments/reagents made from a range of wastes and/or that can be gradually degraded or reused have low impact on the treated soil's physio-chemical and biological properties. Above all, these methodologies also consider costs, carbon emissions, and soil health and capacity following interventions for different land uses. Agricultural practices and objectives have, and continue to prioritize, production optimization over long-term soil health. Given the growing interest in green technologies in soil remediation, the chapter includes the most recent relevant research, with the goal of attracting practitioners, scientists and policymakers from all over the world, particularly those interested in learning more about green and emerging technologies for soil remediation. It would also help authorities pick technology available for degraded lands such as employing agricultural, industrial, and municipal waste amendments (biochar, compost, etc.) for soil immobilization. Moreover, stretching to bioremediation investigations, such as soil washing, leaching, flushing, electronic extraction, phytoextraction employing biodegradable /used /recycled reagents might be of great significance. In addition, soil health and the ability of soil to be reused in various types of land use are essential concepts in the subject (Figure 1). This chapter discusses all of the approaches that can be employed to combat the devastating impacts of soil degradation in the current period, as well as the conditions in which they can be applied.

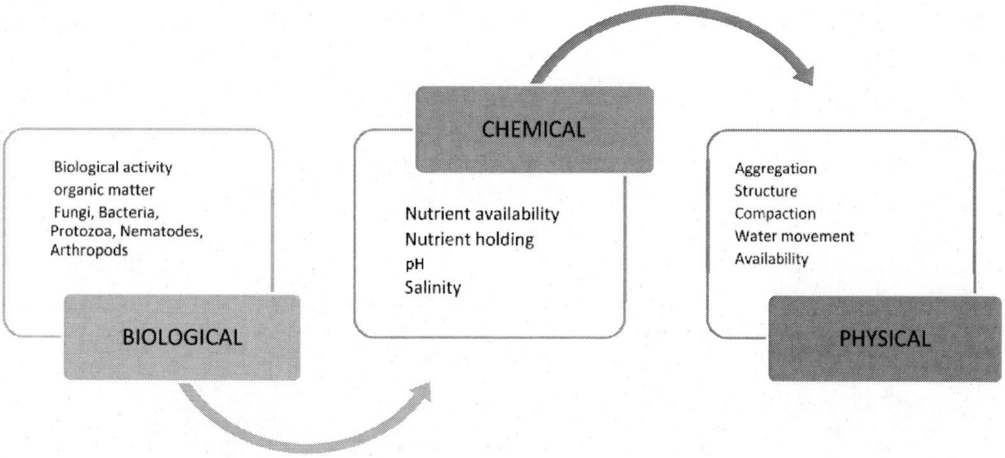

Figure 1. Inter relating factors for determination of Soil health and protection.

Causes of Land/ Soil Degradation

Soil degradation occurs when soil function and usage are lost or diminished. This is further explained in terms of the period of land-use triggered deterioration (short, medium and long-term land-use). There are multiple reasons for several forms of land degradation and several attempts are underway to discover and define essential limiting/indicators for soil health in various agricultural practices (Sánchez-Navarro et al. 2015). Such factors reduce biomass production and as a long-term effect genetic plant reserve also. Nevertheless, these reasons fluctuate from one country to the other as well as between continents. This is due to the fact that various aspects of development occur in different parts of the world. For example, the principal causes of land degradation in countries of Africa may differ from those in a state in the United States, and still again from those in an Australian state. Moreover, causes vary depending on aspects such as agricultural methods and skills (with respect to their intensity), whether the region is developed or underdeveloped and even external problems such as growing population. Most frequent genetic causes of soil deterioration, in addition to the specific causes of land degradation, are as follows:

6. Physical/Chemical Causes
7. Natural/Human Causes
8. Direct Causes/Indirect Causes

At some point, all of these variables may be related. Wind, for example, is a significant source of soil erosion, as is water from rain, both acting as a natural and physical cause. Similarly, soil pollution is primarily seen as a result of human-caused land degradation, including agricultural and industrial chemicals being major sources of chemical contaminants. Likewise human caused Land degradation can be also attributed to a variety of sources, including waste dumping, mining, urbanization, air deposition, and soil erosion. Besides, activities like deforestation, over-cultivation, and overgrazing also contribute to it. Overgrazing, over tillage, over-fertilization (excess nitrogen levels), over-application of pesticides and herbicides, incorrect or overwatering watering, and other components of Industrial Agriculture and/ or intensive or unsustainable farming practices all contribute to the depletion of agricultural ecosystems. Heavy tilling, regular intensive harvests and intensive use of agrochemicals despite having increased yields have staked the cost of long-term sustainability. Employing industrial and intense farming activities disguise short-term decreases in soil productivity and yields, as well as pave the way for long-term unsustainable tendency. Landowners may also overlook environment protection techniques like ground cover and soil/water drains that aid in the preservation of lands and top soil. The growth of industrial agriculture is one of the most harmful types of land degradation. Consequently, the implications of land degradation have been obviously reinforced by industrial agriculture and intensive farming techniques for food production/high rates of consumption of food concerns, as well as inadequate land regulation and farming efficiency.

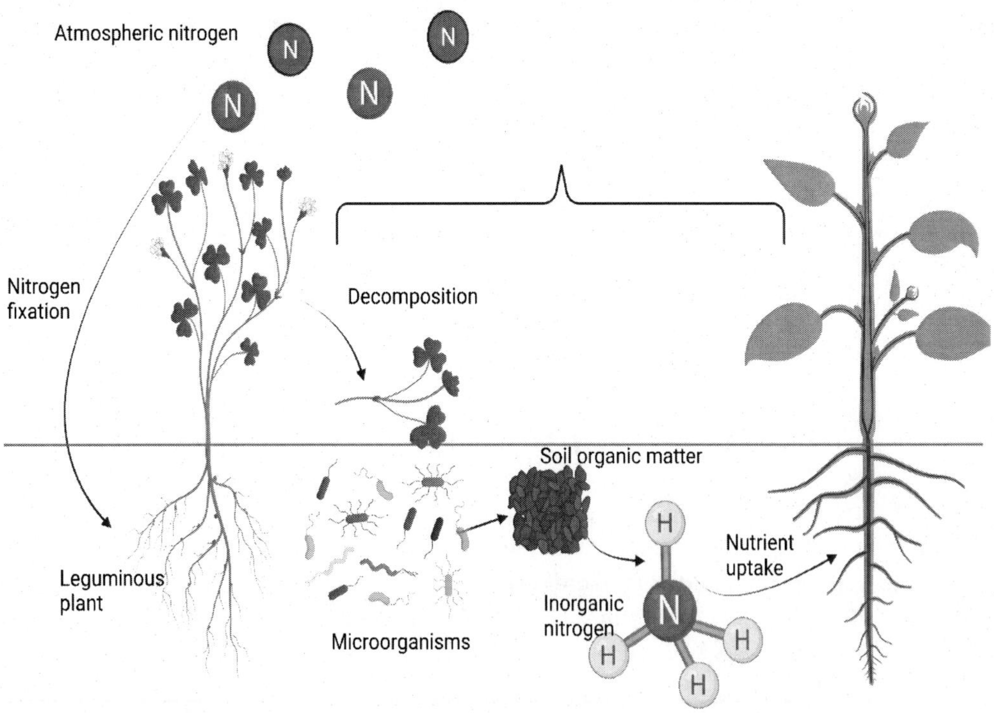

Figure 2. Schematic representation of Nitrogen cycling with a legume green manure.

Other compounding factors include urbanization and human settlement areas such as cities, towns, infrastructure, highways, and so on. At the moment, problems like atmospheric deposition and chemical leaching (through wind, water, etc.) from one place to another, where land/soil becomes contaminated or deteriorated, are being addressed. This includes pesticides transported between one region to another by the airflow and additionally oil can also percolate into land and water systems from roadways and large highways. Another example is poorly managed landfills, where leachate can leak and contaminate soil. This is an obvious case of improper waste management/disposal equivalent to known hard waste, water waste, human sewage, and other sorts of waste infesting soil and the environment. Hazardous chemicals and substances, such as radioactive waste, can also be mismanaged. Extreme weather events (Natural/Man-Made) Land and soil damage can be caused by floods, storms, and other natural disasters. The erosion, salinization, compaction, acidity, and chemical pollution of soils have resulted in mild to severe degrading of the land.

To summarize, industrial effluents, hydrocarbons, insecticides, lead, and other toxic metals are the most commonly involved compounds. Operations which cause other types of soil degradation (erosion, compaction, etc.) may exacerbate pollution effects indirectly, making soil clean-up more difficult. Prominent causes of land degradation, as well as soil loss, vary by location of the world - for example, overgrazing is a big cause in Africa, whereas agricultural practices are a major factor in North America (fewresources.org; ommegaonline.org). Likewise usage of Nitrogen and Phosphorus chemical fertilizers to crop fields exclusively for a long time without the addition of organic matter, as is currently done by Ethiopian farmers, causes soil erosion, degradation of physical, chemical, and biological properties of soils, and a decline in crop responses (Shibabawet al. 2018, 2017; Alemayehuet al. 2020; Tameneet al. 2017) Figure 2.

Quality improving : Soil Green Technologies and Rational for Utilizing Green Technologies

Our understanding of soil science has significant gaps. Before deciding on a green technology, it's critical to assess your farm's soil, climate, and water supply. When using green technology, it's also important to think about planting and termination times and procedures. Before deciding on a green technology, it's critical to assess your farm's soil, climate, and water supply. Improving soil health is both resource intensive and time consuming. By investigating the soil with their roots and sustaining life in the soil, many different species of plants can increase the availability of nutrients. Despite the fact that plants require a wide range of nutrients, nitrogen is often the most limiting factor in crop development and yield. To improve the quality of soils, various treatments are used. Few of them are: -

i. *Use of Organic Farming techniques:* The use of natural agricultural methods to reduce environmental harm is defined as organic farming. Some organic farming practices that assist replenishing the soil include green manure (uprooted or sowed crop sections included or left on topsoil), cover crops, crop rotation, and organic compost, as outlined by Pieters (1927). For soil quality restoration, soil organic matter management, reclaiming degraded soils, and giving plant nutrients, green manure has become a popular environmentally critical agricultural strategy in Asia (Sinha et al. 2009; Kumar et al. 2020). In the last two to three decades, green manure research has resurfaced (Tiwari et al. 1980; Brar and Sidhu, 1997; Blackshawet al. 2001; Larkin and Griffin, 2007; Sinha et al. 2009; Kumar, 2010). Green manure include legume such as pigeon pea, green gram, soybean or ground nut; perennial woody multipurpose legumes trees viz., *Leucaena leucocephala* (Subabul), *Gliricidia sepium* (Gliricidia or Mata Ratón), *Cassia siamea* (Kassod tree) or non-grain legumes like sunn hemp (*Crotalaria sp*), dhaincha (*Sesbania spp*), *Centosemia, Stylosanthes and Desmodium* (Palaniappan,1994). Among others Sunn hemp (*Crotalaria juncea* L.), dhaincha (*Sesbania aculeata L.*), berseem (*Trifolium alexandrinum*) and mungbean (*Vigna radiata*) are most commonly grown as green manure crops (Bistaet al. 2018).

ii. *Green manure and cover crops*: The green manures and cover crops act as mulch, protecting the soil from wind and water erosion. Whereas cover crops are produced to prevent erosion and improve soil, green manures help to stop soil being blown away by wind and rain. The roots hold the earth in place (Schumann et al. 2000). Monocultures like sugarcane can be recovered or improved utilizing green manure crops like *Sesbania rostrata* and *Crotalaria juncea* (Schumann et al. 2000). Further, legumes and other nitrogen fixing plants like cowpea, mungbean, urd-bean, pigeon-pea, chickpea, lentil, field pea, lathyrus, rajmash, temphrasia, groundnut, soybean, woody legume (e.g., *Leucaena, Gliricida, Pongomia and Delonix), dhaincha* (*Sesbania sp.)* and sun-hemp (*Crotalaria sp.)* fixing the atmospheric nitrogen in their root nodules are quite beneficial (Kumar et al. 2013) in this regard.

When integrated into the soil during the succulent stage, plants like dhaincha (*Sesbania aculeata L.*) adds 60-90 kg nitrogen per hectare (Pandey et al. 2008). Since green manures have a higher N content and lower C-N ratio than synthetic nitrogen fertilizers, they tend to improve crop yield (Bhuiyan and Zaman, 1996). Similarly green manuring, for example, increases the availability of rock phosphate in rice (Cavigelli and Thien, 2003). They decompose and add organic stuff to the soil. Generally, soils contain bacteria, actinomycetes, fungus, algae, and protozoa, as well as nematodes, earthworms, and arthropods from the animal kingdom. The mineral nutrition cycling from herbivore, carnivore, and saprophagous food chains is dependent on these bacteria (Witkamp, 1971). The distribution and population dynamics of soil microorganisms are altered by decomposition (Akpor et al. 2006). For instance, during decomposition microbes recycle nitrogen, phosphorus, and potassium (Goyal et al. 1999; Sharma and Ghosh, 2000; Yadav et al. 2000; Singh et al. 2007; Selvi and Kalpana, 2009). Green manure improves the chemical and physical characteristics of the soil by aerating and draining it. Improved soil quality leads to higher agricultural yields (Egodawatta et al. 2011). Also, weeds are controlled with insecticides and green manure (Blackshaw et al. 2001). Green manuring using clover and rye, for example, inhibits the growth of weeds by competing for nutrients, light, and space (Boydston and Hang, 1995). Some green manures show potential to create chemicals that inhibit plant germination, growth, or development. This is known as allelopathy. Allelopathy occurs when plants discharge substances into the soil that affect neighboring or subsequent crops. Sweet clover, buckwheat, and alfalfa are examples. Overall, they are cheap and natural ways to control weeds. Furthermore, green manuring has been demonstrated to boost soil fertility and quality by stimulating soil microbial development and activity (Eriksen, 2005). (Doran et al. 1988). It supplies nutrients rich in organic carbon for the microbial biomass which converts unavailable nutrients in plant leftovers to available nutrients for crops and it also increases soil microorganism diversification (Kumar et al. 2014). Increased organic matter causes beneficial microbial activity to rise (Sikora and Stott, 1996). Green manures are used between crops to assist preserve or increase soil organic matter (Pung et al. 2004). Green manuring has been found to have a significant impact on the management of soil-borne plant diseases, notably in the rhizosphere by many researchers (Bailey and Lazarovits, 2003; Manici et al. 2004; Pung et al. 2004; Janvier et al. 2007; Larkin and Griffin, 2007; Kamil et al. 2009).

Many soil-borne diseases cause crop yield reductions (Blum and Kabana, 2004). The ineffectiveness of chemical fumigants and seed treatments has recently led to the study of organic soil additions, including green manures. Pung et al. (2004) employed *Brassica* green manure crops to regulate soil-borne *Sclerotinia spp* reduce inoculum density by producing various degrees of biofumigant compounds known as isothiocyanates (ITCs) Kumar et al. 2013. Green manure crops, particularly crucifers containing glucosinolates, also help to lower *Pythium spp.* and total fungus population. *Sclerotium rolfsii* germination, sclerotia development and mycelial proliferation were all reduced when kudzu, velvet bean, and pine-bark plant parts were added to soybean and tomato which is recently reported by Lazzeri and Manici (2001) and Blum and Kabana (2004). Moreover, *Rhizoctonia solani, Sclerotium rolfsii,* and *Sclerotinia sclerotiorum* were studied during the breakdown of *Sesbania aculeata* L in soil (Kumar, 2010) with significant reduction in mycelial growth and sclerotia production of these soil borne pathogens.

Earlier Davis et al. (1996) reported a few commonly appliable green manures that includes Austrian winter pea, broccoli, Sudan grass, barley and corn (Davis et al. 2000). It has been reported recently that application of such fresh green manures influence *Rhizoctonia solani* saprophytic growth and damping-off disease in Canola (Yulianti et al. 2006). According to Larkin and Griffin (2007), using *Brassica* crops in crop rotations and also applied as green manure can reduce soil-borne infections/diseases of potato in laboratory, greenhouse, and field studies. Also, in decomposing sun-hemp (*Crotalaria juncea* L.) the green manure lowers soil-borne infections (Yadava, 2001; Kamil et al. 2009). In similar studies Manici et al. (2004) reported *Pythium spp.* population and microbial communities in intensive cropping systems and observed that green manures are useful to reduce the saprophytic and pathogenic population of *Pythium spp.* Furthermore, they encourage beneficial microbial communities in soils. Also, using green manure lessens the severity of bacterial illnesses. For example, using milk vetch as green manure influenced bacterial community structure and nutrient transformation functional groups in soils at a regional scale (Gao et al. 2021). It has also shaped the microbial population in the rice rhizosphere, especially beneficial bacteria like *Acinetobacter* and *Pseudomonas* (Zhang et al. 2017).

iii. *Organic compost*: In comparison to inorganic fertilizers, organic compost is a generally less expensive technique of soil amendment as such commonly used. Composting can help mitigate the harmful impact of intensive vegetable cultivation on soil organic carbon levels (Smukleret al. 2008; Sainjuet al. 2002; Jackson et al. 2004; Evanyloet al. 2008; Willekenset al. 2014; Sharifi et al.2014; Coggeret al. 2016). Compost is made up of decomposing plant and animal matter. Composting has the primary benefit of increasing the organic matter level of soil. Compost contributes carbon that is generally stable (C). It aims to strengthen fertility of the soil, structure and water holding capacity. For example, a study on the effects of organic waste compost on sesame plants growing in sandy soil found that all compost treatments accelerated sesame growth and improved pigment, carbohydrate and mineral content (Abdel-Sabouret al. 1996). Composting is one of the few technologies that can be used at any scale, from domestic composting to large municipal wastes. This even absorbs carbon from the atmosphere and stores it in the soil. Incomplete decomposition and humification of organic matter, on the other hand, would induce an increase of hazardous compounds in the soil and toxic consequences in plants. The addition of immature/non-stable organic solid waste to soil, in particular, may influence plant growth by specific substances or inappropriate growing conditions, such as immobilization/imbalance of essential nutrients, phytotoxicity, the presence of heavy metals, pathogenic bacteria and inorganic salts, all of which can inhibit plant growth (Malińska et al. 2016; Wang et al. 2017). In total, composting lowers the usage of chemical fertilizers, which, if used incorrectly, can contaminate nearby streams or groundwater. It is suggested that we determine the compost's quality by looking at its maturity and stability before its application (Sayara et al. 2020).

iv. *Crop rotation:* Cropping systems are useful in developing sustainable farming systems because they provide spatial and temporal variety. Choosing a crop rotation may necessitate a trade-off between environmental benefits and short-term market opportunities. This agricultural method involves gradually planting various crops in one spot (cropping pattern). Other *Azospirillum spp.* benefits include biological nitrogen fixation (BNF) in the rhizosphere, improved root soil exploration and more effective water and nutrient absorption (Dalla Santa et al. 2004). Forage radish (*Raphanus sativus* var. niger) and sun-hemp (*Crotalaria juncea*) are examples of deep-rooted, fast-growing crops that may successfully collect nutrients (particularly nitrogen) from deep soil layers and make them more easily available for succeeding crops planted or left as a mulch (Douds et al., 1999). Crop fields in Ethiopia's highlands have lost their physio-chemical quality and production due to lack of crop rotation and organic inputs. Overall, research has indicated that this method increases agricultural yields (Talgre et al. 2009, Ndayegamiye et al. 2015; Hayat and Ali 2010). It also prevents soil erosion and promotes soil fertility (Dogliotti, 2004). Crop rotation also helps build micro-habitats that allow various communities with distinct biological roles to coexist while also fostering a healthy, disease-resistant soil. Crop rotation, no-or low-tillage and intercropping cover crops can increase SOM numerically and qualitatively (Kumar et al. 2020).

v. *Soil remediation:* Soil remediation involves the removal of harmful contaminants such as, heavy metals, sewage sludge, coal tar, carcinogenic hydrocarbons, liquors and petroleum from soils. Soil remediation can be achieved using biological techniques. This method is called bioremediation. Some examples of bioremediation techniques include:

- *Phytoremediation:* This low-tech and low-cost clean-up technology uses plants to remove, eliminate, incorporate, or render toxins harmless from soils such as heavy metals, trace elements, organic compounds and radioactive compounds with no adversities. Plants can clean up or remediate contaminated environments in a variety of ways. Some plants may show capability to extract/absorb contaminants from soils which is known as phytoextraction. Other subset technologies include Phyto-transformation, Phyto-stimulation, Phyto-volatilization, Phyto-degradation, Phyto-rhizofiltration and Phyto-stabilization (Dixit R. et al.2015; Schwitzguebel et al. 2001). *Rhazya stricta*, for example, as a heavy metal phytoremediation in contaminated soils (Azab and Hegazy 2020). Likewise, *Pteris vittata* has offered up new avenues for removing this noxious substance from the soil (Ma et al. 2001). Phytodegradation was used to treat organic chemicals such as hydrocarbons using *Helianthus annuus, Brassica campestris, Festuca arundinacea,* pesticide and dye by *Blumeam alcolmii,* and TNT by *Chrysopogon zizanioides* (De-Farias et al. 2009; Park et al. 2011; Schwitzguébel et al. 2006, Kagalkar et al. 2011).

- *Bioaugmentation:* The introduction of genetic engineered microbes into contaminated soils (often bacteria) with the goal of decomposing contaminants is referred to as bioremediation. Though bioaugmentation is a well-known and environmentally safe strategy mostly for boosting hydrocarbon removal from oil-contaminated soil (Ambrazaitienė et al. 2013), it can sometimes fail in soil restoration and disrupt the soil's ecological equilibrium. The effectiveness of this procedure is determined by a variety of parameters, including the soil's physico-chemical qualities and the introduced microbes' capacity to compete successfully with the native soil microflora.

vi. *Land-based treatments:* Techniques such as land farming and composting are examples of this. In the process the contaminated soils are transported to farming locations and overturned and tilled on a regular basis to allow circulation of air while as in composting, microscopic organisms prevalent in organic matter are employed to break down soil pollutants. By and large soil remediation is always a large-scale, expensive activity that requires money or consent from the government or local government.

vii. *Desalinization:* Soil salinization occurs when high levels of soluble salts accumulate in the root zone. till date, more than 3% of the world's total soil resources have been damaged by salt (FAO, 2020). Crop production losses due to salinization range from 18 to 43 percent in arid and semiarid regions of the world (Wichelns and Qadir, 2015; Chang et al. 2019) in a variety of crops have been reported in a number of prior studies (Sima and Ebadi, 2019). The cumulative consequence of crop root soil salinity and submergence in water is said to be worse than the individual effects (Al-Muaini et al. 2019; Zheng et al. 2009). Reducing crop growth and crop yield. The process of soil salinization is aided by:

- Synthesis from high-salt-content parent materials
- In arid places where there isn't enough water to drain salts, little rainfall is a problem.
- Poor soil drainage system
- The drainage system in the soil is inadequate.
- Prolonged contact of soil with saline water resources or chemicals

The following are some of the strategies used to rehabilitate saline soils:

- Drainage can rather be built to remove ions down the soil profile though the process is comparatively expensive and complicated
- Applying water to polluted soils to push salts far beyond the root system to leach out saline soils.
- Bioremediation with salt-tolerant plants/halophytes.
- Gypsum (calcium sulphate dehydrate) is used to treat sodic soils. The sodium in sodium chloride is replaced with calcium when gypsum is added into the sodic soils, lowering the sodium content.

Uses

Green technologies cover a wide range of technologies that aid in reducing human influence on the environment and fostering long-term growth. The key factors for green technology are social equity, economic feasibility, and long-term viability. Other important uses include:

- *Reutilizing/Recycling:* Green technology makes use of non-depleting natural resources. It aids in the management and recycling of wastes. It enables it to be put to useful use. This concept is employed in a variety of applications, including waste management, trash incineration, and much more. Individuals have used a variety of recyclable materials to make plant fertilizer, fuel, and etc.
- *Purification of Water:* Green technology has the intended effect of purifying water. Currently, the major source of concern is the paucity of drinking water supply.
- *Economical farmland methods:* Contour ploughing, terrace farming, perimeter runoff control, keyline design, cover crops/crop rotation, windbreaks, soil conservation farming, green manures, zero tillage/no tillage, agrostology related measures, salinity management, dry farming indigenous crops, liming, natural farming with indigenous microbes etc. are some effective alternatives that are environment friendly and encourages reduced maintenance expenses.
- *Reviving Ecosystems:* Green or clean technology has been utilized to restore ecosystems that have been severely harmed as a consequence of human intervention. Through the use of this technology the ecology is able to re-establish itself and at the same time be protected. This contributes to the survival of many plant and animal species.

Conclusion

The potential reclamation of degraded land is essential to maintain the soil health and sustainable ecosystem. The proposed environmental impact is perhaps the most significant benefit of using green solutions. The current situation is having issues like inflicting significant environmental damage, making it critical for farmers to employ green technology such as green manures and other alternatives. Green technology, in particular, aids in the reduction of carbon footprint, waste reduction, soil and water conservation, as compared to older technologies. The strategy chosen is influenced by a variety of factors, including capital costs, labour costs, time, and availability of land. When taken as a whole, green solutions can help damaged soils improve their operations while lowering expenses. Although green technology is initially more expensive to use, the long-term advantages much outweigh the short-term costs.

References

Abdel-Sabour, M. F., & El-Seoud, M. A. (1996). Effects of organic-waste compost addition on sesame growth, yield and chemical composition. *Agriculture, ecosystems & environment*, *60*(2-3), 157-164.

Akpor, O. B., Okoh, A. I., & Babalola, G. O. (2006). Culturable microbial population dynamics during decomposition of Theobroma cacao leaf litters in a tropical soil setting. *Journal of Biological Sciences (Austria) v. 6 (4) p. 768-774*.

Alemayehu, Y. A., Asfaw, S. L., & Terfie, T. A. (2020). Exposure to urban particulate matter and its association with human health risks. *Environmental Science & Pollution Research*, *27*(22).

Al-Muaini, A., Green, S., Abou Dahr, W. A., Kennedy, L., Kemp, P., Dawoud, M., & Clothier, B. (2019). Water use and irrigation requirements for date palms on commercial farms in the hyper-arid United Arab Emirates. *Agricultural Water Management*, *223*, 105702.

Ambrazaitienė, D., Žukauskaitė, A., Jakubauskaitė, V., Reikaitė, R., Zubrickaitė, M., & Karčauskienė, D. (2013). Biodegradation activity in the soil contaminated with oil products. *Zemdirbyste-Agriculture*, *100*(3), 235-242.

Azab, E., & Hegazy, A. K. (2020). Monitoring the efficiency of Rhazya stricta L. plants in phytoremediation of heavy metal-contaminated soil. *Plants*, *9*(9), 1057.

Bailey, K. L., & Lazarovits, G. (2003). Suppressing soil-borne diseases with residue management and organic amendments. *Soil and tillage research*, *72*(2), 169-180.

Bhuiyan, N. I., & Zaman, S. K. (1996). Use of green manuring crops in rice fields for sustainable production in Bangladesh agriculture. In *Biological nitrogen fixation associated with rice production* (pp. 51-64). Springer, Dordrecht.

Bista, B., & Dahal, S. (2018). Cementing the Organic Farming by Green Manures. *International Journal of Applied Sciences and Biotechnology*, *6*(2), 87-96.

Blackshaw, R. E., Moyer, J. R., Doram, R. C., & Boswell, A. L. (2001). Yellow sweetclover, green manure, and its residues effectively suppress weeds during fallow. *Weed science*, *49*(3), 406-413.

Blum, L. E., & Rodríguez-Kábana, R. (2004). Effect of organic amendments on sclerotial germination, mycelial growth, and Sclerotium rolfsii-induced diseases. *Fitopatologia Brasileira*, *29*, 66-74.

Boydston, R. A., & Hang, A. (1995). Rapeseed (Brassica napus) green manure crop suppresses weeds in potato (Solanum tuberosum). *Weed technology*, *9*(4), 669-675.

Cavigelli, M. A., & Thien, S. J. (2003). Phosphorus bioavailability following incorporation of green manure crops. *Soil Science Society of America Journal*, *67*(4), 1186-1194.

Chang, Q., Zheng, T., Zheng, X., Zhang, B., Sun, Q., & Walther, M. (2019). Effect of subsurface dams on saltwater intrusion and fresh groundwater discharge. *Journal of Hydrology*, *576*, 508-519.

Cogger, C. G., Bary, A. I., Myhre, E. A., Fortuna, A. M., & Collins, D. P. (2016). Soil physical properties, nitrogen, and crop yield in organic vegetable production systems. *Agronomy Journal*, *108*(3), 1142-1154.

Dalla Santa, O. R., Hernández, R. F., Alvarez, G. L. M., Ronzelli Junior, P., & Soccol, C. R. (2004). Azospirillum sp. inoculation in wheat, barley and oats seeds greenhouse experiments. *Brazilian archives of biology and technology*, *47*(6), 843-850.

de Farias, V., Maranho, L. T., de Vasconcelos, E. C., da Silva Carvalho Filho, M. A., Lacerda, L. G., Azevedo, J. A. M., ... & Soccol, C. R. (2009). Phytodegradation potential of Erythrina crista-galli L., Fabaceae, in petroleum-contaminated soil. *Applied biochemistry and biotechnology*, *157*(1), 10-22.

Dixit, R., Malaviya, D., Pandiyan, K., Singh, U. B., Sahu, A., Shukla, R., ... & Paul, D. (2015). Bioremediation of heavy metals from soil and aquatic environment: an overview of principles and criteria of fundamental processes. *Sustainability*, *7*(2), 2189-2212.

Dogliotti, S., Rossing, W. A., & Van Ittersum, M. K. (2004). Systematic design and evaluation of crop rotations enhancing soil conservation, soil fertility and farm income: a case study for vegetable farms in South Uruguay. *Agricultural Systems*, *80*(3), 277-302.

Doran, J. W. (1987). Microbial biomass and mineralizable nitrogen distributions in no-tillage and plowed soils. *Biology and Fertility of Soils*, *5*(1), 68-75.

Douds Jr, D. D., & Millner, P. D. (1999). Biodiversity of arbuscular mycorrhizal fungi in agroecosystems. *Agriculture, ecosystems & environment*, *74*(1-3), 77-93.

Egodawatta, W. C. P., Sangakkara, U. R., Wijesinghe, D. B., & Stamp, P. (2011). Impact of green manure on productivity patterns of homegardens and fields in a tropical dry climate.

Eriksen, J. (2005). Gross sulphur mineralisation–immobilisation turnover in soil amended with plant residues. *Soil biology and Biochemistry*, *37*(12), 2216-2224.

Evanylo, G., Sherony, C., Spargo, J., Starner, D., Brosius, M., & Haering, K. (2008). Soil and water environmental effects of fertilizer-, manure-, and compost-based fertility practices in an organic vegetable cropping system. *Agriculture, ecosystems & environment*, *127*(1-2), 50-58.

Golchin, A., Clarke, P., Oades, J. M., & Skjemstad, J. O. (1995). The effects of cultivation on the composition of organic-matter and structural stability of soils. *Soil Research*, *33*(6), 975-993.

Goyal, S., Chander, K., Mundra, M. C., & Kapoor, K. K. (1999). Influence of inorganic fertilizers and organic amendments on soil organic matter and soil microbial properties under tropical conditions. *Biology and Fertility of Soils*, *29*(2), 196-200.

Janvier, C., Villeneuve, F., Alabouvette, C., Edel-Hermann, V., Mateille, T., & Steinberg, C. (2007). Soil health through soil disease suppression: which strategy from descriptors to indicators? *Soil biology and Biochemistry*, *39*(1), 1-23.

Kagalkar, A. N., Jadhav, M. U., Bapat, V. A., & Govindwar, S. P. (2011). Phytodegradation of the triphenylmethane dye Malachite Green mediated by cell suspension cultures of Blumea malcolmii Hook. *Bioresource technology*, *102*(22), 10312-10318.

Kamil, D., Kumar, R., & SINHA, A. (2009). Effect of green manuring of Crotolaria juncea L. on some soil-borne pathogens. *Indian Phytopathology*, *62*(3), 304-309.

Kumar, R., Mahajan, G., Srivastava, S., & Sinha, A. (2014). Green manuring: a boon for sustainable agriculture and pest management–a review. *Agricultural Reviews*, *35*(3), 196-206.

Kumar, R., Mishra, J. S., Rao, K. K., Mondal, S., Hazra, K. K., Choudhary, J. S., ... & Bhatt, B. P. (2020). Crop rotation and tillage management options for sustainable intensification of rice-fallow agro-ecosystem in eastern India. *Scientific reports*, *10*(1), 1-15.

Kumar, S., & Sukul, P. (2020). Green Manuring and Its Role in Soil Health Management. In *Soil Health* (pp. 219-241). Springer, Cham.

Kumar, V., & Singh, A. P. (2010). Long-term effect of green manuring and farmyard manure on yield and soil fertility status in rice-wheat cropping system. *Journal of the Indian Society of soil science*, *58*(4), 409-412.

Larkin, R. P., & Griffin, T. S. (2007). Control of soilborne potato diseases using Brassica green manures. *Crop protection*, *26*(7), 1067-1077.

Lazzeri, L., & Manici, L. M. (2001). Allelopathic effect of glucosinolate-containing plant green manure on Pythium sp. and total fungal population in soil. *HortScience*, *36*(7), 1283-1289.

Malińska, K., Golańska, M., Caceres, R., Rorat, A., Weisser, P., & Ślęzak, E. (2017). Biochar amendment for integrated composting and vermicomposting of sewage sludge–the effect of biochar on the activity of Eisenia fetida and the obtained vermicompost. *Bioresource technology*, *225*, 206-214.

Manici, L. M., Caputo, F., & Babini, V. (2004). Effect of green manure on Pythium spp. population and microbial communities in intensive cropping systems. *Plant and Soil*, *263*(1), 133-142.

N'Dayegamiye, A., Whalen, J. K., Tremblay, G., Nyiraneza, J., Grenier, M., Drapeau, A., & Bipfubusa, M. (2015). The benefits of legume crops on corn and wheat yield, nitrogen nutrition, and soil properties improvement. *Agronomy Journal*, *107*(5), 1653-1665.

Palaniappan, S., & Siddeswaran, K. (1994). Nitrogen uptake of rice as influenced by green manure, grain legumes and fertilizer N. In *SARP Research Proceedings* (p. 141).

Pandey, D. K., Pandey, R., Mishra, R. P., Kumar, S., & Kumar, N. (2008). Collection of Dhaincha (Sesbania spp.) variability in Uttar Pradesh, biodiversity and agriculture (souvenir). *Uttar Pradesh Biodiversity Board, Lucknow*, 48-51.

Park, S., Kim, K. S., Kim, J. T., Kang, D., & Sung, K. (2011). Effects of humic acid on phytodegradation of petroleum hydrocarbons in soil simultaneously contaminated with heavy metals. *Journal of environmental sciences*, *23*(12), 2034-2041.

Pieters, A. J. (1927). Green manuring. *Willey, New York.*

Pung, H., Aird, P. L., & Cross, S. (2004, February). The use of Brassica green manure crops for soil improvement and soil borne disease management. In *3rd Australian Soil borne diseases Symposium* (pp. 8-11).

Robertson, G. P., & Swinton, S. M. (2005). Reconciling agricultural productivity and environmental integrity: a grand challenge for agriculture. *Frontiers in Ecology and the Environment*, *3*(1), 38-46.

Sakakibara, M., Watanabe, A., Inoue, M., Sano, S., & Kaise, T. (2010, January). Phytoextraction and phytovolatilization of arsenic from As-contaminated soils by Pteris vittata. In *Proceedings of the annual international conference on soils, sediments, water and energy* (Vol. 12, No. 1, p. 26).

Sánchez-Navarro, A., Gil-Vázquez, J. M., Delgado-Iniesta, M. J., Marín-Sanleandro, P., Blanco-Bernardeau, A., & Ortiz-Silla, R. (2015). Establishing an index and identification of limiting parameters for characterizing soil quality in Mediterranean ecosystems. *Catena*, *131*, 35-45.

Sayara, T., Basheer-Salimia, R., Hawamde, F., & Sánchez, A. (2020). Recycling of organic wastes through composting: Process performance and compost application in agriculture. *Agronomy*, *10*(11), 1838.

Schumann, R. A., Meyer, J. H., & Van Antwerpen, R. (2000). A review of green manuring practices in sugarcane production. In *Proc S Afr Sug Technol Ass* (Vol. 74, No. 9, pp. 93-100).

Schwitzguébel, J. P. (2001). Hype or hope: the potential of phytoremediation as an emerging green technology. *Remediation Journal: The Journal of Environmental Cleanup Costs, Technologies & Techniques*, *11*(4), 63-78.

Schwitzguébel, J. P., Meyer, J., & Kidd, P. (2006). Pesticides removal using plants: phytodegradation versus phytostimulation. In *Phytoremediation rhizoremediation* (pp. 179-198). Springer, Dordrecht.

Selvi, R. V., & Kalpana, R. (2009). Potentials of green manure in integrated nutrient management for rice–a review. *Agricultural Reviews*, *30*(1), 40-47.

Sharifi, M., Lynch, D. H., Hammermeister, A., Burton, D. L., & Messiga, A. J. (2014). Effect of green manure and supplemental fertility amendments on selected soil quality parameters in an organic potato rotation in Eastern Canada. *Nutrient cycling in agroecosystems*, *100*(2), 135-146.

Sharma, A. R., & Ghosh, A. (2000). Effect of green manuring with Sesbania aculeata and nitrogen fertilization on the performance of direct-seeded flood-prone lowland rice. *Nutrient cycling in Agroecosystems*, *57*(2), 141-153.

Shibabaw, A., Alemayehu, G., Adgo, E., Germer, J., Asch, F., & Freyer, B. (2017). Growth and biomass yield response of clover (Trifolium decorum) to preceding crop and organic treatment in the highlands of Awi Administrative Zone, Ethiopia. *Ethiopian Journal of Science and Technology*, *10*(3), 151-164.

Shibabaw, A., Gelaw, B., Wang, S. H., & Tessema, B. (2018). Time to sputum smear and culture conversions in multidrug resistant tuberculosis at University of Gondar Hospital, Northwest Ethiopia. *PloS one*, *13*(6), e0198080.

Sikora, L. J., & Stott, D. E. (1996). Soil organic carbon and nitrogen. In 'Methods for assessing soil quality', SSSA Special Publication 49. In *Soil Science Society of America*.

Sima, N. A. K., Ebadi, A., Reiahisamani, N., & Rasekh, B. (2019). Bio-based remediation of petroleum-contaminated saline soils: Challenges, the current state-of-the-art and future prospects. *Journal of environmental management*, *250*, 109476.

Singh, R., Srivastava, P., Singh, P., Upadhyay, S., & Raghubanshi, A. S. (2019). Human overpopulation and food security: challenges for the agriculture sustainability. In *Urban agriculture and food systems: breakthroughs in research and practice* (pp. 439-467). IGI Global.

Sinha, S., Masto, R. E., Ram, L. C., Selvi, V. A., Srivastava, N. K., Tripathi, R. C., & George, J. (2009). Rhizosphere soil microbial index of tree species in a coal mining ecosystem. *Soil Biology and Biochemistry*, *41*(9), 1824-1832.

Smukler, S. M., Jackson, L. E., Murphree, L., Yokota, R., Koike, S. T., & Smith, R. F. (2008). Transition to large-scale organic vegetable production in the Salinas Valley, California. *Agriculture, Ecosystems & Environment*, *126*(3-4), 168-188.

Songjuan, G. A. O., Weidong, C. A. O., & Guopeng, Z. H. O. U. (2021). Bacterial communities in paddy soils changed by milk vetch as green manure: A study conducted across six provinces in South China. *Pedosphere*, *31*(4), 521-530.

Talgre, L., Lauringson, E., Roostalu, H., Astover, A., Eremeev, V., & Selge, A. (2009). The effects of pure and undersowing green manures on yields of succeeding spring cereals. *Acta Agriculturae Scandinavica Section B–Soil and Plant Science*, *59*(1), 70-76.

Tejada, M., Gonzalez, J. L., García-Martínez, A. M., & Parrado, J. (2008). Effects of different green manures on soil biological properties and maize yield. *Bioresource technology*, *99*(6), 1758-1767.

Wang, G., & Zhou, L. (2017). Application of green manure and pig manure to Cd-contaminated paddy soil increases the risk of Cd uptake by rice and Cd downward migration into groundwater: field micro-plot trials. *Water, Air, & Soil Pollution*, *228*(1), 1-15.

Wichelns, D., & Qadir, M. (2015). Achieving sustainable irrigation requires effective management of salts, soil salinity, and shallow groundwater. *Agricultural Water Management*, *157*, 31-38.

Willekens, K., Vandecasteele, B., Buchan, D., & De Neve, S. (2014). Soil quality is positively affected by reduced tillage and compost in an intensive vegetable cropping system. *Applied Soil Ecology*, *82*, 61-71.

Witkamp, M. (1971). Soils as components of ecosystems. *Annual review of Ecology and Systematics*, *2*(1), 85-110.

Yadav, R. L., Dwivedi, B. S., Prasad, K., Tomar, O. K., Shurpali, N. J., & Pandey, P. S. (2000). Yield trends, and changes in soil organic-C and available NPK in a long-term rice–wheat system under integrated use of manures and fertilisers. *Field Crops Research*, *68*(3), 219-246.

Yadava, A. S. (2001). *Studies on decomposing mycoflora of Crotalaria juncea L in soil* (Doctoral dissertation, Ph. D. Thesis).

Yulianti, T., Sivasithamparam, K., & Turner, D. W. (2006). Response of different forms of propagules of Rhizoctonia solani AG2-1 (ZG5) exposed to the volatiles produced in soil amended with green manures. *Annals of applied biology*, *148*(2), 105-111.

Zhang, Y., Brackman, G., & Coenye, T. (2017). Pitfalls associated with evaluating enzymatic quorum quenching activity: the case of MomL and its effect on Pseudomonas aeruginosa and Acinetobacter baumannii biofilms. *PeerJ*, *5*, e3251.

Chapter 7

Sustainable Management and Restoration of the Fertility of Damaged Soils

B. Raghavendra Goud[1,*], M. Raghavendra[2†], P. N. Siva Prasad[3], Veeresh Hatti[4], Hanamant M. Halli[5], G. V. Venkataravana Nayaka[6], G. Suresh[6], K. Sai Maheshwari[6], G. Adilakshmi[6], G. Prabhakara Reddy[7] and Sudhir K. Rajpoot[8]

[1]ICAR-National Rice Research Institute, Cuttack, Odisha, India
[2]ICAR-Indian Institute of Soybean Research, Indore, Madhya Pradesh, India
[3]SBVR Agricultural College, Badvel, ANGRAU, Andhra Pradesh, India
[4]Sardarkrushinagar Dantewada Agricultural University, Palanpur, Gujarat, India
[5]ICAR-National Institute of Abiotic Stress Management, Baramati, Pune, India
[6]Department of Agronomy, S.V. Agricultural College, Tirupati, AP, India
[7]Agricultural College, Mahanandi, ANGRAU, Andhra Pradesh, India
[8]Institute of Agricultural Sciences, BHU, Varanasi (UP), India

Abstract

Soil takes many years to form, however can be damaged in a very short amount of time. With the damage of soil, man's ability decreases to grow food crops, graze animals and to produce fiber and forests. Increasing agricultural development with population pressure and mounting demands created upon the soil ends up in a lot of stress on soil by accrued chance of harm. Human activities such as inappropriate agricultural management, overgrazing and deforestation lead to converting millions of productive lands to nonproductive. Various types of soil damage involve physical degradation (characterized by lowering in pore geometry and continuity, worsening soil's susceptibility to crusting and compaction, decrease in water infiltration, increase in surface runoff, wind and water erosion), chemical degradation (characterized by acidification, salinization, nutrient exhaustion, decreased cation exchange capacity, elevated nutrient toxicities and deficiencies, leaching of essential plant nutrients and industrial waste contamination) and biological degradation (characterized by decline of the soil organic carbon (SOC), biodiversity loss, depletion in soil C sink capability, and elevated greenhouse gas (GHG) emissions from soil into the atmosphere) are becoming ecological threats. Ecological degradation results in disturbance of different agro-ecosystem tasks such as nutrient recycling, purification and infiltration of water, disturbance of the hydrological cycle, and lessening in net biome productivity. Soil degradation can be restored by conversion to a

* Corresponding Author's Email: raghava0160@gmail.com.
† Corresponding Author's Email: raghavendra4449@gmail.com.

In: Ecosystem Services
Editor: Hanuman Singh Jatav
ISBN: 978-1-68507-614-6

restorative land use and practicing of recommended management practices to reduce soil erosion, produce positive SOC and N budgets, increase the soil biodiversity and microbial activity, and improve soil structural attributes. Ameliorating soil quality parameters declines soil degradation with associated improvement in environmental quality. The site-specific techniques of bringing back soil quality includes conservation agriculture, integrated nutrient management and continuous vegetative cover. The restoration of degradation varies according to soil types. Cultivation easily corrects soil crusting and compaction problems, whereas, salinization and alkalization are not readily corrected. Sodic/alkali soils can be corrected by imposition of drainage and application of gypsum. Depleted nutrients can be restored back by the use of organic and inorganic nutrient supplements, and acidity is corrected by liming. Trees are used as wind breaks to control soil erosion by the wind in drier areas of light textured soils. Contour bunds are designed to drain away excess runoff water. No-till farming, tied ridge cultivation and alternate strip cropping, mulching are other effective erosion control techniques. Reloading nutrients is essential for maintaining the productivity of the system. Crop rotation with legumes restore all essential soil nutrients and reduces erosion as the soil is covered with vegetation. Toxic elements can be eliminated by soil remediation using biological techniques and land-based treatments. Therefore, soil management through sustainable management practices and vegetation are crucial in the restoration of degraded lands for effective crop production under starved conditions.

1. Introduction

Land and soils represent the muse for sustainable agricultural development, critical environment capabilities and meals safety. They are key to sustain life on Earth. Soil degradation could be a twenty first century world downside that's particularly severe within the tropics and sub-tropics. Humans face soil degradation problem from the beginning of agriculture, however soil degradation extent and influence on humans is severe currently. It weakens the productive capacity of an ecosystem and affects global climate through changes in water, energy cycles and disturbs carbon and other nutrient cycles. Because of soil degradation influence on agricultural productivity and environment, it ends up in "increase in deforestation, additional use of marginal lands, runoff, soil erosion, pollution of water bodies, emission of greenhouse gases" etc. Therefore, the current rate of soil degradation threatens the capability of future generations to satisfy their wants. As our international population marches steadily towards projections of 9.5 billion in 2050, natural and human evoked soil degradation, if not alleviated, can without doubt increase the potential for negative impacts like disease and malnutrition (Sanchez and Swaminathan 2005). Feeding the exploding world population is extremely tough under soil degradation characterized by soil quality decline and decrease in ecosystem services. Land accessibility for increasing the cultivated area is finite and decreasing. Thus, the sole two viable alternatives for meeting the demands for food and fiber are by: (1) increasing the crop productivity per unit of the area, and (2) restoring degraded/marginal soils. Yield increments on existing harvest terrains will be a fundamental segment to expanded food creation, but by themselves will not suffice (Godfray et al., 2010; Hubert et al., 2010; Ray et al., 2013). Hence, immediate measures should be taken to restore the degraded ecosystems so that areas going out of cultivation can be diverted for increasing agricultural production. Therefore, in this chapter effort has made to discuss in detail related to damaged soils, causes and their restoration process.

2. Soil Damage/Soil Degradation

Soil damage is lessening of its ability to grow crops or natural vegetation under a particular climate (Tinker 1988). Intensive agriculture practices and population pressure have accelerated the demands on soils leading to greater stress on soils, with increased probability of damage. Soil damage with reference to crop production becomes synonymous with soil degradation. Further, soil degradation is defined as a change in the soil quality status resulting in a diminished capacity of the ecosystem to provide goods and services for its beneficiaries (FAO 2019). Therefore; it is the physical, chemical and biological decline in soil quality. Soil degradation further, leads to decline in soil productivity due to "the loss of organic matter, decline in soil fertility, and structural condition, erosion, adverse changes in salinity, acidity or alkalinity, and the effects of toxic chemicals, pollutants or excessive flooding." It is understood as "a change in the soil health status resulting in a diminished capacity of the ecosystem to provide goods and services for its beneficiaries. Deforestation, intensive use of marginal lands, intensive cultivation, indiscriminate use of chemicals, overgrazing and urbanization are resulting in soil degradation. Furthermore, we are able to say soil is degraded once soil cannot perform one or a lot of functions as given by (Lal 1997b) includes

i. Sustain biomass production and biodiversity together with preservation and improvement of the gene pool,
ii. Regulation of water and air quality
iii. Preserve archeological, geologic and astronomical records, and
iv. Support socio-economic structure, cultural and aesthetic values and supply engineering foundation

3. The Extent of Soil Degradation

Past examinations demonstrated that degradation diminished soil ecosystem services by 60 per cent between 1950 and 2010 (Leon and Osorio 2014). Also, GLASOD study assessed that almost 2 billion ha (22.5 per cent) of farming area, pasture, forest and woodland had been debased since the mid-twentieth century. Inline with Oldeman et al. (1990) roughly 2 percent of the soils severely degraded, 7 percent moderately degraded and 6 percent gently degraded. Further, Bot et al. (2000), reported that 26 percent was severely degraded, 21 percent moderately degraded and therefore the remaining 18 percent gently degraded. Accelerated soil degradation has apparently influenced as much as 500 million hectare (Mha) in the tropics (Lamb et al., 2005), and universally 33 per cent of earth's territory surface is littered with some type of soil degradation (Bini 2009). 23 per cent of the broad leaved forest, 19 per cent of the needle-leaved forests and 20-25 per cent of the rangelands are degraded (Bai et al., 2008). In India, soil degradation is calculable to be touching 147 Mha (Bhattacharya et al., 2015) and that is a serious concern for the country that supports 18 per cent of the world's human population and 15 per cent of the animal population with just 2.4 per cent of the global land area. Out of the complete degraded area (147Mha), 94 Mha is from water erosion, 16 Mha from acidification, 14 Mha from flooding, 9 Mha from wind erosion, 6 Mha from salinity, and 7 Mha from a mix of things. Therefore, total land not suited for agriculture constitutes 5.5 per cent.

4. Types of Soil Degradation and Causes

It takes many years to create soil but can be damaged in very short time. Increasing agricultural development and population pressure are increasing the demands made upon soils leading to greater stress on soils, with increased probability of damage. The reasons for soil degradation are natural and human incited. Natural causes include earthquakes, tsunamis, droughts, avalanches, landslides, volcanic eruptions, floods, tornadoes, and wildfires; though human-prompted soil degradation results from land clearing and deforestation, wrong farming practices, ill administration of industrial effluents and wastes, over-grazing, reckless management of forests, surface mining, populated area, and commercial/industrial development. Land use change is the major driver of soil degradation. Improper agricultural practices embrace "over tillage, utilization of hefty machinery, excessive and unbalanced use of inorganic fertilizers, poor irrigation management, chemical overuse, low application of crop residue and/or organic carbon inputs, and poor crop cycle planning." The basic social reasons for soil degradation are land deficiency, decrease in per capita land accessibility, economic pressure on land, land tenancy, poverty, and populace increment. There are several other factors that set-in motion various soil-degradative processes and have various impacts on the soil processes (Table 1).

Table 1. Soil Degradation types, causes and impacts on soil processes (Cheryl Palm et al., 2007)

Type	Causes	Degradation process	Impact on soil processes
Physical	• Deforestation • Burning of biomass • Excessive grazing • Tillage up and down the gradient, excessive animal, human and machine traffic	• Breakdown of soil structure, aggregation and porosity • Crusting and surface sealing • Compaction of top and subsoil • Reduction in extent and strength/stability of aggregates	• Reduction in infiltration capacity • Changes in soil water retention characteristics • Increase in runoff rate and amount • Accelerated erosion by wind and water • Increase in bulk density leading to a decrease in porosity • Waterlogging and anaerobiosis
Chemical	• Irrigation with poor quality water • Inadequate drainage • Little or excess use of fertilizers • Application of industrial, urban wastes	• Salinization • Alkalization • Acidification • Eutrophication • Toxification, contamination with heavy metals, pollution	• An accumulation of base-forming cations • Decreased levels of macronutrients on exchange sites, soil organic matter, and in soil solution • Leaching and runoff of nutrients into the water sources • Excessive buildup of some elements (e.g., Al, Mn, Fe) and heavy metals (Pb and Hg) • An increase in soil borne pathogens
Biological	• Removal and burning of residues • Little or no use of organic inputs • Monoculture • Excessive tillage	• Reduction of soil organic carbon • Depletion of soil biodiversity • Destruction of soil structure	• A reduction in N mineralization, soil aggregation and related properties • Shift in species composition and diversity of favorable soil organisms • Reduction in porosity and infiltration, reduction in activity of soil biota

There are three main types of degradation as listed in Table 1. Each type of soil degradation is further described in detail as follows.

4.1. Physical Degradation

It is the deterioration of physical properties of soil inflicting "reduction in structural attributes as well as pore geometry and continuity, therefore exacerbating soil's condition to crusting, compaction, reduced water infiltration, magnified surface runoff, wind and water erosion, bigger soil temperature fluctuations, and a magnified propensity for desertification."

4.1.1. Causes and the Effects of Soil Physical Degradation

The principle reasons of soil physical degradation are the breakdown of soil structure, wrong land use and soil management practices. All exploitatory practices can eventually bring about degradation and thus cut back soil productivity. In highly mechanized systems, weighty hardware and extreme culturing are significant elements of degradation. In the developing world, land not suitable for cultivation, similar to dry lands or steep parcel of land, is progressively being cropped. The cultivation and agriculture practices in the above land use systems are generally responsible for degradation. Atlast, soil physical degradation brings about reduced crop growth and soil productivity.

4.1.2. Types of Physical Degradation

4.1.2.1. Compaction

Compaction is defined as compression of soil mass into smaller volume, resulting from the reduction of soil pores. Bulk density is increased due to natural and human factors. Expanding mechanization and utilization of bigger and more proficient farm vehicles prompted continuous densification of soils and reduction in productivity. Soil vulnerability to compaction expanded because of inordinate use of inorganic fertilizers, excessive tillage and practices that have prompted a decrease of organic matter in soils. Compaction results in poor infiltration, high evaporative losses, less efficient use of fertilizer, higher runoff, erosion rates, higher fuel costs for tillage and for irrigation. Changes within in the bulk density result in modification in structural properties, thermal and hydraulic conductivity, gaseous transfer characteristics that impact chemical and biological balances. Further, utilization of heavy machinery in present day farming has created 68 Mha compacted arable land (Hamza and Anderson 2005). Compacted soils typically show a degraded structure with low (macro) porosity, and low pore continuity and connectivity (Horn et al., 1995). The structural degradation, which can last for decades (Berisso et al., 2013; Hakansson and Reeder 1994), antagonistically influences ecosystem services of arable soil, including crop productivity (Batey 2009; Hamza and Anderson 2005; Tracy et al., 2011). Low crop productivity on compacted soil results fundamentally from obstructed root frame work and hence restricted access to soil water and nutrients pools (Araki and Iijima 2005; Bengough et al., 2011; Valentine et al., 2012, Colombi et al., 2018; Nosalewicz and Lipiec 2014). There are two principle kinds of soil compaction that can be recognized for practical purposes, in particular the surface layer compaction and subsoil compaction.

1. *Surface layer compaction:* It is the compaction in the upper part of the soil profile, ordinarily the plough layer. Compaction in the surface layer is dynamic and changes fundamentally over the cropping season, expanding with increasing machinery passes over the field and diminishing again with primary tillage for seed bed preparation for the accompanying season. Adequate tillage effectively decreases soil compaction in the surface layer and its belongings.
2. *Sub-soil compaction:* It influences soils past the surface layer at depths >30 cm. Utilization of heavy machinery prompts sub-soil compaction. Swelling-shrinking, freezing-thawing and biological activities can lighten compaction partly. Sub soiling using specially designed equipment can at times reduce sub-soil compaction, however is very energy demanding sometimes, soils become denser than it was before sub-soiling because of the destabilization of the soil caused by the mechanical energy input from the sub-soiling operation. Compaction ought to be viewed as an irreversible, permanent form of degradation.

4.1.2.2. Sealing and Crusting

Seals and crusts are results of downpour and flooding on unprotected soil-surfaces. Under the effect of rain drops and the drenching impact of water, the bonds that hold the particles together become weak and the sealed aggregates will in general self-destruct. Singular particles become isolated. These particles become adjusted and the finer particles will in general be washed into the depressions of the surface. There they structure a slender (1-5 mm) and thick layer that obstructs the soil pores and seals the surface. These seals are normally very elastic. Typical characteristics of soil seals are that they do not break and cannot be taken out from the surface. Soil crusts are a lot thicker than seals (usually 5-20mm) and can be isolated effectively from the soil surface, and they break after drying. Crusts are commonly shaped on soils with high contents of non-swelling clay susceptible to dispersion. Seals that become hard after drying are likewise termed crusts. Soils with a high content of fine or very fine sand, or silt are prone to sealing and crusting. The presence of exchangeable sodium in the soil can improve clay dispersion and hence contribute to seal and crust formation.

4.1.2.3. Soil Erosion and Sedimentation

Soil erosion is a global issue because of economic and environmental impacts. It is the separation or breaking away of soil particles from a land surface by erosive agents like water or wind, and ensuing transportation of the detached particles to another area. Erosion is a natural process and is a basic factor in soil formation from rock parent material. Human activities are responsible for quickening erosion rates, generally by reducing or eliminating plant and residue cover. This exposes the soil to wind and water erosion forces, weakening the soil cohesive forces by tillage disturbance, and increasing erosive agents, particularly by activities that increase surface runoff. Wind erosion leads to desertification. Soil erosion impacts economically due to direct effects on crops or plants on- and off-site. The National Bureau of Soil Survey & Land Use Planning (NBSS & LUP 2005) reported that an area of 119.2 million hectares is suffering from soil erosion in India. The water eroded areas are most widely distributed in India and the severe wind eroded area lies in the extreme western sectors of the country. Lal (1997a) estimated the risks of global annual loss of food production due to accelerated erosion to be as high as 190 Mt yr^{-1} of cereals, 6 Mt yr^{-1} of soybean, 3 Mt $yr^{-1}$of pulses and 73 Mt yr^{-1} of roots and tubers. Soil loss due to erosion is 20-60 t/ha/year, 20 to 40 times higher than the rate of soil formation. GLASOD estimates soil erosion to be the most significant type of soil degradation, since water erosion influences 56 per cent and wind erosion 28 per cent of all land utilized by man. The most significant reasons of water erosion are deforestation (43%), overgrazing (29%), agricultural mismanagement (16%) and of wind erosion, overgrazing (60%), agricultural mismanagement (16%), overexploitation of natural vegetation (16%) and deforestation.

4.1.2.4. The Loss of Soil Structure

Soil structure describes the arrangement of primary particles into aggregates of different sizes and shapes and the associated pore spaces between them. Therefore, a structured soil is heterogeneous; where a degraded, structure less soil is homogeneous. Soil structure significantly influences all processes that take place in the soil (Gomez et al., 1999). Soil structure degradation following intensive agricultural activities, soil compaction, loss of structural stability and the formation of surface crusts give rise to the loss of continuity of elongated transmission pores, which reduces water transport, resulting in increased runoff and soil erosion. Conventional ploughing induced the more relevant modification of soil physical properties resulting in damage to soil structure. The negative aspects associated with this management system are the formation of surface crusts and plough pan at the lower cultivation limit. The formation of the plough pan and the decrease of porosity, in particular the continuous elongated pores, in the surface layers of conventionally tilled soil, besides a reduction of water movement, may also hamper root growth.

4.2. Chemical Degradation

Chemical degradation after erosion is the second most abundant form of soil degradation and as such poses a threat to our finite soil resource, as it tends to render it less usable. Soil chemical degradation has the characters of acidification, salinization, nutrient depletion, reduced cation exchange capacity (CEC), increased Al or Mn toxicities, Ca or Mg deficiencies, leaching of NO_3-N or other essential plant nutrients, or contamination by industrial wastes or by-products. In soils with low activity clays, excessive leaching of cations results in decline in soil pH and base saturation. Buildup of toxic chemicals and nutrient imbalance in the soils also results in chemical degradation of soils. The most important forms of chemical degradation are loss of nutrients and organic matter in South America and Africa and salinization in Asia. The main reasons for chemical soil degradation are agricultural mismanagement (56%), and deforestation (28%). Almost 240 M ha of the world soil is chemically degraded and it accounts 12 per cent of the total area affected by human induced soil degradation (Oldeman (1992). The chemical soil degradation can be assessed by using certain indicative parameters which are included in Table 2.

Table 2. Assessment of chemical soil degradation using chemical indicator parameters (Senjobi et al., 2013)

Indicator	***Degree of degradation (%)**			
	1	**2**	**3**	**4**
Content of the nitrogen element (multiple decrease)	> 0.13	0.10-0.13	0.08-0.10	< 0.08
Content of phosphorus element (mg kg^{-1})	> 8	7-8	6-7	< 6
Content of potassium element (cmol kg^{-1})	> 0.16	0.14-0.16	0.12-0.14	< 0.12
Content of readily soluble salts (increase by %)	< 0.20	0.20-0.40	0.40-0.80	> 0.08%
Content of ESP (increase by % of CEC)	< 10	0-25	25-50	> 50
Content of base saturation (decrease of saturation if more than 50%)	< 2.5%	2.5-5%	5-10%	> 10%
Excess salts (salinization) (increase in conductivity) mmho/cm/yr	< 2	2-3	3-5	> 5

(*1= none to slightly degraded soils, 2= moderately degraded soils, 3= highly degraded soils, 4= very highly).

4.2.1. Soil Acidity

Chemically productive soils have pH range of 6.5 to 7.5. Mineralogical make up (clay minerals, various metal oxides and hydroxides, lime etc), organic matter content of soils and dissolved CO_2 in the aqueous phase determines the soil pH. Dissolution of soil minerals and application of ammonium fertilizers also can lead to soil acidity. Soil acidity limits the plant growth by toxicity and decreases the macronutrient base cation content. Furthermore, the solubility of Fe, Mn, and Al containing minerals are enhanced at low pH level and the toxicity of these elements becomes a most important problem. The activities of soil organisms, including nitrifying bacteria, are severely restricted at pH levels lower than 5.5.

4.2.2. Soil Salinity

Salinity is a common problem in arid and semi-arid areas where evapo-transpiration exceeds rainfall. Under these conditions there is not enough water to wash the soluble salts down the profile below the rooting zone. Thus, soluble salts originating from various sources accumulate in soil profile at certain depths – known as the salic horizon – or at the soil surface, depending on the water regime. If not washed from the soil profile to a drainage system by the leaching fraction of water, the concentration of the chloride, sulfate, carbonate and bicarbonate salts of Na may increase in soil profile and cause salinity in a very short period. Another cause of salinity is the absence of a drainage system or poor drainage especially in the lowlands. A raised water table as a result of an ineffective discharge system is still another cause of salinity. The upward capillary movement of water carries the dissolved salts, previously present at a depth to the rooting zone. The source of soluble salts besides irrigation water are "mineral weathering, fertilizers, salts used on frozen roads, atmospherics transfer of sea spray, and the lateral movement of ground water from salt containing areas." Salinity have an effect on plant growth and development by distressing water and nutrient uptake, and owing to ion toxicity due to an excess amount of Na^+ and Cl^-(Ashraf 1994; Ashraf and Khanum 1997; Negrao et al., 2017). Salinity stress causes elevated levels of ethylene that inhibits root growth and therefore affect a number of plant physiological pathways (Belimov et al., 2002; Sun et al., 2007). The dissolved salts in water increase the osmotic potential, thereby creating the so called physiological drought. Toxicity develops when the ions absorbed from the soil solution accumulate in leaves. As water is lost in transpiration, the concentration of toxic ions increases and causes damage to various degrees, depending on the sensitivity of plants. Salinity affects the mineral nutrition of plants by reducing the availability and uptake of nutrients through the interaction of Na and Cl with nutrient cations and anions and by interfering with transport of elements inside the plant. Salinity is measured as electrical conductivity of soil saturation paste or extract and is reported as deci-Siemens per meter ($dS\ m^{-1}$).

4.2.3. Soil Alkalinity

Accumulation of salts in soil increases the concentration of Na in soil solution above those of Ca and Mg and alters the composition of the replace phase in favor of Na, because the Na salts are the most soluble salts in nature. Akali or sodic soils are degraded by a process called sodification or alkalization, in which there will be an increase of Na ions in the exchangeable sites. The measure of soil alkalinity is exchangeable sodium percentage, which is the ratio of Na_x to cation exchange capacity. This parameter is sometimes expressed as the exchangeable sodium ratio, which is the ratio of Na_x to other exchangeable cations. If the soil solution contains CO_3^{2-} and HCO_3^- in excess of Ca^{2+} and Mg^{2+} ions, then highly soluble Na salts of these anions hydrolyse and the soil pH rises above 8.5. This process is termed as alkalinization. Sodicity although a chemical property, has adverse effects on soil structure. With increase in sodium ions, dispersing action of Na ions reduces the binding effects of Ca and Mg ions on clay particles, the binding effects of divalent Ca and Mg ions on clay particles are overcome by the dispersing action of Na ions. Dispersed particles move with water and quickly clog the soil pores, causing drastic reductions in water and air permeability in sodic non-saline soils.

4.2.4. The Loss of Plant Nutrients

Crop removal, erosion, leaching, and volatilization results in continuous loss of plant nutrients from soils. The rate of loss is determined by the type of vegetative cover, the cropping system practised and climatic conditions of the area. In grasslands and forests, falling plant debris and mineral weathering compensates nutrient losses. On the other hand, in intensive agriculture, much larger amounts of nutrients are removed from the soil with little return in crop residues, in many cases exhausting nutrient reserves in soils. Erosion permanently removes the readily available nutrients from the topsoil, which is the most productive layer. Leaching can be significant for anionic nutrients, especially for nitrate under various farming systems. Basic nutrient cations such as Ca, Mg and K may be leached from soils under acidic conditions.

4.2.5. Soil Pollution

Industrialization and urbanization have negative impact on quality of soil via chemical contaminants, use of polluted waters for irrigation, and deposition of harmful particulates into the atmosphere. These contaminants and their interactions in the environment threaten the ecosystems. The sources of pollution are: (1) the waste waters; (2) agricultural wastes, (3) airborne pollutants, (4) pesticides, (5) the urban wastes: a) sewage sludge; (b) compost (c) fly ash; and (6) industrial wastes: (a) pesticides and fungicides; (b) fertilizers; (c) detergents; and (d) chlorinated solvents. These sources may cause fatal effects and/or irreversible destruction on health of human beings and the environment. The levels of contamination together with their mobility and capability in decomposition and accumulation of chemicals in the soil destroys the natural structure of water, soil and air and harms animals, plants, and micro-organisms. Inactivated enzymes are the measure of heavy metal (divalent) toxicity readily reacting with proteins, amines and sulfhydryl (-SH) groups as well as Hg and Cd replacing Zn in metallic enzymes. Metal toxicity is known to decrease cell membrane permeability and change genetic characteristics of cells, leading to increased risk of cancer. Mercury accumulates in fatty tissues as methyl Hg, whereas cadmium replaces calcium in bones and kidneys destroying their excretory function. Metals such as lead, arsenic, mercury, chromium, cadmium and copper are found at acceptable limits in nature. Further, pollutants and heavy metals such as persistent chlorinated organic compounds and Hg could cause genetic side effects, great care and effort should be paid to prevent the transportation of said substances to soil for sustaining biodiversity. Hence, in the course of the sustainable ecosystem concept, the status of pollutants in soil should be monitored permanently.

4.3. Biological Degradation

Depletion of organic matter, soil organic carbon (SOC) pool, loss of soil biodiversity and soil microbial activity, a decline in soil C sink capacity, and augmented greenhouse gas (GHG) emissions from soil into the atmosphere resulted in soil biological degradation. One of the most brutal penalties of soil biological degradation is that the soil becomes a net source of GHG emissions (CO_2 and CH_4) more willingly than a sink. In addition, accumulation of the soil organic matter, namely humus, in soils starts with the production of biomass and approaches equilibrium dependent on the effects of factors such as the climate, type of vegetation, topography, drainage conditions and soil texture. At equilibrium, when additions by biomass and removal by mineralization are in balance, organic matter contents range from less than 1 per cent in arid regions to over 20 per cent in organic deposits of cool humid climates. Any change in these factors may disturb the system and a new equilibrium towards depletion of soil organic matter results. A highly disturbing factor in this respect is cultivation. Less organic material is returned to soil at harvest in most cropping systems, and tillage hastens the organic matter decomposition in the soil. The conversion of forests and grasslands to crop lands promotes a rapid decomposition of the organic matter present in soils. Soil erosion is another factor that causes significant reductions in soil organic matter content.

5. Impacts of Soil Degradation

Degradation of soil outcome in the loss of critical functions and ecosystem services like production of food, feed, fuel and fiber, ensuring sufficient supplies of clean water, providing a platform for the built environment, acting as a buffer against extreme climatic events, supporting biodiversity, and providing the largest terrestrial store of carbon and nutrients (Janzen et al., 2011). Soil degradation decreases soil productivity by affecting the supply of nutrients and water for plant growth. Deforestation, intensive cropping and overgrazing leads to rapid mineralization of organic matter. Structural deterioration and the erosion cause removal of nutrients from topsoil and deposition in offsite, leading to pollution and eutrophication in water bodies. The direct impact due to soil degradation is depicted in figure 1. Furthermore, soil degradation diminishes the SOC pool, alongside it, plant available N and other essential nutrients like P and S. Carbon sediment is ejected into the atmosphere as CO_2. In addition to harmfully impacting agronomic production, it can also diminish economic growth, particularly in countries where agriculture is the appliance for economic development (Scherr 2001). Over and more the environmental and economic collision, there are also health risks of soil erosion (Guerra et al., 2005) and other degradation processes (Lal 2009).

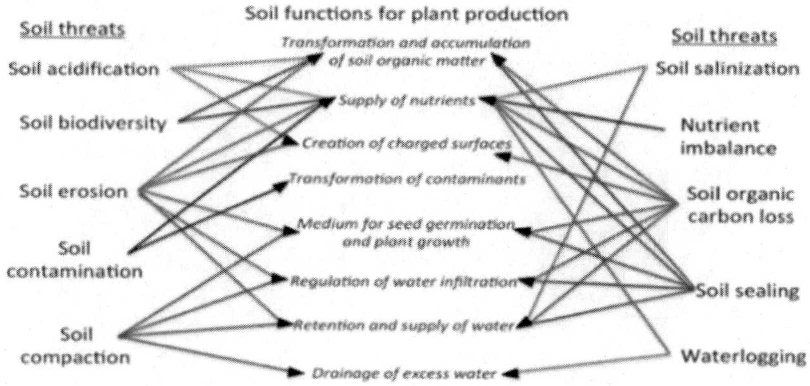

Figure 1. Direct impacts of soil threats on specific functions of soil in relevance to crop production (FAO and ITPS 2015).

6. Degradation: Reversible or Irreversible

All types of soil degradations are able to occur very easily. However, a remedy may be either slow and expensive or completely impracticable. Symptoms of degradation of the topsoil include crusting, low fertility, etc. These problems can usually be alleviated by "increasing the soil organic matter content and adding calcium compounds such as gypsum $CaSO_4$ or lime" if the soil is acidic. Organic matter can be increased by changing crop rotations to include a pasture phase, by keeping plant cover, all the year round, by adding manure, and also by minimizing losses of soil organic matter, which can occur as a result of excessive or high intensity tillage. Organic matter increases soil aggregation and reduces clay dispersion in water. Calcium compounds flocculate the clay, thereby reducing dispersion and increasing soil stability. The building blocks for bigger complex particles, such as aggregates, appear to be clay floccules. The use of such practices over a number of years can produce a soil structure that is better for agricultural production and is more stable. Compaction damage of top soils is usually not permanent as a result of the ameliorative special effects of tillage, biological activity, and cycles of wetting and drying (Figure 2).

Degradation of the subsoil often occurs through compaction. Subsoil tillage may make a temporary improvement, but often the soil will recompact to be as bad as or worse than it was before subsoiling. Soil salinity can be reclaimed by leaching. However, dryland salinity usually cannot be reversed but may be slowed or halted by planting of deep-rooted plants (e.g., salt tolerant trees) that can transpire enough water to stop further water table rise. Sodicity can be reclaimed by displacing the sodium with calcium. However, this is difficult as sodic soils have an extremely low permeability to water. Displacement and leaching may be accelerated by not using pure water, "but using water with a high enough electrolyte concentration" to maintain the clay flocculated and hence high permeability.

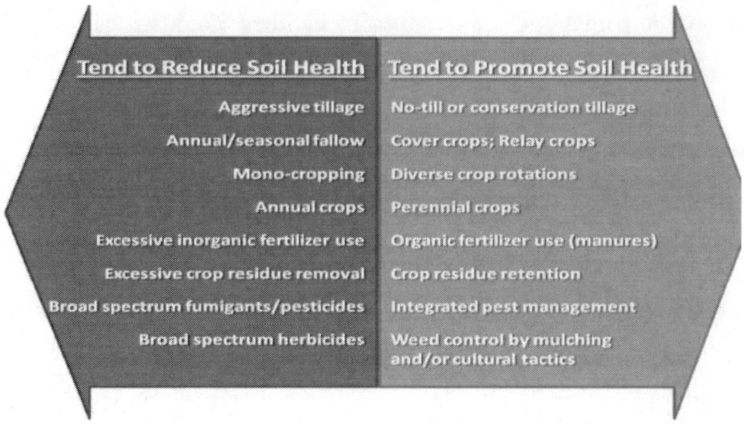

Figure 2. The impact of agricultural practices on soil health.

7. Restoration of Damaged Soils

Restoration refers to the method of repairing the degraded soils (eroded agricultural soils, mined soils) to a condition almost like the predegradation level of capability for sustaining plant growth and maintaining environmental quality. The degraded soils should be restored to meet the increasing food demands, reducing the pressure on prime agricultural soils, improving the value of degraded soils and enhance wildlife habitats, reducing pollution of water resources, increasing economic development and agricultural sustainability and avoiding risks of irreversible soil degradation. Evaluation of the magnitude of soil degradation, identification of degradative processes, and implementation of appropriate mitigation technologies are vital strategies for restoring degraded soils. A reversal may not be possible in strongly degraded soils. Severely eroded, nutrient-depleted, poorly drained, saline and sodic, compacted, and acid soils require restoration. Appropriate land use and judicious management of soil and crop have a beneficial effect on soil resilience and can repair degraded soils. Restoration of degraded soils must follow the following principles: Increase in soil organic matter content, improvement in structural properties of the soil, runoff management, and protection of exposed soil surfaces.

Further, Lal (2015) summarized strategies to reverse degradation and concluded that they include reduction of soil erosion, creating a positive soil C budget, improving nutrient availability, increasing soil biodiversity, and enhancing rhizosphere processes (Figure 3). However, Lal (2015) stressed the importance of managing soil organic C as other soil properties are coupled with soil organic C. There is a strong association between soil restoration, carbon sequestration, food security and biodiversity. The time required to restore degraded soils varies with ecosystem conditions, the severity of degradation, restorative strategies, and climate. A degraded soil without topsoil loss can recover within 3–5 years under appropriate management. Severely degraded soils require as long as 100 years to suit for agriculture (Blanco and Lal 2010). The site-specific and suitable land management practices should be identified to renovate physical, biological, chemical and ecological components of soil quality (Gregory et al., 2012). Measures to restore, mitigate and improve degraded soils is discussed under strategies mentioned (Figure 4) and discussed in detail above.

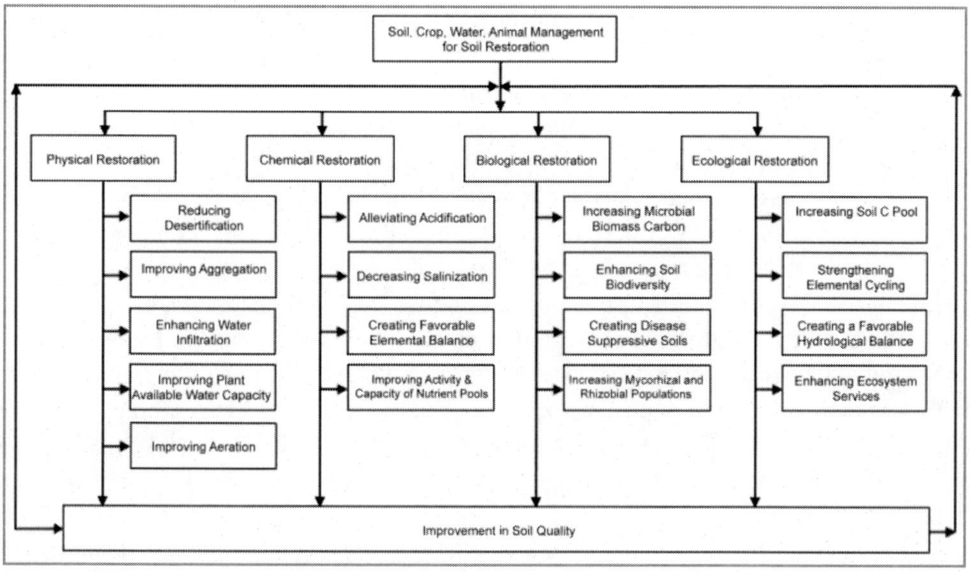

Figure 3. Soil restoration management (Lal 2015).

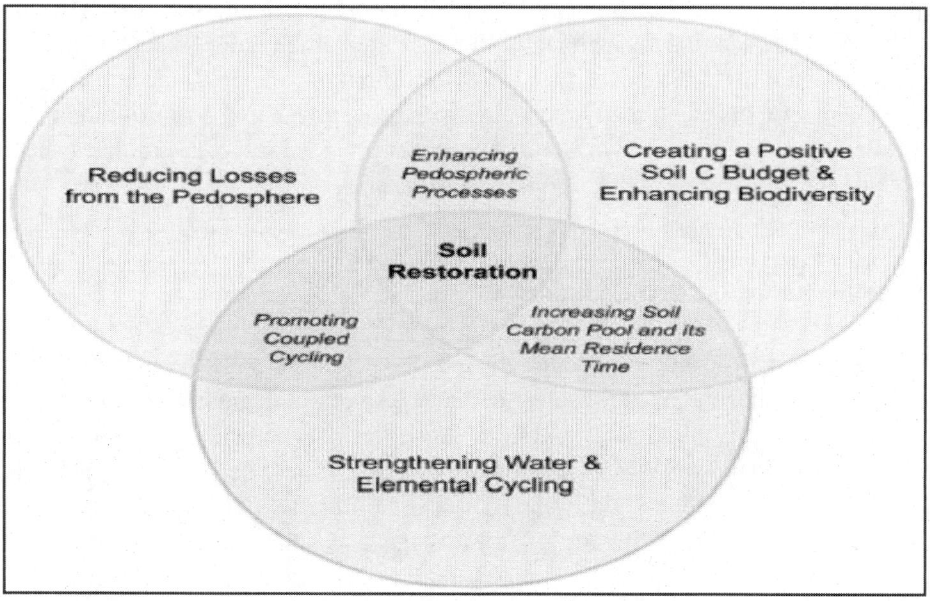

Figure 4. Strategies of restoring and managing soil quality (Lal 2015).

7.1. Management of Soil Erosion

Choosing of appropriate measures to control soil erosion within a tolerable limit is an important component of sustainable soil management (SSM) (FAO 2017). The Voluntary Guidelines for Sustainable Soil Management (VGSSM) (FAO 2017) present four groups of actions that can be taken to control soil erosion. First group is to lower land use changes. The 2nd and 3rd groups of measures are to reduce erosion and involve soil surface protection from erosion, reducing runoff volume and velocity on hill gradients. Measures such as reduced tillage protect the surface and decrease runoff, whereas terrace construction and maintenance reduces runoff. Further, conservation tillage is the most extensively practiced measure (Derpsch et al., 2010) to reduce soil erosion. In reduced tillage, residues are retained on the soil's surface. "No-till reduces soil loss by 60 per cent for regions with temperate climates, but there was no significant difference in soil loss for subtropical and tropical climates" (Mhazo et al., 2016). In temperate climates, precipitation runoff reduced by 33 per cent, but was significantly higher in subtropical and tropical climates. Mulching and other vegetative measures acts as a permanent or semi-permanent protective soil surface cover and is widely used in fire-affected areas, rangelands as well as in agricultural settings. Prosdocimi et al. (2016) observed that mulching reduces the average sediment concentration, soil loss, runoff quantity and depth over to control plots. The soil loss reduction differed with the technique used to assess it. The average decrease in sediment concentration across rainfall simulation studies was 68.9 percent, whereas it was 48.8 percent for run-off plot, silt fence and sediment trap methods. Sediment trapping and terraces can be achieved using vegetative measures such as grass strips, shrub and tree barriers (Mekonnen et al., 2015). These measures are beneficial and can be readily adopted using local grass and shrub species. Terraces are commonly used structural measures to reduce runoff velocity and depth reduction (Arnaez et al., 2015; Mekonnen et al., 2015; Wei et al., 2016). The prevention and control of gullies is a challenge for sustainable soil management, since measures should be adopted throughout the drainage area of the gully (Valentine et al., 2005). Generally, a perennial grass cover will increase infiltration, thereby reducing runoff to the gully, while plant roots can directly reduce gully expansion. Hence, the institution of grass or forest cover in the drainage area will slow gully expansion. Physical measures like check dams, stone bunds and enclosures can also slow gully expansion.

7.2. Creating a Positive Soil C Budget

The single most significant soil quality indicator for nearly all soils throughout the globe is SOM. It is also one of the most prevalent deficiencies seen in degraded soils, contributing to the various chemical, biological, physical properties and processes it influences. Soil degradation subordinate soil organic carbon, as well as it should be maintained at a threshold level of 1.1%-1.5% by soil weight in the root zone in order to avoid soil and environmental degradation risks (Manlay et al., 2007). Soil management options like conservational tillage and multifunctional and integrated cropping rotations that include forage legumes with small grains are suggested to enhance SOM and improve soil quality (Marriott and Wander 2006; Pikul et al., 2008; Singer et al., 2004; Teasdale 2007). Cover crops extend the duration of the photosynthetic capture in annual crop rotations, as a result rising organic C input in the soil and the soil C sequestration – an important practice for re-establishing degraded soils and address rising apprehension regarding global climate changes. Conservation tillage augment surface soil organic carbon content compared to conventional tillage (Cambardella et al., 2000; Franzluebbers, 2010; Johnson et al., 2005).

7.3. Improving Nutrient Availability

Soil fertility provides the base for wholesome food production and resilient and sustainable livelihoods. In the past, mainly inorganic fertilizers were used for improving soil fertility and productivity (Shah and Wu 2019). But, in degraded soils which possess low organic matter, a focus on inorganic fertilizer use alone cannot improve soil fertility. Cultivation of green manure crops and incorporation in the field improve both organic matter and N content of the soil, thereby improving soil fertility (Stewart et al., 2020). The use of manure and residues of crops will increase the organic matter in the soil and thereby fertility of the soil is increased (Madar et al., 2020). Cultivation of cover crops reduces nutrient loss by controlling soil erosion. Provision of drainage in waterlogged soils which otherwise reduce aeration, in turn inhibit microbial activity and mineralization of nutrients. Crop rotation with legumes helps to restore soil nutrients and reduces erosion by keeping the soil covered by vegetation (Meena and Lal 2018).

7.4. Increasing Soil Biodiversity

Soil microorganisms play a key role in improving soil quality and in reduction and restoration of soil degradation. Soil micro-organisms play key role in important ecosystem functions like "organic matter decomposition, nutrient cycling, water quality, regulating CO_2 concentration in the ambience creating disease suppressive soils" etc. The existence of termites, earthworms and other soil biota are widely recognized as important indicators of soil quality in tropical soils (Lal 1987; Ayuke et al., 2012, Madar et al., 2020a). The flourishes of organic matter by incorporating plant residues provide carbon and vigor sources for soil micro-organisms (Raghavendra et al., 2017; Raghavendra et al., 2020). An increase in soil organic matter content improves different soil parameters like soil aeration, temperature, moisture, and aggregate stability, and provides a resilient reserve base for various soil micro-organisms while preserving a rich and diversified source of organic matter and proper supply of nutrients. Farming practices like conservation tillage, cover cropping, cultivation of perennial crops, surface retention of crop residues, crop rotation, the use of organic nutrient resource and implementation of integrated pest management practices modify the overall soil biological community (number, species diversity, biomass and activity) in a typically positive manner (Altieri 1999; Jansa et al., 2006; Moonen and Barberi 2008; Dias et al., 2015; Raghavendra et al., 2018).

7.5. Enhancing Rhizosphere Processes

The rhizosphere is of predominant importance for ecosystem services like carbon and water cycling, nutrient retention, agricultural production, carbon uptake and storage (Adl 2016). Rhizosphere act as an interface between roots and soils and additionally as a centre for interactions among plants, soils and microorganisms. The biological and chemical processes taking place in the rhizosphere determine mobilization and acquisition of soil nutrients and even control NUE by crops (Zhang 2006). Root growth and rhizosphere processes have a good influence on soil nutrient transformation, mobilization, and profitable use in the plants (Hinsinger 2001; Hinsinger et al., 2009; Zhang et al., 2010). Therefore, different rhizosphere properties can be improved by adopting conservation residue retention, tillage practices and addition of organic fertilizer in different crops and cropping systems (Shinwari et al., 2015, Sharma et al., 2016, Madar et al., 2020).

Conclusion

Degradation of fertile soil is proceeding in no time as compared to restoration. All types of soil use destroy soil structure and reduce fertility. Further, speedy population growth and the consequential intensification of land use is one among the majority reasons for increasing soil degradation. Therefore, soil being the most basic natural resource, its sustainable management is essential to achieve global food security. Hence, this could be only possible by minimizing soil erosion, creating positive SOC, nutrient cycling, improved soil fertility, and species miscellany of soil biota under degraded soils. Through intervention of assorted good agriculture management practices, we will be able to further mitigate and restore degraded soils in different ecosystems.

References

Adl, S. (2016). Rhizosphere, food security, and climate change: a critical role for plant-soil research. *Rhizosphere*. 1, 1-3.

Altieri, M. A., 1999. The ecological role of biodiversity in agroecosystems. *Agric. Ecosyst. Environ*. 74, 19–31.

Araki, H. and Iijima, M. (2005). Stable isotope analysis of water extraction from subsoil in upland rice (*Oryza sativa* L.) as affected by drought and soil compaction. *Plant Soil*. 270, 147–157.

Arnaez, J., Lana-Renault, N., Lasanta, T., Ruiz-Flano, P., Castroviejo, J. (2015). Effects of farming terraces on hydrological and geomorphological processes. A review. *Catena*. 128, 122-134.

Ashraf, M. (1994). Breeding for salinity tolerance in plants. *Crit. Rev. Plant Sci*. 13, 17-42.

Ashraf, M. and Khanum, A. (1997). Relationship between ion accumulation and growth in two spring wheat lines differing in salt tolerance at different growth stages. *J. Agron. Crop Sci*. 178, 39-51.

Ayuke, F., Karanja, N., Okello, J., Wachira, P., Mutua, G., Lelei, D., Gachane, C., Hester, R., Harrison, R. (2012). Agrobiodiversity and Potential Use for Enhancing Soil Health in Tropical Soils of Africa. *Soils Food Secur*. 35, 94-134.

Bai, Z. G., Dent, D. L., Olsson, L., Schaepman, M. E. (2008). Proxy global assessment of land degradation. *Soil Use Manag*. 24(3), 223-234.

Batey, T. (2009). Soil compaction and soil management – a review. *Soil Use and Management*. 25, 335–345.

Belimov, A. A., Safronova, V. I., Mimura, T. (2002). Response of spring rape (*Brassica napus var. oleifera* L.) to inoculation with plant growth promoting rhizobacteria containing 1-aminocyclopropane-1-carboxylate deaminase depends on nutrient status of the plant. *Can. J. Microbiol*. 48, 189-199.

Bengough, A. G., McKenzie, B. M., Hallett, P. D., Valentine, T. A. (2011). Root elongation, water stress, and mechanical impedance: a review of limiting stresses and beneficial root tip traits. *J. Exp. Bot*. 62, 59–68.

Berisso, F. E., Schjonning, P., Keller, T., Lamande, M., Simojoki, A., Iversen, B. V., Alakukku, L., Forkman, J. (2013). Gas transport and subsoil pore characteristics: anisotropy and long-term effects of compaction. *Geoderma*. 195–196, 184–191.

Bhattacharyya, R., Ghosh, B. N., Mishra, P. K., Mandal, B., Rao, C., Sarkar, D., Das, K., Anil, K. S., Lalitha, M., Hati, K. M. (2015). Soil Degradation in India: Challenges and Potential Solutions. *Sustainability*. 7, 3528–3570.

Bini, C. (2009). Soil: A precious natural resource. In *Conservation of Natural Resources*; Kudrow, N. J., Ed.; Nova Science Publishers: Hauppauge, NY, USA, pp. 1–48.

Blanco, H. and Lal, R. (2010). Principles of Soil Conservation and Management, C Springer Science+Business Media B.V.

Bot, A. J., Nachtergaele, F. O., Young, A. (2000). Land resource potential and constraints at regional and country levels (L. a. W. D. Division, Trans.) Food and Agriculture Organization of the United Nations, Rome.

Cambardella, C. A., Johnson, J. M. F., Varvel, G. E. (2012). Soil carbon sequestration in central U.S. Agroecosystems. In *Managing Agricultural Greenhouse Gases: Coordinated Agricultural Research through Gracenet to Address Our Changing Climate*; Liebig, M. A., Franzleubbers, A. J., Follet, R. F., Eds.; Academic Press, Elsevier: San Diego, CA, USA, 2012; pp. 41–58.

Cheryl Palm, Pedro Sanchez, Sonya Ahamed, Alex Awiti. (2007). Soils: A Contemporary Perspective. *Annu. Rev. Environ. Resour.* 32, 99–129.

Colombi, T., Torres, L. C., Walter, A., Keller, T. (2018). Feedbacks between soil penetration resistance, root architecture and water uptake limit water accessibility and crop growth – a vicious circle. *Sci. Total Environ.* 626, 1026–1035.

Derpsch, R., Friedrich, T., Kassam, A., Hongwen, L. (2010). Current status of adoption of no-till farming in the world and some of its main benefits. *Intl Journal of Agricultural and Biological Engineering*. 3, 1-26.

Dias, T., Dukes, A., Antunes, P. M. (2015). Accounting for soil biotic effects on soil health and crop productivity in the design of crop rotations. *J. Sci. Food Agric*. 95(3), 447-454.

FAO and ITPS. (2015). Status of the World's Soil Resources (SWSR) – Main Report. Food and Agriculture Organization of the United Nations and Intergovernmental Technical Panel on Soils, Rome, Italy.

FAO, *Soil Degradation*, FAO, Rome, Italy, 2019, http://www.fao.org/soils-portal/soil-degradation-restoration/en.

FAO. (2017). Voluntary guidelines for sustainable soil management. Food and Agriculture Organization of the United Nations. Rome. 26pp.

Franzluebbers, A. J. (2010). Achieving soil organic carbon sequestration with conservation agricultural systems in the southeastern United States. *Soil Sci. Soc. Am. J.* 74, 347–357.

Godfray, H. C. J., Beddington, J. R., Crute, I. R., Haddad, L., Lawrence, D., Muir, J. F. (2010). Food security: the challenge of feeding 9 billion people. *Science*. 327(5967), 812-818.

Gomez, J. A., Giraldez, J. V., Pastor, M., Fereres, E. (1999). Effects of tillage method on soil physical properties, infiltration and yield in an olive orchard. *Soil Till. Res.* 52, 167–175.

Gregory, P., Hester, R., Harrison, R. (2012). Soils and Food Security: Challenges and Opportunities. *Soils Food Secur.* 35, 1-30.

Guerra, A., Marcal, M., Polivanov, H., Lima, N., Souza, U., Feitosa, A., Davies, K., Fullen, M. A., Booth, C. A. (2005). Environment management and health risks of soil erosion gullies in Sao Luiz (Brazil) and their potential remediation using palm-leaf geotextiles. In *Environmental Health Risk II*; WIT Press: Southampton, UK, pp. 459–467.

Hakansson, I. and Reeder, R. C. (1994). Subsoil compaction by vehicles with high axle load extent, persistence and crop response. *Soil Tillage Res.* 29, 277–304.

Hamza, M. A. and Anderson, W. K. (2005). Soil compaction in cropping systems. *Soil Tillage Res.* 82, 121–145.

Hinsinger, P. (2001). Bioavailability of soil inorganic P in the rhizosphere as affected by root-induced chemical changes: a review. *Plant and Soil*. 237, 173-195.

Hinsinger, P., Bengough, A. G., Vetterlein, D., Young, I. M. (2009). Rhizosphere: biophysics, biogeochemistry and ecological relevance. *Plant and Soil*. 321, 117–152.

Horn, R., Domzal, H., Slowihka-Jurkiewicz, A., Van Ouwerkerk, C. (1995). Soil compaction processes and their effects on the structure of arable soils and the environment. *Soil Tillage Res.* 35, 23–36.

Hubert, B., Rosegrant, M., van Boekel, M. A., Ortiz, R. (2010). The future of food: scenarios for 2050. *Crop Science*. 50(Suppl. 1), S-33.

Jansa, J., Wiemken, A., Frossard, E. (2006). *The Effects of Agricultural Practices on Arbuscular Mycorrhizal Fungi;* Geological Society of London: London, UK, 2006.

Janzen, H. H., Fixen, P. A., Franzluebbers, A. J., Hattey, J., Izaurralde, R. C., Ketterings, Q. M., Lobb, D. A., Schlesinger, W. H. (2011). Global Prospects Rooted in Soil Science. *Soil Sci. Soc. Am. J.* 75, 1–8.

Johnson, D. W., Verburg, P. S. J., Amone, J. A. (2005). Soil extraction, ion exchange resin, and ion exchange membrane measures of soil mineral nitrogen during incubation of a tallgrass prairie soil. *Soil Sci. Soc. Am. J.* 69, 260–265.

Lal, R. (1987). Tropical Ecology and Physical Edaphology; John Wiley Sons: Chichester, UK.

Lal, R. (1997a). Degradation and resilience of soils. *Phil. Trans. R. Soc. Lond. B.* 352, 997-1010.

Lal, R. (1997b). Soil Quality and sustainability. Pp 17-30. In "*Rattan Lal, Winfried E. H. Blum, C. Valentin, Bobby A. Stewart. (1997). Methods for Assessment of Soil Degradation. Advances in Soil Science.* Published by Taylor and Francis.

Lal, R. (2009). Soil degradation as a reason for inadequate human nutrition. *Food Sec. 1*, 45–57.

Lal, R. (2015). Restoring Soil Quality to Mitigate Soil Degradation. *Sustainability*. 7, 5875-5895.

Lamb, D., Erskine, P., Parrotta, J. (2005). Restoration of degraded tropical forest landscapes. *Science*. 310, 1628–1632.

Leon, J., Osorio, N. (2014). Role of Litter Turnover in Soil Quality in Tropical Degraded Lands of Colombia. *Sci. World J.13*.

Madar, R., Singh, Y. V., Meena, M. C., Das, T. K., Gaind, S., Verma, R. K. (2020). Potassium and Residue Management Options to Enhance Productivity and Soil Quality in Zero Till Maize–Wheat Rotation. *CLEAN–Soil, Air, Water. 48*(3), 1900316.

Madar, R., Singh, Y. V., Meena, M. C., Das, T. K., Paramesh, V., Al-Mana, F. A., Mohamad A., O Elansary, H. (2020a). Residue and Potassium Management Strategies to Improve Crop Productivity, Potassium Mobilization, and Assimilation under Zero-Till Maize–Wheat Cropping System. *Agriculture*. 10(9), 401.

Manlay, R., Feller, C., Swift, M. (2007). Historical evolution of soil organic matter concepts and their relationships with the fertility and sustainability of cropping systems. *Agric. Ecosyst. Environ.* 119, 217–233.

Marriott, E. E. and Wander, M. M. (2006). Total and labile soil organic matter in organic and conventional farming systems. *Soil Sci. Soc. Am. J.* 70, 950–959.

Meena, R. S., Lal, R. (2018). Legumes and sustainable use of soils. In *Legumes for Soil Health and Sustainable Management* (pp. 1-31). Springer, Singapore.

Mekonnen, M., Keesstra, S. D., Stroosnijder, L., Baartman, J. E. M., Maroulis, J. (2015). Soil conservation through sediment trapping: a review. *Land Degradation and Development*. 26(6), 544-556.

Moonen, A. and Barberi, P. (2008). Functional biodiversity: An agroecosystem approach. *Agric. Ecosyst. Environ.* 127, 7–21.

Negrao, S., Schmockel, S. M., Tester, M. (2017). Evaluating physiological responses of plants to salinity stress. *Ann. Bot.* 119, 1-11.

Nosalewicz, A. and Lipiec, J. (2014). The effect of compacted soil layers on vertical root distribution and water uptake by wheat. *Plant Soil*. 375, 229–240.

Oldeman, L. R. (1992). *Global extent of soil degradation. ISRIC Bi-Annual Report.* Wageningen. The Netherlands. pp. 19-36.

Oldeman, L. R., Hakkeling, R. U., Sombroek, W. G. (1990). *World map of the status of human-induced soil degradation: An explanatory note.* Wageningen: International Soil Reference and Information Centre.

Pikul, J. L. Jr., Johnson, J. M. F., Schumacher, T. E., Vigil, M., Riedell, W. E. (2008). Change in surface soil carbon under rotated corn in eastern south Dakota. *Soil Sci. Soc. Am. J.* 72, 1738–1744.

Prosdocimi, M., Tarolli, P., Cerda, A. (2016). Mulching practices for reducing soil water erosion: a review. *Earth-Sci. Rev.*, 161, 191- 203.

Raghavendra, M., Singh, Y. V., Das, T. K., Meena, M. C. (2017). Effect of crop residue and potassium management practices on productivity and economics of conservation agriculture-based maize (*Zea mays*) - wheat (*Triticum aestivum*) cropping system. *Indian J. Agril Sci.*. 87(7), 855-861.

Raghavendra, M., Singh, Y. V., Gaind, S., Meena, M. C., Das, T. K. (2018). Effect of Potassium and Crop Residue Levels on Potassium Solubilizers and Crop Yield under Maize-Wheat Rotation. *Int J. Curr Microbiol App Sci*. 7(6), 424-435.

Raghavendra, M., Singh, Y. V., Meena, M. C., Das, T. K., Sepat, S., Verma, R. K. (2020). System productivity and economics influenced by residue and potassium management in maize (*Zea mays*)-wheat (*Triticum aestivum*) rotation. *Indian J. Agri. Sci. 90*, 784-789.

Ray, D. K., Mueller, N. D., West, P. C., Foley, J. A. (2013). Yield trends are insufficient to double global crop production by 2050. *PloS One*. 8(6).

Sanchez, P. A. and Swaminathan, M. S. (2005). Cutting world hunger in half. *Science*. 307, 357–359.

Scherr, S. J. (2001). The future food security and economic consequences of soil degradation in the developing world. In *Response to Land Degradation*; Oxford Press: New Delhi, India, pp. 155–170.

Senjobi, B. A., Ande, O. T., Ogunkunle, A. O. (2013). Land Degradation Assessment under Different Uses: Implications on Soil Productivity and Food Security. *Agronomski Glasnik*. 1, 3-22.

Shah, F., Wu, W. (2019). Soil and crop management strategies to ensure higher crop productivity within sustainable environments. *Sustainability*, *11*(5), 1485.

Sharma, S., Kulkarni, J., Jha, B. (2016). Halotolerant rhizobacteria promote growth and enhance salinity tolerance in peanut. *Front Microbiol*. p. 7.

Shinwari, K. I., Shah, A. U., Afridi, M. I., Zeeshan, M., Hussain, H., Hussain, J., Ahmad, O. 2015. Application of plant growth promoting rhizobacteria in bioremediation of heavy metal polluted soil. *Asian J. Multi discip. Stud*. 3, 179-185.

Singer, J. W., Kohler, K. A., Liebman, M., Richard, T. L., Cambardella, C. A., Buhler, D. D. (2004). Tillage and compost affect yield of corn, soybean, and wheat and soil fertility. *Agron. J*. 96, 531–537.

Stewart, Z. P., Pierzynski, G. M., Middendorf, B. J., & Prasad, P. V. (2020). Approaches to improve soil fertility in sub-Saharan Africa. *Journal of Experimental Botany*, *71*(2), 632-641.

Sun, P., Tian, Q. Y., Zhao, M. G., Dai, X. Y., Huang, J. H., Li, L. H., Zhang, W. H. (2007). Aluminum-induced ethylene production is associated with inhibition of root elongation in *Lotus japonicus* L. *Plant Cell Physiol*. 48, 1229-1235.

Teasdale, J. R. (2007). Strategies for soil conservation in no-tillage and organic farming systems. *J. Soil Water Conserv*. 62, 144A–147A.

Tinker, P. B. (1988). Soil damage - physical and chemical Processes. *Ecol. Bull*. 39, 17-22.

Tracy, S. R., Black, C. R., Roberts, J. A., Mooney, S. J. (2011). Soil compaction: a review of past and present techniques for investigating effects on root growth. *J. Sci. Food Agric*. 91, 1528-1537.

Valentine, C., Poesen, J., Li, Y. (2005). Gully erosion: impacts, factors and control. *Catena*. 63(2-3), 132-153.

Wei, W., Chen, D., Wang, L., Daryanto, S., Chen, L., Yu, Y., Lu, Y., Sun, G., Feng, T. (2016). Global synthesis of the classifications, distributions, benefits and issues of terracing. *Earth-Science Reviews*. 159, 388-403.

Zhang, F. S. (2006). Biological processes in the rhizosphere: a frontier in the future of soil science. In: Alfred E. Hartemink (ed.), *The Future of Soil Science*, Wageningen: IUSS International Union of Soil Sciences, pp.155–157.

Zhang, F. S., Shen, J. B., Zhang, J. L., Zuo, Y. M., Li, L., Chen, X. P. (2010). Rhizosphere processes and management for improving nutrient use efficiency and crop productivity: implications for China. In: DL Sparks, ed., *Advances in Agronomy*, vol. 107. San Diego: Academic Press. pp 1–32.

Chapter 8

Agronomic and Economic Assessment of Site Specific Nutrient Management in Crop Production

Gurdeep Singh Malhi[1,*], Manpreet Kaur[2], Angrej Singh[1], Vijay Kant Singh[1], Sat Pal Saini[1] and Hanuman Singh Jatav[3]

[1]Punjab Agricultural University, Ludhiana, Punjab, India
[2]CCS Haryana Agricultural University, Hisar, Haryana, India
[3]Sri Karan Narendra Agriculture University, Jobner, Jaipur, Rajasthan, India

Abstract

India faced the major challenge of feeding its population after independence for which it increased the input consumption in agriculture. The fertilizer consumption has increased because of cultivation of high yielding input responsive varieties. The continuous imbalanced use of fertilizers led to yield stagnation and soil fatigue. It has also added to the emission of greenhouse gases and global warming. Moreover, the leaching of nutrients also causes groundwater contamination and environmental pollution. This necessitates the sustainable cultivation of agricultural crops for which efficient nutrient management is very important. Balanced nutrient application is the key to healthy environment which used need based approach for fertilizer application. Site specific nutrient management (SSNM) is a need based approach for application of nutrients where the fertilizer requirement is estimated to fulfill the deficit between the nutrients required to achieve the target yield and indigenous soil nutrient supply. Plant based and soil based approaches can be used for detection of nutrient deficits some of which are leaf color chart (LCC), chlorophyll meter, optical sensors, nutrient expert etc. These approaches help in on-farm evaluation of nutrients in the crop and correct the nutrient supply if any deficiency symptoms appear. SSNM helps in increasing the nutrient use efficiency (NUE) and helps in better utilization of nutrients by the crops. It ensures timely application of required nutrients at required rate at right place where nutrient is deficit. The SSNM approach increased the agronomic efficiency (AE) and recovery efficiency (RE) of nutrients and thereby increases the crop productivity at a lower rate of nutrient application. This also prevents the environmental pollution and greenhouse gases emission leading to a healthier environment. The increased yield along with reduced input use leads to enhanced economic returns to the farmers making it a viable option for them.

Keywords: net benefits, nutrient use efficiency, productivity, SSNM, sustainability

* Corresponding Author's Email: 89gurdeep.malhi@gmail.com.

In: Ecosystem Services
Editor: Hanuman Singh Jatav
ISBN: 978-1-68507-614-6

1. Introduction

Plants generally have a requirement of seventeen essential elements for growth and development namely carbon (C), hydrogen (H), oxygen (O), nitrogen(N), phosphorus (P), potassium (K), sulfur (S), calcium (Ca), magnesium (Mg), boron (B), chlorine(Cl), copper (Cu), iron (Fe), manganese (Mn), molybdenum (Mo), nickel (Ni), and zinc (Zn). The three macronutrients namely C, H and O are obtained by the plants through water, air or both. The remaining other nutrients are divided into two groups viz. soil-derived macronutrients and soil-derived micronutrients. The soil derived macronutrients are N, P, K, S, Ca and Mg whereas the soil derived micronutrients are B, Cl, Cu, Fe, Mn, Mo, Ni and Zn (Mahler, 2004). These nutrients are classified essential on the basis of essentiality criteria proposed by Arnon and Stout in 1939. The three essentiality criteria are the absence of any element, the plant would be unable to complete its life cycle by distorting its vegetative or reproductive phase, the role of each element is specific and the function of one element in plant growth cannot be played by the other, and the element is directly involved in plant nutrition (Das and Dash, 2019). Moreover, there is interaction between the nutrients in which supply of one nutrient influence the absorption, distribution and function of other nutrients in higher plants. These interactions can either induce deficiencies or toxicity and can even influence growth of the plants (Robson and Pitman, 1983). The agricultural soils of the world are deficient in one or more essential nutrients needed to sustain healthy growth in plants. Moreover, the acidity, salinity, alkalinity, anthropogenic activities, types and nature of farming and soil erosion adds to soil degradation, this necessitates the application of fertilizers and amendments to provide nutrient supply to the plants. Soil fertility is declining all around the world mainly due to intensive cultivation of crops, inadequate application of additional nutrients and deforestation. This low soil fertility significantly puts constraints on the improvement of agricultural production. This necessitates the application of fertilizers for improving agricultural productivity and sustaining the world's growing population. There is projection which states 70 per cent of plant nutrients in 2020 at global level will be from fertilizer application. But this has caused various environmental concerns such as eutrophication of water bodies, ozone depletion, global warming, proliferation of algal blooms and acid rains etc. (Ayoub, 1999). Moreover, it was also reported that use of higher doses of chemical fertilizers doesn't strongly correlate with the growth of agricultural productivity. The agricultural yield has been fluctuating over the years mainly because of improper use of N, P, K nutrients over the years (Patra et al., 2016). NUE of N, P and K is reported to be lower than 50% for N, less than 10% for P and around 40% for K (Baligar et al., 2007). It was also reported that high doses of fertilizers could lead to high NUE only if comprehensive techniques of crop management were used. However, the farmers are getting very low NUE at very high input levels which is causing environmental degradation (Zhang et al., 2000). Climate change is also causing environmental stress such as drought, salinity, heat and cold stress to the plants which is a barrier in proper utilization of nutrients by the plants (Malhi et al., 2021a). The farmers need to adopt recent technologies such as site-specific nutrient management (SSNM), real-time nitrogen management (RTNM), slow release and controlled release fertilizers, use of urease inhibitor and nitrification inhibitor etc. so as to ensure higher fertilizer use efficiency (FUE) (Xiang et al., 2008). Therefore the raising fertilizers utilization rate while reducing the application of fertilizers is a major concern to bring down environmental pollution. Balanced nutrient application has the potential to meet the nutrients requirement for crop growth and development without causing any negative impact on environment (Chen et al., 2021). SSNM is a field specific approach which ensures applications of nutrients as and when needed. This advocates optimum utilization of indigenous nutrients i.e., soil nutrients, plant residues, manures and irrigation water. Fertilizers are then applied to overcome the deficit between the nutrients required to achieve the target yield and soil nutrient supply (Pampolino et al., 2007). The agronomic N use efficiency of SSNM approach is higher than other conventional methods. Moreover, it leads to increased yield and decreased fertilizer application leading to higher returns (Khurana et al., 2008). SSNM is a nutrient smart practice which falls under a broad group of climate smart agricultural activities. It advocates optimum use of nutrients in the arena of climate change to ensure enhanced productivity and returns with minimal environmental implications (Malhi et al., 2021b).

2. Need for Nutrient Management

Although green revolution has helped in attaining self-sufficiency in food production but the intensive use of fertilizers has led to environmental degradation particularly affecting soil, water resources and vegetation. The intensive use of chemical fertilizers has declined soil organic matter. It has also led to water contamination by nitrates and phosphates along with physical and chemical degradation of soils (Singh, 2000). However, the use of chemical fertilizers has played an important role in increasing agricultural productivity and thereby increasing agricultural production. An increase of 0.8 gram Nitrogen per gram of Phosphorus of N/P fertilizer ratio was reported at global level during 1961-2013. It was reported that the fertilizer use rates on per unit area has increased eight times since 1961 and three times since IFA (International Fertilizer Industry Association) and FAO (Food and Agricultural Organization) started surveying the consumption of fertilizer country wise. A shift in hotspots of agricultural N fertilizer application to Asia in 21st century from Europe and USA in 1960s was witnessed. The P fertilizer shows the similar pattern as N fertilizer with one additional hotspot in Brazil (Lu and Tian, 2017). In East-Central China, it was found that there is widespread misapplication of fertilizer and the behavior of farmers shows huge variations in context to fertilizer application. It was reported that lack of scientific knowledge and lack of guidance from extension services that led the farmers to use their own personal judgment and advice from friends and dealers to take decision regarding fertilizer application (Yang and Fang, 2015). The fertilizer consumption in India has increased 322 times in comparison to fivefold increase in the production of cereals form 1950-51 to 2007-08 (Prasad, 2009). India ranks third in fertilizer use over the world but its nutrient application rate is only 85Kg/ha. Moreover, the rainfed area which forms 70% of the total cultivated area has reported to use only 20% of total fertilizers (Mala, 2013). In India, the west zone was reported to consume the maximum amount of fertilizer than rest of the country. West zone was found to use 31116.7 Kilotons (KT) of fertilizer out of which 16621.3 KT is used by Uttar Pradesh only during 2007-11. It was also reported that the fertilizer consumption is increasing at an annual compound growth rate of 9.68 per cent in west zone. Apart from this, it was found the maximum fertilizer was consumed in rice and wheat crop which uses 37% and 24% of total fertilizer consumed in India (Usama and Khalid, 2018). The fertilizer use in India is affected by irrigation, high yielding varieties seeds, size of farm credit etc. (Mala, 2013). In semi-arid tropics of India, soil degradation mainly causes low crop and water productivity. When the soils of farmers' field were evaluated in Madhya Pradesh and Rajasthan, it was found that there are widespread deficiencies of nutrients and soil organic carbon indicating poor soil health in the region (Chander et al., 2012). The increase in fertilizer consumption over the years has been shown in Figure 1. It reveals that there has been tremendous increase in the consumption of fertilizers since green revolution. This has increased the cereal production over the years but the FUE has been decreased which is evident through the figure.

The application of synthetic N and P fertilizers is one of the serious components of anthropogenic inputs in the earth's system. Although nitrogen and phosphorus fertilizers have increased the agricultural productivity but it has dramatically altered the global nutrient budget, greenhouse gases balance and their influence on the climate (Lu and Tian, 2017). The application of nitrogen fertilizers led to the emission of nitrogen containing trace gases. The application of N fertilizers in the intensive wheat systems of Mexico led to emission of nitrous oxide (N_2O) and nitric oxide (NO) which has a huge role in global warming. It was reported that application of lower rates of nitrogen fertilizers in the later periods of crop cycle reduced the nitrogen losses without affecting crop yield and quality (Matson et al., 1998). The application of N fertilizer on a large scale has also increased the nitrate content in the groundwater making it unhealthy for human consumption. The gaseous emission of NH_3 and NO_x resulting from nitrogenous fertilization has deleterious impact on the environment (Prasad, 2009). Apart from this, the continued use of chemical fertilizers also lead to surface run off of phosphorus and nitrogen causing eutrophication of aquatic ecosystem. Nitrogen has received far more attention in this context because of its higher application at global level and N mainly limits the primary production in coastal water and estuaries (Glibert et al., 2005). This necessitates the formulation of new techniques of fertilizer application. The balanced use of nutrients will ensure the optimum utilization of nutrients in the field and therefore decrease the fertilizer consumption. This decreased consumption has the potential to decrease environmental pollution and emission of greenhouse gaseous and associated global warming.

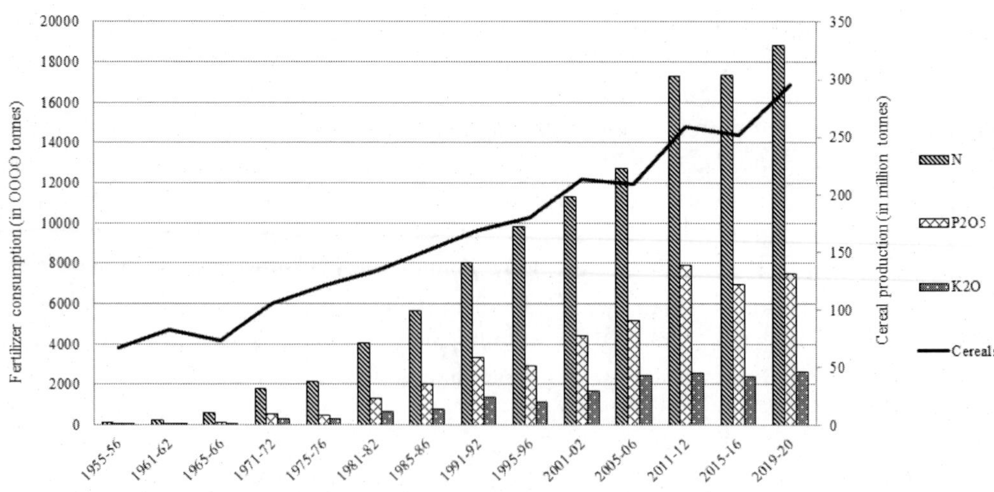

Source: The fertilizer association of India (2020), Agricultural Statistics at a glance (2018) and Ministry of Agriculture & Farmers' Welfare (2020).

Figure 1. Fertilizer consumption and cereal production in India over the years.

3. Site Specific Nutrient Management

The concept of balanced fertilizer is the key to healthy environment. It ensures application of fertilizers to the crops in optimum amounts and exact proportion as needed through appropriate methods, which helps in sustenance of soil fertility along with maintaining crop productivity. It builds the soil physical and biological health. Balance nutrient supply is regarded as best management practice (BMP) which ensures efficient use of nutrients, conservation of resources and maintenance of soil productivity (Dev, 1998). With scientific application of fertilizers, it is possible to reduce pollutant load generated from agricultural runoff (Yang and Fang, 2015). In semi-arid tropics of India, the application of soil test based balanced nutrition was reported to increase the crop yield by 6-40% with benefit cost ratio of 0.81 to 4.28 by increased rainwater use efficiency (Chander et al., 2012). In SSNM, nutrients are applied dynamically to the crops in a field specific approach as and when needed. It allows optimum utilization of available nutrients of soil, residues, manures etc. SSNM is an approach based on plants for effective management of N, P and K so as to provide the crops with required amount of essential nutrients (IRRI). SSNM ensures optimum use of nutrients from soil, residues of plants and manures. Fertilizers are applied in required pattern to overcome the deficit between the crops' demand for nutrients and supply from soil and indigenous sources (Pampolino et al., 2007). SSNM is based on 4R concept dating back to 1988 and is formally attributed to International Plant Nutrition Institute (IPNI) as shown in Table 1.

Table 1. 4R concept of balanced nutrient management

4R	Meaning	Scientific basis
Right product	It advocated application of fertilizers according to crop needs and soil type so as to ensure balance supply of nutrients.	Ensure usage of right fertilizers and combination of organic and inorganic sources of nutrition to ensure balanced supply of nutrients.
Right rate	It ensures the application of required quantity of fertilizers with the indigenous supply of soil nutrients also taken into view. It ensures that the fertilizer is used well and there are no environmental losses associated with it such as runoff, leaching, gaseous emissions, soil degradation etc.	Assessing the nutrient supply from soil sources and computation of fertilizer required to cover the deficit between target yield and indigenous soil nutrient supply
Right time	It ensures the application of nutrients to the crops at the time of need by evaluating the crop nutrient dynamics. It possibly means split application of fertilizers or application of combination of organic and inorganic fertilizers etc.	Applying nutrients in split doses during the critical stages
Right place	It means application of nutrients at optimal soil depth to avoid nutrient losses. It also ensures that the nutrients are kept at optimal distance from the plants. However, this depends on soil type, tillage method employed and type of fertilizer etc.	Recognition of rooting pattern of the crop along with management of spatial variability

Source: Adapted from Richards et al., 2015.

4. Principles of SSNM

1. *Yield gains and fertilizer requirements:* This principle proposes a pre-season computation of balanced fertilizer rates which is done on the basis of difference between the plants' requirement of nutrients and nutrient supplying capacity of soil. This nutrient deficit has a huge dependence on the expected yield increase which is the difference between the yield goal and yield limited by nutrients. These yield gains or increase must be calculated separately for N, P and K. The fertilizer rates are calculated by Witt and Dobermann approach (2002):

   $FN = (GY\text{-}GY_{0N}) \times UN/REN$
   $FP = (GY\text{-}GY_{0P}) \times UP/REP \times 2.292$ [-15%]
   $FK = (GY\text{-}GY_{0K}) \times UK/REK \times 1.2$ [-15%]

   Where FN, FP and FK the recommended fertilizer doses of N, P_2O_5 and K_2O in Kilogram for a hectare land, GY is the desired yield goal (t/ha) and GY_{0N}, GY_{0P} and GY_{0K} (Kg/ha) are grain yield observed in omission plots where fertilizers are not applied. UN, UP and UK are plant uptake requirements for N, P and K for a t grain yield. REN, REP and REK are the expected fertilizer recovery efficiency which is taken as 40-50% for N, 25% for P and 50% for K. A reduction of 15% is done for the fertilizer rates of P and K due to the several other constraints faced by the crop in the season which makes it not possible to achieve the target yield goal.
2. *Use of Omission plots for estimating soil nutrient supplies:* There can be indirect estimation of soil nutrients supply from the nutrient uptake by the plant instead of soil testing. This is termed as plant-based estimation of soil nutrient supply. It constitutes soil nutrient supply from all the indigenous sources under field conditions along possible estimation of nutrient supply by organic manures. The soil nutrient supply can also be estimated by the grain yield from the omission plots with the assumption of an average uptake of 15 Kg N, 2.6 Kg P and 15 Kg K at harvest per ton of grain yield.
3. *Need-based N management:* The application of N fertilizer is generally done in splits. But the number of splits, amount of N fertilizer per split and application time varies significantly within a small recommendation area. However, the farmers are flexible in this approach of fertilizer timing and application and this offers huge potential to coordinate the application of N fertilizers with the demand of the crop. The three most important N management strategies are:
   a. *Location-specific N fertilizer splitting schedules for preventive N management* follows a fertilizer application schedule. These recommendations in extensive use are usually devised according to on-farm and on-station N fertilizer response experiments. However, these fertilizer response experiments have several limitations because of time consuming and costly approach for identification of optimal fertilizer doses for split application. Moreover, these results cannot be extrapolated because of huge variation in N supply by the soil and climatic factors within a recommendation domain. And although, the strategies of corrective N management have huge potential for efficient management of N fertilizer but these recommendations are accurate under certain climatic conditions with low pest infestation.

The split schedules for a specific location can be framed by following first and second principles in which they calculate the N fertilizer requirement of the crop and soil indigenous N supply. The soil indigenous N supply can be estimated with the analysis of current farm yield and farmers' practices regarding N management along with combining local knowledge and information of soil fertility. These locally defined N splitting schedules also take differences in climate, varieties, crop establishment, water application and basal application of N.

b. *Real time corrective N management* with the use of a LCC needs periodic assessment of N status of the plant and the fertilizer application is done if any deficiency symptoms seem to appear. This is a need based approach done without the estimation of soil N supply or computation of pre-season fertilizer rate. LCC provides significant opportunities to the farmers for estimation of N demand in real time to attain efficiency in fertilizer use and high crop yields (Witt et al., 2005). The scientific basis of need based management of N fertilizer was framed with the introduction of chlorophyll meter (SPAD). However, the SPAD was found not very suitable for on-farm testing, so LCC was developed by International Rice Research Institute (IRRI) in collaboration with National Agricultural Research and Experimental System (NARES). LCC is a visual indicator which shows the N deficiency in the plant with its color panels of various shades of green colour. In LCC, an optimal colour identifies whose maintenance is needed throughout the season for attaining high yield. This optimal colour varies with the crop cultivar and establishment method. The reading of leaf colour with LCC is taken after every 10 days and if the colour falls below the optimal or critical value, a pre decided amount of N fertilizer is applied to cure the N deficiency in the plants.

c. *Location specific split schedules combined with LCC* provides methods for both preventive and corrective management of N. The fertilizer requirement of the crop is computed according location specific schedules along with basal N application. These pre decided fertilizer doses are then modified according to plants' requirement in the later growth stages.

4. *Sustainable crop and soil-based P and K management:* The requirement of P and K is very challenging to be estimated because of small fragmented landholdings and huge variation among of P and K nutrients among a small recommendation domain. Moreover, it is not easy to predict the short and long tern response of crop to P and K application and therefore offers a small possibility of correcting P and K deficiencies in the cropping period. The efficiency of P and K nutrients is greatest if these nutrients are applied as basal doses or in early growth stages. So it is very important to develop a fertilizer application plan by the extension worker with the use of nutrient omission plots of the representative farmers' field. A nutrient balance model was constructed to calculate nutrient requirement for target yield by also taking into account the soil nutrient supply, nutrients inputs from irrigation water and nutrient uptake by grain and straw:

$$FP = GY \times UP - StP + (GY - GY_{0P}) \times UP \times 2.292\ [-15\%]$$
$$FK = GY \times UK - StK - WK + (GY - GY_{0K}) \times UK \times 1.2\ [-15\%]$$

Where, StP and StK are estimated inputs of P and K for a tonne of recycled straw. WK is average K supplied with irrigation water during the growing season. GY is the desired yield goal (t/ha) and GY_{0N}, GY_{0P} and GY_{0K} (Kg/ha) are grain yield observed in omission plots where fertilizers are not applied. UP and UK are plant uptake requirements for P and K for a tonne grain yield. There is a reduction of 15% for P and K fertilizer rates because of other constraints faced by the farmers in achieving target grain yield.

5. *Increasing profitability:* The major benefit of improved nutrient management is increased profitability. The application of SSNM may require small amount of additional labour but the labour costs are generally much lower than those of other operations. With the adoption of SSNM in rice fields, it was reported that N use efficiency of crop significantly increased and yield of the crop is also increased. The fertilizer use in a unit area was also reduced which would eliminate the pollution of rice ecosystem (Wang et al., 2007). With the adoption of SSNM, the N_2O emission per unit of grain yield is reduced in those environments where higher yields are possible to be obtained with lower fertilizer application (Pampolino et al., 2007). SSNM also reported to have lower global warming potential than the farmer's fertilization practices in NW India (Sapkota et al., 2014). The increased profitability of SSNM is mainly because of increased yield rather than reduced input cost (Gabinete, 2009; Singh and Singh, 2018).

5. Approaches of Implementation of Site Specific Nutrient Management on Farm

Leaf color chart (LCC): The nutrient status of soil is best judged with the plants grown over there. It gives an idea about the nutrients which are actually taken by the plants from the soil for their growth and development. There are different indicators such as LCC, chlorophyll meter etc. which shows the nutrient status of the plant by which we can plan the fertilizer management of the crop. LCC was developed by International Rice Research Institute (IRRI) in collaboration with Philippines Rice Research Institute (PhilRice) with a Japanese prototype for determining the requirement of nitrogenous fertilizers in rice fields to ensure maximum yield. It is also useful in detection of N deficiency in maize and wheat. It is innovative low cost technique for real time crop need based N management. It a ruler shaped strip having four or more panels having different range for colours from yellowish green to dark green. It subjectively indicates N deficiency by measuring the intensity of leaf colour in relation to N status (IRRI). For South Asian countries, a new four panel LCC was formed and its user guidelines were adjusted according to the local conditions. A critical colour has to be maintained for ensuring optimal plant growth and yield for which LCC provides direction of when to provide fertilizer N to avoid its deficiency (Witt et al., 2005). It was reported that LCC based application of N fertilizer (urea) reduces the emission of nitrous oxide and methane reducing the global warming potential of rice-wheat system by 10.5% (Bhatia et al., 2011). The agronomic parameters of the super rice crop viz. number of spikelets per panicle, percentage of filled grains, grains weight etc. showed significant improvement under AWD (Alternate wetting and drying)-SSNM treatment. AWD and SSNM showed synergistic interaction and increased the N partial factor productivity and water use efficiency of the crop (Liu et al., 2013).

Chlorophyll meter/SPAD meter: It is proximal sensing technique in which the SPAD meter reading is taken to estimate the fertilizer requirement of the crop. The value of SPAD meter depends upon the N concentration of the leaves and these values are compared with critical values to arrive at the nutrient requirement of the crop for optimal yield. However, the values of chlorophyll meter also depend on other factors than nitrogen status (Singh and Ali, 2020). It was reported that SSNM based on chlorophyll meter increase the yield of hybrid rice crop by 8.9-9.3% and agronomic efficiency by 200.9-276.4% compared to Farmers' fertilizer practice which shows the great potential of SPAD meter in fertilizer management (Liu et al., 2004). Nitrogen fertilizer optimization algorithm (NFOA) and nitrogen nutrition index (NNI) was developed using chlorophyll meter or SPAD (Soil and Plant Analysis Development) in paddy and it was found that these adjustment strategies optimized the application of nitrogen by upward and downward adjustment. The highest productivity was observed on application of 264 Kg N/ha. The N recovery efficiency was also found higher for NFOA and NNI by 51-57% than fixed application of N fertilizers (Zhang et al., 2020).

Nutrient expert: It is a precision nutrient management technology guided through decision support system software for enhancing crop yields, environmental quality and agricultural sustainability particularly in south Asian Indo-gangetic plains. It is governed by the principles of SSNM and developed by International Plant Nutrition Institute (IPNI) and International Maize and Wheat Improvement Center (CIMMYT). It helps in framing fertilizer recommendations by taking into view the target yield, agronomic deficiencies and the nutrient supply from indigenous sources (Pooniya et al., 2015). It was first developed for rice and later adapted to maize and wheat (Pampolino et al., 2012). It allows the users to develop the recommendation based on experimental data but it can also use their own experience, farmers' knowledge of local region and farmers' fertilizer practices (Dass et al., 2014). The field specific recommendations of fertilizers based on nutrient expert offers a better opportunity to the farmers of Southern India to manage the nutrient use in maize under increasing prices of fertilizers (Satyanarayana et al., 2013). The SSNM (Site Specific Nutrient Management) according to the NE (Nutrient Expert) and application of FYM has led to increased productivity in the maize wheat cropping system of Southern India along with an improvement in physico-chemical and biological properties of soil leading to long term soil productivity (Pooniya et al., 2015). Location specific nutrient recommendation made according to nutrient expert for wheat in South Asia has led to increased yields and economic returns when compared with farmers' fertilizer practices (Pampolino et al., 2012). The recommendations based on Nutrient Expert have increased the yield of maize by 1.1 t/ha in Odisha and Southern India (Satyanarayana et al., 2014).

Optical sensors: Optical sensors generate vegetative index named NDVI (Normalized Difference Vegetation Index) based on reflectance of leaves. Application of fertilizers according to optical sensors needs local calibration of sensor for nutrient, crop and region which related grain yield with NDVI readings (Richards et al., 2015). These spectral vegetation indices help in detection of photosynthetic efficiency, productivity potential and potential yield. The expected yield measured through NDVI showed strong relation with actual yield. A concept of fertilizer N algorithm is developed by optical sensors using response index and potential yield to compute the fertilizer requirement of the crop (Singh, 2014).

Green seeker is an optical sensor developed by Oklahoma state university, USA and produced by N tech industries Inc., Ukiah, CA, USA for site-specific N management in rice and wheat (Singh, 2017). The green seeker detects the proportion of emitted light which is sensed back by the sensor i.e., crop reflectance and compute NDVI accordingly. NDVI is calculated as:

$$NDVI = (F_{NIR} - F_{Red})/ (F_{NIR} + F_{Red})$$

Where F_{NIR} and F_{Red} is proportion of NIR and red radiations reflected back from sensed area. The NDVI is then used to estimate the expected yield and fertilizer is managed accordingly to achieve the target yield from expected yield (Singh, 2014). The fertilizer management according to green seeker has reported to increase the partial factor productivity and therefore has a potential to enhance N use efficiency (Yao et al., 2012).

6. Agronomic and Economic Implications

Green revolution is considered as golden period in South Asian Agriculture as it helped in boosting agricultural production manifolds and in achieving self-sufficiency in food grains. But the irrational use of chemical fertilizers and pesticides has led to lower factor productivity and stagnation in crop yields. It has caused serious implications to soil physical and biological health. The declined productivity is mainly caused by low NUE of fertilizers which is compelling the farmers to use higher level of fertilizers (Dass et al., 2014). The rice-wheat system of South Asia is facing huge stress because of higher nutrient removal by crops from soil than the replenishment of nutrients in soil with fertilizers (Singh et al., 2014). The nitrogen is the most vulnerable among all the nutrients whose efficiency is therefore attempted to increase by physical manipulation of the size of granules, coating of granules with films, chemicals and amendments to slow the dissolution and denitrification. However, these manipulations weren't economically viable. So the site specific nutrient management with the use of gadgets such as LCC, sensor-based technology etc. came into use which has a huge potential in increasing the NUE (Hedge et al., 2007). The partial factor productivity of Nitrogen is higher in SSNM when compared with state recommendation and farmers' fertilizer practices in Indo-gangetic plains of India (Singh et al., 2014). SSNM is also reported to increase the NUE of N, P and K in the rice crop of Mekong Delta. AEN and REN were found higher in SSNM plots compared to farmers' fertilizer practice (FFP) and extension's fertilizer practices (EFP). AE increased from 10-15 Kg grain/kg applied N in FFP to 11.9-17 Kg grain/kg N in SSNM plots. RE in SSNM plots was also increased to 0.51-0.56 while EFP and FFP have RE ranging from 0.42-0.53. This shows that SSNM enhanced RE and plant uptake of nitrogen. The phosphorus and potassium use efficiency of the rice crop is also improved under SSNM. The AEP (Agronomy Efficiency of Phosphorus) and AEK (Agronomy Efficiency of Potassium) under SSNM was reported to be 25-46 Kg grain/Kg P_2O_5 and 2.4-13.4 Kg grain/Kg K_2O. Consequently, the Recovery Efficiency of Phosphorus (REP) and Recovery Efficiency of Potassium (REK) were also improved under SSNM (Khuong et al., 2007). The Nutrient use efficiency (NUE) of particularly N, P and K was reported to be much higher in SSNM in comparison to FFP. The addition of micronutrients in the schedule of SSNM also enhanced the nutrient use efficiency over the improved state recommendation but extent of increase depends on the cropping system (Gill et al., 2009). The huge doses and improper timing of N fertilizer application are responsible for low nitrogen use efficiency (NUE) of irrigated rice in Zhejiang. With site specific nutrient management, the input of nitrogenous fertilizers is reduced without yield loss in crop (Xie et al., 2007). The site specific nutrient management (SSNM) is reported to increase the grain yield per unit N application or agronomic N efficiency and raise the crop yield (Lijun et al., 2003). The increased uptake of nutrients N, P and K by 8-14% in rice plants on site specific nutrient management was observed in Southeast China. The split application of fertilizers with improved timing increased N recovery efficiency to 0.29 kg/kg in SSNM from 0.18 kg/kg in farmers' fertilizer practices (FFP). The agronomic N use efficiency was also reported to be 80% higher in SSNM than FFP (Wang et al., 2001). SSNM has resulted in increased agronomic N use efficiency in chili crop also. SSNM also resulted in 23.3% increase in the yield along with 25.8% reduction in N losses. However, P losses were not affected significantly (Yi et al., 2018).

The application of site specific nutrient management in southern India, Philippines and southern Vietnam increased the yield of rice crops and reduced the N fertilizer application in some cases. SSNM also increased the partial factor productivity in locations where relatively low level of fertilizers is applied and N use efficiency is low as in Vietnam and Philippines. SSNM increased N use of fertilizer and maintains low emissions of N_2O. The economic analysis as done through focused group discussion showed that there is 0.2 Mg/ha, 0.3 Mg/ha and 0.8 Mg/ha increase in yield as observed in southern Vietnam, Philippines and southern India, respectively with SSNM. The pesticide cost of the farmers is also reduced and annual net benefit of 34 US$/ha/year, 106 US$/ha/year and 168 US$/ha/year were observed in Vietnam, Philippines and India, respectively which is mainly through increased yield rather than reduced input cost (Pampolino et al., 2007). In an experiment conducted at Taraori, Karnal, it was found that SSNM increased the productivity and water use efficiency of sustainable intensification of maize-wheat cropping system with Mungbean along with providing additional net returns of 13.4%, 13.3% and 15.3% when compared with farmers' fertilizer practice (Jat et al., 2018). When the maize-wheat-mungbean cropping system was analyzed in semi-arid and subtropical region of New Delhi for four consecutive years (2012-16), it was found that conservation agriculture based zero tillage and permanent beds practice along with SSNM based balanced application of fertilizers led to significant improvement in water use efficiency, bioenergetics yields, net energy outputs, efficiency and energy & biomass productivity (Parihar et al., 2017). When rice-wheat area of Punjab was evaluated from 2002-03 to 2004-05, it was found that SSNM led to increase in the average wheat yield from 4.2 to 4.8 Mg/ha. It also led to enhanced N, P and K accumulation in the plants by 12-20%. The gross returns above the cost of fertilizer were reported to be 13% higher when compared by farmers' fertilizer practice. The agronomic N use efficiency was reported 63% higher with SSNM as compared to FFP (Khurana et al., 2008). SSNM based fertilizer application in rice i.e., $N_{150}P_{60}K_{120}S_{40}B_5Mn_{20}Zn_{25}$ and wheat i.e., $N_{150}P_{60}K_{120}$ raised the total productivity of cropping system in terms of rice yield equivalent (14.8 t/ha/yr). It also yielded 66.4% and 28.2% higher returns than farmers' practice and state recommended practice, respectively. Total net returns of ₹67033/ha/yr were observed in SSNM which were ₹31681/ha/yr higher than the farmers' practice and ₹16905/ha/yr more than the state recommendations (Mauriya et al., 2013). When SSNM was evaluated in nine locations in the north western region for 2003-04 to 2005-06, the yield of rice (unhusked) was reported to 8.20 t/ha in comparison to 6.95 t/ha and 6.03 t/ha observed with state recommendation and farmers' practice. SSNM raised the average rice yield by 36% over FP. The average yield of succeeding wheat crops was also increased with SSNM. The average wheat yield with SSNM was 4.86 t/ha compared to 3.56 t/ha on FP. There is wheat yield advantage of 41% in SSNM compared to FP. The additional expenditure of US$ 120/ha was reported in rice-wheat cropping system under SSNM which would generate additional gross returns of US$ 852/ha leading to net returns of US$ 732/ha and BCR of 6.1 (Singh et al.,2008).

In Karnataka, application of SSNM increased the yield of maize, rice, wheat and rabijowar over farmers' fertilizer practice by 28.6%, 24%, 17.7%, 32.8% and 35.7%, respectively. While in commercial crops, the yield of seed cotton, dry chilli, sunflower and chickpea was also increased by 27%, 23.6%, 34.8% and 26.4%, respectively. The additional net returns of ₹4057, ₹7950, ₹6778 and ₹3230 on application of SSNM over FFP in maize, rice, wheat and rabijowar were reported. In case commercial crops, these additional net returns were ₹7417, ₹8724, ₹4322 and ₹6838 in cotton, chilli, sunflower and chickpea, respectively (Biradar et al., 2012). In rice-rice cropping system in different locations of India, the additional gross returns of US$ 607/ha/yr and annual net returns of US$ 464/ha/yr was obtained due to 25% increase in productivity (Singh et al., 2009). In upper Gangetic plains of India, the impact of SSNM was analyzed in six-pre dominant wheat based cropping systems from 2007-08 to 2009-10 and it was found that SSNM increased the wheat equivalent yield over state recommendation (SR), improved SR, soil testing laboratory recommendation (STLR) and farmers' fertilizer practice by 19%, 8%, 17% and 29%, respectively. Per hectare additional cost of ₹5097 to ₹7938 under several cropping systems in SSNM, however, additional net returns of ₹13649/ha to ₹58776/ha over SR and ₹25030/ha to ₹68980/ha over FFP were observed (Singh et al., 2015). The three regions of Southeast Asia namely Indonesia, Philippines and Vietnam were evaluated for the site specific nutrient management and it was witnessed that SSNM improved the maize yields by 0.9 to 1.3 Mg/ha around the different sites in each country. SSNM was reported to increase the efficiency of N fertilizer by 53% when compared with farmers' fertilizer practice. The net benefits of the farmers were also increased by 184 US$/ha/crop which are mainly attributed to raised yield rather than reduced input cost (Pasuquin et al., 2010).

In a field experimented conducted in Modipuram during 2001-03, it was found that the value of LCC for maximum yield and nitrogen use efficiency varies (NUE) in different varieties of rice and wheat. In Basmati-370, LCC$\leq$3 produced maximum rice yield and NUE than the recommended fixed time N split doses while in case of Saket-4 and Hybrid 6111/PHB-71, the most optimal value of LCC was 4 and 5, respectively. In wheat, the maximum yield and NUE was observed when LCC$\leq$4 with a requirement of 120 Kg N/ha. It was also found that chlorophyll meter rating and crop growth was not having any significant difference in both rice and wheat for with or without basal dose at 15 days after rice transplanting and 21 days after wheat seeding. This shows that soils have relatively high indigenous N supply. The net returns were found 19 to 31% higher in nutrient management based on LCC than fixed time N application in rice-wheat cropping (Shukla et al., 2004). Use of LCC for N management has increased the rice yield and profit when compared to farmer's practice in Bangladesh. It was reported that an average of 0.1 to 0.7 Mg/ha increase in grain yield was reported across the villages and seasons which is leading to increased net returns of US$ 41 to 65/ha/season. When the recommended dose of P, K, S and Zn were also applied along with N doses according to leaf colour chart, the additional yield increase of $\geq$0.3 Mg/ha were observed in about half the combination of villages. Moreover, if improved crop management is performed such as manual weed control and plant spacing, the yield gains are further increased by $\geq$0.4 Mg/ha in about half the combination of villages (Alam et al., 2005). The application of 30 Kg N/ha when the value of SPAD falls below 37.5 brought application of 90 Kg N/ha in rice in northwestern India and it was found that the subsequent yield obtained by it is equivalent to 120 Kg N applied in a hectare in three splits. The wheat crop on the other hand responded to N fertilizer application at the stage of maximum tillering when the value of SPAD falls lower than 44. The yield of wheat showed significant improvement if 30 Kg N/ha was applied at 42 SPAD value. It was found that N requirement of rice was reduced from 12.5 to 25% with use of chlorophyll meter without any reduction in yield (Singh et al., 2002). In an experiment conducted at Agra, Uttar Pradesh for two consecutive seasons of wheat in 2012-13 and 2013-14, it was found that SSNM increased the wheat yield by 33.8 and 41.8% over state recommendations and farmers' fertilizer practices. SSNM also yielded maximum net returns of ₹62452/ha and highest BC ratio of 2.31 in wheat crop. The quality of produce also improved in terms of protein content under SSNM (Singh and Singh, 2018). The nutrient use efficiency of the crop increases with site specific nutrient management. The balanced application of site specific nutrient management ensures the optimum utilization of nutrients and reduces the emission and leaching losses associated with injudicious use of fertilizers. The agronomic efficiency of nutrients and recovery efficiency of nutrients is thereby increased with SSNM which leads to increased yields. The increased yield with SSNM and reduced fertilizer consumption leads to increased returns and profitability which makes it an economically viable option for the farmers.

Conclusion

The cultivation of high yielding varieties coupled with high amount of fertilizer and pesticide use has resulted in manifold increase in the yield of crops mainly rice and wheat in India since green revolution. However, in recent times it has led to decreased factor productivity along with stagnation of yield which is a major threat to agricultural sustainability. The declining factor productivity and nutrient use efficiency has led to overuse of fertilizers by farmers which caused detrimental effect on chemical and biological properties of soil along with groundwater contamination and environmental pollution. This necessitates focusing on new approaches for rationalized use of nutrients in agriculture such as SSNM. SSNM is a requirement based application of nutrients in the crops to enhance the NUE, crop productivity and profitability. SSNM aims to offer a locally suitable nutrient management practice modified according to the field and season specifications of the crop. The principles of SSNM includes balanced application of fertilizers according to the crop requirement, estimation of root nutrient supplies on the basis of plants, fertilizer N management based on needs, sustainable management of P and K, raising profitability. The recommendations of SSNM are done either on the basis of plants analysis or soil cum plants analysis. Plants analysis based SSNM has five steps which involves selection of the yield goal (target yield) to be achieved, evaluation of nutrient requirement of the crop, estimation of nutrient supply from the soil or indigenous nutrient supply, calculation of nutrient rates of fertilizers and dynamic correction of nitrogen rates. However, in soil cum plant based SSNM, the availability of nutrients in the soil, demand of crop for the nutrients and RE of nutrient application is kept in view for developing the schedule of fertilizer use. A number of diagnostic tools such as chlorophyll meter or SPAD meter, LCC etc. are used for application of need based N fertilizer doses. SSNM increases the partial factor productivity of nitrogen and reduced the proportion of nitrogen losses from applied fertilizers.

SSNM has the potential to increase the crop yield and NUE while reducing the N_2O emission and associated global warming. Moreover, increased net benefits are associated with SSNM as it increases the yield while reducing the cost of cultivation. There is still scope for working on nutrient dynamics, soil organic matter, soil physical and biological activities for development of more suitable methods for fertilizer recommendations.

References

Agricultural statistics at a glance. (2018). *Directorate of Economics and Statistics. Ministry of Agriculture & Farmers Welfare.* Government of India.

Alam, M. M., Ladha, J. K., Khan, S. R., Foyjunnessa, Rahid, H., Khan, A. H., Buresh, R. J. (2005). Leaf color chart for managing nitrogen fertilizer in lowland rice in Bangladesh. *Agronomy Journal.* 97, 949-959.

Arnon, D. I., Stout, P. R. (1939). The essentiality of certain elements in minute quantity for plants with special reference to copper. *Plant Physiology.* 14, 371-375.

Ayoub, A. T. (1999). Fertilizers and the environment. *Nutrient Cycling in Agroecosystems.* 55, 117-121.

Baligar, V. C., Fageria, N. K., He, Z. L. (2001). Nutrient use efficiency in plants. *Communication in soil science and plant analysis.* 32, 921-950.

Bhatia, A., Pathak, H., Jain, N., Singh, P. K., Tomer, R. (2011). Greenhouse gas mitigation in rice-wheat system with leaf color chart-based urea application. *Environmental Monitoring and Assessment.* 184, 3095-3107.

Biradar, D. P., Aladakatti, Y. R., Basavanneppa, M. A. (2012). Enhancing the productivity and economic returns of field crops with balanced nutrient application through site specific nutrient management approach. *Proc. Agro-informatics and Precision Agriculture.* Pp: 146-151.

Chander, G., Wani, S. P., Sahrawat, K. L., Kamdi, P. J., Pal, C. K., Pal, D. K., Mathur, T. P. (2012). Balanced and integrated nutrient management for enhanced and economic food production: case study from rainfed semi-arid tropics in India. *Archieves of Agronomy and Soil Science.* 59, 1643-1658.

Chen, Y., Hu, S., Guo, Z., Cui, T., Zhang, L., Lu, C., Yu, Y., Luo, Z., Jin, Y. (2021). Effect of balanced nutrient fertilizer: A case study in Pinggu district, Beijing, China. *Science of the total environment.* 754, 142069, https://doi.org/10.1016/j.scitotenv.2020.142069.

Das, R., Dash, B. P. (2019). Principles of plant nutrition. In: *New dimension of crop sciences.* Nigam R, Singh J, Rathi MS, Kumar M, Singh A and Sharma AK (eds.) Weser books, Germany.102-109.

Dass, A., Suri, V. K., Choudhary, A. K. (2014). Site-specific nutrient management approaches for enhanced nutrient-use efficiency in agricultural crops. *Research & Reviews: Journal of Crop Science & Technology.* 3, 1-6.

Dev, G. (1998). Balanced fertilizer use increases crop yield and profit. *Better Crops International.* 12, 25-27.

Gabinete, G. (2009). Site specific nutrient management (SSNM): Profitability to farmers, current level of adoption, and constraints to wider adoption in Iloilo province, Philippines. *IFA crossroads Asia Pacific 2009, Kota Kinabalu, Malyasia.*

Gill, M. S., Shukla, A. K., Singh, M. P., Tomar, O. K., Kumar, R., Majumdar, K., Tiwari, K. N. (2009). Evaluation of Nutrient management options for yield, economics and nutrient use efficiency. *Better Crops,* 12-15.

Glibert, P. M., Seitzinger, S., Heil, C., Burkholder, J. M., Parrow, M. W., Codispoti, L. A., Kelly, V. (2005). The role of eutrophication in the global proliferation of harmful algal blooms. *Oceanography.* 18, 198-209.

Hegde, D. M., Babu, S. N. S., Qureshi, A. A., Murthy, I. Y. L. N. (2007). Enhancing nutrient-use efficiency in crop production-a review. *Indian Journal of Agronomy.* 52, 261-274.

Jat, R. D., Jat, H. S., Nanwal, R. K., Yadav, A. K., Bana, A., Choudhary, K. M., Kakraliya, S. K., Sutaliya, J. M., Sapkota, T. B., Jat, M. L. (2018). Conservation agriculture and precision nutrient management practices in maize-wheat system: Effects on crop and water productivity and economic profitability. *Field Crops Research.* 222, 111-120.

Khuong, T. Q., Huan, T. T. N., Tan, P. S., Buresh, R. (2007). Effect of site specific nutrient management on grain yield, nutrient use efficiency and rice production profit in the Mekong delta. *Omonrice.* 15, 153-158.

Khurana, H. S., Phillips, S. B., Singh, B., Alley, M. M., Dobermann, A., Sidhu, A. S., Singh, Y., Pend, S. (2008). Agronomic and economic evaluation of site-specific nutrient management for irrigated wheat in northwest India. *Nutrient Cycling in Agroecosystems.*82, 15-31.

Lijun, L., Dazhi, S., Cuilian, L., Zhiqin, W., Jianchang, Y., Qingsen, Z. (2003). Effects of real-time and site-specific nitrogen managements on rice yield and nitrogen use efficiency. *ZhongguoNong ye kexue= ZhongguoNongyeKexue.* 36, 1456-1461.

Liu, L., Chen, T., Wang, Z., Zhang, H., Yang, J., Zhang, J. (2013). Combination of site-specific nitrogen management and alternate wetting and drying irrigation increases grain yield and nitrogen and water use efficiency in super rice. *Field Crops Research.*154, 226-235.

Liu, L., Sang, D., Liu, C., Yang, J., Zhu, Q. (2004). Effects of real-time and site-specific nitrogen management on rice yield and nitrogen use efficiency. *Agricultural Sciences in China.*262-268.

Lu, C., Tian, H. (2017). Global nitrogen and phosphorus fertilizer use for agriculture production in the past half century: shifted hot spots and nutrient imbalance. *Earth Syst. Sci. Data*, 9, 181–192, https://doi.org/10.5194/essd-9-181-2017.

Mahler, R. L. (2004). Nutrient plants require for growth. University of Idaho. *College of Agricultural and Life Science. CIS 1124.*

Mala, P. (2013). Fertilizer scenario in India. *International Journal of Social Science & Interdisciplinary Research.*2, 62-72.

Malhi, G. S., Kaur, M., Kaushik, P. 2021b. Impact of Climate Change on Agriculture and Its Mitigation Strategies: A Review. *Sustainability*. 1318. https://doi.org/10.3390/su13031318

Malhi, G. S., Kaur, M., Kaushik, P., Alyemeni, M. N., Alsahli, A. A., Ahmad, P. 2021a. Arbuscularmycorrhiza in combating abiotic stresses in vegetables: An eco-friendly approach. *Saudi Journal of Biological Sciences.*1465, doi: 10.1016/j.sjbs.2020.12.001.

Matson, P. A., Naylor, R., Ortiz-Monasterio, I. (1998). Integration of environmental, agronomic and economic aspects of fertilizer management. *Science.*280, 112-115.

Mauriya, A. K., Maurya, V. K., Tripathi, H. P., Verma, R. K., Shyam, R. (2013). Effect of site-specific nutrient management on productivity and economics of rice (*Oryza sativa*)-wheat (*Triticumaestivum*) system. *Indian Journal of Agronomy.*58, 282-287.

Ministry of Agriculture and Farmers' Welfare. (2020). Third advanced estimates of production of foodgrains, oilseeds and other commercial crops for 2019-20.

Pampolino, M. F., Manguiat, I. J., Ramanathan, S., Gines, H. C., Tan, P. S., Chi, T. T. N., Rajendran, R., Buresh, R. J. (2007). Environmental impact and economic benefits of site-specific nutrient management (SSNM) in irrigated rice system. *Agricultural Systems.*93, 1-24.

Pampolino, M., Majumdar, K., Jat, M. L., Satyanarayana, T., Kumar, A., Shahi, V. B., Gupta, N., Singh, V. (2012). Development and evaluation of nutrient expert for wheat in South Asia. *Better Crops,* 96, 29-31.

Parihar, C. M., Jat, S. L., Singh, A. K., Majumdar, K., Jat, M. L., Saharawat, Y. S., Pradhan, S., Kuri, B. R. (2017). Bio-energy, water-use efficiency and economics of maize-wheat-mungbean system under precision-conservation agriculture in semi-arid agro-ecosystem. *Energy.*119, 245-256.

Pasuquin, J. M., Witt, C., Pampolino, M. (2010). A new site-specific nutrient management approach for maize in the favorable tropical environments of Southeast Asia. *19th World Congress of Soil Science, Soil Solutions for a Changing World.* Pp: 4-7.

Patra, S., Mishra, P., Mahapatra, S. C., Mithun, S. K. (2016). Modelling impacts of chemical fertilizer on agricultural production: a case study on Hooghly district, West Bengal, India. *Modeling Earth System and Environment.*2, 1-11.

Pooniya, V., Jat, S. L., Choudhary, A. K., Singh, A. K., Parihar, C. M., Bana, R. S., Swarnlakshmi, K., Rana, K. S. (2015). Nutrient expert assisted site-specific-nutrient-management: An alternative precision fertilization technology for maize-wheat cropping system in South-Asian Indo-Gangetic Plains. *Indian Journal of Agricultural Sciences.*85, 996-1002.

Prasad, R. (2009). Efficient fertilizer use: The key to food security and better environment. *Journal of Tropical Agriculture.*47, 1-17.

Richards, M. B., Bahl, K. B., Jat, M. L., Lipinski, B., Monasterio, I. O., Sapkota, T. (2015). Site specific nutrient management: Implementation guidance for policymakers and investors. Practice Brief. *Global Alliance for Climate-Smart Agriculture.*1-10.

Robson, A. D., Pitman, M. G. (1983) Interactions between Nutrients in Higher Plants. In: Läuchli A., Bieleski R. L. (eds) Inorganic Plant Nutrition. *Encyclopedia of Plant Physiology* (New Series), vol 15. Springer, Berlin, Heidelberg. https://doi.org/10.1007/978-3-642-68885-0_6.

Sapkota, T. B., Majumdar, K., Jat, M. L., Kumar, A., Bishnoi, D. K., McDonald, A. J., Pampolino, M. (2014). Precision nutrient management in conservation agriculture based wheat production of Northwest India: Profitability, nutrient use efficiency and environmental footprint. *Field Crops Research.*155, 223-244.

Satyanarayana, T., Majumdar, K., Dutta, S., Jat, M. L., Pattanayak, S. K., Biradar, D. P., Aladakatti, Y. R., Pampolino, M., Johnston, A. (2014). Nutrient Exprt-Maize: A tool for increasing crop yields and farm profit. *Better Crops,* 7-10.

Satyanarayana, T., Majumdar, M., Pampolino, M., Johnson, A. M., Jat, M. L., Kuchanur, P., Sreellatha, D., Sekhar, J. C., Kumar, Y., Maheswaran, R., Karthikeyan, R., Velayutahm, A., Dheebakaran, G., Sakthivel, N., Vallalkannan, S., Bharathi, C., Sherene, T., Suganya, S., Janaki, P., Baskar, R., Ranjith, T. H., Shivamurthy, D., Aladakatti, Y. R., Chiplonkar, D., Gupta, R., Biradar, D. P., Jeyaraman, S., Patil, S. G. (2013). Nutrient Expert^{TM}: A tool to optimize nutrient use and improve productivity of maize. *Better Crops.* 97, 21-24.

Shukla, A. K., Ladha, J. K., Singh, V. K., DWivedi, B. S., Balasubramanian, V., Gupta, R. K., Sharma, S. K., Singh, Y., Pathak, H., Pandey, P. S., Padre, A. T., Yadav, R. L. (2004). Calibrating the leaf color chart for Nitrogen Management in different genotypes of rice and wheat in a systems perspective. *Agronomy Journal.*96, 1606-1621.

Singh, B. (2014). Site specific and need based management of nitrogen fertilizers in cereals in India. In: *Advances in fertilizer technology: Biofertilizers* (vol. 2) by Sinha S, Pant KK and Bajpai S (576-605) Studium press LLC, USA.

Singh, B., Ali, A. M. (2020). Using hand-held chlorophyll meters and canopy reflectance sensors for fertilizer nitrogen management in cereals in small farms in developing countries. *Sensors.*20, 1127, https://doi.org/10.3390/s20041127

Singh, B., Singh, Y., Ladha, J. K., Bronson, K. F., Balasubramanian, V., Singh, J., Khind, C. S. (2002). Chlorophyll meter and leaf color chart based nitrogen management for rice and wheat in northwestern India. *Agronomy Journal.*94, 821-829.

Singh, H. (2017). Precision nitrogen management through green seeker. *Journal of Emerging technologies and innovative research.*4, 433-436.

Singh, R. B. (2000). Environmental consequences of agricultural development: a case study from the green revolution state of Haryana, India. *Agriculture, Ecosystems & Environment.*82, 97-103.

Singh, S., Singh, V. (2018). Maximizing wheat (*Triticumaestivum*) productivity and profitability using site specific nutrient management strategy. *Annals of Plant and Soil Research.*20, 103-106.

Singh, V. K., Dwivedi, B. S., Tiwari, K. N., Manjumdar, K., Rani, M., Singh, S. K., Timsina, J. (2014). Optimizing nutrient management strategies for rice-wheat system in the Indo-Gangetic plains of India and adjacent region for higher productivity, nutrient use efficiency and profits. *Field Crops Research.*164, 30-44.

Singh, V. K., Shukla, A. K., Singh, M. P., Majumdar, K., Mishra, R. P., Rani, M., Singh, S. K. (2015). Effect of site-specific nutrient management on yield, profit and apparent nutrient balance under pre-dominant cropping systems of Upper Gangetic Plains. *Indian Journal of Agricultural Sciences.*85, 335-343.

Singh, V. K., Tiwari, K. N., Gill, M. S., Sharma, S. K., Dwivedi, B. S., Shukla, A. K., Mishra, P. P. (2008). Economic viability of site-specific nutrient management in rice-wheat cropping system. *Better Crops.* 92, 28-30.

Singh, V. K., Tiwari, R., Sharma, S. K., Dwivedi, B. S., Tiwari, K. N., Gill, M. S. (2009). Economic viability of rice-rice cropping as influenced by site-specific nutrient management. *Better Crops.* 93, 6-9.

The Fertilizer association of India. (2020). All-India consumption of fertilizer nutrients 1950-51 to 2019-20. https://www.faidelhi.org/general/con-npk.pdf (accessed on 10 Apr 2021).

Usama, M., Khalid, M. A. (2018). Fertilizer consumption in India and need for its balanced use: A review. *Indian Journal of Environmental Protection.*38, 564-577.

Wang, G., Dobermann, A., Witt, C., Sun, Q., Fu, R. (2001). Performance of site-specific nutrient management for irrigated rice in Southeast China. *Agronomy Journal.*93, 869-878.

Wang, G., Zhang, Q. C., Witt, C., Buresh, R. J. (2007). Opportunities for yield increases and environmental benefits through site-specific nutrient management in rice systems of Zhejiang province, China. *Agricultural Systems.*94, 801-806.

Witt, C., Dobermann, A. (2002). A site-specific nutrient management approach for irrigated, lowland rice in Asia. *Better Crops International.* 16, 20-24.

Witt, C., Pasuquin, J. M. C. A., Mutters, R., Buresh, R. J. (2005). *New leaf color chart for effective nitrogen management in rice.* 89, 36-39.

Xiang, Y., Yun, J., Ping, H., Zao, L. (2008). Recent advances on the technologies to increase fertilizer use efficiency. *Agricultural Sciences in China.*7, 469-479.

Xie, W., Wang, G., Zhang, Q., Guo, H. (2007). Effects of nitrogen fertilization strategies on nitrogen use efficiency in physiology, recovery, and agronomy and redistribution of dry matter accumulation and nitrogen accumulation in two typical rice cultivars in Zhejiang, China. *Journal of Zhejiang University Science B.* 8, 208-216.

Yang, X., Fang, S. (2015). Practices, perceptions, and implications of fertilizer use in East-Central China. *Ambio.*44, 647-652. DOI 10.1007/s13280-015-0639-7.

Yao, Y., Miao, Y., Huang, S., Gao, L., Ma, X., Zhao, G., Jiang, R., Chen, X., Zhang, F., Yu, K., Gynp, M. L., Bareth, G., Liu, C., Zhao, L., Yang, W., Zhu, H. (2012). Active canopy sensor-based precision N management strategy for rice. *Agronomy for Sustainable Development.*32, 925-933.

Yi, Bo., Gu, C., Li, J., Abbas, T., Di, H. (2018). Effects of different fertilization regimes on nitrogen and phosphorus losses by surface runoff and bacterial community in a vegetable soil. *Journal of Soils and Sediments.*18, 3186-3196.

Zhang, F. S., Ma, W. (2000). The relationship between fertilizer input level and nutrient use efficiency. *Soil and Environmental Sciences.* 2, 154-157.

Zhang, K., Yuan, Z., Yang, T., Lu, Z., Cao, Q., Tian, Y., Zhu, Y., Cao, W., Liu, X. (2020). Chlorophyll meter-based nitrogen fertilizer optimization algorithm and nitrogen nutrition index for in-season fertilization of paddy rice. *Agronomy Journal.*112, 288-300.

Chapter 9

The Risk Associated with Crop Ecosystem Management and Pesticides Pollution

Manoj Kumar Chitara[1,*], Rajesh Pratap Singh[1], Prince Kumar Gupta[1], Dhruv Mishra[2], Surendra Singh Jatav[3] Sachin Sharma[3] and Hanuman Singh Jatav[4]

[1]Department of Plant Pathology, College of Agriculture,
G. B. Pant University of Agriculture and Technology, Pantnagar (Uttarakhand), India
[2]Department of Biological Sciences, College of Basic Sciences and Humanities,
G. B. Pant University of Agriculture and Technology, Pantnagar, (Uttarakhand), India
[3]Department of Soil Science and Agricultural Chemistry,
Institute of Agricultural Sciences, Banaras Hindu University, Varanasi, India
[4]Department of Soil Science and Agriculture Chemistry,
Sri Karan Narendra Agriculture University, Jobner, Jaipur, India

Abstract

Pesticides are large groups of chemical compounds including herbicides, insecticides, fungicides, nematicides, rodenticides, and plant growth regulators etc., commonly used for crop protection against pests. Indiscriminate utilization of pesticides could adversely impose the risk to food safety, the environment and the living population. Farmers are ignoring the risk associated with pesticides used, safety guidelines, and protective directives in crop management. The improper utilization of pesticides caused the degradation of the quality and fertility of the soil as well as disturbing the nutrient cycling, which leads to heavy metal deposition and toxicity. Excessive exposure of the pesticide to the insect population will be created the development of insect pest resistance, resurgence and decline of the population of the natural enemies. Disposal of unwanted pesticides waste is responsible for the pollution of the water reservoir, groundwater, and pond water sources. The bio-magnification of contaminant cause health associated risk in human beings and animal at different levels of the food chain. In this chapter, we are discussing in detail the risk associated with pesticide application and crop management.

Keywords: pesticide pollution, risk, bio-magnification, heavy metal toxicity

---

* Corresponding Author's Email: manojchitara01@gmail.com.

In: Ecosystem Services
Editor: Hanuman Singh Jatav
ISBN: 978-1-68507-614-6

1. Introduction

Globally, in the agriculture production sector, pesticides are widely used to control the pest population at optimum level as well as improve the yield and quality of the produce by employing the pesticide treatment at various time intervals in crop management (Damalas and Eleftherohorinos, 2011). For crop management program, need for pesticides is originated from the concern that an agro-ecosystem is an open or artificial system where part of the energy derived from the sun is removed by harvesting. Consequently, the sustainability of such a system depends on additional energy, artificially provided by farmers through common agricultural practices (i.e., fertilization, irrigation, pesticide treatments, etc.). Optimal use of pesticides gives benefits to the crops but inadequate use causes severe losses in respect to environmental risk and exposure to pesticides poses a continuous health hazard, especially in the agricultural working environment. Most of the pesticides exhibit a high degree of toxicity because they are developed to kill crop pests and as a result generate some risk of harm. Pesticide risk increases the frequency of damage in terms of human and animal health and their surrounding environment (Damalas and Eleftherohorinos, 2011; Power, 2010), it also reduces the population of the microorganisms in the soil etc. (Kalia and Gosal, 2011). In some instances, accidental intake of pesticide-contaminated food causes illness in humans and animals (Pimentel et al., 1992). During pesticide application in the form of spray on the crop hampered the beneficial insect's viz. predators and parasites as well as continue application causes resistance against pesticides (Dutcher, 2007). Few studies were reported that pesticide application in the field causes surface and groundwater pollution (Azizullah et al., 2011).

Pesticides include insecticides, herbicides, fungicides, nematicides and other chemical substances used for pest controls, contains some toxic ingredients such as phosphorous, chlorine, oxygen, nitrogen, sulfur, and bromine in addition to heavy metals such as lead, arsenic, mercury, copper and sulfates (Greaver et al., 2012; Men et al., 2018) which interferes with the natural environment and cause the harmful effects (Edwards, 2013). After application of the pesticides to the crops, may enter through the runoff from the agricultural fields with irrigation water or another source such as rain, drained water etc. to the natural sources of water like a pond, lake, river thereby contaminating (Rasmussen et al., 2011) as well as affects the health of animals and human beings after drinking the contaminated water (Mekonen et al., 2016)causing the various type of neuropsychiatric disease, cancer and teratogenicity among consumer of pesticides contaminated water (Patel and Sangeeta, 2019). Several soil microbes' are directly and indirectly involved in the breakdown of the organic material, nutrient cycling and helps the plants in the absorption of the minerals etc., these beneficial microbial population faces heavy consequences due to pesticides pollution (Jacobsen and Hjelmso, 2014) as well as it hampers the all microbial related process in the soil, resulting reduces the natural properties of the soil. Here in this chapter, we are discussing the risk due to pesticides application and pesticide pollution.

2. Pesticide Pollution

Pesticides are chemical compounds that are used in a crop management program for suppressing or destroying the pest at the optimum level, which include insecticides (suppress the insects), herbicides (suppress the weeds), fungicides (suppress the fungi), nematicides (suppress the nematode) and rodenticides (suppress the rodents) etc. (Hassaan and El Nemr, 2020). Extensive uses of the pesticides cause discomfort or pollution of the environment as well as it enters in the food web of the ecosystem causes adverse effects on the associated living organisms commonly termed as pesticide pollution (Özkara et al., 2016). The pesticides common ingredients are oxygen (O_2), chlorine (Cl), sulphur(S), phosphorous (P), bromine (Br), nitrogen (N_2) etc. and heavy metals including arsenic (As), copper sulphate ($CuSO_4$), lead (Pb) and mercury(Hg)etc. These all pesticides are being toxic at certain concentrations when it comes in contact with natural sources viz. surface land, water reservoir (pond, lake, river and ocean) plants, animals and microbial communities etc.

Pesticides can interfere with the natural environment which hurts its component at four major levels, first at organism population level particularly a person or life-affecting its birth rate, numbers, growth and mortality, at second level it affects the community of the plants, microbes, animals interactions and diversity of the species in the food web, third at ecosystem level in which affects its functions such as primary and secondary productivity, organic matter breakdown and nutrient cycling and forth at the landscape level, in which it changes in spatial heterogeneity of plants and soil organisms, material transfer of soil and nutrients, and hydrologic transfers of nutrients (Figure 1) etc. (Edwards, 2002).

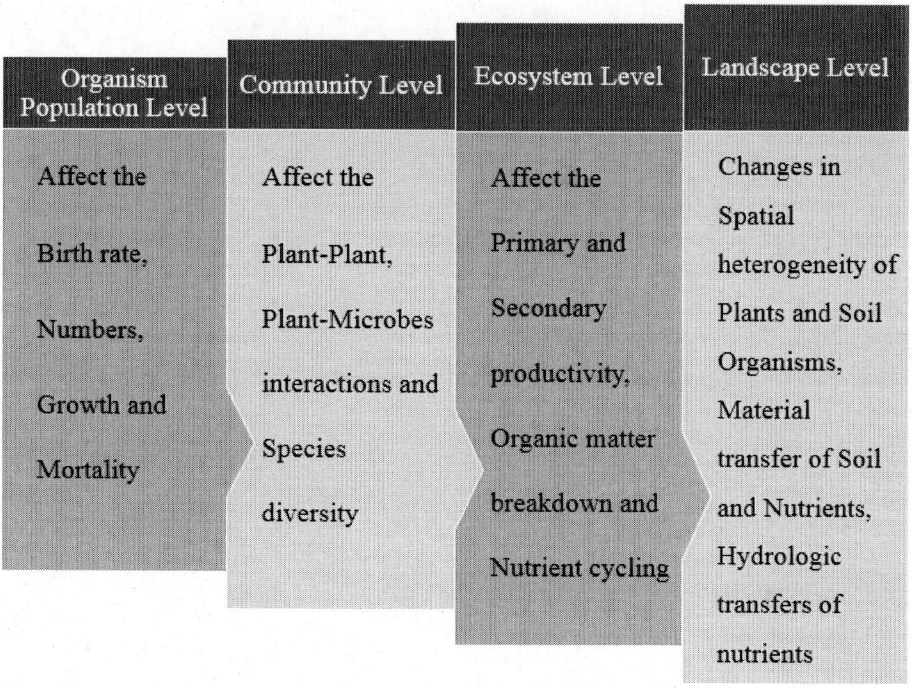

Figure 5. Negative interaction of pesticides with the components of the environment at different levels.

3. Risks Associated with Pesticide Use

Pesticides associated risk may have several prospectives including the effects on the non-target species as well as diversity and the richness of the plant, animal and microorganisms of the terrestrial (Sánchez-Bayo, 2011) and aquatic ecosystem (Muturi et al., 2017). The chemical composition of the pesticides is very toxic, which is capable to disturb the structure and functions of these terrestrial and aquatic communities as well as affecting their living and nonliving component. The chlorinated hydrocarbon derivatives include the class of organochlorine includes DDT, DDD, Dicofol, Eldrin, Dieldrin, chlorobenziate; organophosphate includes Dimefox, Mipafox, Methyl parathion, Phosphomidon; carbamates include carbaryl, dimethan, methan, pyrethroids includes allethrin, cypermethrin; phenyl amides include propanil, dipropanil, phenoxy alkaloids, triazines, benzoic acid, phthalimide, dipyridyl etc., in which some pesticides get dissolve in water and enter groundwater, streams, rivers and lakes hence causing harm to live component viz. plant, animal and microorganism (Jayaraj et al., 2016) as well as affect the quality of the nonliving component viz. water reservoir and land surface of the ecosystem (Kookana et al., 1998) etc. Most of the pesticides after application volatilize within hours or a few days (Garcia et al., 2014). These volatilized pesticides evaporate in the air and may harm the non-target organism. Indiscriminate utilization of pesticides is responsible for the extinction of many species (Helfrich et al. 2009).In contrast, some fat-soluble pesticides may enter the bodies of terrestrial and aquatic animals by the process of biomagnification, in which the fat-soluble pesticides get absorbed through fatty acid of the animals hereafter the pesticide may persist for a longer time in the food chain (Wang et al., 2012; Zhou et al., 2018).

4. The Risk Associated with Living Organisms

The living organisms on the earth include terrestrial and aquatic organism, which is directly or indirectly exposed to pesticides. The pesticide application can pose risk to both terrestrial and aquatic living organisms, but it depends on the amount of the pesticides likely to be exposed, and also affects the health of organisms at the very minute level (Sarwar, 2015). Globally 3 lakh death-occur due to pesticide poisoning every year (Sabarwal et al., 2018). The high toxic pesticides exposure to living organisms causes very severe health issues. The pesticide associated risk may vary with the time and level of exposure to the pesticide. Based on the time and level of exposure, the possibility of the risk of the pesticides is divided into (i) short term & high-level exposure, (ii) long term & high-level exposure and (iii) long term & low-level exposure.

4.1. Risk Associated with Terrestrial Organisms

Pesticides exposure to terrestrial organisms causes the lethal effect. In the case of the plants, the herbicide volatilization or glyphosate drifting during agricultural operation in the standing crops causes significant damage (Huang et al., 2010). Insecticide application of the pyrethroids, for example, cyfluthrin and lambda-cyhalothrin, are highly toxic to social insects such as honey bees etc (Johnson et al., 2006). Neonicotinoids insecticides such as clothianidin affect the activity of the honey bee and imidacloprid at low concentration affects the foraging behaviour (Yang et al. 2008) as well as reducing the learning capacity of the honey bee (Decourtye et al. 2003). The pesticides accumulation such as DDT and its metabolites are responsible for the extinction of many bird species due to accumulation in which is causing the death of the birds (Hellou et al., 2013). Pesticides affect the nervous system of the organisms (EFSA, 2019) such as loss of memory, loss of coordination, slow response to stimuli, loss of visual ability, violent mood, the general pattern of behaviour, and less activity of the motor skill (Sarwar, 2015) as well as carcinogens, causing the hormonal imbalance of the endocrine system, asthma, allergy and hypersensitivity in the body (Van Maele-Fabry et al., 2010; Falzone et al., 2016)while others may irritating to the skin or eyes for example organophosphates and carbamates etc. (Jira et al., 2012).

Pesticides also affect the micro and macro flora of the soil. Application of pentachlorophenol, bisphenol A, DDT, methyl parathion and chrysin reduces the efficiency of the microbes associated with the crop plants such as reducing the nitrogenase activity of the symbiotic microorganism and overall yield of the alfalfa crop (Fox et al., 2007) as well as influences the vital process of the microorganism, for example, it affects the biochemical reaction such as mineralization of organic matter, nitrification, denitrification, ammonification, redox reactions and methanogenesis etc. (Hussain et al., 2009). Application of the herbicides or insecticides are involved in reducing the soil microbial communities for example application of the butachlor or carbofuran reduces the population of the *Azospirillum* (Lo, 2010). Pesticides application such copper oxychloride (Helling et al., 2000), malathion (Espinoza-Navarroand Bustos-Obregón, 2005), acetachlor (Xiao et al., 2006), chlorpyriphos (Zhou et al., 2007), cypermethrin (Shiping et al., 2008), benomyl (Riepert et al., 2009) affect the growth and reproduction of the earthworms (Yasmin and D'Souza, 2010).

4.2. Risk Associated with Aquatic Organisms

The aquatic organism is living in the natural water reservoir such as ponds, lakes, sea etc. The pesticide contamination to the aquatic organism may receive through pesticide drift with water, surface runoff, leaching and during mosquito control etc. The aquatic animal including pieces (small and large fishes), amphibians (frogs), reptiles (turtles) and mammals (blue whales) etc. adversely affected by pesticide pollution. These are exposed to the pesticides via dermally (through the skin), breathing (through the gills) and orally (through the mouth) (Helfrich et al., 2009). The pesticide-contaminated water reduces the activity and function of the vital system of the aquatic animals. Amphibians exposed to chlorpyriphos, diazinon, endosulfan and malathion etc. affects the survival and growth of these animals (Hua and Relyea, 2014). Many pesticides and their related chemicals block the secretion of the action of a hormone in fishes, causing dysfunction and abnormal development of the reproduction system in the amphibians and reptiles as well as organs deformities in the frogs (Khan and Law, 2005). A researcher artificially exposed the pesticide in the lab condition to the aquatic species, reduced the species richness and biomass as well as destroyed the aquatic species e.g., application of the insecticides Sevin and Malathionreducedthezooplankton diversity by 15 and 30% and herbicide Roundup application, eliminated the tadpoles species (Relyea, 2005). The aquatic microbial communities including the majority of the Proteobacteria followed by Bacteroidetes and Firmicutes tested against herbicides (Atrazine and Glyphosate) and insecticides (Carbaryl, Malathion and Permethrin) reduced the aquatic microbial diversity, richness and altered the structure of the microbial community (Muturi et al., 2017).

5. Impact of Pesticide Use on Living Organisms and Their Environment

The toxic effect of the pesticide varied according to the time, if the toxicity of the pesticide appear within the days or immediately then it comes under the acute effect of the pesticide or if it may take several months or years then it comes under the chronic effect of the pesticide. Long term and indiscriminate utilization of pesticides cause several health-related problems for humans and animals in terms of disease and also affect the surrounding environmental activity (Rahman et al., 2015; Sharma and Singhvi, 2017). Infants and children are most prone to the injurious effects of pesticides due to the non-specific nature and inadequate application of pesticides (Goldman, 1995). Pesticide use has increased over the past decades, which causes several adverse effects on the living organism and its surrounding environments (Table 1).

Table 1. The toxic effect of the pesticide on the living species

Sources	Species	Pesticides	Toxic effects
		Insect species	
Guillaume et al., 2017	Honey bee	Fipronil	Reduceddrone fertility
Brunner et al., 2001	*Colpoclypeusflorus* and *Trichogramma platneri*	Organophosphate and carbamate	Causing severe toxicity in the parasitoid
Samuelson et al., 2016	Bumblebee	Thiamethoxam	Impairment of spatial working memory
Cloyd and Bethke, 2011	Predatory insect and mites	Neonicotinoid (Acetamiprid, Clothianidin, Dinotefuran, Imidacloprid and Thiamethoxam)	Inhibit foraging and egg-laying
		Animal species	
Narra, 2016	Fresh water catfish (*Clariasbatrachus*)	Endosulfan, Carbofuran, Methyl parathion and Cypermethrin	Reduced growth, hepatosomatic index and survival performance, inhibition of brain acetylcholinesterase, gills Na+/K+ ATPase activities, and abnormal behaviour
Astiz et al., 2009	Male WistarRat (*Rattusnorvegicus*)	Dimethoate, zineb and glyphosate alone and in combination	Affect the liver and brain by inducing oxidative stress
Yasmin and D'Souza 2007	Earthworm (*Eisenia fetida*)	Carbendazim, dimethoate, and glyphosate alone and in combination	Significantly reduced growth and reproduction
Yang et al., 2019	Human	Triazophos	Negative effect on the endocrine system
Bridges, 2000	Southern Leopard Frog (*Ranaspheno cephala*)	Carbaryl	Causing developmental deformities
		Plant species	
Firdos et al., 2018	*Vigna radiata* (L.) Wilczek (Greengram)	Mancozeb, Chlorpyrifos, Anchor and Metribuzin	Significantly reduced growth, photosynthetic pigments, increased enzymatic activity
Petit et al., 2012	Plants	Copper fungicide	Inhibit photosynthesis, Significantly reduced growth and yield

Pesticide exposure to the human body through ingestion, inhalation or penetration through skin etc. causes acute and chronic health effects such as neurotoxicity, lung damage, chemical burns and infant methemoglobinemia (Sankhla et al., 2018) and majorly it affects through consumption of contaminated food (Schafer and Kegley, 2002).

5.1. Acute Effects of Pesticides

Annually, in the world, 26 million cases are reported of acute poisoning of pesticides (Richter, 2002). Acute effects mean the pesticide can cause some short term or immediate health effects such as headache, watering in the eye, burning sensation in eye and skin, dizziness, irritation of the nose and throat, skin itching, appearance of the rash and blisters on the skin, diarrhoea, abdominal pain, nausea and vomiting, blurred vision, blindness etc. have appeared (Konradsen, 2007; Rastogi et al., 2010). The exposure of the pesticide in short term can be significantly more than average because the consumed food in a single time can be very large compared with long-term consumption and the food may have an ample range of residue than normal (Hamilton et al., 2004). In the case of the organophosphate, acute exposure can affect both the central and peripheral nervous system as well as it evidence that increasing prevalence of the neurologic symptoms may cause neurologic dysfunctions (Rastogi et al., 2010).

5.2. Chronic Effect of Pesticides

The chronic effect of the pesticides means the small number of pesticide doses repeated over the period. It often causes the lethal effect, which may take a longer period for appearing even for years (Alavanja et al., 2004). These are long term effects that cause damage to multiple body organs, for example, disruption of the endocrine system, diabetes, cancer (Porta, 2001), birth defects, reproductive disorders, a neurological disorder such as Alzheimer's and amyotrophic lateral sclerosis (ALS), Parkinson's disease (Betarbet et al., 2000) developmental toxicity, immune toxicity etc (Mostafalou et al., 2013).

Management of Pesticide Pollution

The main aim of pesticide use is the control of the pest population at the optimum level, which does not harm the crop and its surrounding environment. For this, we can use the pesticide at the recommended dose and never use it in excess or indiscriminate amount. Another way is the farmers can use some environmentally friendly pesticides instead of the highly toxic chemical, for example, bio-pesticides e.g., *Trichoderma*sp(Keswani et al., 2017; Gupta et al., 2020), *Pseudomonas* sp.,(Yadav et al., 2013), *Mycorrhiza*(Parihar et al., 2020; Gautam et al., 2021) and endophytes (Mishra et al., 2021) plant extract/botanicals e.g., neem-based products (Yadav et al., 2018), natural enemies for the killing of the target pest e.g., Egg parasitoid *Trichogramma* spp. (Mills, 2009; Gupta and Sumit, 2010; Al-Samarrai et al., 2012) etc.

Farmers should adopt the integrated pest management (IPM) strategies for pest control which include the low use of pesticides, sustainable utilization of the farm product and its waste in pest control, for example, a crop from early growth stage to the harvesting attacked by several pests, for this, we can apply all the measures including cultural, mechanical, physical, chemical and biological management of the pest at optimum level (Stenberg, 2017). Recent IPM strategies include some modern tools which are based on epidemiological studies which mean, these tools are developed based on several years of studies of the pest activity and pattern, so we can easily decide the application pattern of all IPM technique before the pest has arrived on the crop. For example in the case of the apple scab(Singh et al., 2016) and epidemiological tools PREMISE are very helpful in management at the farm (Heijne et al., 2013), VIGICULTURES for an early warning system (Singh et al., 2010) andIndoblight for Potato (Pundhir, 2011; Sharma et al., 2018)etc for crop pest management.

Conclusion and Future Prospects

Agricultural uses of the hazard on pesticide have imposed the risk to human health, environment and raised concerns about the safety of pesticides. It is not possible to eliminate the effect of the pesticide, but we can minimize its toxic effects by alternative cropping approaches, using the pesticides in the appropriate amount, using well-equipped spray material, production of safe and environment-friendly pesticide formulations etc. Similarly, less toxic pesticide formulation application is appropriate in agriculture production. Less biodegradable pesticides have been banned in many countries. Most of the farmers are unaware of the toxic effect of pesticides. They have little or no information about the right way of the pesticide use, their level of poisoning, hazards and safety measures to be taken before use of those pesticides, so it should be made aware by providing training to the farmers and safe handling of pesticides. In future, the sustainable management of pests and diseases by the application of pesticides coupled with cultural and biological means will be more beneficial. It will also help in conserving environmental sustainability and safety. It should be needed, to assimilate the studies of different disciplines including agriculture, chemistry toxicology, ecology etc. to proper understand the direct and indirect effects of pesticides pollution on the environment.

References

Alavanja, Michael CR, Jane A. Hoppin, and Freya Kamel. "Health effects of chronic pesticide exposure: cancer and neurotoxicity." *Annu. Rev. Public Health* 25 (2004): 155-197.

Al-Samarrai, Ghassan, Harbant Singh, and Muhammad Syarhabil. "Evaluating eco-friendly botanicals (natural plant extracts) as alternatives to synthetic fungicides." *Annals of Agricultural and Environmental Medicine* 19, no. 4 (2012).

Astiz, Mariana, María JT de Alaniz, and Carlos Alberto Marra. "Effect of pesticides on cell survival in liver and brain rat tissues." *Ecotoxicology and environmental safety* 72, no. 7 (2009): 2025-2032.

Azizullah, Azizullah, Muhammad Nasir Khan Khattak, Peter Richter, and Donat-Peter Häder. "Water pollution in Pakistan and its impact on public health—a review." *Environment international* 37, no. 2 (2011): 479-497.

Betarbet, Ranjita, Todd B. Sherer, Gillian MacKenzie, Monica Garcia-Osuna, Alexander V. Panov, and J. Timothy Greenamyre. "Chronic systemic pesticide exposure reproduces features of Parkinson's disease." *Nature neuroscience* 3, no. 12 (2000): 1301-1306.

Bridges, C. M. "Long-term effects of pesticide exposure at various life stages of the southern leopard frog (*Ranasphenocephala*)." *Archives of environmental contamination and toxicology* 39, no. 1 (2000): 91-96.

Brunner, Jay F., John E. Dunley, Michael D. Doerr, and Elizabeth H. Beers. "Effect of pesticides on *Colpoclypeusflorus* (Hymenoptera: Eulophidae) and *Trichogrammaplatneri* (Hymenoptera: Trichogrammatidae), parasitoids of leafrollers in Washington." *Journal of Economic Entomology* 94, no. 5 (2001): 1075-1084.

Cloyd, Raymond A., and James A. Bethke. "Impact of neonicotinoid insecticides on natural enemies in greenhouse and interiorscape environments." *Pest Management Science* 67, no. 1 (2011): 3-9.

Damalas, Christos A., and Ilias G. Eleftherohorinos. "Pesticide exposure, safety issues, and risk assessment indicators." *International journal of environmental research and public health* 8, no. 5 (2011): 1402-1419.

Decourtye, Axel, Eric Lacassie, and Minh-Hà Pham-Delègue. "Learning performances of honeybees (Apismellifera L) are differentially affected by imidacloprid according to the season." *Pest Management Science: formerly Pesticide Science* 59, no. 3 (2003): 269-278.

Dutcher, James D. "A review of resurgence and replacement causing pest outbreaks in IPM." *In General concepts in integrated pest and disease management*, pp. 27-43. Springer, Dordrecht, 2007.

Edwards, C. A., ed. *Environmental pollution by pesticides*. Vol. 3. Springer Science & Business Media, 2013.

Edwards, Clive A. "Assessing the effects of environmental pollutants on soil organisms, communities, processes and ecosystems." *European Journal of Soil Biology* 38, no. 3-4 (2002): 225-231.

Espinoza-Navarro, Omar, and Eduardo Bustos-Obregón. "Effect of malathion on the male reproductive organs of earthworms, Eiseniafoetida." *Asian Journal of Andrology* 7, no. 1 (2005): 97-101.

European Food Safety Authority (EFSA), Federica Crivellente, Andy Hart, Antonio F. Hernandez-Jerez, Susanne HougaardBennekou, Ragnor Pedersen, Andrea Terron, GerritWolterink, and Luc Mohimont. "Establishment of cumulative assessment groups of pesticides for their effects on the nervous system." *Efsa Journal* 17, no. 9 (2019): e05800.

Falzone, Luca, Andrea Marconi, Carla Loreto, Sabrina Franco, Demetrios A. Spandidos, and Massimo Libra. "Occupational exposure to carcinogens: Benzene, pesticides and fibers." *Molecular medicine reports* 14, no. 5 (2016): 4467-4474.

Fatma, Firdos, Aisha Kamal, and Alka Srivastava. "Exogenous application of salicylic acid mitigates the toxic effect of pesticides in *Vignaradiata* (L.) Wilczek." *Journal of Plant Growth Regulation* 37, no. 4 (2018): 1185-1194.

Fox, Jennifer E., Jay Gulledge, Erika Engelhaupt, Matthew E. Burow, and John A. McLachlan. "Pesticides reduce symbiotic efficiency of nitrogen-fixing rhizobia and host plants." *Proceedings of the National Academy of Sciences* 104, no. 24 (2007): 10282-10287.

Garcia, Lucas, Carole Bedos, Sophie Génermont, Pierre Benoit, Enrique Barriuso, and Pierre Cellier. "Modeling pesticide volatilization: testing the additional effect of gaseous adsorption on soil solid surfaces." *Environmental science & technology* 48, no. 9 (2014): 4991-4998.

Gautam, Kamini, Chhavi Sirohi, N. Raju Singh, Yourmila Thakur, Surendra Singh Jatav, Kiran Rana, Manoj Chitara, Rajendra Prasad Meena, Ashish Kumar Singh, and Manoj Parihar. "Microbial biofertilizer: Types, applications, and current challenges for sustainable agricultural production." In *Biofertilizers,* 3-19. Woodhead Publishing, 2021.

Goldman, Lynn R. "Children--unique and vulnerable. Environmental risks facing children and recommendations for response." *Environmental Health Perspectives* 103, no. suppl 6 (1995): 13-18.

Greaver, Tara L., Timothy J. Sullivan, Jeffrey D. Herrick, Mary C. Barber, Jill S. Baron, Bernard J. Cosby, Marion E. Deerhake et al. "Ecological effects of nitrogen and sulfur air pollution in the US: what do we know?." *Frontiers in Ecology and the Environment* 10, no. 7 (2012): 365-372.

Gupta, Suman, and A. K. Dikshit. "Biopesticides: An ecofriendly approach for pest control." Journal of Biopesticides 3, no. Special Issue (2010): 186.

Gupta, Prince Kumar, Manoj Kumar Chitara, and Sanjay Kumar Singh. "Antagonism of Native and Commercial *Trichoderma* spp. against *Fusarium solani* Isolates Causing Root Rot of Papaya (*Carica papaya* L.)." *International Journal of Current Microbiology and Applied Sciences* 9, no. 2 (2020): 1260-1269.

Hamilton, Denis, ÁrpádAmbrus, Roland Dieterle, Allan Felsot, Caroline Harris, Barbara Petersen, Ken Racke et al. "Pesticide residues in food—acute dietary exposure." *Pest Management Science: formerly Pesticide Science* 60, no. 4 (2004): 311-339.

Hassaan, Mohamed A., and Ahmed El Nemr. "Pesticides pollution: Classifications, human health impact, extraction and treatment techniques." *The Egyptian Journal of Aquatic Research* (2020).

Heijne, Bart, W. Hennen, and J. Buurma. "PREMISE, a prototype of an empirical model of the epidemiology of apple scab for economic assessment of IPM tools." *PREMISE, a prototype of an empirical model of the epidemiology of apple scab for economic assessment of IPM tools*. 91 (2013): 229-234.

Helfrich, Louis Anthony, Diana L. Weigmann, Patricia A. Hipkins, and Elizabeth R. Stinson. "Pesticides and aquatic animals: a guide to reducing impacts on aquatic systems."https://vtechworks.lib.vt.edu/handle/10919/48060 (2009) pp. 7.

Helling, B., S. A. Reinecke, and A. J. Reinecke. "Effects of the fungicide copper oxychloride on the growth and reproduction of *Eiseniafetida* (Oligochaeta)." *Ecotoxicology and environmental safety* 46, no. 1 (2000): 108-116.

Hellou, Jocelyne, Michel Lebeuf, and Marilynn Rudi. "Review on DDT and metabolites in birds and mammals of aquatic ecosystems." *Environmental Reviews* 21, no. 1 (2013): 53-69.

Hua, Jessica, and Rick Relyea. "Chemical cocktails in aquatic systems: Pesticide effects on the response and recovery of >20 animal taxa." *Environmental Pollution* 189 (2014): 18-26.

Huang, Yanbo, S. J. Thomson, B. V. Ortiz, K. N. Reddy, W. Ding, R. M. Zablotowicz, and J. R. Bright. "Airborne remote sensing assessment of the damage to cotton caused by spray drift from aerially applied glyphosate through spray deposition measurements." *biosystems engineering* 107, no. 3 (2010): 212-220.

Hussain, Sarfraz, Tariq Siddique, Muhammad Saleem, Muhammad Arshad, and Azeem Khalid. "Impact of pesticides on soil microbial diversity, enzymes, and biochemical reactions." *Advances in agronomy* 102 (2009): 159-200.

Jacobsen, CarstenSuhr, and Mathis HjortHjelmsø. "Agricultural soils, pesticides and microbial diversity." *Current Opinion in Biotechnology* 27 (2014): 15-20.

Jayaraj, Ravindran, PankajshanMegha, and PuthurSreedev. "Organochlorine pesticides, their toxic effects on living organisms and their fate in the environment." *Interdisciplinary toxicology* 9, no. 3-4 (2016): 90-100.

Jira, David, StanislavJanousek, Jiri Pikula, Frantisek Vitula, and Kristina Kejlova. "Toxicity hazard of organophosphate insecticide malathion identified by *in vitro* methods." *NeuroendocrinolLett* 33 (2012): 3-9.

Johnson, Reed M., Zhimou Wen, Mary A. Schuler, and May R. Berenbaum. "Mediation of pyrethroid insecticide toxicity to honey bees (Hymenoptera: Apidae) by cytochrome P450 monooxygenases." *Journal of economic entomology* 99, no. 4 (2006): 1046-1050.

Kairo, Guillaume, Yannick Poquet, Haïthem Haji, Sylvie Tchamitchian, Marianne Cousin, Marc Bonnet, Michel Pelissier, André Kretzschmar, Luc P. Belzunces, and Jean-Luc Brunet. "Assessment of the toxic effect of pesticides on honey bee drone fertility using laboratory and semifield approaches: A case study of fipronil." *Environmental toxicology and chemistry* 36, no. 9 (2017): 2345-2351.

Kalia, Anu, and S. K. Gosal. "Effect of pesticide application on soil microorganisms." *Archives of Agronomy and Soil Science* 57, no. 6 (2011): 569-596.

Keswani, Chetan, Kartikay Bisen, Manoj Kumar Chitara, Birinchi Kumar Sarma, and Harikesh Bahadur Singh. "Exploring the role of secondary metabolites of *Trichoderma* in tripartite interaction with plant and pathogens." In *Agro-environmental sustainability*, pp. 63-79. Springer, Cham, 2017.

Khan, M. Zaheer, and Francis CP Law. "Adverse effects of pesticides and related chemicals on enzyme and hormone systems of fish, amphibians and reptiles: a review." *Proceedings of the Pakistan Academy of Sciences* 42, no. 4 (2005): 315-323.

Konradsen, Flemming. "Acute pesticide poisoning–a global public health problem." *Dan Med Bull* 54, no. 1 (2007): 58-59.

Kookana, Rai S., S. N. R. S. N. R. Baskaran, and R. Naidu. "Pesticide fate and behaviour in Australian soils in relation to contamination and management of soil and water: a review." *Soil Research* 36, no. 5 (1998): 715-764.

Lo, Chi-Chu. "Effect of pesticides on soil microbial community." *Journal of Environmental Science and Health Part B* 45, no. 5 (2010): 348-359.

Mekonen, S., Argaw, R., Simanesew, A., Houbraken, M., Mekonen, Seblework, RobaArgaw, AkliluSimanesew, Michael Houbraken, David Senaeve, ArgawAmbelu, and Pieter Spanoghe. "Pesticide residues in drinking water and associated risk to consumers in Ethiopia." *Chemosphere* 162 (2016): 252-260.

Men, Cong, Ruimin Liu, Fei Xu, Qingrui Wang, LijiaGuo, and ZhenyaoShen. "Pollution characteristics, risk assessment, and source apportionment of heavy metals in road dust in Beijing, China." *Science of the total environment* 612 (2018): 138-147.

Mills, Nick. "Egg parasitoids in biological control and integrated pest management." In *Egg parasitoids in agroecosystems with emphasis on Trichogramma*, pp. 389-411. Springer, Dordrecht, 2009.

Mishra, Dhruv, Ajay Kumar, Supriya Tripathi, Manoj Kumar Chitara, and Preeti Chaturvedi. "Endophytic fungi as biostimulants: an efficient tool for plant growth promotion under biotic and abiotic stress conditions." In *Biostimulants for Crops from Seed Germination to Plant Development*, pp. 365-391. Academic Press, 2021.

Mostafalou, Sara, and Mohammad Abdollahi. "Pesticides and human chronic diseases: evidences, mechanisms, and perspectives." *Toxicology and applied pharmacology* 268, no. 2 (2013): 157-177.

Muturi, Ephantus J., Ravi KiranDonthu, Christopher J. Fields, Imelda K. Moise, and Chang-Hyun Kim. "Effect of pesticides on microbial communities in container aquatic habitats." *Scientific reports* 7 (2017): 44565.

Narra, Madhusudan Reddy. "Single and cartel effect of pesticides on biochemical and haematological status of Clariasbatrachus: A long-term monitoring." *Chemosphere* 144 (2016): 966-974.

Özkara, Arzu, DilekAkyıl, and MuhsinKonuk. "Pesticides, environmental pollution, and health." In Environmental Health Risk-Hazardous Factors to Living Species. *IntechOpen,* 2016.

Parihar, Manoj, Manoj Chitara, Priyanaka Khati, Asha Kumari, Pankaj Kumar Mishra, Amitava Rakshit, Kiran Rana et al. "Arbuscular mycorrhizal fungi: Abundance, interaction with plants and potential biological applications." In *Advances in Plant Microbiome and Sustainable Agriculture*, pp. 105-143. Springer, Singapore, 2020.

Patel, Seema, and Sushree Sangeeta. "Pesticides as the drivers of neuro psychotic diseases, cancers, and teratogenicity among agro-workers as well as general public." *Environmental Science and Pollution Research* 26, no. 1 (2019): 91-100.

Petit, Anne-Noëlle, Florence Fontaine, ParulVatsa, Christophe Clément, and Nathalie Vaillant-Gaveau. "Fungicide impacts on photosynthesis in crop plants." *Photosynthesis research* 111, no. 3 (2012): 315-326.

Pimentel, David, Herbert Acquay, Michael Biltonen, P. Rice, M. Silva, J. Nelson, V. Lipner, S. Giordano, A. Horowitz, and M. D'amore. "Environmental and economic costs of pesticide use." *BioScience* 42, no. 10 (1992): 750-760.

Porta, Miquel. "Role of organochlorine compounds in the etiology of pancreatic cancer: a proposal to develop methodological standards." *Epidemiology* 12, no. 2 (2001): 272-276.

Power, Alison G. "Ecosystem services and agriculture: tradeoffs and synergies." *Philosophical transactions of the royal society B: biological sciences* 365, no. 1554 (2010): 2959-2971.

Pundhir, V. S. "Plant Disease Forecasting (Late Blight Forecasting)." *Centre of Advanced Faculty Training in Plant Pathology* (2011): 116-121.

Rahman, KM Atikur, and Sankar Chandra Debnath. "Agrochemical use, environmental and health hazards in Bangladesh." *International Research Journal of Interdisciplinary & Multidisciplinary Studies* 1 (2015): 75-79.

Rasmussen, Jes J., Annette Baattrup-Pedersen, Peter Wiberg-Larsen, Ursula S. McKnight, and Brian Kronvang. "Buffer strip width and agricultural pesticide contamination in Danish lowland streams: Implications for stream and riparian management." *Ecological Engineering* 37, no. 12 (2011): 1990-1997.

Rastogi, S. K., S. Tripathi, and D. Ravishanker. "A study of neurologic symptoms on exposure to organophosphate pesticides in the children of agricultural workers." *Indian journal of occupational and environmental medicine* 14, no. 2 (2010): 54.

Relyea, Rick A. "The impact of insecticides and herbicides on the biodiversity and productivity of aquatic communities." *Ecological applications* 15, no. 2 (2005): 618-627.

Richter, E. D. "Acute human pesticide poisonings." *Encyclopedia of pest management* 1 (2002): 3-6.

Riepert, F., J. Römbke, and T. Moser. "Earthworm Reproduction Tests." In *Ecotoxicological Characterization of Waste*, pp. 171-176. Springer, New York, NY, 2009.

Sabarwal, Akash, Kunal Kumar, and Rana P. Singh. "Hazardous effects of chemical pesticides on human health–Cancer and other associated disorders." *Environmental toxicology and pharmacology* 63 (2018): 103-114.

Samuelson, Elizabeth EW, Zachary P. Chen-Wishart, Richard J. Gill, and EllouiseLeadbeater. "Effect of acute pesticide exposure on bee spatial working memory using an analogue of the radial-arm maze." *Scientific Reports* 6 (2016): 38957.

Sánchez-Bayo, Francisco. "Impacts of agricultural pesticides on terrestrial ecosystems." *Ecological impacts of toxic chemicals.* Bentham Science Publishers Ltd, USA (2011): 63-87.

Sankhla, Mahipal Singh, MayuriKumari, Kirti Sharma, Ravindra Singh Kushwah, and Rajeev Kumar. "Water contamination through pesticide and their toxic effect on human health." *International Journal for Research in Applied Science & Engineering Technology* (IJRASET) 6, no. 1 (2018): 967-970.

Sarwar, Muhammad. "The dangers of pesticides associated with public health and preventing of the risks." *International Journal of Bioinformatics and Biomedical Engineering* 1, no. 2 (2015): 130-136.

Schafer, Kristin S., and Susan E. Kegley. "Persistent toxic chemicals in the US food supply." *Journal of Epidemiology & Community Health* 56, no. 11 (2002): 813-817.

Sharma, Nayana, and RituSinghvi. "Effects of chemical fertilizers and pesticides on human health and environment: a review." *International Journal of Agriculture, Environment and Biotechnology* 10, no. 6 (2017): 675-680.

Sharma, Priyanka, B. K. Singh, and R. P. Singh. "Prediction of potato late blight disease based upon weather parameters using artificial neural network approach." In 2018 9th International Conference on Computing, *Communication and Networking Technologies* (ICCCNT), pp. 1-13. IEEE, 2018.

Sine, Mehdi, Emmanuel Morin, DanièleSimonneau, Michel Brochard, Gérald De Cosnac, and Hervé Escriou. "VIGICULTURES–An early warning system for crop pest management." In Scientific and Technical Information and Rural Development. IAALD XIIIth World Congress, Montpellier, pp. 26-29. 2010.

Singh, K. P., J. Kumar, A. Singh, R. K. Prasad, R. P. Singh, and D. Prasad. "Maturation, ascospores discharge pattern and relevance of mills criteria for predicting apple scab infection period in India." *Plant Pathology Journal* (Faisalabad) 15, no. 3 (2016): 108-123.

Stenberg, Johan A. "A conceptual framework for integrated pest management." *Trends in plant science* 22, no. 9 (2017): 759-769.

Van Maele-Fabry, Geneviève, Anne-Catherine Lantin, Perrine Hoet, and Dominique Lison. "Childhood leukaemia and parental occupational exposure to pesticides: a systematic review and meta-analysis." *Cancer Causes & Control* 21, no. 6 (2010): 787-809.

Wang, De-Qing, Ying-Xin Yu, Xin-Yu Zhang, Shao-Huan Zhang, Yu-Ping Pang, Xiao-Lan Zhang, Zhi-Qiang Yu, Ming-Hong Wu, and Jia-Mo Fu. "Polycyclic aromatic hydrocarbons and organochlorine pesticides in fish from Taihu Lake: Their levels, sources, and biomagnification." *Ecotoxicology and environmental safety* 82 (2012): 63-70.

Xiao, Nengwen, Bobin Jing, Feng Ge, and Xianghui Liu. "The fate of herbicide acetochlor and its toxicity *to Eiseniafetida* under laboratory conditions." *Chemosphere* 62, no. 8 (2006): 1366-1373.

Yadav, R. N., RenuYadav, MdMahtab Rashid, A. I. Khan, and M. K. Chitara. "*In vitro* evaluation of botanicals against rice blast caused by *Pyriculariaoryzae*." *Trends in Biosciences* 11, no. 24 (2018): 3274-3276.

Yadav, Sudheer Kumar, Anushree Dave, Ankita Sarkar, HarikeshBahadur Singh, and Birinchi Kumar Sarma. "Co-inoculated biopriming with *Trichoderma, Pseudomonas* and *Rhizobium* improves crop growth in *Cicerarietinum* and *Phaseolus vulgaris*." *International Journal of Agriculture, Environment and Biotechnology* 6, no. 2 (2013): 255-259.

Yang, E. C., Y. C. Chuang, Y. L. Chen, and L. H. Chang. "Abnormal foraging behavior induced by sublethal dosage of imidacloprid in the honey bee (Hymenoptera: Apidae)." *Journal of economic entomology* 101, no. 6 (2008): 1743-1748.

Yang, Fang-Wei, Yi-Xuan Li, Fa-ZhengRen, JieLuo, and Guo-Fang Pang. "Assessment of the endocrine-disrupting effects of organophosphorus pesticide triazophos and its metabolites on endocrine hormones biosynthesis, transport and receptor binding in silico." *Food and Chemical Toxicology* 133 (2019): 110759.

Yasmin, Shahla, and Doris D'Souza. "Effects of pesticides on the growth and reproduction of earthworm: a review." *Applied and Environmental soil science* 2010 (2010).

Yasmin, Shahla, and Doris, D'Souza. "Effect of pesticides on the reproductive output of Eiseniafetida." *Bulletin of environmental contamination and toxicology* 79, no. 5 (2007): 529-532.

Zhou, Shanshan, Yongqiang Pan, Lina Zhang, Bin Xue, Anping Zhang, and Meiqing Jin. "Biomagnification and enantiomeric profiles of organochlorine pesticides in food web components from Zhoushan Fishing Ground, China." *Marine pollution bulletin* 131 (2018): 602-610.

Zhou, Shiping, Changqun Duan, Fu, Hui, Yu-hui Chen, Xue-hua Wang, and Ze-fen Yu. "Toxicity assessment for chlorpyrifos-contaminated soil with three different earthworm test methods." *Journal of Environmental Sciences* 19, no. 7 (2007): 854-858.

Zhou, Shiping, Duan, Changqun, Wang, Xuehua, Wong Hang Gi Michelle, Y. U. Zefen, and F. U. Hui. "Assessing cypermethrin-contaminated soil with three different earthworm test methods." *Journal of Environmental Sciences* 20, no. 11 (2008): 1381-1385.

Chapter 10

Nanotechnology: An Efficient Tool in Plant Nutrition Management

Sachin Sharma[1,*], Surendra Singh[1], Ayush Bahuguna[1], Bharti Yadav[1], Akshita Barthwal[2], Raghu Nandan Singh Khatana[2], Astha Pandey[1], Rajneesh Thakur[3] and Hanuman Singh Jatav[4]

[1]Department of Soil Science and Agricultural Chemistry, Institute of Agricultural Sciences, Banaras Hindu University, Varanasi (U.P), India
[2]Department of Soil Science and Agricultural Chemistry, Naini Agricultural Institute, SHUATS, Prayagraj (U.P.), India
[3]Dr. Yashwant Singh Parmar University of Horticulture and Forestry, Nauni, Himachal Pradesh, India
[4]Sri Karan Narendra Agriculture University, Jobner, Jaipur, Rajasthan, India

Abstract

The increasing global population accelerates world food demand with limited land resources. It shifts the research towards higher productivity with higher fertilizer use on large scale but with high economic inputs. This promotes the use of Nanotechnology in fertilizer industries which offers higher production with the desired customization in nutrient values, high reactivity, changeable surface area, high nutrient use efficiency because of its different physic-chemical properties. Nanoparticles facilitate a smart delivery system of nutrients with the targeted cellular approach in plants as well as in the rhizosphere. The nanopolymers and nano fertilizers also benefit the plant in stress resistance, crop nutrient quality and improve water use efficiency. However, the behavior of nanoparticles to surpass the food chain, its concentration and toxicity in edible products is still a topic of research. The present chapter deals with their mechanisms, controlled release pattern, particle morphology, chemical kinetics, interaction with soil colloids as well as the effect on the microbiome with lesser amount application in the field and proves to replace the current conventional fertilizers with sustainability and minimal environmental risks.

Keywords: nanoparticles, nanotechnology, nanopolymers, controlled release, smart delivery

* Corresponding Author's Email: sssachin726@gmail.com

In: Ecosystem Services
Editor: Hanuman Singh Jatav
ISBN: 978-1-68507-614-6

1. Introduction

Agriculture is continuously facing many challenges either natural or anthropic in nature like climate change, natural calamities, excess use of fertilizers and pesticides, soil pollution. The current population of 7.7 billion in the world in the present scenario thereby intensified demands for increased food production by 2050 (Chen and Yada 2011). It faces many critical issues with the reduction in cultivable land with economic development. Many technologies have been developed that increase production with fewer inputs but they are not commercially feasible and may increase disruptions between developed and developing countries (Prasad et al., 2012). Therefore the requirement of such a technology that can revolutionize the agriculture and food industry with prospects of sustainable development and nanotechnology proved to be the efficient tool to fulfill demands in all scenarios.

Nanotechnology is a vast field of modern science which deals with the physical, chemical and biological properties of matter considered at the Nanoscale (1-100nm) and their various aspects for the welfare of human beings (Holdren, 2011).Nanotechnology's main prospective in agriculture is to provide timber, fuel, food and also to fulfill the needs of the population by increasing production. Nano fertilizers a tool of nanotechnology can be utilized for the efficient delivery of fertilizers and plant nutrients (Prasad et al., 2012).The mechanism involves smart delivery, controlled release, encapsulation, chemical bonds and others (Sawant et al., 2006).The mechanism itself protects it from chemical degradation, runoff, leaching losses and alleviates environmental issues. Nano fertilizers proved to eliminate conventional fertilizers which have low use efficiency which not only increases cost but also pollution (Wilson et al., 2008). The applied fertilizer is also subjected to various types of losses such as leaching volatilization, a fixation that reduces their efficiency (Naderi et al., 2012, Afify et al., 2019). Based on recent fertilizer use efficiency studies, it is identified that fertilizer efficiency nitrogen is 30-40% in rice and 50-60% in other cereals, while the efficiencies of fertilizer phosphorus, potassium, and sulfur are 15-20%, 60-80% and 8-12% respectively in most of the crops (Stewart and Roberts 2012).

2. Nanoparticles and Their Classification

Nanoparticles, which are particles between 1 and 100 nanometers in size and consist of carbon, metal, metal-oxide, or organic materials, are the most core element of nanotechnology. When compared at a larger scale, nanoparticles have unique physical, chemical, and biological features (Ahmed and Kibret 2014). Nanomaterials with a large surface area could be used as nanocoatings (100nm) to ensure controlled release as well surface protection (Brady and Weil 1999).

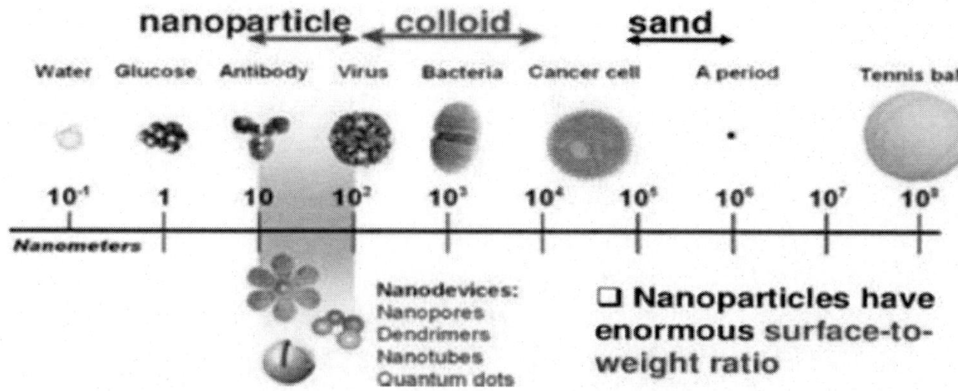

Figure 1. The particle size of nanoparticles in comparison of soil colloid and sand.

It can precisely detect and deliver the accurate quantity of nutrients to crop thereby reducing the effect in soil. The synthesis of nano-fertilizer is done by fortification of nutrients with nano-dimension singly or in combination on to various adsorbent materials, nano-fertilizer can be defined as nanoparticles that sum up materials that slowly delivers nutrients to crop (Cho et al., 2013). A nanoparticle can either be zero-dimensional where the length, breadth and height are fixed at a single point Ex-nanodots, One-dimensional nanoparticle where it can possess only one parameter Ex-graphene, Two dimensional where it has length and breadth Ex-carbon nanotubes and three dimensional is having all the parameters length, breadth and height Ex-gold nanoparticles (NPs) (Guo et al., 2018).

2.1. Types of Nanoparticles

2.1.1. Three Types of Nanoparticles Based on Their Origin

Based on their origin there are three types of nanoparticles:

Natural nanoparticles-

Naturally occurring particles

a) Volcanic dust

b) Lunar dust

c) Terrestrial dust storms

d) Mineral composites

e) Photochemical reactions

f) Forest fires

Incidental nanoparticles-

Man-made/industrial particles

a) Petrol/diesel exhaust

b) Coal combustion

c) Welding fumes

d) Industrial exhaust etc.

Engineered nanoparticles-

a) Carbon based NPs

b) Metal based NPs

c) Magnetic NPs

d) Dendrimers NPs

e) Composite NPs

- Natural nanoparticles

Natural nanoparticles naturally occurring particles have existed since before life began on earth. Ex-soil organic colloids, polysaccharides, humic materials and peptidoglycan, clay, ocean salt, silicate clay minerals and sesquioxides.

- Anthropogenic/Man-made nanoparticles

i. Engineered nanoparticles or manufacturers – particles prepared by humanbeings due to possessing specific nanotechnological properties.
ii. Accidental nanoparticles – particles produced accidentally during the manufacturing and several anthropogenic processes like- electricity generation, diesel-burning, welding, etc. (Kah et al., 2018)

2.1.2. Three Types of Nanoparticles Based on Their Component

Based on their component there are three types of nanoparticles.

2.1.2.1. Organic Nanoparticles

These nanoparticles are bio-degradable, non-toxic and sensitive to thermal and electromagnetic radiation such as heat and light particles Ex-micelles ferritin and liposomes (Tiwari, Bihari and Sen 2008).

2.1.2.2. Inorganic Nanoparticles

These particles are not made up of carbon. Nanoparticles made of metals and metal oxides are included.

a) Metal-based – Metal-based materials are produced to nanometric sizes from metals, either destructively or creatively. Metals that are commonly used for nanoparticles synthesis are Al, Cd, Co, Au, Fe, Pb, Ag and Zn (Salavati, Davar and Mir 2008).
b) Metal oxide-based – These nanoparticles are created to affect the properties of their respective metal-based, and they are mostly synthesized due to their increased reactivity and efficiency. Commonly synthesized are Aluminium oxide, Cerium oxide, Iron oxide, Silicon dioxide, Zinc oxide (Tai, Chang and Liu 2007).

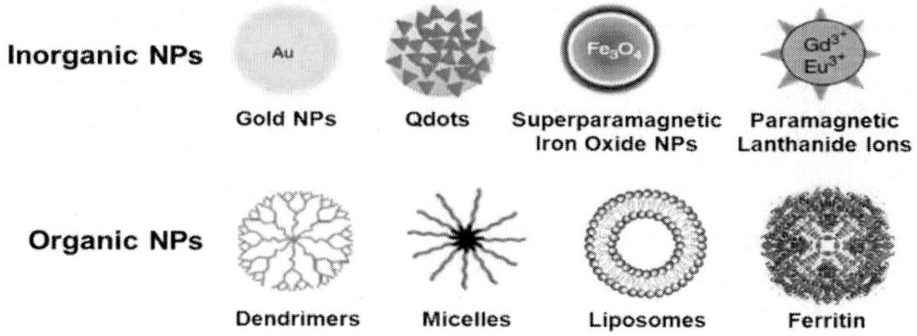

Figure 2. Representing organic and inorganic nanoparticles.

2.1.2.3. Carbon-Based Nanoparticles

The nanoparticles are completely based on the formation of carbon, they can be categorized into fullerenes, graphene, carbon nanotubes, carbon nanofibers and carbon black (Bhaviripudi et al., 2007).

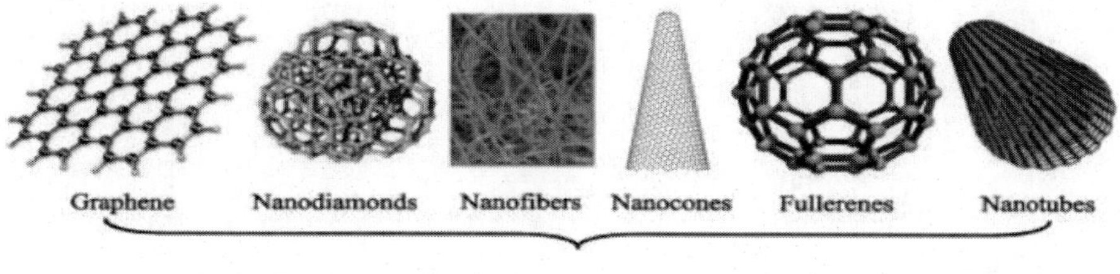

Figure 3. Representing carbon-based nanoparticles.

3. Nanofertilizers and Their Classification

3.1. Nanofertilizer

Nanofertilizers are prepared from conventional fertilizers, bulky materials for fertilizers, or extracted from plant parts or different plants by covering them with nanomaterials for control as well as a slow release of nutrients for the establishment of soil fertility, productivity and quality of agricultural products (Zulfiqar et al., 2019). The combined nutrients can be released in a nano sized form in a controlled manner to improve the efficiency of crop plants along with the minimum impact on the environment. The use of nanofertilizers in a correct way can feed plants slowly in such a manner that increases the nutrient use efficiency, prevents leaching, minimizes volatilization and diminishes overall environmental risks (Linquist et al., 2013). Nanofertilizers improve the bioavailability of nutrients owing to the high specific surface area, mini size and more reactivity (Solanki et al., 2015).

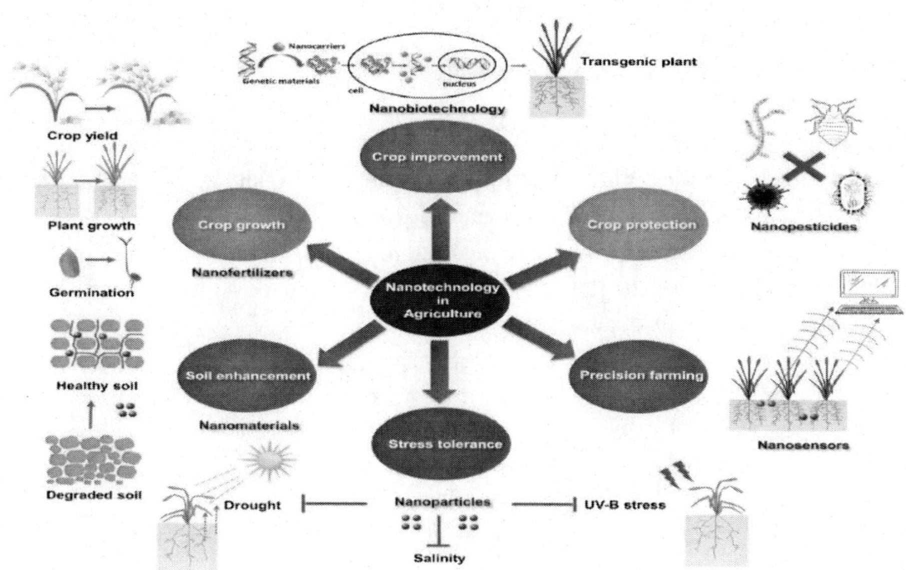

Figure 4. Impact of the use of nanofertilizers in agriculture.

The encapsulation of nutrients with nanomaterials can be done inthe following ways (Chan and Wei 2018, Iqbal 2019).

a) Entrapped within the nanomaterials
b) Coated with a layer of nanomaterials
c) Delivered in the form of nano-emulsions

3.2. Types of Nanofertilizer

Nanofertilizers have been classified into three groups

i. Nanoformulation of macronutrients
ii. Nanoformulation of micronutrients
iii. Nutrients-loaded nanomaterials

3.2.1. Nanoformulation of Macronutrients

This classification includes all the primary and secondary elements for identifying as a label of nanofertilizer.

a) *Nitrogen nanofertilizer* – Zeolite-based nitrogen nanofertilizer indicates higher accumulation of N with reduced leaching and volatilized losses of N and exhibition of better soil pH, moisture (Rajonee et al., 2016).
b) *Phosphorus nanofertilizer* – The use of nanohydroxyapatitenanofertilizer enhance crop growth and seed yield and chemical contents in plants, as well as encapsulated biodegradable P, found that it can be accumulated through leaves which eliminates the requirement of P bonding in the soil (Soliman et al., 2016).
c) *Potassium nanofertilizer* – Application of 150 ppm Nano-K is most effective in increasing the leaf area with no. of leaves, harvest index, grain yield, biological yield, K percentage, disease and pest resistance and chlorophyll content (Ghahremani et al., 2014, Gerdini 2016).
d) *Calcium nanofertilizer* – Spraying of 500 mg/L of Nanocalcium significantly reduces fruit cracking with improvement in fruit quality and quantity and also influences the increase in no. of flowers (Seydmohammad 2020, Ranjbar et al., 2019).
e) *Magnesium nanofertilizer* – Concentration of 500 ppm of Magnesium hydroxide nanofertilizer gives an improvement in growth and development with 100% seed germination (Shinde et al., 2018).
f) *Sulfur nanofertilizer* – Applying sulfur nanofertilizer to the soil enhances sulfur metabolism and limits the deleterious effects of Mn stress (Ragab and Saad-allah 2020).

3.2.2. Nanoformulation of Micronutrients

a) *Iron nanofertilizer* – Iron chelated nanofertilizer is highly stable and provides a slow release of iron in the abroad pH range. Application of iron oxide nanofertilizer at a concentration of 30μM significantly increase growth parameter, photosynthetic pigments and total protein content (Askary et al., 2017).

b) *Zinc nanofertilizer* – Out of all method,the seed priming method of application of zinc nanofertilizer (≤100 mg/kg) efficiently enhance Zn uptake, chlorophyll content with increased no. of leaves flowers and lateral branches (Seydmohammad et al., 2020, Narendhran et al., 2016, Sharifi et al., 2016, Khanm et al., 2018, Munir et al., 2018, Moghaddasi et al., 2017).

c) *Copper nano fertilizer* – Application of biosynthesized copper nanofertilizer at doses up to 20μg/mL improved mitotic index in dividing cells, improvement in stress tolerance, improves the level of protein in starch degradation and increases root length, etc. (Yasmen et al., 2017).

d) *Manganese nano fertilizer* – By using 0.05 mg/L lengthen the roots and shoots and improve the photosynthesis process with increase fresh and dry biomass and causing improved salt tolerance in crops (Preetha and Balakrishnan, 2017).

e) *Boron nano fertilizer* – At 90mg/L concentration of boron nanofertilizer indicated increased plant height, no. of pods total seed yield and forage quality in calcareous condition (Ibrahim and Taherian et al., 2019).

f) *Molybdenum nano fertilizer* – With the application of Molybdenum nanofertilizer causes disease resistance in legumes and improved microbial activity in the rhizosphere (Thomas et al., 2017, Taran et al., 2014).

3.3. Biofertilizers-Based Nano Fertilizer

Biofertilizer-based nanofertilizer is the coexistence of biocompatible nanomaterial and biological sources which focus on slow and gradual nutrient release over a long span for promoting crop yield and productivity (Duhan et al., 2017). Nanocarriers ensure the long-range availability of fertilizers at different stages of plant growth. It betters the field performance and reduces the economic expenditure. The PGPR is beneficial for soil nutrient content via nitrogen fixation, formation of siderophores and phosphate solubilization (Thirugananasambandan 2019).

3.3.1. NanoFertilizer: Methods of Preparation

Nanotechnology used for making nanofertilizer provides a clean environment by providing safer air, water and clean renewable energy for sustainable energy. It's being tested for a variety of different applications to improve the object's efficiency and performance while also lowering the cost so that it's affordable to everyone. Because of its efficiency and environmental friendliness, it has a bright future. Aggregates and colloids not only endanger the security of soil and water bodies, but they may also prove tough to remove (Zulfiqar et al., 2019).

To get a proper understanding of NPs for the delivery of fertilizers to plant. There is a need for a scientific study to evaluate the nanocarriers, nanocomposites and nano-encapsulated fertilizers on crop plants and their environmental consequences. The ultimate target is the production of eco-friendly nanofertilizers which will release their nutrient in a regulated manner in response to different biotic and abiotic signals. The priority of research should include the mechanism of smart delivery of fertilizers in a controlled-release manner as well as to find out the optimum soil condition and other aspects in the environment which determine the behavior and fate of slow-controlled-release fertilizers through the novel formulations of nanotechnology (Solanki et al., 2015).

*Likewise: Worlds' first nano urea introduced by IFFCO which is an indigenously developed fertilizer is in liquid form and promised to revolutionize the fertilizer for farmers in India.

3.3.2. Manufacturing of NanoFertilizers

Nanofertilizers can be prepared by different approaches:

a) Physical (top-down)
b) Chemical (bottom-up)
c) Biological (biosynthesis of natural source)

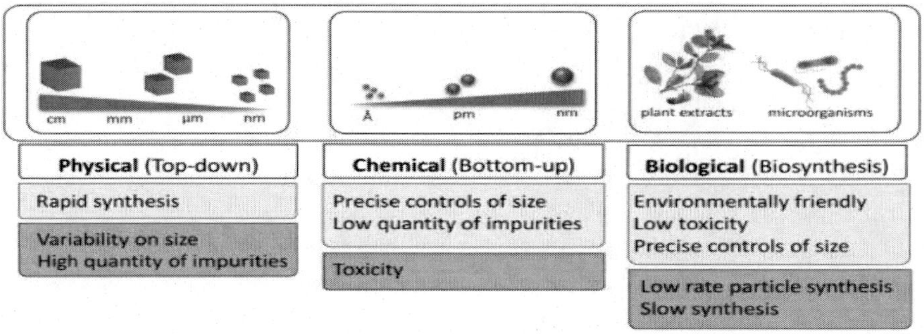

Figure 5. Representing the different methods for the production of nanofertilizer.

a) Physical (Top-down)

The top-down approach is based on the size breaking down of relatively big particles into smaller ones of the nanoscale by mechanical attrition that forms variability in size. It is very easily synthesized, but having some disadvantages in this approach is less control on the size of nanoparticles as well as it causes a greater quantity of impurities. (Zulfiqar et al., 2019)

b) Chemical (Bottom-up)

A bottom-up method starts with molecules in solution and moves via association of these molecules to form nanoparticles using a chemical reaction. It is a chemically controlled synthetic process so that managing the size of particles is possible and therefore it causes a low quantity of impurity and toxicity to the environment. (Subramanian and Tarafdar 2011)

For e.g.,

Preparation of the nanooxide fertilizer using green microwave-assisted hydrothermal method.

↓

$Fe(NO_3)_3.9H_2O$, $Mn(NO_3)_2.4H_2O$, $Zn(NO_3)_2.6H_2O$ dissolved in 50 ml water to obtain 0.2 Msalt concentration.

↓

Adjust pH to 11 by NaOH solution and transfer to 100mL Teflon autoclave vessel.

For Mn3O4, ZnO, and Fe2O3 oxides, transfer to a 750 w advanced microwave synthesis lab station for 3 minutes at 100°, 160°, and 180°C.

Oxides are washed three times, dried at 100°C, ground, and stored in a desiccator under anhydrous calcium chloride.

c) Biological (Biosynthesis)

In the biological method the synthesis of bioagents or natural sources such as plants, fungi, yeast and bacteria. This process hasa slow rate of synthesis, particle size is controlled by this method. Fertilizers are formed and eco-friendly so they cause less toxicity to the environment.

For eg-

0.5 Kg of Banana peel was washed by the tap water and blended using high -speed mechanical blender

The obtained viscous slurry was mixed with the potassium hydroxide and stirred for 1 min. and kept for boiling for 30 min.

Cool it at room temp. and filter it to get clear brown filtrate, heat it again at 70°C with continuous stirring at 300 rpm.

Add urea and citric acid (5% solution) dropwise till 5 pH.

Now obtained sludge is dried at 105°C and grinded to a fine powder for use.

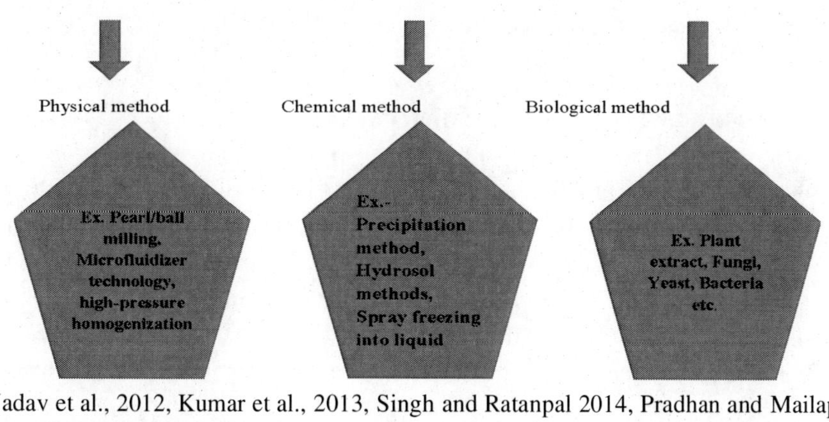

Source: Yadav et al., 2012, Kumar et al., 2013, Singh and Ratanpal 2014, Pradhan and Mailapalli 2017, Ingale and Chaudhari 2013, El-Ghamry et al., 2018.

Figure 6. Sources for preparation of nanofertilizers.

4. Mode of Action and Chemistry of Nanofertilizers

The application of mineral fertilizer in excess hasa huge impact on soil and groundwater. Excess minerals get leached down in soil or contribute to air harming sustainability and crop production (Chandini et al., 2019). Nanofertilizers being ecofriendly are one of the alternatives to mineral fertilizers, capable to increase soil fertility, improve yield, reduce pollution and increasing microbial activities (Ahmed et al., 2012). Nanotechnology can be considered as a powerful solution in the agriculture sector to increase the efficiency of fertilizers and pesticides, reduce pollution and is also used in food processing technology (Baruah and Dutta, 2009).

Nano-fertilizers provide control release of agrochemicals, reduce toxicity in soil and plant, site targeted delivery as well as maximize nutrient efficiencies of the utilized fertilizer (Cui et al., 2010). High surface area to volume ratio, high solubility due to small size, specific targeting, high mobility and low toxicity is the innovative characters that provide advantages of nano-fertilizers (Sasson et al., 2007). Nanotechnology is progressively shifting from experimental areas to practical applications. In the following heads,the mechanism of action of nano-fertilizer can be understood.

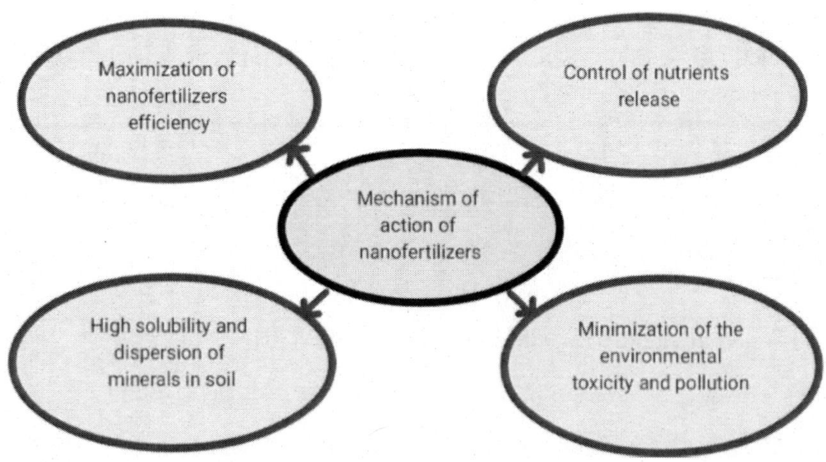

Figure 7. Mechanism of action of nanofertilizers.

4.1. Maximization of Nanofertilizers Efficiency

The nutrient use efficiency in agriculture is 20-50% for N, 10-25%for P and 30-40% for K fertilizers. It indicates the need for food production to be much more efficient (Tarafdar et al., 2013). Fertilizers that are typified in nano-particles have increased accessibility and uptake of nutrients. The utilization of nano-fertilizers has a positive effect on the effectiveness of fertilizers and supplements uptake by plants from soil (Liscano et al., 2000).

Navarro et al., (2008) accredited the high proficiency of the nanofertilizers to

i. Reactivity of nano-materials with other compounds is higher than those of ordinary ones. It is due to the small size and high surface area which provide more sites for reaction.

ii. Enhancement of nutrients penetration and plant uptake, due to the small size of nano-particles, specific surface area and particle number per unit is increased that causes increase in the contact surface between nano-fertilizers and plants (Lin and Xing, 2007). The critical factor to determine the entry of nano-particle through cell-wall is the diameter of plant cell wall pores which range between 5-20 nm (Fleischer et al., 1999). The penetration of nanofertilizer is easier because of their small size i.e., less than 100 nm. Plant cell walls pores are larger than the nano-particle aggregates which makes it easier for nano-particles to enter the plant through the cell wall and reach the plant cell membrane (Moore, 2006).

iii. To enhance uptake, active nano-particles can cause enlargement of pore size and induce new pores in the plant cell wall (Juarez-Maldonado et al., 2019). Nair et al., (2010) in their studies emphasized that nano-particle uptake by plant cells occurs through binding to the carrier protein, through aquaporin, endocytosis, or ion channels.

Nano zeolites and clays are normally available minerals. They are used to increase the efficiency of fertilizers use. These networks of nano-particles are filled with macro-nutrients such as nitrogen, phosphorus, potassium, calcium, or minor or trace elements that act as a slow-release source of nutrients in the soil (Chinnamuthu Boopathi, 2009). Nano-zeolites are important in agriculture due to their ability to capture, store and slow-release nutrients (Leggo, 2000). Urea fortified NaO-zeolite has been considered a slow releaser of nitrogen fertilizers as the release of nitrogen from zeolite is slower than in the ionic form.

4.2. Control of Nutrients Release

Doses of traditional fertilizer applied are not used totally for the crop. Only a few parts reach the targeted plant parts. The remaining fates to leaching, runoff, evaporation, hydrolysis by soil moisture, and degradation by soil flora (Preetha and Balakrishnan, 2017). About 40-70% of nitrogen, 80-90% of phosphorus, and 50-90% of potassium fertilizers are lost within the soil and not able to reach the plant (Saigusa, 2000). More fertilizers and pesticides will be applied to the soil to compensate for the fertilizer lost. It will affect the balance of nutrients adversely.

Nano-fertilizers are one of the ways to overcome the problem of excessive use of fertilizers. Slowly-released nanofertilizer s can be considered a great alternative to traditional fertilizers. They provide a moderate rate of discharge of supplementsduring the development of crops. Plants can absorb most of the nutrients without much loss (Huiyuan et al., 2018). Nano-particles strongly bond the nutrients when fertilizer is coated on the surface of nanomaterials because of their higher surface tension than that of conventional fertilizers. This coating increases the efficiency of fertilizers by controlling the release of these fertilizers (Brady and Weil, 1999).

Nitrogen fertilizer is very important but because of its high solubility nature,it causes severe damage to the plants and surroundings. Leggo, (2000), in his study, highlighted that there was a considerable increase in nitrogen uptake with efficient controlled release when nanoporous zeolite was used with urea. Urea modified hydroxyapatite (HA) nanoparticles were synthesized for controlled release of nitrogen through the crop growth period and it was found to have a slower release of nitrogen reaching 60 days of plant growth when compared to traditional fertilizers which reach upto 30 days only (Kottegoda et al., 2011).

Soluble fertilizers are enveloped inside nano-materials in controlled-release nanofertilizer s. It prevents the exposure of water-soluble materials from the surroundings. The soluble part is released into soil mainly through diffusion (Huiyuan et al., 2018). Nanofertilizers release nitrogen 50 days slower when compared to conventional mineral fertilizers (Subramanian and Rahale, 2009). To encourage the slow release from fertilizers, a nono-membrane can be used to coat fertilizer particles (Tarafdar et al., 2012). Nanocomposite containing macronutrients like nitrogen, phosphorus and potassium are widely in use as they increase the uptake and utilization of nutrients by the crop plants.

4.3. Minimizationof the Environmental Toxicity and Pollution

Excessive use of fertilizer and pesticides had harmed flora and fauna by creating an imbalance of nutrients in the soil and causing environmental pollution. Continuous and large-scale application of fertilizer reduces the fixation of nitrogen, increases the bioaccumulation of pesticides, increases pest and pathogen resistance, and destroys the natural habitat of organisms. This creates a huge environmental imbalance which causes sustainable and economic losses (Tilman et al., 2002).

Nanofertilizers reduce the rate at which fertilizers are lost in the soil. It decreases the use of conventional mineral fertilizers, which causes contamination of soil. Encapsulated nanofertilizer s like nano-clays and zeolites increase the efficiency of using fertilizers, which in turn enhance plant health and soil fertility and also reduce environmental pollution and agro-ecological degradation (Manjunatha et al., 2016). It is important to check the use of mineral fertilizers, provide nutrients in optimum amount and decrease the damages of the environmental contamination. These objects can be accomplished by using nanofertilizer s.

4.4. High Solubility and Dispersion of Minerals in the Soil

Nano-mineral micronutrients increase the solubility of insoluble nutrients and dispersion in soil. It will reduce the absorption and fixation of nutrients in the soil. All these together lead to an increase in the bioavailability of nutrients and will also increase nutrient and will also increase nutrient uptake efficiency (Guru et al., 2015). The use of nano-sized rock phosphate increases the availability of phosphorus to the plants. It reduces the fixation of phosphorus in soil. There is no phosphorus obsession due to iron, silica acid, or calcium in the soil, thus it leads to increased phosphorus accessibility within the soil. ZnO in form of nano-particles have a higher dissolution rate than the bulk ones. This character of ZnO nano-particles of high solubility can improve their efficiency as novel fertilizer.

4.5. Mobility of Nanofertilizer in Soil

Soil is a sink for nanoparticles released in the environment. It is essential to reveal the potential impact of nanofertilizer s on the food chain and groundwater to understand the transport behavior of nanofertilizer in soil systems. Interaction between nanofertilizer s and solid surfaces decides the mobility of nano-particles (Doshi et al., 2009). This interaction can be influenced by the environmental and physicochemical properties of particles. Soil is a porous medium as it contains macropores and micropores. Micropores are very small pores, consist of a network of humic substances and soil particles (Kretzschmar and Schafer, 2005). Nanoparticles are enough small to fit in the micropores and clog them. This will affect the mobility of nanoparticles. When nano-particles form aggregate they become too large to fit in micropores, this makes them remain in macropores (Kumar et al., 2012).

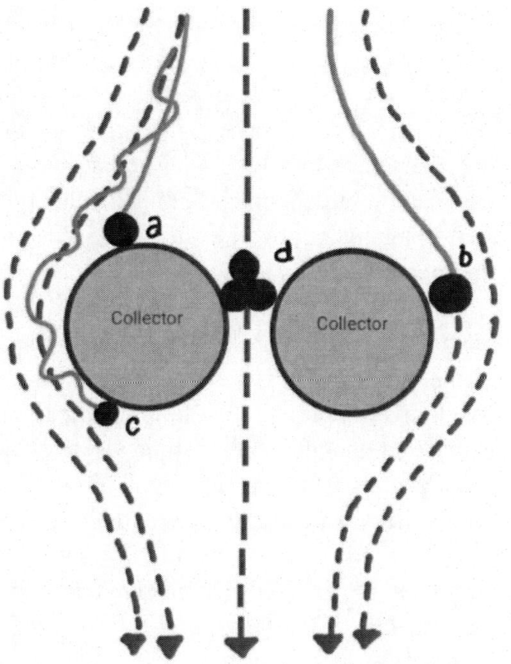

Figure 8. Transport of nanoparticles in porous media through infiltration mechanism. The purple spheres, dotted lines and thick lines are NPs, fluid streamlines and trajectory paths respectively. The NPs can attach to the collector through: (a) Gravitational sediment; (b) Interception; (c) Brownian diffusion; (d) Large aggregates retained by small pores. [modified from Moore and Mondal (2016)].

Nano-particles adsorb to soil and sedimental particles very strongly due to their high surface areas (Oberdorster et al., 2005). Nano-particle when sorbed to the immobile particle its mobility gets hindered, while sorption to the mobile particle may build up its mobility. The shape of the nanoparticle and collector surface hasa considerable effect on the adhesion of nanoparticles to the surface. The factor that controls the stability of nano-particles also tends to control the mobility properties of nanoparticles. Sedimentation and diffusion are also affected by gravity and Brownian motion (Tourinho et al., 2012). Brownian motion governs the transport of nanoparticles in porous media. When nano-particle agglomerate and aggregate gravitational force predominates, then the particles are more likely to interact with the surfaces of soil particles (Dunphy Guzman et al., 2006). The interaction between metal-based nanoparticles and soil surfaces is also dependent on the surface charge of both nanoparticles and the soil. The difference between charges on nanoparticles and soil has an impact on electrostatic attraction or repulsion and aggregation among particles and between particles and soil (Tourinho et al., 2012).

The mobility of nanoparticles depends on the rate of their infiltration or capture by stationary surfaces. Particle infiltration theory explains the behavior of engineered nanomaterials in soil. There are mostly three mechanisms for the capture of colloidal material in porous media: (i) gravitational sediment; (ii) interception; (iii) Brownian diffusion (Schrick et al., 2004). The sedimentation and interception mechanism works for larger particles and aggregates. For the particles of nanoscale, diffusion is the dominant mechanism. The high diffusivity of nanoparticles leads to a higher number of collisions with the surface of the soil.

Nanoparticles when aggregates may get screened by the soil pores due to physical screening if the particle size is greater than thatof pore size through which fluid is flowing. The capture or infiltration of nanoparticles by collector surfaces is determined mainly by chemical colloidal interaction between particles and surfaces. This interaction is regulated by solution chemistry and the chemical characteristics of particles and surfaces. Physical parameters like fluid velocity, particle size, water temperature and grain size also play important roles in the infiltration of engineered nanomaterials. Physic-chemical properties of engineered nanomaterials like size, morphology, and surface properties can significantly influence their mobility in porous media. The transport of nanoparticles is more complicated than microscale particles because of their nature to aggregate, which depends on the particle size.

Surface charge is an important property that can determine the mobility of engineered nanomaterials in porous media (Darlington et al., 2009). Soil particles are normally negative and positively charged engineered nanomaterials will be electrostatically attracted to the soil surface. Engineered nanomaterials having a negative charge will have more mobility due to stronger electrostatic repulsion between nanoparticles and soil and nanoparticles themselves. Therefore, various methods have been used to modify the surface properties of engineered nanomaterials to control the mobility of engineered nanomaterials in porous media.

Solution pH and ionic strength also affect the capture of engineered nanomaterials by porous media, thus affecting the transport of nanoparticles. Solution pH controls the solubility of nanoparticles and their surface charges, to the extent, it controls electrostatic interactions between nanoparticles and between nanoparticles and porous media (Doshi et al., 2008).

5. Factors That Influence Smart Delivery Systems of Nanofertilizer

After release into the environment, the soil is a major sink for nanofertilizers and a potential source of nanofertilizers in grains. It's crucial to comprehend the nanofertilizer transport behavior in natural soil systems to comprehend their possible impact on the food chain and groundwater. The chemical-colloidal interaction between particles and the surface, which is regulated by solution chemistry, is used to estimate particle filtration by collector surfaces (e.g., pH, coexisting organic matters and ionic strength) chemical and physical properties of particles and surfaces.The behavior of nanofertilizer s during their transportation in natural soil systems is critical for determining their possible effects on groundwater, human health, and the environment.

5.1. Interaction with Soil Colloids

Soils are a natural complex matrix with mineral particles and colloids in the pore water, and there are issues concerning nanofertilizer adsorption and binding. Brownian motion into soil pores is the most important factor affecting nanoparticle migration in soil. The shape of nanoparticles and collectors, as well as many variables that alter the environment surrounding the particles, influence the interaction between soil molecules and nanoparticles. As a result, soil conditions can either increase or decrease nanoparticle mobility in soils (Fang et al., 2009).

The interaction between particles and solid surfaces determines how nanofertilizer transportation through soils (Darlington et al., 2009) is controlled by the particles' physiochemical and environmental properties (Tourinho et al., 2012). Nanofertilizers have a large surface area and a great potential for adsorption to soil and sediment particles (Oberdorster et al., 2005). The interaction of nanofertilizers with soil surfaces is influenced by the surface charge ofthe nanofertilizer and the soil. To decrease the uncontrolled release in the soil environment, nanofertilizer s are blended with various materials such as hydrogen and polymers such as Chitosan (Kashyap, 2015). These materials cause nanofertilizers to clump together and form complexes with mineral nanoparticles derived from soil clay.

The size, chemistry, and structure of the nanofertilizer core determine its reactivity. The combination of an inorganic core and organic/inorganic shell determines the reactivity of nanofertilizers.

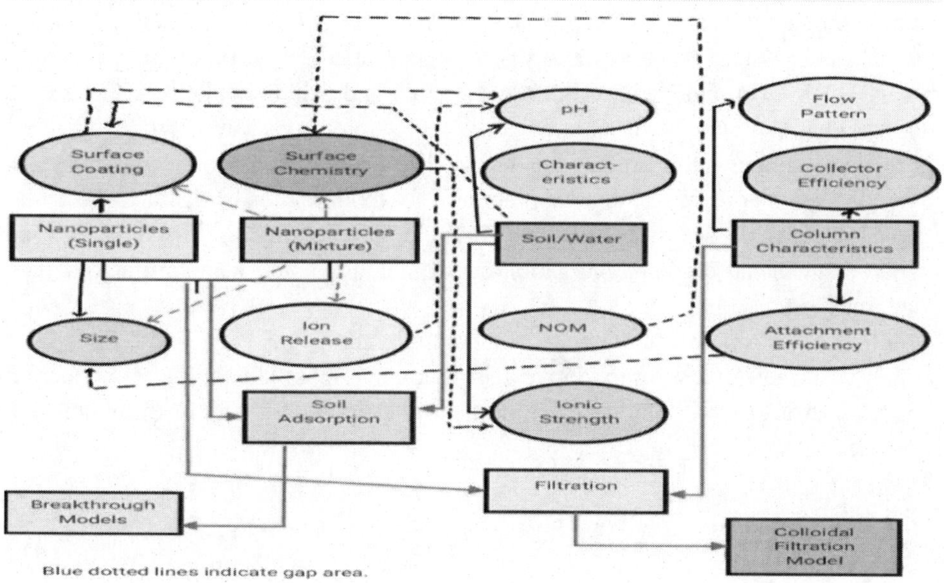

Figure 9. Factors affecting the transportation of nano fertilizer in porous media.

The shell can be organic or inorganic, with organic chemistry imparting interfacial contact. As a result, the particle surface coating plays a crucial role in modulating particle reactivity, as well as defining its solubility and selectivity against the targeted target. Metal-based nanofertilizers have a variety of physical qualities, including size, shape, and chemical properties such as acidic/basic surface properties and metal solubility.

The transformation, which involves processes such as agglomeration, sorption to surfaces, and ionic metal dissolution, will control their destiny performance and ecotoxicity in the environment. Nanofertilizers enter the soil in three different ways.

- By dissolved state
- By absorption of solid nanoparticles
- By the direct entry of nanofertilizer

5.2. Interaction with Organic Matter

The agricultural soil has been enhanced with a variety of nutrients. Agricultural soils are high in organic carbon, which can contribute to the organic surface coating and influence nanofertilizer mobility in the soil. Soil with dissolved organic matter can adsorb nanofertilizer on its surface. At ambient pHs, humic substances are negatively charged, and the sorption of nanofertilizer causes the overall particle-humic conglomerate to become negatively charged (Gosh et al., 2008), resulting in particle stability in solution, reduced aggregation and settling. (Fang et al., 2009; Ben Moshe et al., 2010).Nanofertilizers can be stabilized by natural organic materials. After being released into the soil solution, nanofertilizers may interact with decomposed organic matter in particle form, causing them to be suspended and transported rather than aggregate.

The decomposed organic particles adhere to the surface of nanofertilizer and modify their surface physicochemical properties and improve their stability. Soil organic matter includes compounds such as organic acids, Chitosan, dextran and starch and others. Certain organic compounds coat the surface of nanofertilizer and cause aggregation while others induce desegregation. Although the organic matter modifies the surface of the nanofertilizer, the outcome of the nanofertilizer depends on charge modification. Organic matter induces reduction of nanofertilizer size likely resulting in their ability to lessen particle-particle interaction that can cause homoaggregation or particle interaction with soil factors that cause heteroaggregation.

Stabilization of nanofertilizer due to the coating by humic acid could result in minimizing the particle surface available for oxidative dissolution into ions. The long-term effect of humic acid on nanoparticles and stated that humic acid play important role in the mobility of the nanoparticles.

5.3. Ionic Strength

The ionic strength of the soil solution impacts nanofertilizer distribution in porous media by causing aggregation or stabilization (Cornelis et al., 2012; Fang et al., 2016). Nano fertilizer deposition has been reported to increase with increasing solution strength in both monovalent and divalent salt solutions. Jiang et al., (2012) examined the transport behavior of ZnO nanoparticles in quartz sand in divalent and monovalent salt solutions at varying ionic strengths. Higher ionic strength resulted in increased van der Waals forces over repulsive electrostatic interactions, resulting in improved aggregation. The electrical double layer will be compressed as the ionic strength rises, lowering the repulsive barrier between nanoparticles. As your ionic strength grows, you'll be able to do more things. By developing ionic strength diffuse layer between collector and nanoparticles is reduced, thus reducing repulsion and zeta potential (Jones et al. 2012).

For nanofertilizer retention in porous media, the Derjaguin, Landau, Verwey, and Overbeek (DLVO) theory can be applied. The DLVO theory addresses the interaction energy between nanoparticles and collector surfaces as well as the stability of nanomaterials in suspension and is used to investigate nanoparticle adsorption on soil particles

5.4. pH

Understanding the influence of pH on nanofertilizer suspension is critical to understanding the behavior of nanofertilizer in the soil. The zeta potential of colloids is mostly controlled by pH. When the pH is less than spec, the colloidal surface is positively charged, and the zeta potential increases as the pH fall below zero. When the pH is higher than zero, the colloidal surface has a positive charge, and the zeta potential decreases. Nanoparticles with a low zeta potential are more prone to flocculate, while those with a greater zeta potential are more stable to the suspension of nanoparticles. The pH of the soil is a critical component in regulating the behavior of nanofertilizers in the soil.

Nanofertilizers are rapidly converted into ionic species in acidic environments, whereas alkalinity promotes nanofertilizer aggregation. At alkaline pH, they clump together, lose their nano-specific features, and the rate of further dissolution into ions slows. Surface charge, which is a function of pH, determines the responsiveness of nanofertilizers. In acidic (pH 4.5-5.5) vs. Basic (pH 4.5-5.5) settings, CuO and ZnOfertilisers produced different reactions. Alkaline (pH 7.8-8.1) soils showed positive growth responses, while acidic soils showed negative growth responses (Watson et al., 2015; Anderson et al., 2017).

5.5. Temperature

Soil temperature has a big impact on fertilizer effectiveness because it affects fertilizer solubility, cation exchange capacity, and the plant's ability to absorb and use nutrients. Too high a temperature distorts the structure, reducing nanofertilizer availability, while too low a temperature inhibits nutrient solubility, therefore the ideal temperature is preferable for its stability.

6. Ecological Issues withNanoFertilizers

Nanomaterials may trigger tissue injury either by a single or combined mechanism at the target site. Interactions of nanomaterials with various biological components like cells, fluids and protein binding and translocation. Enzyme malfunctions may develop as a result of structural alterations on the surface of nanoparticles. Nanoparticle surface characteristics are acknowledged to be one of the most critical elements governing their stability and mobility as colloidal suspensions on adsorption, aggregation, and deposition (Tantawy 2015).

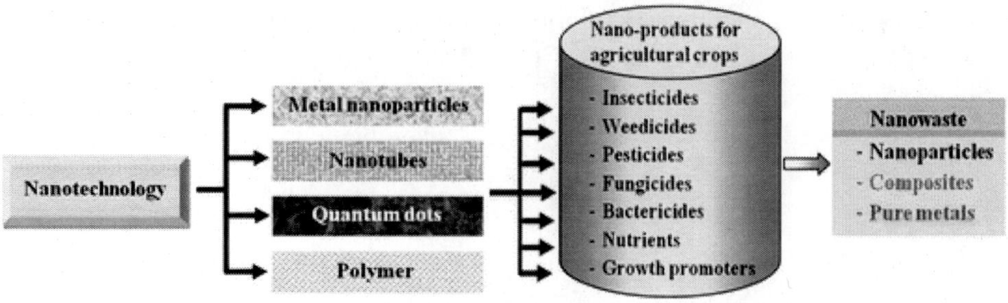

Figure 10. Manufacturing of nano-products by nanotechnology to the outlet of nanowaste (Thul et al., 2013).

6.1. Impact of Nanomaterials on the Rhizospheric Environment of Soil

The inorganic NPs like TiO_2, SiO_2 and ZnO gives a toxic effect on bacteria and other organisms thus uptake of manufactured nano-CeO_2 nanomaterials into roots and root nodules arefound to eliminate N_2 fixation potentials and impaired growth of the plant (Priester et al., 2012, Fan et al., 2014). The Nano TiO2 has a significant influence on rhizobium-legume found that nano TiO_2 exert morphological changes in bacterial cells. Soil bacteria may be exposed to NPs applied directly to land or treated biosolids containing mobile NPs (Ge et al., 2012, Holden et al., 2013, Zeliadt 2013). Microbicidal AgNPs negatively impact the growth of plants that kills the soil microbes and the activity of soil enzymes such as soil protease, catalase and peroxidase were significantly reduced. Physico-chemical characteristics of NPs (e.g., shape, size and surface charge) and soil (e.g., pH, ionic strength, organic matter and clay content) after physicochemical processes results in NPs dissolutions, agglomeration and aggregation in which CuNPs can be considered the least mobile in comparison to Fe3O4, CuO, TiO_2 and ZnONPs. Their behavior in soil controls mobility and their bioavailability to soil organisms which may interact with beneficial soil microbes and extend the impact on their survival (Du et al., 2011, Adams et al., 2006, Suresh et al., 2013).

6.2. Impact of Nanomaterials on the Accumulation in Plants

The accumulation and subsequent availability of nanoparticles in food crops transfer of NPs into the food chain through edible plants is a great concern (Yin et al., 2011, Servin et al., 2013). Adsorption onto the root surface, integration into the cell wall, and absorption into membranes are the most common interactions of NPs with plant roots. The interaction of NPs with the toxic, organic compounds can both amplify and alleviate the toxicity of compounds in plants (Zhao et al., 2013, Kureapa et al., 2010, Ma et al., 2013), For e.g., nanoTiO_2 shows toxic effect in green algae which have a similar cell wall to plants, the large amount of nZVI coated on plant root surface as irregular aggregates and some penetrate several layers of epidermal cells of root cells of the plant, etc (Shi et al., 2014, Hu et al., 2014, Zhai et al., 2008).

6.3. Impact of Nanomaterials on the Accumulation in Water Bodies

The level of NPs in soil and water is increasing due to the growing consumer products that contained NPs on waste streams. The organic carbon concentration is high in agricultural soil likely a layer made of organic materials and resulted in NPs mobility through soil have the engineered nanoparticles' fate and bioavailability in soil wherein the ENM bioavailability is determined by salinity, texture, pH, concentration and type of mobile organic molecules, and saturation level (Boxall et al., 2007, Klaine et al., 2008). The leftover leachates from excess and after use of NPs ultimately find the way and accumulate in aggregates and colloids in their diverse types to the water bodies. These aggregates and colloids having NPs will generate additional anthropogenic waste (nanowaste) in the agroecosystem that is should be in continuous monitoring of the fate of nanoproducts leftover nanowaste products and soil composition (Biswas 2005,Cornelis et al., 2012), For e.g., Arsenic is well-known groundwater contaminant and exists as both arsenate oxyanions at neutral pH under oxidizing conditions and arsenite under mildly reducing conditions at pH below 9.2, both can be remediated by using INP (iron nanopolymer) along with their corrosive products that tend to form aggregate during oxidative corrosion to iron oxide/hydroxide. These corrosive products restrict effective transport as well as delivery of INP through porous media which is normally essential for in situ groundwater remediation (Zhao et al., 2013, Collins et al., 2012).

Conclusion

Nanomaterials are accumulated over time in the form of aggregates and colloids are not only a threat to soil and water bodies but may be impossible to remediate thus nanocarriers could be designed in such a manner that can anchor the plant roots and surrounding soil structure and organic matter. The ultimate target is the production of eco-friendly nanofertilizers which will release their nutrients in a regulated manner in response to biotic and abiotic signals as well as to find out optimum soil conditions and other variables in the environment through the novel formulation of nanotechnology.

References

Adams L. K., Lyon D. Y., Alvareez P. J. (2006) Comparative eco-toxicity of nanoscale TiO_2, SiO_2, and ZnO water suspensions. *Water Res* 40:3527-3532.

Afify, R.R., El-Nwehy, S.S., Bakry, A.B., and M. E., A.E.A. (2019). Response of peanut (*Arachis hypogaea L.)* crop grown on newly reclaimed sandy soil to foliar application of potassium nano-fertilizer, *Sciences,* 9, pp.78-85.

Ahmed, M., and Kibret, M. (2014). Mechanism and applications of plant growth-promoting rhizobacteria: Current perspective. *J. King Saud Univ.Sci.,* 26, pp.1-20.

Ahmed, S., Niger, F., Kabir, M. H., Chakrabarti, G., Nur, H. P. and Imamul Huq, S.M. (2012). Development of Slow Release Nano Fertilizer. *Proceedings of the InternationalWorkshop on Nanotechnology*, Dhaka, Bangladesh. 1(45): 21-23.

Anderson, Anne, Joan McLean, Paul McManus, and David Britt. (2017) "Soil chemistry influences the phytotoxicity of metal oxide nanoparticles." *International Journal ofNanotechnology* 14: 15-21.

Askary, M., Amirjani, M.R., and Saberi, T. (2017). Comparison of the effects of nano-ironfertilizer with iron-chelate on growth parameters and some biochemical properties of *Catharanthusroseus. J. Plant. Nutr.,*40,pp.974-982.

Baruah, S. and Dutta, J. (2009). "Nanotechnology applications in sensing and pollution degradation in agriculture." *Environmental Chemistry Letters*. 7: 191-204.

Ben-Moshe, Tal, Ishai Dror, and Brian Berkowitz. (2010). "Transport of metal oxide nanoparticles in saturated porous media." *Chemosphere* 81: 387-393. Collins, Daniel, Todd Luxton, Niraj Kumar, Shreya Shah, Virginia K.

Bhaviripudi S., Mile E., lii S. A. S., Zare A. T., Dresselhaus M. S., Belcher A. M. and Kong J. (2007) CVD Synthesis of single-walled Carbon Nanotubes from Gold Nanoparticles Catalysts 1516-7.

Biswas P., Wu C. Y. (2005) Nanoparticles and the environment. *J Air Waste Manag Assoc* 55:708-746.

Boxall A. B., Chaudhry Q., Sinclair C. (2007) Current and future predicted environmental exposure to engineered nanoparticles. Central Science Laboratory, Department of theEnvironment and Rural Affairs, London, UK.http://randd.defra.gov.uk/Document.aspx?Document= CB01098_6270_FRP.pdf.

Brady, N.C. and Weil, R.R. (1999).*The Nature and Properties of Soils.* 12th Edition, PrenticeHall Publishers, London. 1-9: 453-536.

Chan, J. and Wei, X. (2018).Controlled-released fertilizers as a means to reduce nitrogen leaching and runoff in container-grown plant production. In *Nitrogen in Agriculture*, pp. 33-52. IntechOpen.

Chandini, K.R.; Kumar, R. and Prakash, O. (2019). "The Impact of chemical fertilizers on our environment and ecosystem." *Research Trends in Environmental Sciences*. 69-86.

Chinnamuthu, C.R. and Boopathi, P.M. (2009). "Nanotechnology and Agroecosystem." *Madras Agricultural Journal.* 96: 17-31.

Chen, H., & Yada, R. (2011). Nanotechnologies in agriculture: new tools for sustainable development. *Trends in Food Science & Technology*, *22*(11), 585-594.

Cho E. J. Holback H., Liu K. C., Albouelmagd S. A., Park J. and Yeo Y. (2013) Nanoparticle characterization: state of the art, challenges and emerging technologies.

Collins D., Luxton T., Kumar N. (2012). Assessing the impact of copper and zinc oxide nanoparticles on soil: a field study. *PLoS ONE* 7:e42663.

Cornelis G., Thomas C. D., McLaughlin M. J., et al., (2012) Retention and dissolution ofengineered silver nanoparticles in natural soils. *Soil Sci Soc Am J* 76:891-902.

Cornelis, Geert, Casey Doolette Madeleine Thomas, Mike J. McLaughlin, Jason K. Kirby, Douglas G. Beak, and David Chittleborough. (2012) "Retention anddissolution of engineered silver nanoparticles innatural soils." *Soil Science Society of America Journal* 76: 891-902.

Cui, H.X., Sun, C.J., Liu, Q., Jiang, J. and Gu, W. (2010). Applications of nanotechnology in agrochemical formulation, perspectives, challenges and strategies. In*: International conference on Nanoagri*, Sao Pedro, Brazil. 28-33.

Darlington, T. K., Neigh, A. M., Spencer. M. T., Nguyen, O. T., Oldenburg, S. J. (2009). "Nanoparticle characteristics affecting environmental fate and transport through soil." *Environ Toxicol Chem* 28:1191–1199.

Darlington, Thomas K., Arianne M. Neigh, Matthew T. Spencer, Oanh TN Guyen, andSteven J. Oldenburg. (2009). "Nanoparticle characteristics affecting environmental fate and transport through the soil." *Environmental Toxicology and Chemistry: AnInternational Journal* 28: 1191-1199.

Doshi, R., Braida, W., Christodoulatos, C., Wazne, M., O'Connor, G. (2008). "Nano-aluminum: transport through sand columns and environmental effects on plants and soil communities." *Environ Res* 106:296–303.

Du W, Sun Y, Ji R et al., (2011) TiO_2 and ZnO nanoparticles negatively affect wheat growth and soil enzyme activities in agricultural soil. *J Environ Monit* 13:822-828.

Duhan, J.S., Kumar, R., Kumar, N., Kaur, P., Nehra, K., and Duhan, S. (2017). Nanotechnology: The new perspective in precision agriculture. *Biotechnol. Rep.,*15,pp. 11-23.

Dunphy Guzman, K. A., Finnegan, M. P., Banfield, J. F. (2006). "Influence of surface potential on aggregation and transport of Titania nanoparticles." *Environ Sci Technol* 40:7688– 7693.

El-Gharmy, A., Mosa, A.A., Alshaal, T., and El-Ramady, H. (2018). Nano fertilizers vs. biofertilizers: New insights. *Environ. Biodiverse. Soil Security,* 2, pp.51-72.

Fan R., Huang Y. C., Grusak M. A., (2014) Effects of nano-TiO_2 on the agronomically-relevant *Rhizobium*-legume symbiosis. *Sci Total Environ* 466-467:503-512.

Fang, Jing, Keke Zhang, Peide Sun, Daohui Lin, Bing Shen, and Yan Luo. (2016) "Co- transport of Pb2+ and TiO2 nanoparticles in repacked homogeneous soil columnsunder saturation condition: Effect of ionic strength and fulvic acid." *Science of theTotal Environment* 571: 471-478.

Fang, Jing, Xiao-quan Shan, Bei Wen, Jin-Ming Lin, and Gary Owens. (2009). "Stability of titania nanoparticles in soil suspensions and transport in saturated homogeneous soil columns." *Environmental pollution* 157: 1101-1109.

Fleischer, A., O'Neill, M.A. and Ehwald, R. (1999). "The pore size of non-graminaceous plantcell walls is rapidly decreased by borate ester cross-linking of the pectic polysaccharide rhamnogalacturonan II." *Plant Physiology*. 121: 829-838.

Ge Y., Schimel J. P., Holden P. A. (2012) Identification of soil bacteria susceptible to TiO_2 andZnO nanoparticles. *Appl Environ Microbiol* 78:6749-6758.

Gerdini, F.S., (2016). Effect of nano potassium fertilizer on some parchment pumpkin (*Cucurbitapepo)* morphological and physiological characterstics under draught conditions. *Int. J. Farm Alli. Sci.,* 5, pp.367-371.

Ghahremani, A., Akbari, K., Yousefpour, M., and Ardalani, H. (2014). Effects of nano-potassium and nano calcium chelated fertilizers on qualitative and quantitative characteristics of *Ocimumbasillicum. Int. J. Pharm. Res. School.,* 3, 00167.

Ghosh, Saikat, Hamid Mashayekhi, Bo Pan, Prasanta Bhowmik, and Baoshan Xing. (2008) "Colloidal behavior of aluminum oxide nanoparticles as affected by pH and natural organic matter." *Langmuir* 24: 12385-12391.

Guo, H., White, J. C., Wang, Z., and Xing, B. (2018). Nano-enabled fertilizers control there lease and use efficiency of nutrients. *Curr. Opin, Environ. Sci. Health, 6*, pp.77- 83 Hasan S 2015 A Review on Nanoparticles: Their synthesis and types biosynthesis:4 9-11.

Guru, T., Veronica, N., Thatikunta, R. and Reddy, S.N. (2015). "Crop Nutrition Management with Nano fertilizers." *International Journal of Environmental Science and Technology*. 1(1): 4-6.

Holden P. A., Nisbet R. M., Lenihan H. S., et al., (2013). Ecological nanotoxicology: integrating nanomaterial hazard consideration across the subcellular, population, community, andecosystems levels. *Acc Chem Res* 46(3): 813-822.

Holdren J. P. (2011). *The national nanotechnology initiative strategic plan report at subcommitteeon nanoscale science, engineering and technology of committee on technology*. National science and technology council (NSTC), Arlington.

Hu C., Liu X., Li X. et al., (2014) Evaluation of growth and biochemical indicators of Salvinia natans exposed to zinc oxide nanoparticles and zinc accumulation in plants. *Environ Sci Pollut Res Int* 21(1):732-739.

Huiyuan, G., Jason, C.W., Zhenyu, W. and Baoshan, X. (2018). "Nano-enabled fertilizers tocontrol the release and use efficiency of nutrients."*Current Opinion in Environmental Science and Health*. 6: 77-83.

Ibrahim, N.K., and Al Farttoosi, H.A.K. (2019). Response of mung bean to boronnano particles and spraying stages (*Vigna Radiata* L,).*Plant Archives*,19, pp. 712-715.

Ingale, A.G. and Chaudhari, A.N. (2013). Biogenic synthesis of nanoparticles and potential applications: An eco-friendly approach. *J.Nanomed. Nanotechnol.,* 4, pp. 1-7.

Iqbal, M.A. (2019). Nano-fertilizers for sustainable crop production under changing climate:changing climates: Aglobal perspective. In *Sustainable Crop Production.* IntechOpen.

Jiang, Xujia, Meiping Tong, and Hyunjung Kim (2012) "Influence of natural organic matter on the transport and deposition of zinc oxide nanoparticles insaturated porous media." *Journal of colloid and interface science* 386: 34-43.

Jones, E. H., & Su, C. (2012). Fate and transport of elemental copper (Cu0) nanoparticles through saturated porous media in the presence of organic materials. *Water research*, *46*(7), 2445-2456.

Juárez-Maldonado, A., Hortensia, O., América, B.M., Susana, G., Álvaro, M.M., Cabrera-Dela Fuente, M. Sandoval-Rangel, A., Cadenas P., G. and Benavides-Mendoza, A. (2019) "Nanoparticles and Nanomaterials as Plant Biostimulants." *International Journal of Molecular Sciences.* 20(1): 162.

Kah, M., Kookana, R.S., Gogos, A., and Bucheli, T. D. (2018). A critical evaluation of nanopesticides and nanofertilizers against their conventional analogues. *Nat. Nanotechnol*., 13 pp. 677-684.

Kashyap, Prem Lal, Xu Xiang, and Patricia Heiden. (2015). "Chitosan nanoparticle based delivery systems for sustainable agriculture." *International journal ofbiological macromolecules* 77: 36-51.

Khanm, H., Vaishnavi, B.A., and Shankar, A.G., (2018). Rise of nano-fertilizer era: Effect of nano scale zinc oxide particles on the germination, growth and yield of tomato (*Soalnum lycopersicum). Int. J. Curr. Microbiol. Appl. Sci.,* 7, pp. 1861-1871.

Klaine S. J., Alavarez P. J., Batley G. E. (2008) Nanomaterials in the environment: behaviour, fate, bioavailability, and effects. *Environ Toxicol Chem* 27;1825-1851.

Kottegoda, N., Munaweera, I., Nadeesh, M.A. and Karunaratne, V. (2011). "A green slow- release fertilizer composition based on urea-modified hydroxyapatite nanoparticles encapsulated wood." *Current Science*. 101(1): 73-78.

Kretzschmar, R. and Schafer, T. (2005). "Metal retention and transport on colloidal particles in the environment." *Elements* 1:205–210.

Kumar, R., Rawat, K. S., Mishra, A. K. (2012). "Nanoparticles in the soil environment and their behaviour: an overview." *J App Natural Sci* 4(2):310–324.

Kumar, S., Dilbaghi, N., Rani, R., Bhanjana, G., and Umar, A. (2013). Novel approach for enhancement of drug bioavailability. *Rev. Adv. Sci.Eng.,* 2, pp.133-154.

Leggo, P.J. (2000). "An investigation of plant growth in an organo-zeolitic substrate and its ecological significance." *Plant and Soil*. 219: 135-146.

Lin, D. and Xing, B. (2007). "Phytotoxicity of Nanoparticles: Inhibition of Seed Germination and Root Growth." *Environmental Pollution*. 50: 243-250.

Linquist, B. A., Liu, L., van Kessel, C., and van Groenigen, K. J. (2013) Enhanced efficiency nitrogen fertilizers for rice systems: Meta-analysis of yield and nitrogen uptake. *Field Crops Res.*, 154 pp. 246-254.

Liscano, J.F., Wilson, C.E., Norman, R.J. and Slaton, N.A. 2000. "Zinc availability to rice from seven granular fertilizers." *AAES Research Bulletin*. 963: 1-31.

Ma X., Gurung A., Deng Y. (2013) Phytotoxicity and uptake of nanoscale zero valent iron (nZVI) by two plant species. *Sci Total Environ* 443:844-849.

Mani, P. K., & Mondal, S. (2016). "Agri-nanotechniques for Plant Availability of Nutrients."*Plant Nanotechnology, 263–303.* doi:10.1007/978-3-319-42154-4_11

Manjunatha, S.B., Biradar, D.P. and Aladakatti, Y.R. (2016). "Nanotechnology and itsapplications in agriculture: A review." *Journal of Farm Sciences*. 29: 1-3.materials. *Water research* 46: 2445-2456.

Moghaddasi, S., Fotovat, A., Khoshgoftarmanesh, A.H., Karimzadeh, F., Khazzaei, H.R., and Khorassani, R. (2017), Bioavailability of coated and uncoated ZnO nanoparticles to cucumber in soil with or without organic matter. *Ectox. Enviro. Safe.*144, pp.543-551.

Moore, M. (2006). "Do nanoparticles present ecotoxicological risks for the health of the aquatic environment?."*Environment International*. 32: 967-976.

Munir, T., Rizwan,M., Kashif, M., Shahzad, A., Ali, S., Zahid, R., Alam,M.F.E., andImran. (2018). Effect of zinc oxide nanoparticles on the growth and Zn uptake in wheat (*Triticumaestivum* L.) by seed priming method. *Dig. J. Nanometer.Bios.,* 13, pp.315-323.

Naderi A., (2012). The role of nanotechnology in improving the use efficiency of nutrients and chemical fertilizer. *Nanotechnology* 11.12 16-32.

Nair, R., Varghese, S.H., Nair, B.G., Maekawa, T., Yoshida, Y. and Kumar, D.S. (2010)."Nanoparticulate material delivery to plants." *Plant Science*. 179: 154-168.

Narendhran, S., Rajiv, P., and Sivaraj, R.(2016). Influence of zinc oxide nanoparticles on growth of *Sesamum indicum* L., in zinc deficient soil. *Int. J.Pharm. Pharm. Sci.,* 8,pp.365-371.

Navarro, E., Baun, A., Behra, R., Hartmann, N.B., Filser, J., Miao, A.J., Quigg, A., Santschi P.H. and Sigg, L. (2008)."Environmental behavior and ecotoxicity of engineered nanoparticles to algae, plants, and fungi." *Ecotoxicology*. 17: 372-386.

Oberdorster, G., Oberdorster, E., Oberdorster, J. (2005). "Nanotoxicology: an emerging discipline evolving from studies of ultrafine particles." *Environ Health Perspect*.113(7):823–839.

Oberdörster, Günter, Eva Oberdörster, and Jan Oberdörster. (2005). "Nanotoxicology: an emerging discipline evolving from studies of ultrafine particles." *Environmental health perspectives* 113: 823-839.

Pradhan, S. and Mailpalli, D.R. (2017). Interaction of engineered nanoparticles with the agri- environment. *Agric. Food Chem.,* 65, pp.8279-8294.

Prasad R., Bagde U. S., Varma A. (2012) Intellectual property rights and agricultural biotechnology: an overview. *Afr J Biotechnol* 11(73):13746–13752.

Preetha, S. and Balakrishnan, N. (2017) "A review of nano fertilizers and their use and functions in soil." *International Journal of Current Microbiology and Applied Science*. 6(12): 3117-3133.

Priester J. H., Ge Y., Mielke R. E. (2012) Soybean susceptibility to manufactured nanomaterials with evidence for food quality and soil fertility interruption. *PNAS* 109:E2451-E2456.

Rajonee, A.A., Nigar, F., Ahmed, S., and Huq, S.I. (2016). Synthesis of nitrogen nano fertilizer and its efficacy. *Can. J. Pure Appl. Sci.,* 10, pp.3913-3919.

Ranjbar, S., Ramezanian, A., and Rahemi, M. (2019). Nano-calcium and its potential to improve 'Red Delicous' apple fruit characterstics. *Hortic, Environ.Biotech.*pp.1-8.

Saigusa, M. and Ombódi, A. (2000). "Broadcast application versus band application of polyolefin-coated fertilizer on green peppers grown on andisol." *Journal of Plant Nutrition*. 23: 1485-1493.

Salavati-nisari M., Davar F. and Mir N. (2008) *Synthesis and Characterization of metallic copper nanoparticles via thermal decomposition polyhedron*. 27 3514-8.

Sasson, Y., Levy-Ruso, G., Toledano, O. and Ishaaya, I. (2007). *Nanosuspensions: emergingnovel agrochemical formulations*. In: Ishaaya, I.; Horowitz, A.R. and Nauen R. (eds.). *Insecticides design using advanced technologies.*Springer, Berlin.1-39.

Sawant R., Hurley J., Salmaso S., Kale A., Tolcheva E., Levchenko T., Torchilin V. (2006) "SMART" drug delivery systems: double-targeted pH-responsive pharmaceutical nanocarriers. *Bioconj Chem* 17(4):943–949.

Schrick, B., Hydutsky, W., Blough, J. L., Mallouk, T. E. (2004). "Delivery vehicles forzerovalent metal nanoparticles in soil and groundwater." *Chem Mater.* 16:2187–2193.

Servin A. D., Morales M. I., Castilo-Micheal H. (2013) Synchroton verification of TiO_2 accumulation in cucumber fruit: a possible pathway of TiO_2 nanoparticle transfer from soil into the food chain. *Environ Sci Technol* 47(20):11592-11598.

Seydmohammadi, Z., Roetin, Z., and Rezvanipour, S. (2020). Accelearting the growth and flowering of *Eustomagrandiflorum* by foliar application of nano-Zn and nano-CaCo3. *Plant Physiol. Rep,* 25,pp.140-148.

Sharifi, R., Mohammadi, K., and Rokhzadi, A. (2016).Effect of seed priming and foliar application with micronutrient on quality of forage corn (*Zeamays.). Environ. Exp. Biol.* 14, pp.151-156.

Shi J., Peng C., Yang Y. (2014) Phytotoxicity and accumulation of copper oxide nanoparticles to the Cu-tolerant plant *Elsholtzia splendens. Nanotoxicology* 8:179- 188.

Singh, G., and Rattanpal, H. (2014). Use of nanotechnology in horticulture: A review. *Int. J. Agric. Sci. Vet. Med,* 2, pp.34-42.

Solanki, P., Bhargava, A., Chhipa, H., Jain, N., and Panwar, J. (2015) Nano-fertilizers and their smart delivery system. In *Nanotechnologies in Food and Agriculture,* pp 81-101. Springer, Cham.

Solanki, Priyanka, Arpit Bhargava, Hemraj Chhipa, Navin Jain, and Jitendra Panwar.(2015) "Nano-fertilizers and their smart delivery system." In *Nanotechnologies infood and agriculture*, pp. 81-101. Springer, Cham.

Soliman, A.S., Hassan, M., Abou-Elella, F., Ahmed, A.H., and El-Feky, S.A. (2016). Effect of nano and molecular phosphorous fertilizers on growth and chemical composition of baobab (*Adansoniadigitat L.). J. Plant Sci.,* 11, pp.52-60.

Subramanian and Tarafdar (2011) "Nanotechnology: interdisciplinary science of applications." *African Journal of Biotechnology* 12.3(2013).

Subramanian, K.S. and Rahale, S. (2009). "Synthesis of nanofertiliser formulations for balanced nutrition." *Proceedings of the Indian society of soil science-platinum Jubilee celebration*. IARI Campus, New Delhi. 85.

Suresh A. K., Pelletier D. A., Doktycz M. J. (2013) Relating nanomaterial properties and microbialtoxicity. *Nanoscale* 5:463-474.

Taherian, M., Bostani,A., and Omidi, H.(2019). Boron and pigment content in alfalfa affected by nano fertilization under calcareous conditions. *J. TraceElem. Med. Bio.,* 53,pp. 136-143.

Tai C. Y., Tai C., Chang M. and Liu H. (2007) Synthesis of Magnesium Hydroxide and Oxide Nanoparticles using spinning Disk reactor 5536-41.

Tantawy A. S. (2015)."Nano silicon application improves salinity tolerance of sweet pepperplants." *International Journal of Chem Tech Research* 8.10(2015): 11-17.

Tarafdar, J.C., Raliya, R. and Rathore, I. (2012). "Microbial synthesis of phosphorous nanoparticle from tri-calcium phosphate using Aspergillus tubingensis TFR-5." *Journal of Bionanoscience*. 6: 84-89.

Tarafdar, J.C., Sharma, S. and Raliya, R. (2013). "Review nanotechnology: interdisciplinary science of applications." *African Journal of Biotechnology*. 12(3): 219-226.

Taran, N.Y., Gonchor, O.M., Lopatko, K.G., Batsmanova, L.M., Patyka, M.V., and Volkogon, M.V. (2014). The effect of colloidal solution of molybdenum nanoparticleon the microbial composition in rhizospher of *Cicerarientinum* L. *Nanoparticle Res. Lett.*., 9, pp.289.

Thirugnanasambandan, T.(2019). Advances and trends in nanobiofertilizers, *SSRN,* p, 59. https://doi.org/10.2139/ssrn.3306998.

Thomas, E., Rathore, L., and Trafdar, J.C., (2017). Bioinspired production of molybdenum nanoparticles and its effect on chickpea (*Cicerarientinum* L.). *J. Bionanosci.,*11, pp.153-159.

Thul, S. T., B. K. Sarangi, and R. A. Pandey. (2013): "Nanotechnology in agroecosystem: implications on plant productivity and its soil environment." *Expert Opin Environ Biol* 2:2Ā7.

Tilman. D., Knops, J., Wedin, D. and Reich, P. (2002). Plant diversity and composition: effects on productivity and nutrient dynamics of experimental grasslands. In: Loreau, M.; Naeem Inchausti, P.S. (eds.). *Biodiversity and ecosystem functioning,* Oxford University Press, Oxford. 21-35.

Tiwari D. K., Bihari J. and Sen P. (2008) *Application of Nanoparticles in Waste WaterTreatment* 3 417-33.

Tourinho, P. S., Van Gestel, C.A.M., Lofts, S., Svendsen, C., Soares, A.M.V.M., Loureiro, S.(2012). "Metal based nanoparticles in soil: fate, behavior, and effects on soil invertebrates." *Environ Toxico Chem*. 31(8):1679–1692.

Tourinho, Paula S., Cornelis A. M. Van Gestel, Stephen Lofts, Claus Svendsen, Amadeu M. V. M. Soares, and Susana Loureiro. (2012). "Metal-based nanoparticles in soil: Fate,behavior, and effects on soil invertebrates." *Environmental Toxicology and Chemistry* 31: 1679- 1692.

Watson, Jean-Luc, Tommy Fang, Christian O. Dimkpa, David W. Britt, Joan E. McLean, Astrid Jacobson, and Anne J. Anderson. (2015). "The phytotoxicity of ZnO nanoparticles on wheat varies with soil properties." *Biometals* 28:101-112.

Yadav, T.P., Yadav, R.M., and Singh, D.P.(2012). Mechanical milling: A top down approach forthe synthesis of nanomaterials and nanocomposites. *Nanosci. Nanotechnol.,*2, pp.22-48.

Yasmen, F., Raja, N.I., Razzaq, A., and Komatsu, S., (2017). Proteomic and physiological analyses of wheat seeds exposed to copper and iron nanoparticles. *Biochimicaet Biophysica Acta-Proteins Proteomics,*1865, pp.28-42.

Yin L., Cheng Y., Espinasse B. (2011). More than the ions: the effects of silver nanoparticles on *Loliummulti florum. Environ Sci Technol* 45:2360-2367.

Zeladit N. (2010) Silver beware: antimicrobial nanoparticles in soil may harm plant life. *Sci Am*, Aug 9, 2010. http://www.scientificamerican.com/article/silver-beware-antimicrobial-nanoparticles-in-soil-may-harm-plant-life/.

Zhai L., Gu F. X., Chan J. M. (2008). Nanoparticles in medicine: therapeutic application and developments. *Clin Pharmacol Ther* 83:761-769.

Zhao L., Hernandez-ViezcasJ. A., Perleta-Videa J. R. (2013a) ZnO nanoparticles fate in soilandzinc bioaccumulation in corn plants (*Zea mays)* influenced by alginate. *EnvironSci.Processes Impacts* 15:260-266.

Zhao L., Sun Y., Hernandez-Viezcas J. A. (2013b) Influence of CeO_2 and ZnO nanoparticles on cucumber physiological markers and bioaccumulation of Ce and Zn:a life cyclestudy: *J Agric Food Chem* 61(40), 11045-11051.

Zulfiqar, F., Navarro, M., Ashraf, M., Akram, N.A., and Munne-Bosch, S. (2019). Nanofertilizer use for sustainable agriculture: advantages and Limitations. *Plant Sci*., 110270.

Zulfiqar, Faisal, Míriam Navarro, Muhammad Ashraf, Nudrat Aisha Akram, and Sergi Munné-Bosch. (2019): "Nanofertilizer use for sustainable agriculture: advantages andlimitations." *Plant Science* 289 110270.

Chapter 11

Climate Change Effects on Biodiversity, Ecosystems, Ecosystem Services in an Indian Context

Varsha Kumari[1,*], Priyanka Kumawat[2], D. K. Gothwal[2] and Bhuri Singh[2]

[1]Department of Plant Breeding and Genetics, SKN Agriculture University, Jobner, Jaipur, India
[2]Department of Agronomy, SKN Agriculture University, Jobner, Jaipur, India

Abstract

Climate change appears as remarkable environmental threat globally. It is challenge which evokes awareness of researchers, common people, politicians and authority in India and whole world. Climate change has severe impact on ecosystems in India like natural resources, water, flora, faunal species, agriculture, forest vegetation and industries. Biodiversity represents total biological diversity of existing in ecosystem. Changes in climatic situation have adverse impact on biodiversity. Biodiversity performs a key part in regulation of climatic condition. Conservation of biological diversity from genome to biosphere is essential to withstand adverse and threatening climatic condition by human kind. Preservation of biodiversity will upgrade ecosystems which strengthen ecosystem services in course of high climatic pressure. This chapter highlights effects of climate change on biodiversity, ecosystems and ecosystem services in Indian context.

Keywords: biodiversity, climate change, ecosystems, ecosystem services, global warming

1. Introduction

Climate change invokes to long-term remarkable weather variations of specific zone over length of time. It is most ubiquitous growing environmental challenge which poses threat to natural ecosystem and biodiversity worldwide. It includes recent temperature rise results to massive shift in weather patterns called as global warming driven by release of heat-trapping greenhouse gases, aerosols and clouds, changes in land surface, solar and volcanic eruptions etc. Greenhouse gases includes carbon-dioxide, carbon monoxide, methane, nitrous oxide, O_3 chlorofluorocarbon etc. Although, era of climate change also existed previously, since beginning of 20th century change of climate is ascribed to human activity directly or indirectly

* Corresponding Author's Email: varshakumari.pbg@sknau.ac.in.

In: Ecosystem Services
Editor: Hanuman Singh Jatav
ISBN: 978-1-68507-614-6

which had unrivalled effect on composition of the global atmosphere cause global scale change on climate system (Chaturvedi et al., 2011).

Human activity behind climate change composed of, emission of carbon dioxide into the air, burning fossil fuels, expansion of agriculture by deforestation without reforestation, extensive use of fertilizers, industrialization as like producing cement is one of the major contributors to climate change. As CO_2 concentrations doubles in the atmosphere, earths temperatures raise by approximately 2 to 6°C. The CO_2 and many heat-trapping gases cumulatively called as "CO_2 equivalent." Fossil fuels as like coal, oil, gas etc. contains locked away carbon dioxide for billions of years in the earth. Burning of fossil fuels releases accumulated carbon dioxide directly into the air. Likewise, forest tree store and absorb carbon dioxide. Cutting and burning of trees release carbon dioxide into atmosphere (Swaminathan and Keshavan, 2012). Similarly, rearing animals and plantation also have enormous role in increasing parabola of greenhouse gases. Nitrous oxide used as fertilisers and methane produced by animals are worse and more potent than carbon dioxide. All these factors cause more than 3% of total carbon dioxide release and carbon dioxide level in atmosphere arose by 42% and nitrogen concentrations on land are substantially greater during 20th and 21st century. Additionally, extensive use of fertilizers accumulated nitrogen and phosphorous in soil changes growth and survival of crops and trees which in turn regulates dry forest ecosystem leads to fire and drought (Varma et al., 2018 & Figure 1).

Figure 1. Depiction of various human activities behind climate change.

Climate change ultimately uplift global land and sea temperatures, alter precipitation patterns and increases frequency of heat waves, rainfall intensity and droughts like extreme events (Lewis and Maslin, 2015). Global warming trailing on vehicle of climate change drives to shift in ocean currents due to melting of ice leads to high sea level globally which affects and threatens human livelihood in coastal margins. Mass migration and mitigation from densely populated coastal region to other places due to threat of life losses could be a potential cause of concern today. It is also increasingly clear that monsoon cycle and melting of ice from the

Himalaya Mountains in India are sensitive to shift in climatic conditions ultimately has a major impact on surface and groundwater.

Human beings are benefitted from natural environmental resources by procuring food, fuelwood, water, sheltering material, raw substances, non-timber forest produces and many more called as ecosystem services. Climate change alters human-natural ecosystem interactions by affecting flow of ecosystem services (Kataki and Konwer 2002; Smith, 2007; Mooney et al., 2009; Narsimlu et al., 2013).

India composed of 10 ecoregions and act as prime centre of two global hotspots for boidiversity *viz.*, Indo-Malayam and Western Ghats, where Eastern Himalayas, North-east and Andaman Islands are included in Indo-Malayam region. 50,000 species of plants and 85,000 species of animals are estimated in different habitats in India (Olson and Dinerstein 1998; Myers et al, 2000).

Biodiversity is vigorously depleted in recent times in India because of the deforestation, destruction of natural habitat, entry of invasive species, overgrowing population, increasing pollution and severe exploitation of resources, rigorous poaching, cutting and clearing of forest for development purposes, conversion of forest and grasslands area into residential place etc. Deforestation has a huge impact on the biodiversity and biodiversity loss has alarming impact on ecology and environmental condition in present, past and will sustain in future. Environmental changes are counted under climate change (Kumar and Verma, 2017) which destruct climate regime ultimately results to global warming and ecosystem imbalances.

Climate change has powerful impacts on biodiversity pattern, ecosystem, ecosystem services and ecological functions. Altered climate change drives to change distribution and functioning of ecoregions of Indian subcontinent and replacement of arid and dry vegetation with wetter forms (Rasquinha and Sankaran 2016; Byers et al., 2018). This chapter comprises of overview of climate change and its impacts on biodiversity and summaries key insights into threats posed by climate change to ecosystems and ecosystems services in India.

2. Impact of Climate Change in India

Climate change implies alteration in mean temperature, degree of precipitation, weather behavior, amount of precipitation and length of sunlight which either straightly or obliquely assigned to human interference which changes the constitution of the atmosphere globally. Changed temperatures will bring about the large difficulty to floral, faunal species and humankinds. Big list of plant and animal species existed in India accounts for biological resources due to its vast regime of climatic condition. Wide aggregates of national parks, sanctuaries, biosphere reserves bear distinctive biological diversity.

It is extensively confirmed that climate change and global warming had remarkable influence in Indian subcontinent. High rainfalls in summer month of June-August and low precipitation in October-December month of winter are actually predicated. However, the weather event is so erratic which trigger too much rainfall in September and October in some part and in March-April in other part of country. Some states of India suffer severe flood due to extreme rainfall pattern and dry spells in other part of country. Sometime there is huge flush of shower in one place but dry spell in other place at distance of only 500 meters. Winter and summer rainfall declined by 25% in 2050 and temperature would increase by 3 to 6°C by 2100, has been forecasted in the Indian subcontinent (Vinod and Chopra, 2009).

India is idiosyncratic for biodiversity, diverse climate range, landscape pattern, physical geography etc. Climatic condition of the country is unique and varied due to diversity in temperature, rainfall reaction from region to region. As the mercury raises 50°C in desert in summer season, it is going to fall to – 40°C in winter season in Leh of Jammu and Kashmir, Union territory. Cherrapunji and Mawsynram (Meghalaya) receives 11,777 mm and 11, 872 of average annual rainfall respectively which is highest in the world whereas, Jaisalmer district of Rajasthan receives only 15-30 mm of rainfall of subcontinent (Bajracharya et al., 2007).

Largest aspect of worry in present time is aggregation of greenhouse gases especially carbon-dioxide and carbon monoxide and leading to depletion of ozone layer and changing of climate globally. Changes in the earth's climate during recent decades had notable consequences on the mountains and glaciers which promote glacier retreat (Kireet, 2005). Melting of Shiva Lingam which is naturally made up of snow in divine cave of Amaranth (Jammu and Kashmir) is signal of rising temperature in the Himalayas. It is predicted that Mumbai city is going to be drop below sea till 2050 due to increasing sea level as of increasing temperature, climate change and melting of ice.

2.1. Threats to Biodiversity via Climate Change

Biodiversity is explained as variability of life on earth and ecological niche where they survive. It is measure of diversity within genus, within species, between the species and at ecosystems level and can encompasses evolutionary ecological and cultural process that sustain life. It includes enormous variety of life on earth from genome to ecosystem and ecological processes and encompasses endangered and threatened species of microorganism, fungi, invertebrates, plants and animals (Pereira et al., 2010). Biodiversity is foundation of survival of humankind and furnish expansive direct and indirect benefits to human being through natural ecosystems, economic means and array of essential services and plays an eminent part in modulating ecosystem stability as well as functioning (Singh, 2002).

Changes to climate system presently affects different scales of biodiversity such as genes, genomes, species, biodiversity, community, ecosystem and ecosystem services which correspondingly alter major sectors covering water resources, agricultural output, natural ecosystems include as forestry and vegetation, health and industrial areas. Mere alteration in pattern of climate has measurable influence on biodiversity which cause shift in habitat and niche of organism which in turn act as principal threat to survival presenting a force towards vulnerability. Climate change exacerbates biodiversity loss and rise risk of extinction of species having very less population, restricted habitat and climatic ranges. Millennium Ecosystem Assessment foretells that change in climate is the key threat to ecological and biological diversity (Anonymous, 2007; Bellard et al., 2012).

Biological responses towards change in climate system vary from positive to negative among populations. Positive responses increase growth rate results in expansion, on the other hand negative responses leads to widespread reduction in growth rates and ranges. Shifts of many species beyond their geographical ranges brings out gathering of species, create inter and intraspecific interactions and finally extinction in severe scenarios. (Butchart et al., 2010; Montoya and Raffaelli, 2010; Barnosky et al., 2011; Chen et al., 2011; Dawson, 2011; Geyer et al., 2011).

Increased temperature due to climate change affects normal life cycle and forces plant and animal species to leave their natural habitat and shift to another or being replaced by other species. As for example, Sal forest are replaced by Teak forests and conifer trees are replaced by the deciduous forest in central India. According to earlier studies, rise in temperature by 5°C induces movement of forest by 3.00 km/year that is multiple times more than normal forest motion. Invasive alien species are vital reason of threat and extinction on islands and in freshwater as being resistant to climatic changes and variations. Variation in temperature patterns and precipitation ensue persistent droughts and floods makes indigenous species susceptible to various diseases, pests and abiotic stress (Tibbetts, 2007).

Variable ecosystem, climate, edaphic, cultural, topographical and geological conditions harbors enormous diversity at biological level in India which is contributing to immense range of diversity of habitats as forest land, grasslands, wetlands, coastal areas, marine and deserts ecosystems. In word of abundance, India is recognized as mega-diverse country with fifth position in reptiles, seventh in mammals and tenth in bird specifies its richness in biodiversity. India harbours four biodiversity hotspots *viz*., Himalaya, Indo-Burma, Western Ghats and Sri Lanka and Sundaland, out of 34 hotspot centers recorded worldwide. Approximately, 1,00000 fauna species have been identified in India. Country has 420 threatened animal species which is nearly 5.0% of total endangered animal species globally (Subodh and Lokesh, 2018; Sadguru and Seema, 2019).

Little change in climate posed to severe loss of biodiversity condition results in extinction of animal species. Golden frogs and harlequin toads populations are extincted due to climate change. Decline in Arctic ice cover has put Polar bears in red zone of extinction. Similarly, reduction in planktons due to climate change is chief reason of extinction of Atlantic whale as planktons are their major food of Nilgiri tahr is endemic of Nilgiri Hills in Tamil Nadu and Kerala are at the edge of extinction. Sex ratio of sea turtle disrupt as increased temperature induced to produce a greater number of female turtles. Climate change pattern also have severe impact on animals. Amphibian chytrid fungus which causes chytridiomycosis disease are intensifying due to higher temperatures and leaving amphibians diminishing day by day. Loss of biodiversity also affects the fishing and hunting practices by local populations creating an implication on their food source (Sukumar et al., 1995; McCarthy et al., 2001).

3. Climate Change Effects on Ecosystem

Climate change had a quite fair impact on ecosystem and biodiversity as for example diversity in rainfall, excessive wetting and drying, lengthy warm period, water deficit is growing physiological and hydrological pressure on ecosystem. Extended heatwaves and long-lived acidification in the sea has enhances strain on ecosystem and species. Overfishing, significant threat and loss to faunal species, alien and invasive species, habitat destruction is some of the events which used to magnify the susceptibility of ecosystem to climate change.

Forecasting change in ecosystem loss is very tough as it is so unexpected and uncertain. Adaptive ecosystem management should be implemented through research and monitoring under uncertainty. Lengthy monitoring performs a critical function also. Long-term forest monitoring is a vital proof about carbon sink of biosphere which is limiting factors of climate change (Hoffmann et al., 2019; Platts et al., 2019).

3.1. Agriculture

Agriculture is backbone of civilization which provides food, fodder, fuel, shelter material and many more to drive vehicle of livelihood. Variation in rainfall, rising heat stress, irregular weather pattern, pest and disease proliferation are predominant after consequences of climate change have severe impact on curtailing breeding cycle of crop which ultimately leads to declined production, productivity and growth of plants (Verma 2017; Verma 2018). Climate change has impact on sustainable and unsustainable agriculture system.

3.2. Coastal and Marine Ecosystem

It has been investigated that more than 60% of earth's floor is protected *via* ocean consisting of specific ecosystem comprises of coral reef, grass beds, mangroves etc. Change in climate pattern leads to leap in level of sea margin, extended coastal erosion, over floods, excessive surges of storms, ingress of salinity, extended temperature of sea surfaces, acidification of sea etc. Increasing ocean volume presents intense danger to marine ecosystems results to disruption of survival pattern and ecological niche of various species and poses a big risk to marine and wetland ecosystem. Coastal regions which are susceptible to climate change in India are Goa, Maharashtra, Sunderban, Kutch of Gujarat etc. Sunderban is biggest low-lying natural mangrove ecosystem globally, disbursed over 3,860 square miles including land and water both (Rathore and Jasrai, 2013). Recorded hike in ocean over forty years is liable to affects mangrove ecosystems and tiger habitat notably. Species composition and distribution is virtually be seriously be stricken by modifications of temperature and climate change (Loucks et al., 2010; Rathore and Jasrai, 2013).

3.3. Himalaya Ecosystem

Himalayan temperatures are rising at a rate of 1.0°C annually that is drastically greater than average rise of 0.8°C per decades globally. Mosquitoes are seen in Lhasa and Tibet cities first time placed at 3490 meters above sea and flies are spotted at Mount Everest in Nepal due to temperature loft as cold temperature save people earlier from feasible spread of vector-borne illness and dangerous threats. Mosquitoes, flies and insects are carriers of fatal diseases like malaria, dengue, cholera etc.

3.4. Island Ecosystem

Islands are ample reservoir of biodiversity which has wide importance economically. Change in climate at recent times has threatened and subsequently shifted more than 30% of island species towards verge of extinction.

3.5. Water Ecosystem

It comprised of fresh water ecosystem which includes 1.0% of total earths surface, however, it assists more than 5% of full existing species. They are wealthy supply of food, earning, employment and biodiversity. Irregular climatic situations such as temperature and precipitation lead to alter in physiology, cyclic and seasonal natural phenomenon and migration pattern of inland birds, fishes and water organisms.

3.6. Forest Ecosystem

Forest covers 31% (one third) of total earth surface and act as dwelling place of 15% (two third) of total terrestrial species. Forest lands are biodiversity hotspots for much number of species. Until now, half of the forest has been cleared up for procuring food, wood and shelter places and also due to industrialization. Migration of tree species to elevated latitude, extended pest and insect population, ingression of alien and invasive genus, expanded growth of few of forest and wild fire due to greenhouse effect have consequently changes forest constitution. Moreover, many animal species and primate are at edge of extinction due to all these modifications.

3.7. Dry Lands Ecosystem

Dry lands and grassland ecosystem have diversified range of crops and livestock and consist of numerous localized species. Growing threat of wild fire could shift biodiversity of species.

3.8. Impact on Humans

Elevated temperature, snow melting, natural calamities like cyclones, droughts, floods are numerous after effects of change in climatic conditions which displaces human being from their dwellings. Climate changes have foremost significance on human well-being. Generally, hot environment increased the multiplication, survival and growth rate of insects, pest vectors. One degree enhance in temperature is predicted to ten degree enhance in prevalence of insect, pest and diseases like cholera and typhoid, malaria, dengue, plague etc. Disease curve is rising swiftly due to broadening horizon of greenhouse gases through emission of carbon. Carbon emission expanded to risky degree forming life of animals susceptible to contagious disease inclusive of bacterial illnesses. Growing sea level has drowned various islands and could make shelter less to billions of people ultimately force them to become refugee soon (Bhatt et al., 2018).

4. Impact of Climate Change on Ecosystem Services in Indian Context

Generally, ecosystem services confer to profit attained by human being from nature (Daily 1997). Ecosystem services are consequences of interactions between physical factors of ecosystem such as climate, temperature, moisture, structure of biological communities, biotic and biotic factors, nutrient management etc. Millennium Ecosystem Assessment has categorized ecosystem services broadly into various types as cultural, provisional, regulating, cultural and supporting types of services (Millennium Ecosystem Assessment 2005). Provisioning ecosystem services covers food, fuel and timber like material obtained from nature world, regulating ecosystem services includes regulates air and water quality and atmospheric constitution. Whereas, cultural services serve spiritual and aesthetic values procured from nature by humans. Benefits derived from ecosystem are vast still ecosystem disservices like economic threat, loss of personal safety to all the living includes human and wild life exist (Wallace 2007; Fisher and Turner 2008; Dunn 2010; Lele et al., 2013).

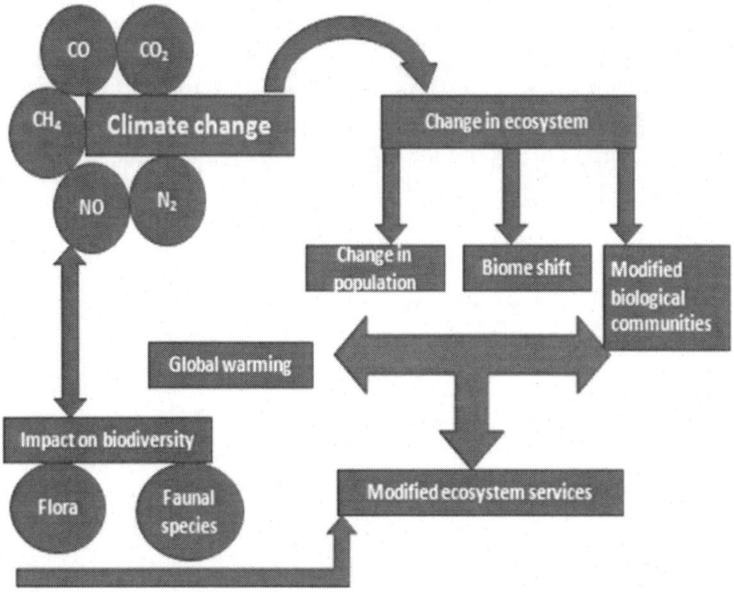

Figure 2. Schematic diagram showing effects of climate change on biodiversity, ecosystem and ecosystem services.

Understanding of biological hierarchy and interplay between individuals of biosphere is most important to identify mechanism through which climate affects ecosystem services. Ascending from individual to population to community consists of species and finally to biosphere is hierarchy of biological organization. Competition and interaction between and within individuals of different trophic level contributes to ecosystem operation and services in India. As like population and growth of forest tree stand giving fuel and wood could strongly influence accessibility of fuel and ultimately livelihood. Similarly, microorganism present in soil can have impact on nutrient availability for plant growth and development. Further, community of soil microbes and tree species help in carbon absorption metabolism. Balancing water supply and wetland quality, biomass requirement and existence of rural people are fulfilled completely by tree-grass ecosystems. Ecosystem functions and interactions are ultimate region behind supply of ecosystem services. Thus, benefits procured from biosphere by human beings are results of ecosystem functioning. For instance, fuel woods are produced by photosynthetic activity of tree species which is determined by temperature, resources and abiotic factors. Similarly, biomass production by forest stand is governed by deducting gain minus loss through photosynthesis and respiration (Konwer et al., 2001; Kataki and Konwer, 2002; Pan et al., 2011; Campbell et al., 2012; Mitsch et al., 2015; Upadhaya et al., 2017; Perović et al., 2018).

Major drivers as like ecosystem fragmentation, nutrient deposition and degradation have greater impact that post danger to ecosystem services such as drought may exacerbate loss in stored carbon resources. Fragmentation and drought synergistic amplified effects on carbon ultimately threat to trees in forest. Fragmentation of niche and habitat can also have serious impact via spread of contagious diseases and water manoeuvre in catchment areas. Similarly, nitrogen and phosphorous accumulation from industries and agriculture can modulate survival of tree leads to fire responses of dry forest (Laurance et al., 2000; Nepstad et al., 2007, Gosain et al., 2011; Madhusoodhanan et al., 2016; Rulli et al., 2017; Varma et al., 2018).

Environment in India is primarily human-dominated where climate change is sole stressor to ecosystem services. Therefore, it is important to establish and integrates research connection between climate change and ecosystem services to understand how climate change will be shaping tract of ecosystem services in future in India (Figure 2).

Conclusion

Climate change is a pervasive and growing threat to biodiversity, ecosystems, and ecosystem services in India. Ecosystems and biodiversity underpin important services to people, thus these changes impact provisioning, regulating, supporting and cultural services, with implications for human wellbeing. Supporting, regulating and provisioning ecosystem services contribute to food security in the India and the fate of the nation's food production are very likely to depend on the interplay of these services and how the agriculture and fishery sectors respond to climate stresses. Moving forward, evaluations of effectiveness and demonstrative case studies of adaptation success stories are needed to promote and guide climate-smart management.

References

Anonymous, (2007). *Biodiversity and Climate Change: Convention on Biological Diversity*: www.biodiv.org accessed on 30-7-2010.

Bajracharya, S. R., Mool, P. K. and Shrestha, B. R. 2007. Impact of Climate Change on Himalayan Glaciers and Glacial Lakes: *Case Studies on GLOF and Associated Hazards in Nepal and Bhutan*. Kathmandu: ICIMOD.

Barnosky A. D., Matzke, N., Tomiya, S., Wogan, G. O. U., Swartz, B., Quental, T. B., Marshall, C., McGuire, J. L., Lindsey, E. L., Maguire, K. C., Mersey, B. and Ferrer, E. A. (2011). Has the Earth's sixth mass extinction already arrived? *Nature*, 471, 51-57.

Bellard, C., Bertelsmeier, C., Leadley, P., Thuiller, W. and Courchamp, F. (2012). Impacts of climate change on the future of biodiversity. *Ecology Letters,* 15, 365-377.

Bhatt, J. R, Das, A. and Shanker, K. (eds.). 2018. *Biodiversity and Climate Change: An Indian Perspective.* New Delhi, India: Ministry of Environment, Forest and Climate Change, Government of India.

Butchart, S. H. M., Walpole, M., Collen, B., van Strien, A., Scharlemann, J. P. W., Almond, R. E. A., Baillie, J. E. M., Bomhard, B., Brown, C., Bruno, J., Carpenter, K. E., Carr, G. M., Chanson, J., Chenery, A. M., Csirke, J., Davidson, N. C., Dentener, F., Foster, M., Galli, A., Galloway, J. N., Genovesi, P. D., Gregory,

R., Hockings, M., Kapos, V., Lamarque, J. F, Leverington, F., Loh, J., McGeoch, M. A., McRae, L., Minasyan, A., Hernández Morcillo, M., Oldfield, T. E. E., Pauly, D., Quader, S., Revenga, C., Sauer, J. R., Skolnik, B., Spear, D., Stanwell-Smith, D., Stuart, S. N., Symes, A., Tierney, M., Tyrrell, T. D., Vié, J. C. and Watson, R. (2010). Global biodiversity: indicators of recent declines. *Science*, 328, 1164-1168.

Byers, E., Gidden, M., Leclère, D., Balkovic, J., Burek, P., Ebi, K. and Greve, P. et al., (2018). Global exposure and vulnerability to multi-sector development and climate change hotspots. *Environmental Research Letters* 13(5), 055012.

Campbell, A. J., Biesmeijer, J. C. Varma, V. and Wäckers. F. L. (2012). Realising multiple ecosystem services based on the response of three beneficial insect groups to floral traits and trait diversity. *Basic and Applied Ecology*, 13(4), 363–370.

Chaturvedi, R. K., Gopalakrishnan, R., Jayaraman, M., Bala, G., Joshi, N. V., Sukumar, R. and Ravindranath, N. H. (2011). Impact of climate change on Indian forests: A dynamic vegetation modeling approach. *Mitig Adapt Strateg Glob Change*, 16**,** 119–142.

Chen, I. C., Hill, J. K., Ohlemuller, R., Roy, D. B. and Thomas, C. D. (2011). Rapid range shifts of species associated with high levels of climate warming. *Science*, 333,1024-1026.

Daily, G. C. (ed.). 1997. *Nature's services*. Washington, DC: Island Press.

Dawson, T. P. (2011). Beyond predictions: Biodiversity conservation in a changing climate *Science*, 332, 664-664.

Dunn, R. R. (2010). Global mapping of ecosystem disservices: the unspoken reality that nature sometimes kills us. *Biotropica,* 42(5), 555–557.

Fisher, B. and Turner. R. K. (2008). Ecosystem services: classification for valuation. *Biological Conservation*, 141(5), 1167–1169.

Geyer, J., Kiefer, I., Kreft, S., Chavez, V., Salafsky, N., Jeltsch, F. and Ibisch, P. L. (2011). Classification of Climate-Change-induced stresses on biological diversity. *Conservation Biology*, 25, 708-715.

Gosain, A. K., Rao, S. and Arora, A. (2011). Climate change impact assessment of water resources of India. *Current Science*, 101(3), 356–371.

Hoffmann, S., Irl, S. D. H. and Beierkuhnlein, C. (2019). Predicted climate shifts within terrestrial protected areas worldwide. *Nature Communications*, 10, 4787.

Kataki, R. and Konwer. D. (2002). Fuelwood characteristics of indigenous tree species of north-east India. *Biomass and Bioenergy*, 22(6), 433–437.

Kireet, K. (2005). Receding glaciers in the Indian Himalayan region. *Current Science,* 88 (3), 10.

Konwer, D., Kataki, R. and Deka. D. (2001). Fuel-wood characteristics of some indigenous tree species of north-east India. *Indian Journal of Forestry*, 24(3), 316–319.

Kumar, A. and Verma A. K. (2017). Biodiversity loss and its Ecological impact in India. *International Journal on Biological Sciences*, 8(2), 156-160.

Laurance, W. F., Delamônica, P., Laurance, S. G., Vasconcelos, H. L. and Lovejoy, T. E. (2000). Rainforest fragmentation kills big trees. *Nature*, *404*(6780), 836–836.

Lele, S., Springate-Baginski, O., Lakerveld, R., Deb, D. and Dash. P., (2013). Ecosystem services: origins, contributions, pitfalls, and alternatives. *Conservation and Society,* 11(4), 343–358.

Lewis, S. L. and Maslin, M. A. (2015). Defining the anthropocene. *Nature*, 519(7542), 171–180.

Loucks, C., Barber-Meyer, S., Hossain, A. A., Barlow, A. and Chowdhury, R. M. (2010). Sea level rise and tigers: predicted impacts to Bangladesh's Sundarbans mangroves. *Climate Change*. 98 (1-2), 291-298.

Madhusoodhanan, C. G., Sreeja, K. G. and Eldho, T. I. (2016). Climate change impact assessments on the water resources of India under extensive human interventions. *Ambio*, 45(6), 25–741.

Mitsch, W. J., Bernal, B. and Hernandez, M. E. (2015). Ecosystem services of wetlands. *International Journal of Biodiversity Science, Ecosystem Services and Management,* 11(1), 1–4.

McCarthy, J. J., Canziani, O. F., Leary, N. A., Dokken, D. J. and White K. S. (2001). Climate Change 2001: Impacts, Adaptation, and Vulnerability. IPCC. Cambridge University Press, UK.

Millennium Ecosystem Assessment. 2005. Ecosystems and human well-being: biodiversity synthesis. Washington, DC: Island Press.

Montoya, J. M. and Raffaelli, D. (2010). Climate change, biotic interactions and ecosystem services. P*hilosophical Transactions of the Royal Society B*-Biological Sciences, 365, 2013-2018.

Mooney, H., Larigauderie, A., Cesario, M., Elmquist, T., Hoegh-Guldberg, O., Lavorel, S., Mace, G. M., Palmer, M., Scholes, R. and Yahara, T. (2009). Biodiversity, climate change, and ecosystem services. *Current Opinion in Environmental Sustainability*, 1(1), 46–54.

Myers, N., Mittermeier, R. A., Mittermeier, C. G., da Fonseca, G. A. B. and Kent, J. (2000). Biodiversity hotspots for conservation priorities. *Nature*, 403(6772), 853–858.

Narsimlu, B., Gosain, A. K. and Chahar. B. R. (2013). Assessment of future climate change impacts on water resources of upper Sind river basin, India using SWAT model. *Water Resources Management,* 27(10), 3647–3662.

Nepstad, D. C., Tohver, I. M., Ray, D., Moutinho, P. and Cardinot, G. (2007). Mortality of large trees and lianas following experimental drought in an Amazon forest. *Ecology*, 88(9), 2259–2269.

Olson, D. M. and Dinerstein, E. (1998). The Global 200: a representation approach to conserving the earth's most biologically valuable ecoregions. Conservation Biology 12(3), 502–515.

Pan, Y., Birdsey, R. A., Fang, J., Houghton, R., Kauppi, P. E., Kurz, W. A., Phillips, O. L. et al., (2011). A large and persistent carbon sink in the world's forests. *Science,* 333(6045), 988–993.

Platts, P. J., Mason, S. C., Palmer, G., Hill, J. K., Oliver, T. H., Powney, G. D, Fox, R. and Thomas, C. D. (2019). Habitat availability explains variation in climate-driven range shifts across multiple taxonomic groups. *Scientific Reports*, 9, 15039.

Perović, D. J., Gámez-Virués, S., Landis, D. A., Wäckers, F., Gurr, G. M., Wratten, S. D., You, M. S. and Desneux, N. (2018). Managing biological control services through multi-trophic trait interactions: review and guidelines for implementation at local and landscape scales. *Biological Reviews*, 93(1), 306–321.

Pereira, H. M., Leadley, P. W, Proenca, V., Alkemade, R., Scharlemann, J. P. W., Fernandez-Manjarres, J. F, Araujo, M. B., Balvanera, P., Biggs, R., Cheung, W. W. L., Chini, L., Cooper, H. D., Gilman, E. L., Guenette, S., Hurtt, G. C., Huntington, H. P., Mace, G. M., Oberdorff, T., Revenga, C., Rodrigues, P., Scholes, R. J., Sumaila, U. R. and Walpole M. (2010). Scenarios for global biodiversity in the 21st century. *Science*, 330: 1496-1501.

Rasquinha, D. N. and Sankaran, M. (2016). Modelling biome shifts in the Indian subcontinent under scenarios of future climate change. *Current Science*, 111(1), 147–156.

Rathore, A. and Jasral, Y. T. (2013). Biodiversity: Importance and Climate change Impacts. *International Journal of Scientific and Research Publications*, 3(3), 1-5.

Rulli, M. C., Santini, M., Hayman, D. T. S. and D'Odorico, P. (2017). The nexus between forest fragmentation in Africa and Ebola virus disease outbreaks. *Scientific Reports*, 7(1), 41613.

Sadguru, P. and Seema, S. (2019). Impact of Climate Change on Biodiversity: An Overview. *International Journal of Biological Innovations*, 1(2), 60-65.

Singh, J. S. (2002). The Biodiversity Crisis. *Current Science*, 82(6), 638-647.

Smith, P. J. (2007). Climate change, mass migration and the military response. *Orbis*, 51(4), 617–633.

Subodh, K. S. and Lokesh, C. D. (2018). *Climate Change and India: Impacts on Biodiversity.* Chapter 11,. indd, 499.

Sukumar, R., Suresh, H. S. and Ramesh, R. (1995). Climate change and its impact on tropical montane ecosystems in southern India. *Journal of Biogeography*, 22, 533-536.

Swaminathan, M. S. and Kesavan, P. C. (2012) Agricultural Research in an Era of Climate Change. *Agricultural Research*, 1(1), 3–11.

Tibbetts, J. (2007). Health effects of climate change. *Environmental. Health Perspectives*, 115, 196-203.

Upadhaya, K., Watham, T., Bordoloi, N. and Kataki. R. (2017). Trees as sources of livelihood and fuelwood: a case study of an eastern Himalayan village. *Energy Sources, Part A: Recovery, Utilization, and Environmental Effects*, 39(4), 398–405.

Varma, V., Catherin, A. M. and Sankaran, M. (2018). Effects of increased N and P availability on biomass allocation and root carbohydrate reserves differ between N-fixing and non-N-fixing savanna tree seedlings. *Ecology and Evolution*, 8(16), 8467-8476.

Verma, A. K. (2017). Multiple effects of Unsustainable Agriculture. *International Journal on Agricultural Sciences*, 8(1), 24-26.

Verma, A. K. (2018). Unsustainable Agriculture, Environmental Ethics and Ecological Balance. *HortFlora Research Spectrum*. 7(3), 239-241.

Vinod, K. and Chopra, A. K. (2009). Impact of climate change on biodiversity of India with special reference to Himalayan region-An overview. *Journal of Applied and Natural Science*, 1(1), 117-122.

Wallace, K. J. (2007). Classification of ecosystem services: problems and solutions. *Biological Conservation*, 139(3–4), 235–246.

Chapter 12

Overview of Soil Fertility from Past To Present

Eetela Sathyanarayana[1,*], J. Bharghavi[1], Suddakanti Saranya[2], Santhosh Kumar Manchala[1], Mandla Rajashekhar[1], Jatoth Veeranna[1], Banda Rajashekar[1], T. Prabhakar Reddy[1], Saideep Thallapally[3], Krishnananda Pralhad Ingle[4], Alamuru Krishna Chaitanya[5], Asik Duta[6], Alavala Umarajashekhar[1], Kumari Sunita[7] and Hanuman Singh Jatav[8]

[1]Professor Jayashankar Telangana State Agricultural University, Hyderabad, India
[2]GD Goenka University, Gurugram, Haryana
[3]Forest College and Research Institute, Mulugu, India
[4]College of Agriculture, Guntur, Andhra Pradesh, India
[5]Lovely Professional University, Punjab, India
[6]ICAR-Indian Institute of Pulses Research, Kanpur, Uttar Pradesh, India
[7]Deen Dayal Upadhyaya Gorakhpur University, India
[8]Department of Soil Science and Agricultural Chemistry,
Sri Karan Narendra Agriculture University, Jobner, Jaipur, India

Abstract

Soil is an essential base for farming and crop production as it is the media for the plant growth and advancement. Its fertility can be evaluated by testing the soil, analyzing the plant, and symptoms of deficiency in plants. The deterioration of fertility of the soil is major problems of the ecosphere, is the most vital matter since expansion of the "sedentary system of agriculture." The continues declining in fertility status of soil is measure challenge to obtain global food security. Conventionally, fertility of the soil is controlled via organic matter in the form of manures, chemical fertilizers, and various natural substance in maximum enduring systems of agriculture. The soil fertility can be improved with the application of manure and fertilizers in judicious way. . Diminishing in the fertile nature of soil is the prime constriction in enhancing the yearly harvest of produce causing decline in production. Thus, to uphold productivity of the crop, fertility and health of the soil are crucial fundamental enhancers to the growth of the sustainable farming that was oriented by biochemical responses inside the earth, quantity, and accessibility of vital nutrients of the plant, in addition to the mechanism of nourishment reduction plus replacement inside the soil.

---

* Corresponding Author's Email: sathyasoil92@gmail.com.

In: Ecosystem Services
Editor: Hanuman Singh Jatav
ISBN: 978-1-68507-614-6

Keywords: agriculture, soil fertility, soil health, soil productivity, plant nutrients, fertilizers

1.1. Introduction

Soil plays a diverse role in plant development which directly influences the harvests of the produce and its quality. Assessment of the fertility of soil let know the considerably around the potent of production. Providentially fertility may be tackled by agriculturalists and farmers by handling the nutritional status of the plant (Flynn et al., 2004). "Soil" is regarded as the substantial centers in agricultural science for the reason that it is the means of development and growth of the plant. From "*Atharva Veda* (1200-1000 B.C.), the soil has been thought of as a mother giving foods" for the society. The formation, existence, & maximum loss of whole civilizations almost dependent upon the efficiency of soil to furnish the food, feed, additional crucial commodities to the individuals (Mueller et al., 2010). Productivity of the soil can be defined by way of "agricultural soils" capability to produce plants and or crop (yield), is dependent on soil fertility" (Basak, 2000). Productiveness is the inherent capability of the soil in a purpose to deliver vital nourishment in adequate quantities to guarantee optimal crop efficiency, comprising the maximum commercial revenue and least deprivation to the environment (Basak, 2000; Biswas and Mukherjee, 1994; Foth, 1990; Prasad and Power, 1997).

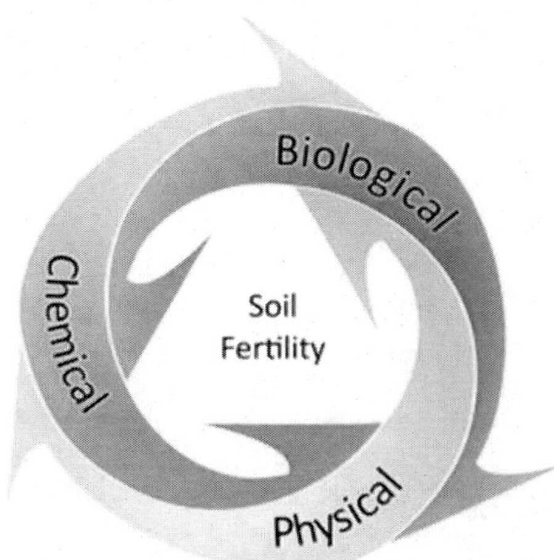

Figure 1. "Soil fertility is originated from the interactions among physical, chemical and biological processes (Abbott and Murphy, 2003). Illustration created by Kara Skye Gibson."

Deterioration of fertile soil is one among the global greatest serious challenges, was the subject of concern since from the advancement of the "sedentary" farming approach prior before ten thousand years (Hartemink, 2003). Reduction in the productiveness is a substantial obstacle among maximum tropical zones (Akinrinde, 2004) for overutilization of agrarian soil sources (Knudsen et al. 2006) as 95 parts in growth of populace is ever-growing in such hot and humid territories (Hartemink, 2003). Conventionally, fertility of the soil may be tackled in maximum everlasting farming systems by usage of synthetic and organic fertilizers (Hartemink, 2003; Foth, 1990). Fertility of the soil is derived by prolonged treatment with manures which are organic or by using of chemical and synthetic fertilizers (Akinrinde, 2004). Deteriorating the fertility was the foremost constriction to enhance the harvests of crop which results in high production (Buresh et al., 1997). Consequently, for upholding this elevated efficiency of crop, managing of productiveness and healthiness of soil are very crucial leading strategies for the advancement of sustainability in agriculture (Prasad and Power, 1997), which is dealt by biochemical responses inside the crust, quantity of accessible or inaccessible nutrients, phenomenon of replenishment and soil nutrient (Prasad and Power, 1995).

1.2. Historical Facts Concerning Fertility of the Soil

Concerning productiveness of the soil, it is derived along the advancement through agricultural science. Through Ancient roman times, many soil management practices for enhancing the fertility of the soil remained in usage. These encompassed nourishing by organic manures, reclamation by liming, rotations of crop season to season, and following to develop the allocation of accessible nutrients for the crop. Substantial development has been started in the early mid of the 19th era to the early twentieth era for assessing fertility and development in conjunction through the supplement by fertilizers besides manures. Jean Baptiste Boussingault (1802-1882) computed nutrients inside the organic manures utilized for fields and thereby formulated "balance sheets" demonstrating contributions & productivities within the reaped produce. At that time, "Justus von Liebig (1803-1873)" proposed the "principle of law of minimum", that says the harvest produced from a crop is directly proportional to the essential nutrients of plant in the least amount. With 1855, it is recognized that (i) plants necessitate equally phosphorus (P) and potassium (K), (ii) non leguminous plants needs a source of nitrogen (N) for plant development, and (iii) soil fertility be able to be retained through inorganic stimulants i.e., fertilizers and organic manures. Advanced, several extensive experiments on fertilizers were recognized everywhere in the domain persuasively exhibited extraordinary profits as of the usage of animal waste such as organic manures plus the synthetic chemical fertilizers, subsequent green manuring, and reduction in the soil productiveness with associated produce falls because of unceasing harvesting exclusive of the incorporation by nutrients, organics, and synthetic chemical nourishment through the fertilizers.

In consideration of Past, within the country like India, classifications of soil based on fertility or nature as "urvara (fertile) and usara (sterile)". The sub classification of fertile soils was as per the crop based and suitability. The promptest source of manuring to augment fertility of the soil can be outlined as concerns the eras of "*Atharva Veda* (1500 - 500 B.C.)".

Maximum Foreigners recognized such inherent natural productiveness of the soils pertaining to India prior to 20th era. The fertility status of the soil in India based on the Royal Commission on Agriculture 1926 was as follows: (i) shortage of Nitrogen, Phosphorus and organic matter in red & black soils of peninsular India, but potassium and lime remained satisfactory, and (ii) "contents of N, P and organic matter in alluvial soils were low and variable; K was adequate and P though not plentiful was less deficient than in other Indian soils" (Abrol and Nambiar, 2008). Subsequent to shortage of food while World War II and famine of the Bengal which was happened in the year 1943, technical, technological potentials of farming advancements were sightseen, besides it was noticed that the mean increase in rice yield increased by 50 percent, 10 percent by utilizing the improved varieties, diversities and cultivars and 40 percent increase via manuring. Comparable surges existed in various crops as mentioned by (Randhawa, 1979). During the years of 1950s, several everlastings "manure plot experiments" were introduced, persuasively manifesting those fertilizers would not constrain productivity of the soil if used carefully and judiciously. Research on soil fertility was started in the 1960s by launching great resilient rice and wheat. During 1970s, the prominence was placed on balance of nutrients and fertility of soil and its supervision in "multiple cropping schemes." Research on micronutrients was acknowledged substantial consideration around 1980s after shortage of Zn which was recognized in cropping schemes (Abrol and Nambiar, 2008).

With the advancements in sustainable farming, fertility degradation is a foremost alarm. Sustaining optimal fertility of the soil, Farmers in India have remained practicing an agricultural scheme that safeguards uncertain and steady yields. The improved inhabitants take directed to familiarizing extraordinary-yielding seeds diversities, exhaustive surplus fertilizers, insecticides, and widespread tillage. India is positively influenced with deprivation of land and rainfed areas are mostly affected.

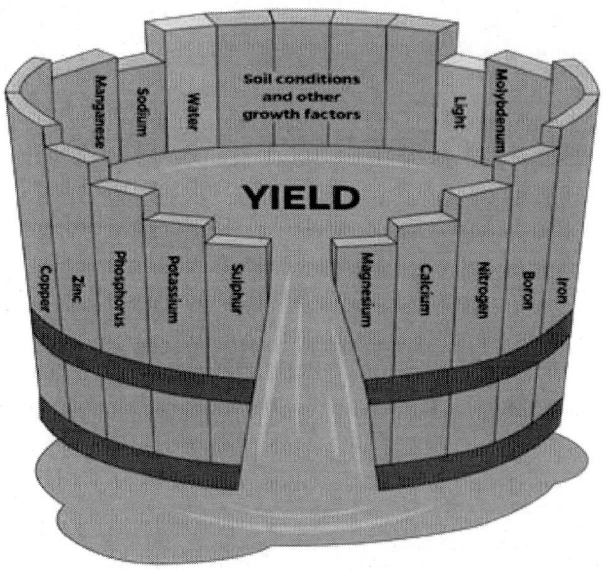

Source: https://www.acgmaterials.com/liebigs-law-minimum/.

Figure 2. "Liebig's law of the minimum demonstrating the impact of trace elements deficiency on digester performance."

The 2015 year declared as "International Year of Soil" and the succeeding aims selected must be attained:

- To create responsiveness amongst people along with decision creators the strength of the soil among anthropoid existence.
- Delivering proper training for the society in the areas of soil's function in enhancing insecurity of the food, important ecology amenities, poverty alleviation, adaption to change in climate, and overall sustainability in agriculture.
- Endorse conferring the strategies besides agendas for "sustainable management and soil resources fortification".

- To implement sustainable managing actions and advance the beneficial soils for several users and populace clusters.
- Proper-prearranged progression of the Universal Goals ingenuities should be agreed.
- Care must be handled for speedy ability augmentation for the soil info assembly there by supervising among state, nationwide, and universal stages (Patel, 2016).

The situation must be reminisced in a way ensuring all shareholders allied with farming globally are performing to one's own potential to enhance the resources of soil and production of management practices in scientific approaches by protection of soil, preservation and soil conservation, and justifiable efficiency in production. The Governments to attain this objective must [i] seek worldwide methodological cooperation and support its investment in research and development [ii] stimulate research and development on soil concentrating on recognized breaks and main concerns and collaborations with related productive, ecological and societal expansion activities [iii]improve the amount and eminence of soil data and info: data assemblage (generation), investigation, authentication, reportage, supervising and amalgamation with further branches [iv] complement approaches, capacities, and indices to reach the development and fortification among the available soil sources. Preceding of all, start policies and agendas to instruct agriculturalists and generate responsiveness amongst farmers to encourage soil integration and landscape reformation, water, flora, and ecological management for permitting sustainable agrarian creation which may be the upright for ecology and revenues of the farmers (Patel, 2016).

To Rejuvenate the health of soil, "Integrated Soil Fertility Management (ISFM)" have been implemented. The engrossment by mineral fertilizers along with the biological material (residues of the crop, dung, and green manuring) besides the inculcation of turning in crops also by using intercrops with legumes so that atmospheric nitrogen can be fixed thereby improving the health of the soil (Mundt, 2002; Srinivasarao et al., 2012; Kumar, 2017). Present day research was observed that implications of the united inputs and strategies may approximately duple crop harvests contrasted to application of fertilizers individually (Agegnehu and Bekele, 2005; Vaje, 2007; Dercon and Hill, 2009). Soil fertility is a major decline in India and several countries which is a serious delinquent (Sanchez et al., 1997; Bationo et al., 2006; Sanginga and Woomer, 2009; Vanlauwe et al., 2010).

1.3. History of Fertility and Quality Concepts of Soil

Scientific impression regarding the productiveness ascended since 1850s, accenting agriculture to utilize soil for enhancing the productivity. Which is concurred from the commencement of the "period of minerals" (1840-1940), that started by the initial systematic demo regarding dry matter origin through the minerals, which lead the assumption of H_2 originates by H_2O, carbon arising by CO_2 and various nutrients via salts present in aquatic forms (Manlay et al., 2007). Within the preliminary stage with fertility of soil commencement, farming sciences henceforth generally emphasized the function of soil's physical and biochemical attributes to sustenance plant growth. Since then, fertility of the soil has become a stock for agronomy and soil disciplines. The predominant method reflected that fertility of the soil depends on about intrinsic abilities ensuing since the appearance of formation of soil. The indicated method directed for effort on the categorization of soil and study tables, wherever the concept of capability of land covered with fertility one.

The notion that fertility of the soil has remained discussed by way of a "piece of basic acquaintance," and communal and governmental circumstances were progressed. It's restriction contrasts extensively, from works such that definite produce or production is corresponding or entirely illustrative of productiveness to works delivering relatively extra or fewer complex descriptions depends on various amalgamations of the numerous features comprising properties, environment, exertion, and societal frameworks (Patzel et al., 2000). Till late 1980s, there remained nope strong consecutive development to this theory. Although various elucidations co-occurred. For instance, promptest definitions specified as "soil fertility is a product of soil and manpower" (von Wulffen, 1847). Additional terminologies and scientific statements although passing over many periods concurrently. In Authors systematic assessments, they depicted as thought regarding the fertile nature of the soil might-not be grabbed among any particular practical definition. Initially, it states towards a outlook where at no time occurs. Secondly, it may not overcome the "trade-off correlation" amongst vividness and extensiveness owing over the imperative aspects contravening the accepted disciplines (Patzel et al., 2000).

1.4. Characteristic Features of Soil

Soil is the maximum and crucial environment to the numerous existing creatures which intermingle and remain accountable for living over this earth. Existing creatures of inside the soil regulate "penetration of water," "inert density," and "nutrient cycling." Microbes such as bacteria and fungi interact by carbon-based material inside the mass of soil for breaking into tiny molecules; earthworms help in digesting the organic material within the soil, thereby also reutilize essential elements, and make topsoil more enrich for the plant growth. Besides the existing live microbes, the soil also consists various elements and minerals. Element namely Carbon is the vital attribute, regulating various biological activities like "soil structure progress, aquatic storing, and nutrient cycling." Organic carbon present inside the soil can be accessible in the mentioned forms, namely, "living carbon, labile carbon, and fixed carbon." Microorganisms, earthworms, crop radicle portions, etc. are categorized in "living carbon." "Labile carbon" comprises constant amalgams like "humates and glomalins." "Sequestered carbon" comprises "fixed carbon" in addition to whole existing biomass. The water from the rain is infiltrated easily and retention of soil is good if there is good amount of organic carbon inside the crust. However, it requires nearly 500 years for forming topsoil, erosion of soil occurs at an abundant quicker proportion. Through deliberately selecting extra enriched nutrient foods, the earth's physical, biotic, as well as the inorganic health (Nitrogen, phosphorus, and potassium) may be measured (Patel, 2016).

1.5. "Soil Is Our Life"

The utmost composite bio-organic resource present over this earth is "the soil" as, soil is keystone of all the ecological sustainability and many systems of foods etc. Soil generates 95%

nourishment required for us and, besides, providing space for living humans and vital ecological amenities required to control supply of the water, conservation of ecology and biodiversity, changes in climate, and sequestration of the organic carbon. It maintains and sustains small and marginal farmer's maintenance. "Soil" is the slender organic coat present over the crust of the earth and is an alive medium. Soil is also the repository of organic carbon, H_2O, various mineral materials, besides the atmosphere among variable extents establishing the plant growth system. Due to its existing natural resource of nutrients, it provisions plant development which is considered as vital for crop growth. Healthful soil is main foundation for "food, feed, fuel, fiber, water, and medicinal substances" required for wellness of humans. The carbon pool is due to the soil mass, which is vital in mitigation and acclimatization with change in climate. In the period of shortage for irrigations, soils serve as the indispensable components for suitable loading and circulation. Food and Agricultural Organization has assessed that a part out of the three proportion of world cultivated soils are deteriorated because of soil erosion, acidification, salinity, pollution, decline of soil organic material, vital nutrients, and various methods caused because of unmanageable land sustainable approaches. Unless novel systematic methods are investigated and implemented, the total productive area & barrens lands per capita for the year "2050 may be solitarily a quarter of the level in 1960." Hence, politicians, Policy-decisions people and agriculturalists should escalate functions of soil and evaluate the hazards and risks which are facing currently.

1.6. Soils Status in India

Approximately 18% of the global people and 15% of the livestock populace is raised in India, owning only 2% of the global topographical range and 1.5% of forest and pastureland. The geographical area in country like India is around 328.7 MHA and out of that 142 MHA is regarded as the cultivable net area. Among this net area 57 MHA i.e., around 40 percent of the area is irrigated land mass and left over 85 MHA i.e., is 60 percent remained as rainfed. "Out of 328.7 MHA about 120.4 MHA (37%) suffer from various kinds of land degradation; water and wind erosion (94.9 MHA), water logging (0.9 MHA), soil alkalinity/sodicity (3.7 MHA), soil acidity (17.9 MHA), soil salinity (2.7 MHA) and mining and industrial waste (0.3 MHA)." Rigorous cultivation and "mining of nutrients" although exhausted fertility and insufficiencies of minor and secondary elements, diminishing level of the water table and eminence. All these may lead to the erosion of soil and deprivation and to minimize only good practices, skills, and consciousness could carry source anywhere adjacent to "natural soil" (Anonymous, 2016; Patel, 2016).

1.7. Sustaining and Management of Soil Fertility

Health of the soil shows a important function for sustainability which is within the prevalent hazard since deterioration progressions. Sustainability in agriculture starts from the soil such as "reducing soil erosion and improving water-holding capacity, nutrient balances, physical structure of soil, and presence of organic matter inside the soil." Health of the soil should be improved by the utilization of soil erosion resisting pulse crops, green manuring and cover crops, amalgamation with the capacity to supply phosphorus among the rotations of crops, utilizing organic compost waste also the dung from animals which are manures, implementation of zeroth tillage, and utilization of mineral or chemicals wherever desirable (Kibblewhite et al., 2008). Farm men now a days currently acclimatizing knowledges thereby enhancing the fertilizer use efficiency and presence of organic portion, water seepage, rainwater seepage, besides few agriculturalists are eradicating contour terraces. A communal technique was similarly formed while healthiness of the soil may be enriched through augmented carbon-based organic material. These material in the soil consists of carbon, and soils exceeding the level of the surface biomass would regard as the "carbon sinks" or "sites for carbon sequestration". For example, agriculturalists working on the zeroth tillage in addition to usage of legume crops for green manuring or those utilizing of the "cover crops" leading the enhancement in carbon proportions of the soil (Pretty et al., 2003).

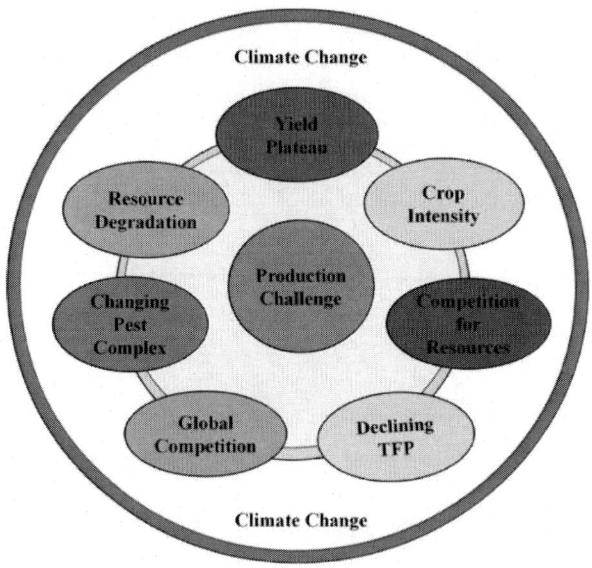

Source: Sendhil et al. 2017.

Figure 3. "Challenging reasons for sustainable agriculture system."

Soils which are fertile having maximum-grade physical attributes for enhancing growth of the roots may be required mainly for sustainability in agriculture. Nevertheless, from the time when 1945, around 17% of the mass under vegetation was endured "human-induced soil degradation" & "loss of productivity," often from insufficient nourishment by chemical fertilizers and management of water, erosion of soil, and condensed fallow stages (Oldeman, 1992). Constant harvesting and insufficient supplement of nutrients detached in reaped resources or vanished by erosion of soil, leaching or emissions in the form of gases reduce fertility of soil and might cause reduction in content of carbon matter, frequently to partial or to a lesser extent original level (Matson et al., 1998). Tilling of the soil accelerates the carbon matter decomposition eventually, leading to supply of mineral nutrients. Erosion of the soil will be more on vertical slanty slopes wherever windbreaks have been blank, absence of vegetative coverage in spring season, and substantial technology is intricated in preparation of the land (Naylor, 1996). The hazards by degradation of land on crop production may occasionally be remunerated by augmented fertilizer supply, hydrating, or drenching by irrigation, and control over the diseases, growing production budgets (Naylor, 1996). Timely rotation of the crop season to season, condensed tillage, utilization of cover crops, uncultivated fallow times, composting and manuring, and well-adjusted fertilizer treatment, can benefit and maintain to rejuvenate the fertility of soil.

1.8. Management of Soil Fertility

"Soil" is the vital function of farming sectors which regulate as the media to different process like "physical, chemical, and biological activities" which improve health and crop production. While topsoil can restore certainly amongst uninterrupted ecologies, unremitting implanting of crops in barren systems leading to degradation and decline in plant nutrient availability. Eventually, reduced fertility of the soil ensuing from continuing farming is the main constriction towards production (Tening et al., 1995, 2013; Nottidge et al., 2005). Subsequently, elevated input farming system based upon chemical amendments to enhance nutrition of the plant and fertility of the soil (Singh 2000; Saha et al., 2008; Hepperly et al., 2009). Chemical and synthetic fertilizers may be utilized for enhancing the nutrient uptake by the plant and owing to fast accessible of nutrients which are required for the plant, nevertheless their enduring usage sooner or later reduces the soil's "physical, chemical, and biological" attributes (Albiach et al., 2000; Thy and Buntha, 2005). In the intervening time, simply encouraging chemical fertilizers in deteriorated soils deprived of mixing biological inputs is unmanageable and may slowly enhance or deteriorate quality of the soil by speeding up carbon loss.

Moreover, deprived and deteriorating fertility of the soil which constrains farming productivity, pests of the crops and microbes causing the diseases too cause substantial decline in production which frequently necessitate the usage of antifungals and chemical pesticides in arable lands. Nevertheless, dependence on antifungals or fungicides and chemical insecticides causes to ecological contamination by pollutions, high cost of production, and resistance against insects and microbial diseases (Shelton et al., 2000; Susila et al., 2003; Sarfraz and Keddie, 2005; Xu et al., 2010). By means of insecticides which are correspondingly toxic for "non-target" creatures, through potent noxiousness over humans as well as the atmosphere (Reinecke and Reinecke, 2007; Fountain et al., 2007; Kumar et al., 2012; Ghananand et al., 2011; Fosu-Mensah et al., 2016 Mostafalou and Abdollahi, 2013). These all had directed to increasing concerns and/or the capability to explore for integrated sustainable methods which increase fertility of the soil and protection of the crops without jeopardizing and thereby creating the sustainable environment (Suge et al., 2011). Congruently, "botanicals" regarded as the budget friendly substitute managing approaches for crops pest control (Liu et al., 2007; Sayyed et al., 2008; Tanyi et al., 2017 Li et al., 2008). Recently, "ISFM" is encouraged as the budget friendly substitute and justifiable approach for advancing health and efficiency of the soils (Scoones and Toulmin, 1998; Sanginga and Woomer, 2009; Vanlauwe and Giller, 2006; Vanlauwe et al., 2010). "ISFM" provisions the usage through chemical fertilizers or manures, enhanced germplasm, besides inter- cropping with pulse crops. (Sanginga and Woomer, 2009; Vanlauwe et al., 2010). Inside "ISFM" framework, residue mulches by natural substances and green manuring are vital strategies for fertility enrichment and there by encouraging supportable advancements. The above strategy is to tackle the fertility status and sustenance mulching with wastes is the judicious and budget friendly choice for augmenting the health of the soil there by permitting the process of sustainable development.

1.9. Evaluation of the Soil Fertility

Assessing fertility of the soil is a process of evaluating its volume of innate & remains of nutrients which can be accessible for usage through rising produces in specific soil and the fertilizer requirement to augment the productivity of crop (Sanchez et al., 1997). Hence, assessment of fertility of the soil is a "tool forward."

Assessing amount of requirement in chemical for the soil as well as plant; (b) Attaining consistent and economical recommendation of fertilizer which is guaranteeing accurate kinds and amounts of synthetic fertilizer chemicals are applied; (c) Inspection of fertilizer residues, and (d) Reducing contamination of water and soil by the usage of high quantities of synthetic fertilizers. However, fertility of the soil assessment is a potent means to sustenance great production in yields by justifying the usage of nutrients, its existing influence on agriculture application is hidden. For doing in an appropriate and farmer-based provision, the usage must be imperious to (1) Increase canopy of the fertility assessment outside important nutrients like "nitrogen, phosphorus and potassium" including pH of the soil. (2) Improve directions of fertilizers for yielding better harvests, concerning maximum scarce elements, and manipulating useful optimistic nutrient relations.

1.10. Tools for Evaluating the Fertility of the Soil

There are approximately three tools that help in assessing the fertile nature of the soil, namely "Plant tissue analysis," "Soil analysis," "Visual symptoms of nutrient deficiency."

1.10.1. Using Visual Symptoms of Nutrient Deficiencies to Determine Fertilizer Needs

"Deficiency Symptoms" which could been seen are the potent ways of evaluating tools for knowing the status of nutrient amiability in the plants. However, diagnosis of the entire plant cannot be based on the visualizing the given individual symptoms. Wade (2010) contended as maximum of the typical symptoms for deficiency like "chlorosis, tip burn, and necrosis" was due to extra deficiencies besides single nutrient. Nevertheless, their recognition is exceptionally beneficial for evaluating the presence of nutrients.

In maximum no of cases, deficiency of nutrients can significantly decrease harvests deprived of revealing some apparent symptom which is regarded as "hidden hunger" whereby a shortage has a undesirable consequence exclusive of being predictable. Although, if deficiency of the nutrient analysis is initially completed, actual activity could generally be reserved (Wade, 2010). The major uses of easily obtained visual indicative indications can deliver an instant signal for status of the nutrients (Wade, 2010), besides wherever indications may not necessitate the validation, strategy of any type is not necessary (Wallace, 1943).

Foremost drawback is when the "visual symptoms do not develop until after there has been a significant effect on plant growth, yield, and growth" (Wade, 2010); indications with shortage in nutrients may be curbed by supplementary aspects, like climate phenomenon's, saltiness, & non-consistency of nutrients, microbes, also insect attacks (Wallace, 1943). However, an alert agriculturalist may acquire to utilize the visual approaches faster with high usage with experience. That type of agriculturalist should be aware of required conceptual skills and knowledge on deficiency of nutrients, as depicted in Table 1, as well as the applied information of identifying indications of the disease when recognized over the plants in standing crop.

Table 1. Description of nutrient deficiency symptoms in crops

S. No	Nutrient element	Deficiency symptoms in crops
1	Phosphorus	Dark-green to blue-green coloration appears on the mature leaves. The root development is restricted, the plant is spindly and stunted. In the acute case, red, purple, or brown leaf coloration reveals, and maturity is delayed. There is insufficient development of seed and fruit
2	Nitrogen	General chlorosis starts on the older leaves (appearance of light-green to pale-yellow color). Old leaves scorch or drop depending on the degree of deficiency, and the tips are first affected. Growth becomes stunted, and plants may have an etiolated appearance. In the severe case, flowering is significantly reduced. Deficient crops have lower protein content.
3	Potassium	Slow and little growth of plants. Chlorosis along the leaf margins followed by scorching and browning of tips of older leaves. These symptoms gradually progress inward. Stalks are weak, and plants lodge easily. Terminal and lateral buds may die ("*dieback*"). Shriveled seeds of fruits
4	Magnesium	Interveinal or marginal chlorosis, mainly of older leaves, is often accompanied by the development of various pigments. Chlorosis may also begin in patches or pouches, which later merge and spread to the leaf margins and tips. With an acute deficiency, the affected tissue may dry up and die. Leaves are usually small, brittle in the final stages, and curve upwards at the margin. Twigs weak and prone to fungus attack, usually premature leaf drop These are first noticed in the meristematic regions and the young leaves. The young leaves of new plants are often distorted, small, and abnormal. Leaves may be cup-shaped and crinkled, and the terminal buds deteriorate with some breakdown of petioles. Root growth is severely impaired. Roots get rotten. Buds and blossoms shed prematurely. Stem structure weakened
5	Calcium	These are first noticed in the meristematic regions and the young leaves. The young leaves of new plants are often distorted, small, and abnormal. Leaves may be cup-shaped and crinkled, and the terminal buds deteriorate with some breakdown of petioles. Root growth is severely impaired. Roots get rotten. Buds and blossoms shed prematurely. Stem structure weakened
6	Sulphur	Younger leaves turn uniformly yellowish or chlorotic. Plants are spindly and grow poorly. Flower production is often indeterminate. Stems are stiff, woody, and small in diameter
7	Boron	Growing tips are often damaged and may die. The leaves may become distorted, curling, and becoming brittle. Stems become rough and cracked, often with corky ridges of spots. Flowers do not form. Roots are stunted and prone to bacterial infection. Brown's heart in root crops is characterized by dark spots on the root's thickest part or splitting at the center. Fruits such as apples develop "internal and external cork" symptoms
8	Zinc	Deficiency symptoms often appear on the 2nd or 3rd, fully mature leaves from the top of plants. In some varieties, leaves may be blue-green or dark-green. Flowering and fruiting are much reduced under conditions of severe zinc deficiency, and the entire plant may be stunted and misshapen
9	Copper	Leaves may become chlorotic (yellowing) or deep blue-green. Curling of leaf blades or rolling up of margins. Flowering and fruiting are curtailed. Ear production in cereals is restricted, and grains are poorly set. Young shoots often die back; after that, new shoots emerge from multiple buds further back

Source: Marschner (1995).

1.10.2. Plant Tissue Analysis

Investigation of tissues of plant is one of the laboratories aim of crops total fundamental portion or definite portions of the plant (Reuter and Robinson, 1997; Steinhilber and Salak, 2010). This analysis help in finding solution for different objectives, checking the position of the plants, and resolving problematic and effected palces. This directs as main source for enrichment mainly to fruit orchards (Steinhilber and Salak, 2010). Plant tissue analysis is the only means if crop is adequately nurtured throughout the peak phase of growth (Flynn et al., 2004). There should be misperception between "Tissue Analysis" and "tissue testing. The fore one indicated to take samples of sap from the fresh plants and investigating it on site. Analysis is always done on laboratory prepared dried plant tissue (Steinhilber and Salak, 2010). Invisible deficiencies are identified in some tissues of the plants (Flynn et al., 2004; Steinhilber and Salak, 2010; Cleveland et al., 2008; Walsh and Steinhiber, 2005), authorize noticeable deficiency indications besides to know nutrient toxicity stages. Though typically utilized as an analytical technique to correct the nutrient complications, analyzing tissues of young plant would permit the correct application of fertilizers in the constant environment conditions (Flynn et al., 2004; Cleveland et al., 2008). It is used for monitoring status of nutrients and to detect problems of nutrient which exists (Cleveland et al., 2008; Flynn et al., 2004) and retain a brilliant yearly record about use of nutrient and necessities of crops in several ecological circumstances.

The "sampling method" of tissue is important for our achievement. Each crop they have a different procedure. Specific tissue at specific stage of growth is selected from the plants (Steinhilber and Salak, 2010; Flynn et al., 2004; Cleveland et al., 2008; Walsh and Steinhilber, 2005) to avoid soil adulterations (Cleveland et al., 2008). If any of the recommended steps are not utilized, then it may lead to ambiguity in the results leading to decreased production of crop. However, plants from fields infected with pests, crop weeds, insects, disease casual organisms, moisture stress, and physical damage there is no advantage of doing analysis (Flynn et al., 2004). As Steinhilber and Salak (2010) postulated that for successful analysis, a small error cause detrimental effect. Table 2 displays intricate methods needed for detecting how, where & when, at the time many samples are accessible, comprising field crops, ornamental trees and plantations of nuts. If analysis was done, critical values of yield are compared with the nutrient substances and finally assessment is done by detecting the excess or shortage of nutrients. Adequacy levels, vital values for numerous species were given in the Table 3. The least boundary of "sufficiency range" signifies the critical level under which appropriate doses of fertilizers to be utilized. The higher range shows the limit beyond harmfulness indicating toxicity.

Testing of soils is performed for finding how much quantity of nutrient is present with their use to the plant natural life. Numerous approaches are established, and the key to success is to calibrate the method. Calibration of soil test suggests establishing a correlation among relative crop response plus soil test values (Agboola and Ayodele, 1987). For efficacious extensive-standing soil and crop nutrient managing strategy sampling of soil should be done properly (Potash and Phosphate Institute, 2004). This is done to estimate the requirement of applying lime or fertilizer prior to planting (Reisenauer et al., 1983). By this we can know nutrient levels specially. But cannot show if plants can take it or not. pH of soil can be assessed by this soil test (Kidder, 1993).

Table 2. Tissue sampling techniques for specific plants

S. No	Crop	When to sample	Where to sample	Number to sample
Field Crops				
1	Alfalfa	Early bloom	Top 6 inches or a topmost third of the plant	12-30
2	Canola	Before seed set	Recently mature leaf	60-70
3	Clover	Before bloom	Upper 1/3rd of plant	30-40
4	Sweet corn / Corns	Seedling stage or before tasseling or tasseling to silking	All above-ground portions first fully developed leaf from the top of the plant Leaf opposite and below the ear	15-20 15-20 12-20
5	Cotton	Full bloom	Recently mature leaf from the main stem	40-50
6	Grasses/forage mixes	Stage of the best quality (before seed emerges)	Upper 4 leaves	30-40
7	Peanuts	Before or at bloom	Recently mature leaves	40-50
8	Small grains (barley, oats, wheat, rye, rice)	The seedling stage before heading	All above-ground portions 4 uppermost leaf blades	25-40 25-40
9	Sorghum (milo)	Before or at heading	2nd leaf from the top of the plant	20-30
10	Soybeans	Before or at bloom	Recently mature, trifoliate leaves from the top of the plant	20-30
11	Sugar beets	Midseason	Recently mature leaf at center of the whorl	25-30
12	Sunflowers	Before heading recently mature leaf	Before heading recently mature leaf	20-30
Vegetable Crops				
1	Asparagus	Maturity	Fern, 18-30 inches above ground line	10-30
2	Beans	Seedling stage or before or at bloom	All above-ground portions Recently mature leaf	20-30 20-30
3	Broccoli	Before heading	Recently mature leaf	12-20
4	Brussels sprouts	Midseason	Recently mature leaf	12-20
5	Celery	Midseason	Outer petiole of recently mature leaf	12-20
6	Cucumbers	Before fruit set	Recently mature leaf	12-20
7	Head crops (cabbage, cauliflower)	Before heading	Recently mature leaf at center of the whorl	12-20
8	Leaf crops (such as lettuce, spinach)	Midseason	Recently mature leaf	12-20
9	Melons	Before fruit set	Recently mature leaf	12-20
10	Peas	Before or at bloom	Leaves from 3rd node from top	40-60

Table 2. (Continued)

Field Crops				
S. No	**Crop**	**When to sample**	**Where to sample**	**Number to sample**
11	Peppers	Midseason	Recently mature leaf	25-50
12	Potatoes	Before or at bloom	3rd to 6th leaf from growing tip	25-30
13	Sweet potatoes	Midseason or before root enlargement	3rd to 6th leaf from tip center or mature leaves	20-30 25-35
14	Root/bulb crops (such as carrots, beets, onions)	Midseason before root or bulb enlargement	Recently mature leaf	20-30
15	Tomatoes (field)	Mid bloom	3rd to 4th leaf from growing tip	15-20
16	Tomatoes (trellis or indeterminate)	Mid bloom from 1st to 6th cluster stage	Petiole of leaf below or opposite top cluster	12-20
Fruit and Nut Crops				
1	Apples, pears, almonds, apricots, cherries, prunes, plums	Midseason (June-July)	Leaves from the current season's nonfruiting, nonexpanding spurs	50-100
2	Peaches and nectarines	Midseason (June-July)	Mid shoot leaflets/leaves	25-100
3	Grapes	At bloom	Petioles or leaves adjacent to basal clusters at bloom	50-100
4	Pecans	Midseason	Mid shoot leaflets/leaves	25-60
5	Pistachios	Mid- to late season (August)	Terminal leaflets from nonfruiting shoots	25-60
6	Raspberries	Midseason	Recently mature leaves from laterals of primocanes	30-50
7	Strawberries	Midseason	Recently mature leaves	25-40
8	Walnuts	(June-July)	Terminal leaflets/leaves from nonfruiting shoots	25-40

Source: Flynn et al., 2004.

Table 3. Sufficiency range of nutrient elements in some crops

S. No	Crop	Source/amount of fertilizer (bags)						
		Nitrogen (N)	**Phosphorus (P)**	**Potassium (K)**	**Calcium (Ca)**	**Magnesium (Mg)**	**Zinc (Zn) (ppm)**	**Boron (B) (ppm)**
1	Maize (whole plants, less than 30 cm tall)	3.5 - 5.0	0.30 - 0.50	2.50 - 4.0	0.30 - 0.70	0.15 - 0.45	20 - 60	5 - 25
2	Maize (leaf below the whorl prior to tasseling. Ear leaf at tasseling collected before silks turn brown)	Leaf below the whorl: 3.0 - 3.50	0.25 - 0.45	Leaf below the whorl: 2.0 - 0.50 Ear Leaf: 1.75 - 2.25	0.25 - 0.50	0.13 - 0.30	15 - 50	4 - 25
3	Cotton: (upper mature leaves on vegetative stems taken prior to bloom or when first squares appear).	3.50 - 4.50	0.30 - 0.50	1.50 - 3.0	2.0 - 3.0	0.30 - 0.90	20 - 200	20 - 60
4	Groundnut (upper mature leaves taken before or at bloom stage)	3.50 - 4.50	0.25 - 0.50	1.70 - 3.00	1.25 - 2.00	0.30 - 0.80	20 - 60	20 - 60

Source: Plank and Kissel (2008) 1.10.3. Soil Testing/Soil Analysis.

Purposes for testing of soil are (1) to define the status of existing nutrients within soils, (2) To specify farmer about deficiency and toxicity of various mineral nutrients for crop (3) To determine fertilizer needs (4) To prompt the outcomes in such a manner that could allow the assessment of recommendation for fertilizers which is budget friendly.

1.11. Soil Fertility and Its Security: Future Challenges and Hopes

In soil sciences fertility plays a vital role and is a major issue. Without soil and its fertility, it is not possible to achieve any agriculture safety. Soils provide necessity such as water, nutrient, and air to grow. Consequently, soil capability must be to enhance the growth and the changing "sun's energy into biomass energy." Prime source of contentment to human needs, including security for humans, health, creation of humans. Soil's fertility is linked to produce or harvest quality, growth of the animal and capacity for conservation of biodiversity (Blum, 2014).

Fertility generally deliberated in various prehistoric eras, age-old writings, religious superstitions globally (e.g., Warkentin, 2006; Winiwarter and Blum, 2006; Blum, 2014). Fertility sustains as well. Typically, fertility of soil conservation comprises applying diverse soil "amendments, bio stimulators, and fertilizers" such as natural, synthetic, and bio-fertilizers, tillage and ploughing practices are used to retain a least of 30% residue of the crop on the surface of crust. Moreover, erosion, soil pollution and soil degradation such are evaded which effect fertility of soil. Numerous hazards regarding the fertility of the soil are still prevailing which comprises, "soil sealing, pollution, urban sprawl, soil erosion, salinization, climate changes, the loss of organic matter, loss of biodiversity, floods, and landslides". We can state fertility of the soil is a reserve and a forecaster to life as well, helps to growth of population to a great degree with all the good quality in lives and best index maintaining the community health (Blum et al., 2010; Blum and Nortcliff, 2011; Blum, 2014).

As we know that security of the soil has handy relation to fertility of soil, and it tells us protection of soil happening beginning with fertility. these dimensions are include (1) capacity of soil for yielding enough and good quality nourishment, clean drinking water is also produced (Bouma et al., 2017), (2)chemical, physical are properties of soil that influence quality of nutrients in farming produce (McBratney et al., 2017), (3) human health is promoted by soil by providing healthy food (Mc Bratney et al., 2017), (4) soils and neighboring ecosystem interact with each other or connectivity by soil (Carre et al., 2017; Kim et al., 2017) and (5) soil and water quality improvement, conservation leads codification of soil (Amundson, 2017; Koch, 2017).

More than a few studies are available about the fertility and universal safety with emphasizing on "root-microbe relations" in diverse soil environments (Somenahally, 2017), various relations amongst security of soil and practices for its sustainable management (Murphy 2017), other relations are concerning health of mankind and protection of soil (Brevik et al., 2017), applicability of dissimilar assessments in soil estimation systems to security of soil comprising the safety of feed, aquatic source, & energy, better human life, weather variation decline and biodiversity safety (Liniger et al., 2011; McBratney et al., 2014), "soil security" plays major part for agronomic production (Koch, 2017), etc. by this we know, it could be concluded that soil safety and security needs fulfilled mainly by maintenance and then improve resources what we need like drinking water cattle feed, food for humans, and crude oil, looking for production of food and energy in a sustainable manner and also adapting to changing climate, biodiversity maintenance, mortal health, and work in ecosystems. Based on soil security and fertility many ideas and goals are developed or designed to attain different socio-economic and environmental dimensions. Also, goals should secure soil and fertility to solve other global issues. We should agree to make soil security a standard universal goal (Morgan et al., 2017).

Conclusion

The quote *"feed the soil to feed the human"* is famous currently. A request to the world is to take care of fertility of the soil so it can provide good and quality food (i.e., security for food). Soils provide us with biofuels or energy (energy security), harmless and sufficient food (safety of human feed), justifiable, low, required water use (water security), and soil protection. All around the universe fertility for soil, its well-being is acknowledged as a significant distress. For farmers benefit government started man programs. At that time, scientists are vexing to improve fertility of the soil by using various techniques and tools and its healthiness and to reclaim lost land. Success can be attained by using proper. By doing this we can improve food production and economic yield of crop which in turn helps in pulling the people away from poor and suffering conditions of society.

Referencess

Abbott, L. K., & Murphy, D. V. (Eds.). (2003). *Soil biological fertility: a key to sustainable land use in agriculture*. Springer Science & Business Media.

Abrol, I. P. (1990). Fertility management of Indian soils-a historical perspective. In *Transactions 14th International Congress of Soil Science, Kyoto, Japan, August 1990, Volume V.* (pp. 203-207).

Agboola, A., & Ayodele, O. (1987). Soil test calibration for upland rice in south western Nigeria. *Fertilizer research*, *14*(3), 227-234.

Agegnehu, G., & Bekele, T. (2005). On-farm integrated soil fertility management in wheat on Nitisols of central Ethiopian highlands. *Ethiopian Journal of Natural Resources*, *7*(2), 141-155.

Akinrinde, E. A. (2004). *Soils: Nature, Fertility Conservation and Management*. AMS Publishing, Ibadan.

Albiach, R., Canet, R., Pomares, F., & Ingelmo, F. (2000). Microbial biomass content and enzymatic activities after the application of organic amendments to a horticultural soil. *Bioresource technology*, *75*(1), 43-48.

Amundson, R. (2017). Whose security is important? Communicating environmental riskaboutsoil to a diverse audience. In *Global Soil Security* (pp. 383-387). Springer, Cham.

Anonymous (2016) Annual report of Department of Agriculture, Cooperation and Farmers Welfare, Ministry of Agriculture Farmers Welfare, pp 194.

Basak, R. K. (2000), *Soil Testing and Recommendation: A Text Book*, Kalyani Publishers, Kolkata-09.

Bationo, A., Hartemink, A. E., Lungu, O., Naimi, M., Okoth, P. F., Smaling, E. M. A., & Thiombiano, L. (2007). *African soils: their productivity and profitability of fertilizer use.*

Bationo, A., Waswa, B., Kihara, J., & Kimetu, J. (2007). Advances in integrated soil fertility management in sub Saharan Africa: challenges and opportunities. *Nutrient cycling inagroecosystems*, 1-2.

Bessou, C., & Marichal, R. (2015). Soil fertility, evolving concepts and assessments. ACIAR.

Biswas, T. D. & Mukherjee, S. K. (1994). *Textbook of Soil Science*, Tata McGraw-Hill Publishing Company Limited, New Delhi.

Blum W (2014) Soil fertility. In: Michalos AC (ed) *Encyclopedia of quality of life and well-being research.* Springer Science + Business Media, Dordrecht, pp 6212-6214. https://doi.org/10.1007/978-94- 007-0753-5.

Blum, W. E. H., Gerzabek, M. H., Hackländer, K., Horn, R., Rampazzo-Todorovic, G., Reimoser, F., & Zehetner, F. (2010). Ecological consequences of biofuels. *Soil quality and biofuel production*, 63-92.

Blum, W. E., & Nortcliff, S. (2011). Soilsforthe production of Food and/orRenewable Energy-The Future Perspectives. *Annals of Agrarian Science*, *9*(4), 1-9.

Bouma, J., Van Ittersum, M. K., Stoorvogel, J. J., Batjes, N. H., Droogers, P., & Pulleman, M. M. (2017). Soil capability: exploring the functional potentials of soils. In *Global soil security* (pp. 27-44). Springer, Cham.

Brevik, E. C., Steffan, J. J., Burgess, L. C., & Cerda, A. (2017). Links between soil security and the influence of soil on human health. In *Global soil security* (pp. 261-274). Springer, Cham.

Carre, F., Caudeville, J., Bonnard, R., Bert, V., Boucard, P., & Ramel, M. (2017). Soil contamination and human health: a major challenge for global soil security. In *Global soil security* (pp. 275-295). Springer, Cham.

Cleveland Brenda, Michelle Mc Guinn and Catherine Stokes (2008). *Plant Tissue Analysis: Measuring Nutrients to Optimize Fertilizer Efficiency and Protect Environmental Quality.* NC Department of Agriculture & Consumer Services.

Dercon, S., & Hill, R. V. (2009, October). Growth from agriculture in Ethiopia: Identifying key constraints. In *IFPRI's ESSP-II policy conferenceAccelerating agricultural development, economic growth and poverty reduction in Ethiopia, Hilton Hotel, Addis Ababa* (pp. 22-24).

Field, D. J., Morgan, C. L. S., & McBratney, A. B. (2017). *Global soil security: progress in soil science series.* Springer International Publishing Switzerland, DOI, 10, 978-3.

Flynn, R., Ball, S. T., & Baker, R. D. (2004). *Sampling for plant tissue analysis*, Guide A-123. New Mexico State University, New Mexico.

Fosu-Mensah, B. Y., Okoffo, E. D., & Mensah, M. (2016). Synthetic pyrethroids pesticide residues in soils and drinking water sources from cocoa farms in Ghana. *Environment and Pollution*, *5*(1), 60.

Foth, H. D. (1990). *Fundamentals of Soil Science*, 8th Edition, John Wiley & Sons, New York.

Fountain, M. T., Brown, V. K., Gange, A. C., Symondson, W. O., & Murray, P. J. (2007). The effects of the insecticide chlorpyrifos on spider and Collembola communities. *Pedobiologia*, *51*(2), 147-158.

Ghananand, T., Prasad, C. S., & Lok, N. (2011). Effect of insecticides, bio-pesticides and botanicals on the population of natural enemies in Brinjal ecosystem. *Vegetos*, *24*(2), 40- 44.

Hartemink, A. E. (2003). *Soil fertility decline in the tropics: with case studies on plantations*. Cabi.

Hepperly, P., Lotter, D., Ulsh, C. Z., Seidel, R., & Reider, C. (2009). Compost, manure and synthetic fertilizer influences crop yields, soil properties, nitrate leaching and crop nutrient content. *Compost Science & Utilization*, *17*(2), 117-126.

Kibblewhite, M. G., Ritz, K., & Swift, M. J. (2008). Soil health in agricultural systems. *Philosophical Transactions of the Royal Society B: Biological Sciences*, *363* (1492), 685-701.

Kidder, G. (1993). Methodology for calibrating soil tests. In *Proceedings*.

Kim, S. C., Lim, K. J., & Yang, J. E. (2017). The measurement of soil security in terms of human health: examples and ideas. In *Global Soil Security* (pp. 297-303). Springer, Cham.

Knudsen, M. T., Halberg, N., Olesen, J. E., Byrne, J., Iyer, V., & Toly, N. (2006). Global trends in agriculture and food systems. In *Global development of organic agriculture- challenges and prospects* (pp. 1-48). CABI Publishing.

Koch, A. (2017). Soil security for agricultural productivity: the policy disconnect and a promising future. In *Global Soil Security* (pp. 425-435). Springer, Cham.

Kumar, P. (2017). Integrated soil fertility management: converting subsistence farming to productive farming. *Newsreach. Nov-Dec*, 19-23.

Kumar, R., Kranthi, S., Nitharwal, M., Jat, S. L., & Monga, D. (2012). Influence of pesticides andapplication methods on pest and predatory arthropods associated with cotton. *Phytoparasitica*, *40*(5), 417-424.

Laboski, C. A., & Lamb, J. A. (2003). Changes in soil test phosphorus concentration after application of manure or fertilizer. *Soil Science Society of America Journal*, *67*(2), 544-554.

Li M, Gao X, Gao Z, Zhao W, Su Z (2008) Insecticidal activity of extracts from forty-eight plants including Xanthium sibiricum Patrin. Huanjing Xuebao Jinan 17:33-37.

Liniger, H. P., Studer, R. M., Hauert, C., & Gurtner, M. (2011). *Sustainable land management in practice: guidelines and best practices for sub-Saharan Africa*. FAO.

Liu, S. W., Ji, M. S., Zhao, L., Wei, S., Wang, G. J., Li, X., & Li, L. (2007). Preliminarystudy on bioactivity of two plant extracts against three kinds of pests. *Modern Agrochemicals*, *6* (6),27.

Manlay, R. J., Feller, C., & Swift, M. J. (2007). Historical evolution of soil organic matter concepts and their relationships with the fertility and sustainability of cropping systems. *Agriculture, Ecosystems & Environment*, *119*(3-4), 217-233.

Marschner, H. (1995). *Mineral nutrition of higher plants*. 2nd. *Edn. Academic Pres.*

Matson, P. A., Naylor, R., & Ortiz-Monasterio, I. (1998). Integration of environmental, agronomic, and economic aspects of fertilizer management. *Science*, *280*(5360), 112-115.

McBratney, A. B., Morgan, C. L., & Jarrett, L. E. (2017). The value of soil's contributions to ecosystem services. In *Global Soil Security* (pp. 227-235). Springer, Cham.

McBratney, A., Field, D. J., & Koch, A. (2014). The dimensions of soil security. *Geoderma*, *213*, 203-213.

Morgan, C. L., McBratney, A. B., Field, D. J., Koch, A., Bouma, J., & Carre, F. (2017). Synthesis: Goals to achieve soil security. In *Global soil security* (pp. 457-462). Springer, Cham.

Mueller, L., Schindler, U., Mirschel, W., Shepherd, T. G., Ball, B. C., Helming, K., & Wiggering, H. (2010). Assessing the productivity function of soils. A review. *Agronomy for Sustainable Development*, *30*(3), 601-614.

Mundt, C. C. (2002). Use of multiline cultivars and cultivar mixtures for disease management. *Annual review of phytopathology*, *40*(1), 381-410.

Murphy, B. (2017). Testing the links between soil security, sustainable land management practices and land evaluation. In *Global Soil Security* (pp. 87-97). Springer, Cham.

Nafiu, A. K., Abiodun, M. O., Okpara, I. M., & Chude, V. O. (2012). Soil fertility evaluation: a potential tool for predicting fertilizer requirement for crops in Nigeria. *African Journal of Agricultural Research*, *7*(47), 6204-6214.

Nafiu, A. K., Abiodun, M. O., Okpara, I. M., & Chude, V. O. (2012). Soil fertility evaluation: a potential tool for predicting fertilizer requirement for crops in Nigeria. *African Journal of Agricultural Research*, *7*(47), 6204-6214.

Naylor, R. L. (1996). Energy and resource constraints on intensive agricultural production. *Annual review of energy and the environment*, *21*(1), 99-123.

Nottidge, D. O., Ojeniyi, S. O., & Asawalam, D. O. (2005). Comparative effects of plant residues and NPK fertilizer on soil properties in a humid Ultisol. *Nigerian Journal of Soil Science*, *15*, 9-13.

Nziguheba, G., Palm, C. A., Buresh, R. J., & Smithson, P. C. (1998). Soil phosphorus fractions and adsorption as affected by organic and inorganic sources. *Plant and soil*, *198*(2), 159-168.

Oldeman, L. R. (1992). Global extent of soil degradation. In *Bi-Annual Report 1991- 1992/ISRIC* (pp. 19-36). ISRIC.

Patel, A. (2016). Addressing soil health management issues in India. *International Journal of Research-GRANTHAALAYAH*, *4*(12), 110-123.

Patzel, N., Sticher, H., & Karlen, D. L. (2000). Soil fertility-phenomenon and concept. *Journal of Plant Nutrition and Soil Science*, *163*(2), 129-142.

Plank CO, Kissel DE (2008). *Plant Analysis Handbook for Georgia*: *cooperative Extension/University of Georgia college of Agriculture*, Athens.

Potash and Phosphate Institute (2004). Agric Briefs: Pay Attention to Soil Sampling Details. Georgia. p. 5.

Power, J. F., & Prasad, R. (1997). *Soil fertility management for sustainable agriculture*. CRC press.

Pretty, J. N., Morison, J. I., & Hine, R. E. (2003). Reducing food poverty by increasing agricultural sustainability in developing countries. *Agriculture, ecosystems & environment*, *95*(1), 217-234.

Randhawa, M. S. (1979). *History of the Indian Council of Agricultural Research*, 1929-1979.

Reinecke, S. A., & Reinecke, A. J. (2007). The impact of organophosphate pesticides in orchards on earthworms in the Western Cape, South Africa. *Ecotoxicology and Environmental Safety*, *66*(2), 244-251.

Reisenauer HM, Quick J, Voss RE, Brown AL (1983). Chemical soil tests for soil fertility evaluation. In HM Reisenauer, (ed.), *Soil and plant tissue testing in California*, pp 39-41. University of California Division of Agriculture and Natural Resources, Oakland, CA. Bulletin 1879. Report on Soil and Water Baseline Survey for National Fadama III Project in Nigeria (2010).

Reuter, D., & Robinson, J. B. (Eds.). (1997). *Plant analysis: an interpretation manual*. CSIRO publishing.

Saha, S., Mina, B. L., Gopinath, K. A., Kundu, S., & Gupta, H. S. (2008). Organic amendments affect biochemical properties of a subtemperate soil of the Indian Himalayas. *Nutrient cycling in Agroecosystems*, *80*(3), 233-242.

Sanchez, P. A., Shepherd, K. D., Soule, M. J., Place, F. M., Buresh, R. J., Izac, A. M. N., & Woomer, P. L. (1997). Soil fertility replenishment in Africa: an investment in natural resource capital. *Replenishing soil fertility in Africa*, *51*, 1-46.

Sanginga, N., & Woomer, P. L. (Eds.). (2009). *Integrated soil fertility management in Africa: principles, practices, and developmental process*. CIAT.

Sarfraz, M., & Keddie, B. A. (2005). Conserving the efficacy of insecticides against Plutella xylostella (L.) (Lep., Plutellidae). *Journal of Applied Entomology*, *129*(3), 149-157.

Sayyed, A. H., Saeed, S., Noor-Ul-Ane, M., & Crickmore, N. (2008). Genetic, biochemical, and physiological characterization of spinosad resistance in Plutella xylostella (Lepidoptera: Plutellidae). *Journal of Economic Entomology*, *101*(5), 1658-1666.

Scoones, I., & Toulmin, C. (1998). Soil nutrient balances: what use for policy?. *Agriculture, Ecosystems & Environment*, *71*(1-3), 255-267.

Sendhil, R., Kumar, A., Singh, S., Chatrath, R., & Singh, G. P. (2017). Framework for doubling the income of wheat producers' by 2022: Trends, pathway and drivers. *Indian Journal of Economics and Development*, *13*(2a), 1-8.

Shelton, A. M., Sances, F. V., Hawley, J., Tang, J. D., Boune, M., Jungers, D., & Farias, J. (2000). Assessment of insecticide resistance after the outbreak of diamondback moth (Lepidoptera: Plutellidae) in California in 1997. *Journal of Economic Entomology*, *93*(3), 931-936.

Singh, B., & Singh, Y. (2015). Soil fertility: Evaluation and management. *Soil science: An introduction. Indian Soc. Soil Sci*, 649-669.

Singh, R. B. (2000). Intensive agriculture during the green revolution has brought significant land and water problems relating to soil degradation over exploitation of ground water and soil pollution due to the uses of high doses of fertilizers and pesticides. *Agric. Ecosyst. Environ*, *82*, 97-103.

Srinivasarao, C., Venkateswarlu, B., Lal, R., Singh, A. K., Vittal, K. P. R., Kundu, S., & Singh, S. P. (2012). Long-Term Effects of Soil Fertility Management on Carbon Sequestrationina Rice-Lentil Cropping System of the Indo-Gangetic Plains. *Soil science society of Americajournal*, *76*(1), 168-178.

Suge, J. K., Omunyin, M. E., & Omami, E. N. (2011). Effect of organic and inorganic sourcesof fertilizer on growth, yield and fruit quality of eggplant (*Solanum Melongena* L). *Archives of Applied Science Research*, *3*(6), 470-479.

Susila W, Sumiartha K, Nemoto H, Kawai S (2003). The effect of insecticides on populations of diamondback moth, Plutella xylostella (Lepidoptera: Yponomeutidae) and itsparasitoid, Diadegma semiclausum (Hymenoptera: Ichneumonidae) in cabbage. *J Int Soc Southeast Asian Agric Sci* 9:132-138.

Tanyi, C. B., Ngosong, C., & Ntonifor, N. N. (2017). Comparative effects of Piper guineense emulsion and cabbage-tomato intercropping for controlling cabbage pests and improving performance. *Journal of Agriculture and Ecology Research International*, 1-12.

Tening, A. S., Omueti, J. A. I., Tarawali, G., & Mohamed-Saleem, M. A. (1995). Potassium status of some selected soils under different land-use systems in the subhumid zone of Nigeria. *Communications in soil science and plant analysis*, *26*(5-6), 657-672.

Thy, S., & Buntha, P. (2005). Evaluation of fertilizer of fresh solid manure, composted manure or biodigester effluent for growing Chinese cabbage (*Brassica pekinensis*). *Livestock Res. Rural Dev*, *17*(3), 149-154.

Vanlauwe, B., & Giller, K. E. (2006). Popular myths around soil fertility management in sub-Saharan Africa. *Agriculture, ecosystems & environment*, *116*(1-2), 34-46.

Vanlauwe, B., Bationo, A., Chianu, J., Giller, K. E., Merckx, R., Mokwunye, U., & Sanginga, N. (2010). Integrated soil fertility management: operational definition and consequencesforimplementation and dissemination. *Outlook on agriculture*, *39*(1),17-24.

Von Wulffen C. 1847. Entwurf einer Methodik zur Berechnung der Feldsysteme. Verlag Veit & Co.: Berlin.

Wade, B. (2010). Symptoms of Deficiency in Essential Minerals. *Plant Physiology, 5th Edition by Taiz Lincoln and Zeiger Eduardo*.

Wallace, T. (1943). The diagnosis of mineral deficiencies in plants by visual symptoms. A colour atlas and guide. *The diagnosis of mineral deficiencies in plants by visual symptoms. A colour atlas and guide.*

Walsh C, Steinhilber P (2005). Nutrient Management for Tree Fruits and Small Fruits. NM-5. University of Maryland Department of Natural Resource Sciences and Landscape Architecture, College Park, MD, 20742.

Warkentin, B. P. (Ed.). (2006). *Footprints in the soil: People and ideas in soil history*. Elsevier.

Winiwarter, V., & Blum, W. E. (2006). Souls and soils: a survey of worldviews. *Warkentin, B. (Hg.)(2006): Footprints in the Soil. People and Ideas in Soil History, Elsevier: Amsterdam/Oxford*, 107-122.

Xu, Q. C., Xu, H. L., Qin, F. F., Tan, J. Y., Liu, G., & Fujiyama, S. (2010). Relay-intercropping into tomato decreases cabbage pest incidence. *Journal of Food, AgricultureandEnvironment*, *8*, 1037-1041.

Chapter 13

Stubble Burning in Rice-Wheat Cropping System: A Threat to the Ecosystem

Sachin Kumar[1,*], Saurbh Soni[2], Surinder Singh Rana[3] and Ranbir Singh Rana[1]

[1]Centre for Geo Informatics Research and Training,
CSK HPKV, Palampur, India
[2]Department of Entomology, [3]Department of Agronomy,
CSK HPKV, Palampur, India

Abstract

In rice-wheat cropping system use of combine harvester in order to harvest preceding croptimely and sowing of next crop has worsened the problem of crop residue as it leaves behind a large amount of rice residue to be burnt in the open fields. This study detail the amount of pollution being caused by rice residue burning and its adverse impact on human health and soil and air quality. It is estimated that paddy straw burning results in nutrient losses which negatively affects the nutrient budget in the soil. The burning of crop residue results in the emission of smoke which added methane, nitrogen oxide and ammonia, can cause severe atmospheric pollution and also contributes to the increased ozone pollution. Although, open field burning of remained residue is a simple and cost-effective method, but it's after effects are more hazardous in the long run and contribute to global warming and climate change. Hence, this chapter illustrate the ill effects of the residue burning and also impart knowledge and awareness among masses.

Introduction

The agricultural production and trading play a vital role in the overall economic growth of a country like India, whose around 70 per cent of the population involve direct or indirectly on the farming and related entrepreneurship activities with approximately 58 per cent of the total employment generation. With ever-increasing population, the growing demand for food is increasing day by day which results to over exploitations of the resources available. India, the second largest agriculturally based economy has year round crop cultivation because of diverse agro-ecological circumstances cause to generate a huge amount of agricultural waste, including crop residues. About 500 million tonnes (Mt) of crop residues is produced either on

* Corresponding Author's Email: schnagri@gmail.com.

In: Ecosystem Services
Editor: Hanuman Singh Jatav
ISBN: 978-1-68507-614-6

farm or off farm annually (Anonymous, 2016 (a); Devi et al., 2017). Due to the lack of appropriate sustainable management practices, around 140 million tons of crop waste as residue is burned each year is burned in field primarily to clear the field, causing excessive particulate matter emanation result to poor air quality and its levels. Agriculture crop residue burning generates severe pollution of air, water and soil on the local and regional scale. The Rice-wheat cropping system is most widely adopted system practiced on approximately 11 million hectares (Mandal et al., 2018) which is most prevalent in north-western and northern states including Indo-Gangetic plains (IGP) and states mainly Punjab, Haryana, Uttar Pradesh, Bihar, Madhya Pradesh and West Bengal which occupies around 53 per cent of the total area and near about 42 per cent of the total agricultural area of the India (Bimbraw, 2019). In Indo-Gangetic plains, during wheat and rice harvesting periods, crop residue burning (stalks and stubble) is the main source of atmospheric pollution (Venkataraman et al., 2006), human health and poor crop productivity (Auffhammer et al., 2006). The major reasons for burning agricultural crop residues in field include unavailability of labour during the peak period of crop harvesting, high removal cost of the residues and use of combines in mainly rice-wheat cropping system especially in the Indo-Gangetic plains (IGP). Out of total on-field crop residues during 1997, 43% was rice and 23 per cent wheat (Koopmans and Koppejan, 1997). The amount of agricultural waste produced depends upon the activities used, geographical conditions, cultural and management factors of a particular region. However, a huge amount of these residues is not employed and further remains in the agricultural farms as waste material, which in fact far-reaching trouble for the farmers (Singh et al., 2018). Although, post-harvest processing of crop produce through milling and packaging also produces a considerable amount of waste as residue. This incident is spread universally and also a significant cause of poor air quality throughout the globe. From different studies, it is found that about 25 per cent of nitrogen, 25 per cent phosphorus, 75 per cent of potassium and 50 per cent of sulphur uptake by cereal crops are kept in residues which make them rewarding sources of nutrients. It is calculated that paddy straw burning results in annual nutrient losses to the tune of 59,000 tonnes of nitrogen, 20,000 tonnes of phosphorus, 34,000 tonnes of potassium and 3.85 million tonnes of organic carbon at the aggregate which as result unfavorably affects the nutrient status of the top layer of the soil and adversely affect the cultivation of the succeeding crop. Residue burning causes the emission of smoke into the atmosphere, which can cause severe atmospheric pollution. Crop residue burning in open conditions resulted in the emission of about 25 per cent of organic matter, black carbon, and carbon monoxide (CO), 9-13 per cent of fine tiny particulate matter and carbon dioxide and about 1per cent of sulphur dioxide (SO_2) (Venkataraman et al., (2006). Carbon dioxide emissions are the major contributor to current global warming, which is followed by black carbon emissions (Ramanathan and Carmichael, 2008). These hazardous gases emission contributes to respiratory diseases to human and animals, eye irritation, chronic bronchitis, emphysema, asthma, and decreased lung function (Kumar and Kumar, 2016). In this chapter, we are interested in highlighting the (i) Crop residue production in India (ii) Status of on-farm stubble burning in India (iii) Impact of stubble burning regarding the environment, soil and human health (iv) Challenges of using crop residue for conservation agriculture (v) Crop residue management plan at the local and regional level, and (vi) Identifying research and governmental policy and plans of crop residue management for profitable, healthy and sustainable agriculture.

Crop Residue Production in Major Crops

The crop residue is the material that remained left after the harvest of the crop (leaves, stalk and roots), pruning or processing of crop (husks, seeds, bagasse and roots) and byproduct after processing of agriculture-based industry. The area under rice-wheat cropping system in the north-west states of India is approximately 3.79 million hectares (Lohan et al., 2018). Crop residue produced by major crops in different states is described in Table 1 (Anonymous, 2014 (a).

Table 1. Total crop residue production (tonnes) in different states of India during 2014

State/UT	Rice	Wheat	Coarse cereals	Others (Pulse, oilseeds, sugarcane, cotton and jute)	Total
Uttar Pradesh	14.3	31.3	6.1	57.5	109.2
Punjab	13.0	19.6	0.9	3.5	36.9
Bihar	7.5	5.0	3.8	6.8	23.1
Haryana	4.7	14.7	1.4	4.7	25.4
Madhya Pradesh	13.7	17.6	5.1	18.1	45.0
Rajasthan	0.4	12.2	12.9	9.5	35.1
Himachal Pradesh	0.1	0.9	1.4	0.1	2.5
West Bengal	17.2	1.2	1.1	5.1	24.6
All India	122.6	110.3	71.3	212.2	516.3

In India, cereals are the major source of surplus residues, which are often burnt in-situ in the field itself (Devi et al., 2017). Pulses, fiber crops, oilseed, and sugarcane are the other high surplus residue producing crops. Rice is contributed to 121 million tonnes of the annual crop residue, followed by wheat 114 million tonnes. Data related to the total quantity of crop residue generated in India as per different studies is described in Table 2.

Table 2. Total quantity of crop residue generated in India as per different studies

Total quantity of crop residue generated	Claimant and year
184, 902 Gg	Agarwal et al., 2008
133, 138 Gg	Garg, 2008
350×10^6 Kg year^{-1}	Mandal et., 2004
334 million tonnes	Gupta et., 2004
500 million tonnes	Bimbraw, 2019

Status of On-Farm Stubble Burning in India and Its Cause

Stubble burning is the consciously setting fire to the left straw stubble after harvesting of the crops. Recently, with the use of mechanical harvesters large quantities of crop residue remained in the field which is the major cause of stubble burning at the farm level. There are two major reasons for crop residue burning in the farm fields. a) There is a brief time period after the harvesting of paddy and sowing of wheat, at the end of the *kharif* season. b) Second, the removal of the paddy stalk and stubbles which remained left after the harvest is a tedious and labour-intensive job. However, between the harvesting of paddy and sowing of wheat unavailability of the labour can be a major cause of stubble burning and maybe the only easiest way that the farmer has is burning the crop residue right on the field itself. Sidhu and Beri, (2005) reported that total paddy stubble production in Punjab reached 18.8 million tonnes, out of which 15 million tonnes were burnt in open field conditions. They further found that the major reason for the burning of paddy straw in open fields was the frequent use of combined harvester, which leads to about 80 per cent of stubble burning in the fields themself. However, Singh et al., (2008) concluded that in Punjab about 17 million tonnes of paddy straw are produced every year, out of which 90 per cent are burnt in open fields. The scrape of an enormous amount of crop yield remains in the field, which is further trouble for the farmers, finally leads to stubble burning (Singh et al., 2018).

Table 3. State-wise status of residue burning in India (Jain et al., 2014)

States	Residue burned (Mt year^{-1})
Uttar Pradesh	22.38
Punjab	13.30
Andhra Pradesh	12.60
Maharashtra	10.96
Haryana	6.85
Tamil Nadu	5.57
Bihar	5.21
Madhya Pradesh	3.86
Assam	2.65
Himachal Pradesh	0.25
Arunachal Pradesh	0.16

Burning of left out residues is cheaper, easiest and requires less effort for most of the farmers. Lopes et al., (2020) reported that the primary reason for choosing crop residue burning is that this method is cheapest as compared to other physical practices followed by the use of a combine harvester for harvesting of paddy crop. He further quoted that labor availability and their cost is expensive, which results to burning of stubbles in the field. Jain et al., (2014) studied about the crop residues burnt in various states of India and concluded that maximum residue was burnt in Uttar Pradesh followed by Punjab and Andhra Pradesh. In Himachal Pradesh only 0.25 Mt year^{-1} residues are burned (Table 3).

Farmer's Views on Stubble Burning

- Increased use of machines particularly combine harvesters
- Inaccessibility of the labor during peak hours of crop harvesting. Also, high labor wages between the harvesting of paddy and sowing of wheat crop
- Residue burning is quickly clear the field with left stubbles in the field
- Burning in the field itself is cheap than other removal practices that need more money to clear their fields for the sowing of the next crop

- There is a brief time span between the harvesting of paddy and sowing of succeeding crop result in stubble burning at the field level.
- Stubble burning in the field destroy weeds, including herbicide-resistant and other harmful pests, which reduces the need for chemical control
- Long duration required for compost preparation

Table 4. Primary reason for picking crop residue burning (Lopes et al., 2020)

Reason of residue burning	Percentage (%)
The cheapest method	35.89
Combine harvester	18.39
High Labour cost	16.37
No other use	13.44
Machines are not available	6.98
Expensive to hire machines	3.45
No farm pits are available	2.55
Tradition	1.58
Small landholding	1.28

From the study, maximum per cent of rice and wheat stubble was burnt i.e., 81 per cent and 45 per cent respectively, although only 7 per cent of rice stubble and 45 per cent of wheat stubbles animal feed. Compared to the wheat crop, low rice stubble consumption as a feed for animals can be seen which was because of the high silica content in the rice crop (Singh et al., 2018). The Ministry of Agriculture and farmers' welfare, GOI quoted that the increase in the on-farm crop residue burning is because of the shortage of human labour at peak farm work hours (Anonymous, 2019 (a).

Table 5. Utilization pattern of straw and stubbles by the farmers of Punjab (Singh et al., 2018)

Sr. No.	Pattern of end usage	Percentage of entire rice stubble generated	Percentage of entire wheat stubble generated
	Animal feed	7	45
	Soil amalgamation	1	1
	Burning	81	45
	Rope production	4	0
	Miscellaneous	7	7

Why This Is a Key Problem only in the Northern States? (Saini et al., 2018)

Northern states viz. Punjab, Haryana, Uttar Pradesh, Uttarakhand and Bihar are the main wheat bowl of the country. However, states in south India are also involved in wheat production. Mainly in the northern states, including IGP has the urgency to remove the stubble to make it ready for the next crop. The crucial time for the maturity of the wheat crop is in mid-April, when the temperature is around or more than $35^{0}C$ in the northern states. For full maturity and maximum yield, wheat crop requires 140-150 days. In this case, farmer has no choice but to sow the crop in the first fortnight of November, so that the full duration crop can be grown.

Because of this brief time period for sowing and harvesting of paddy, and after this ready the field for wheat crop in just about 20 days. So, farmers have a choice of – *Stubble Burning.*

Adverse Impact of Stubble Burning

Typically, surplus crop stubble is burnt in open field in-situ results in the emission of harmful greenhouse gases (CO_2, CO, CH_4, N_2O, SO_2), along with particulate matters, aerosols, hydrocarbons, radioactive gases and other detrimental volatile organic compounds (Gupta and Sahai, 2005; Lal, 2006; Agarwal et al., 2006) which transform regional and global atmosphere chemistry (Crutzen and Andreae, 1990).

Table 6. Amount of gases emission from burning of 98.4 Mt of crop residue (Jain et al., 2014)

Greenhouse Gas	Amount (Mt)	Greenhouse Gas	Amount (Mt)
CO	8.57	NH_3	0.12
CO_2	141.15	NMVOC*	1.46
SO_x	0.037	NMHC**	0.65 Mt
NO_x	0.23	PM***	1.21

* Non-methane volatile organic compounds, ** Non-methane hydrocarbons, *** Particulate matter.

Effect of Stubble Burning on Environment and Air Quality

India is one of the most polluted countries in the world as out of the top 10 most polluted cities in the world, 7 are here in India which is because of the enhanced anthropogenic activities like stubble burning which is one of the one of the main causes of air pollution. C which is present in rice straw which on burning emits as CO_2, CO and CH_4 in the amount of 70, 7 and 0.66 per cent respectively, while N in straw is emitted as 2.09 per cent of N_2O which leads to depletion of the ozone layer. National Remote Sensing Agency in Punjab reported that during May 2005, burning of wheat crop residue contributed to about 113 Gg of CO, 13 Gg of PM_{10}, 12 Gg of $PM_{2.5}$, 8.6 Gg of NO_x, and 1.33 Gg of CH_4. On the other hand, burning of paddy stubbles contributed to 261 Gg of CO, 30 Gg of PM_{10}, 28.3 Gg of $PM_{2.5}$, 19.8 Gg of NO_x, and 3 Gg of CH_4 during October 2005 (Badarinath et al., 2006). Gupta et al., 2004 reported that burning of one ton of straw/stubbles emitted 1460 kg of CO_2, 199 kg of ash, 60 kg of CO, 3 kg of particulate matter and 2 kg of SO_2. Singh et al., (2008) described the major pollutants that are emitted during the burning of crop residue (Table 7).

Lohan et al., (2018) reported that the burning of crop residue in the north-west part of India including Punjab and Haryana emit about 20 per cent of elemental and organic carbon towards the total country share of emission from agricultural stubble burning. Badarinath et al., (2008) reported that around 5,500 km^2 of wheat crop area was burnt after the harvesting of the crop with an average of 5.94 t ha^{-1} biomass in the field. While in paddy, nearly 12,685 km^2 of area was burnt during that period. Mandal et al., 2004 estimated that on-field burning of rice and wheat stubbles cause emission of CH_4, CO, N_2O and NO_2 was 110, 2306, 2 and 84 Gg respectively.

Table 7. Emission of major pollutants owing to crop residue burning

Category	Pollutant	Source
Gases	CO	Incomplete combustion of organic matter
	NO_2, N_2O	Oxidation of N_2 in air at high temperature
	Ozone (O_3)	Secondary pollutant, form due to Nitrogen Oxide and Hydrocarbon
	CH_4	Incomplete combustion of organic matter
	PAH_s*	Incomplete combustion of organic matter
Particulate	SPM** (PM_{100})	Incomplete combustion of in organic material, particle on burnt soil
	PM_{10}, FPM*** ($PM_{2.5}$)	Condensation after combustion of gases and incomplete combustion of organic matter

*Polycyclic aromatic hydrocarbons, ** Suspended particulate matter, ***fine particulate matter

Table 8. Total emission by burning of rice and wheat crop (Badarinath et al., 2008)

Crop	Total emission (Gg)				
	CO	NO_x	CH_4	PM_{10}	$PM_{2.5}$
Rice	261	19.8	3	30	28.3
Wheat	113	8.6	1.33	13	12

The burning of crop residues in Delhi NCR region emitted particulate matter (PM) which is 17 times than from all other sources of pollution such as garbage burning, vehicle emissions, construction work and industries (Anonymous, 2017 (a). Singh et al., (2015) reported that maximum PM_{10} concentration was observed during harvesting of the rice crop (increase in 33.3 percent) due to the open field burning of paddy residue as compared to pre-harvesting and post-harvesting. Talhelm, (2019) reported that PM are the particles like construction dust, organic dust, airborne bacteria and coal particles from power plants in the air and categorized as $PM_{2.5}$ and PM_{10} have the difference of size (aerodynamic diameter) and chemical composition. $PM_{2.5}$ are fine, with an aerodynamic diameter <2.5 μm whereas, PM_{10} are coarse particulate matter with aerodynamic diameter < 10 μm. Light weight PM are more dangerous because of longer suspension in the air and can travel a longer distance with the wind (Singh and Panigrahy, 2011). The burning of rice residue contributes to around 60 to 390 mg m^{-3} of $PM_{2.5}$ annually in the Patiala district of Punjab (Lohan et al., 2018). During the period of October, crop residue burning generates smoke in Northern states of India including Punjab and Haryana which during November month with the onset of cooler weather getting mixed with fog, industrial pollution and dust and thereafter form a thick haze. This causes a major pollution problem in several cities like New Delhi, Lucknow and Kanpur (Allen and Voiland, 2017). Particulate matter of different sizes gets worsened by the weather like temperature, rainfall, light intensity, relative humidity as the particles are of light weight remains suspended for a longer in air and causes smog.

According to World Health Organization (WHO), the standard for acceptable levels of $PM_{2.5}$ in the air is 10 µg m^{-3} whereas according to India's National ambient air quality index, its permissible level is set at 40 µg m^{-3}. Delhi (NCR Region) recorded a mean value of 98 µg m^{-3} during the month of November, which is twice more than the Indian acceptable levels and ten times higher than the WHO permissible level (Zehra, 2017).

Impact of Crop Stubble Burning on Soil Health and Quality

The soils of Punjab, Haryana and other parts of the IGP are usually low in nitrogen, low to medium in phosphorus and medium to high in potassium content. A rice-wheat cropping system removes over 300 kg ha^{-1} of nitrogen, 30 kg ha^{-1} of phosphorus and 300 kg ha^{-1} of potassium from the soil. The burning of crop residues in open fields has a negative impact on the soil fertility, microbial population and their diversity, defacement of soil structure and results in poor drainage and removes the amount of nutrients present in the soil. The burning of residues elevates heat cause a rise in temperature of the upper soil layer, which causes death and permanently diminishes bacterial and fungal populations. Gupta et al., (2004) concluded that residue burning gives rise to the subsoil temperatures to about 33.8-42.2°C at 10 mm depth which on a long-term basis can even affect up to 15 cm of the top layer of the soil. Periodic stubble burning kills the micro flora and fauna and reduces or loses of nitrogen and carbon potential of the soil (Kumar et al., 2015). Crop residue burning also kills soil-borne pests and pathogens which are otherwise deleterious. Stubble burning results in the reduction of soil organic carbon to the atmosphere as carbon dioxide (CO_2), nitrogen equilibrium in the soil transforms because of which nitrogen in the soil is converted to nitrate. It is estimated that 10 t ha^{-1} crop yield takes off 730 kg NPK from the soil. In a studied, Jain et al., (2014) concluded that approximately 80-90 per cent nitrogen, nearly 25 per cent of phosphorus, 20 per cent of potassium and 50 per cent of sulphur present in crop residues are lost in the environment because of burning as various harmful gaseous and particulate matters which resulted to atmospheric pollution (Table 9).

Table 9. Nutrient losses (Mt $year^{-1}$) because of burning of residues (Jain et al., 2014)

Crop residues	N loss	P loss	K loss	Total
Rice	0.236	0.009	0.200	0.45
Wheat	0.079	0.004	0.061	0.14

Anonymous (2014, b), estimated that burning of one ton of rice straw reduces the nitrogen in an account of 5.5 kg, 2.3 kg of phosphorus, 25 kg of potassium and 1.2 kg of sulphur besides, total amount of organic carbon. Tung et al., (2004) reported about the impact of rice straw burning on the soil and concluded that maximum post-burning temperature was in the upper 0-2 cm layer of the soil under intensive burning which caused a sudden increase of soil temperature.

Table 10. The effect of different crop residue management practices on the soil parameters (Mandal et al., 2004)

Parameters of the soil	Crop residue management		
	Incorporated	Removed	Burnt
pH	7.7	7.6	7.6
EC(dSm^{-1})	0.18	0.13	0.13
Organic C (%)	0.75	0.59	0.69
Available N (kg $ha^{-1)}$	154	139	143
Available P (kg ha^{-1})	45	38	32
Available K (kg ha^{-1})	85	56	77
Total N (kg ha^{-1})	2501	2002	1725
Total P (kg ha^{-1}	1346	924	858
Total K (kg ha^{-1})	40,480	34,540	38,280

Effect of Stubble Burning on Human Health

The burning of crop stubbles in open fields results in the emission of harmful gases, aerosols, particulate matter and dust particles which leads to respiratory diseases like Bronchitis, Asthma and eye irritation. These harmful gases affect the life of animals, birds and other insects on and below the earth, which in critical stages can cause the death of animals. Gadde et al., (2009) reported that crop stubble burning in open conditions emits harmful chemicals which are toxic and are potential carcinogens. Most times smoke from burning causes poor visibility which results to road accidents. More than critical levels of CO and CO_2 in the human blood convert normal hemoglobin into deadly hemoglobin and cause a problem of loss of consciousness and arrhythmias. The burning of crop residues also has adverse effects on the health of milk producing animals which on further can decrease the yield of the milk from animals. Breathe in fine particulate matter ($<PM_{2.5}$) prompts asthma and bronchial attack. Small children and pregnant women are more susceptible to adversity from the smoke produced because of the burning of crop waste. According to Singh et al., (2008), rice-wheat growing areas of Punjab showed an increase in 10 per cent in the number of patients within 20-25 days of the stubble burning period every season when exposed to air pollution because of the burning of rice waste. An article in Times of India reported about the suffering percentage because of smoke which is generated because of stubble burning (Table 11).

Table 11. Household members suffering because of stubble burning

Problem	Suffering (%)
Watery eyes	81.25
Coughing	34.09
Respiratory allergies	11.93
Asthma	10.80
Bronchial problems	1.14
Nose/throat irritation	1.65

* Aggregate of three surveyed villages.

Management and Utilization of Crop Stubbles

Across India, crop residue is being used differently depending upon the type of cultivated crop from region to region, the number of crops per year and socio-economic status of the farmers of that area, etc. For example, in the northern part of India, a large volume of crop residue is burnt on-farm, whereas in southern India, rice residues are used for boilers for parboiling rice or for domestic fuel. Likewise, in sugarcane left over crop wastes are used for cattle feeding or burnt in the farm itself for ratoon crop. Similarly, groundnut residues (cake) are burnt in lime and brick kilns or used by cattle as protein supplements. Growers use crop residues for fodder, fuel, packaging and cattle sheds purposes. Pathak et al., (2010) reported that residues from cotton, pulses, oilseed crops, rapeseed and mustard stalks, coconut shells and jute residues as domestic fuel purposes. Cereals are the highest stubbles producing crops grown across India which contribute to surplus residues often burnt in the field itself. Crop stubbles can be used or managed as given below:

1. Off-farm utilization
   Straw removal
2. *In-situ* management
   - Partial straw retention
   - Full straw retention

1. Off-Farm Utilization

Off-farm usage of crop stubbles is a better option than burning of waste material on-farm as it addresses the problem of air pollution. But off-farm use of crop waste is only the second best option as it also leads to depletion of soil fertility as the result of removal of residues, which are otherwise the source of NPK in the soil.

Composting of Residues for Manure

The residues from various crops can be used for compost preparation by using waste material as animal bedding and then place this bedding heaping in the dung pit. Each kg of straw/stubbles soaks about 2-3 kg of animal urine from the shelter. This is an excellent source of nutrients and can make compost on the farm itself. Paddy residues from one hectare give approximately 3.2 tons of manure or farmyard manure (FYM) which is rich in nutrients.

Energy/Power Generation in Case of Paddy Straw

Biomass from the field after the crop harvest can be efficiently utilized as a source of energy based biogas power production. Generation of energy from crop waste substitute for fossil fuels. It also helps in the reduction of CO_2 content in the atmosphere (Anonymous, 2017 (b).

Ethanol Production from Crop Residues

The conversion of lignocelluloses biomass of crop waste into bio based alcohol production is of immense importance. Ethanol can be blended with gasoline as a fuel extender and octane enhancing agent. It can be a neat fuel in internal combustion engines.

Biomethanation

Biomass from rice straw or waste can be converted into biogas, which is a mixture of carbon dioxide (CO_2) and methane (CH_4) and used as fuel. It is found that biogas of 300 $m^3\ t^{-1}$ of dry rice straw yields an excellent quality of gas with 55-60 per cent of methane and the extracted slurry as manure. In this process, crop residues are utilized in a non-destructive way to draw out high-quality fuel gas.

Gasification of Crop Waste

Gasification is a process in which gas is generated in thermo-chemical partial combustion of residues. The waste crop residues from the field can be used in the gasifiers for the generation of gas. One ton of crop biomass can be used for the generation of 300 kWh of electricity.

Fast Pyrolysis

Fast pyrolysis of crop residues is a process in which the temperature of biomass to be raised to 400-500°C within a few seconds, which as result yields a dark brown viscous liquid commonly called bio-oil having a calorific value varies from 16-20 MJ kg^{-1}.

Biochar

Biochar is produced from slow pyrolysis, which is heating in the absence of oxygen. It is a high carbon material that can play a major role in the long-term storage of carbon and has got advantages in terms of its efficiency as an energy source. It is used as a fertilizer when mixed with soil and has the ability to stabilize and reduce emissions of harmful gases in the atmosphere. In soil, biochar enhances the fertility capacity, water retention capability of the soil and increases mineral and carbon delivery to roots of the plants.

Fodder for Animal and Bedding Purposes

The use of rice residue as fodder for animals is relatively low in Punjab as compared to the wheat stubble because of the high silica content in rice stubbles, which is not good for animal health. Therefore, crop stubble is treated with urea before animals feed. Badve, (1991), concluded that treatment of crop residues with 4 per cent urea and 45-50 per cent moisture enhances its nutritive value, crude protein content, increasing digestibility and palatability of the feed material.

Cultivation of Mushroom

Maize crop residues (husk, cob and stalk) can be supplemented with rice bran as growth media could be exploited as a sustainable substrate for the cultivation of mushroom. Paddy waste can be used for the production of *Volveriella volvacea*, *Agaricus bisporus* and *Pleurotus* spp. The environmental and economic parameters make these on-farm crop residues a viable alternative for mushroom cultivation (Adjapong et al., 2015). One kg of paddy straw yields around 120-150g of *Volveriella volvacea*, 300 g of *Agaricus bisporus* and 600 g of *Pleurotus* spp.

Use of Rice Residue in Paper Production and Board Manufacturing

The paddy waste is also being used in combination with wheat stubbles in 40:60 ratios for paper production. The technology is already being operational in some paper mills, which are meeting

60 per cent of their energy requirement through this waste utilization method. Paddy straw is also ideal raw material for pulp board manufacturing.

2. *In-situ* Management

Harvesting of the crop with combine harvester is the major cause of residue burning as it left scattered residues behind in the field itself, which farmers find somehow difficult to remove from the field. Because of this, farmers have only two alternatives, either burning open in the field itself or in-situ incorporation of the wastes of crop stubble. Even though the crop stubble has various alternate uses, but the farmers don't prefer in-situ incorporation of the paddy stubbles as it takes time to decompose in the soil which may delay the sowing of the wheat. Advantages of incorporation of crop stubbles:

- Improves soil health
- Environment friendly

Sidhu and Beri, (2005) resulted that in-situ incorporation is the best alternative to the burning of rice residue in the field. Singh et al., (1996) concluded that the in-situ incorporation of paddy straw three weeks before sowing significantly increased wheat yield under clay loam soil conditions but not on sandy loam soil. Crop residues incorporation increased the organic carbon content of the soil by 14-29 per cent. Incorporating rice stubble in the soil has favourable effects on the soil's physical, biological and chemical properties such as pH, organic carbon, water holding capacity, particle density and bulk density of the soil.

The Government of India launched a Central Sector Scheme, 'Promotion of agricultural mechanization for In-situ management of crop residue in the Northern state of India such as Punjab, Haryana, Uttar Pradesh and Delhi' in order to tackle air pollution and to subsidize machinery required for in-situ management of the crop. The scheme includes different machinery for individual farmers and through a custom hiring center (CHC). However, in 2018-19, 7.93 Mt of rice straw residue (16.9%) in Haryana and nearly 20.17 Mt of rice straw residue (50%) in Punjab were burnt in-situ (Anonymous, 2019, a). In the above scheme, turbo happy seeder has been included with 50 per cent subsidy to the farmers, which increases wheat production under the rice-wheat cropping system.

3. Straw Retention- Partial

Direct drilling in standing stubbles with zero-till drill.

4. Straw Retention-Full

Happy seeder and rotary disc drill are used for direct seeding in the next crop after rice. Happy seeder is used for planting in standing paddy stubbles manufactured by the Department of Farm Machinery and Power Engineering (PAU, Ludhiana).

Advantages of Happy Seeder

- Direct sowing of the wheat crop without soil disturbance in paddy stubbles, which enable it to keep more organic content, nutrients and moisture.
- There is no need to burn rice stubble in the field before planting wheat, therefore reducing air pollution.

Saves money as less time is needed for carrying out field operations, which reduces fuel consumption and labor costs. Dhillon, (2016) compared happy seeder technology with normal sowing in the wheat crop during 2014-15 and concluded that higher yield, total and net returns were recorded under happy seeder at Punjab.

Disadvantages of Straw Incorporation

- Requires extra time and energy
- Delay in sowing by 2-3 weeks
- No or low yield increase in early years
- Immobilization of nitrogen
- Termite problem is the major threat

(Saini et al., 2018) reported practices that can be used for residue management that includes the establishment of a marketplace for crop residue, outreach and public awareness programs through campaigns, wall writings, newspapers articles, TV news etc. Rice stubbles in-situ incorporation simultaneously increase macro and micro flora micro flora population (bacteria, actinomycetes and fungi) as compared to direct burning on the soil which as a result increase temperature which killed the beneficial biological organisms.

Government Policies and Initiatives for Restricting Agriculture Stubble Burning

- Ban on stubble burning
- Agricultural diversification
- Satellite based monitoring mechanism through State Remote Sensing Agency (SRSA)
- National Air Monitoring Programme
- Promotion of zero tillage
- Currently, the most cost-effective and beneficial way to manage the residue is to use the "zero-till sowing machine" called Happy Seeder

Ban on Stubble Burning

Crop residue burning is a crime and there are provisions of some laws by different organizations are: Section 144 of the Civil Procedure Code (CPC) to ban the burning of paddy by various states government. The Air Prevention and Control of Pollution Act (1981); the Environment Protection Act (1986); The National Tribunal Act (1995); and The National Environment Appellate Authority Act (1997). Particularly, in the states of Punjab, Haryana, Rajasthan and Uttar Pradesh strict laws have been taken by the National Green Tribunal (NGT) to limit the burning of waste crop material at the field level itself. NGT orders fine for stubble burning as follow:

Table 12. Provision of fine for burning

Fine for each incident	Burning
Rs. 2,500	Less than 2 acre
Rs. 5,000	2-5 acre
Rs. 15,000	More than 5 acre

Agricultural Diversification

Diversification is the need of the hour and different state Governments have been increasingly felt to shift the farmers into new ventures like vegetables and fruits cultivation, oilseeds and pulses, sugarcane production instead of rice-wheat crop rotation. However, shifting to a short duration paddy crop in Punjab (harvest by September) can reduce the problem of burning which will provide a longer window to the farmers for preparation and wheat sowing (Anonymous, 2019 (b). The Government of Punjab reported that the state's farmers have successfully diversified nearly 2.28 lakh hectares area out of the traditional paddy cultivation during current *kharif* (2020) sowing season. Also, the state has achieved an increase of 5000 hectares area under pulses cultivation and 1000 hectares area under sugarcane cultivation.

Promotion of Zero Tillage

Zero Tillage Technique is promoted by Punjab state Government since 2001-02 in areas where wheat is sown immediately after harvesting of the rice. In this system crops sowing is done with minimum soil disturbance by using a happy seeder which is a novel approach to combine stubble/mulching and seed drilling functions in one machine. With this technique, conservation of soil moisture along with management of waste residue benefits the farmers with timely sowing, reducing soil erosion and run-off reduce deep percolation and improving soil health (physical, chemical and biological characteristics) by incorporating plant nutrients. Keeping crop residues in the field for the long run builds up soil organic matter and as a result decreases farming's carbon footprint. However, increase in the farmer's profit in terms of the monetary amount by Rs. 2,200-3,000/- per hectare by saving 80 per cent of diesel consumption for wheat sowing in one pass only.

Other Government Initiatives

- Delhi government is working with IARI, New Delhi to prepare a fermented liquid solution to convert crop stubbles into manures. Four capsules can be mixed with a liquid solution prepared with jaggery and gram flour and sprayed over one hectare of land. The mixture softens hard straw and turns it into manure.
- Besides the above scheme, Central Government allocates separate funds to states for crop residue management under "*Sub-mission on agriculture mechanization*" by the Ministry of Agriculture and farmers Welfare.
- Punjab government disbursed INR 19 crore among 29,343 non-basmati cultivating small and marginal farmers (INR 2500 per acre) who did not burn paddy residue (PTI 2019, b) during 2019-20.

Conclusion

The Rice-wheat cropping system is the most widely accepted and predominant cropping system of northern India. The burning of paddy straw is a significant concern in India and has taken the attention of policy makers, government agencies and researchers. In-situ burning of crop residues on a large scale is a matter of great contention which results in atmospheric pollution, deteriorating soil health and health risks of human and animals. The burning of farm waste can be trimmed by *in-situ* management (incorporation and surface retention), off-farm usages (straw removal). Paddy straw can be used for sustainable energy production and to reduce emissions of GHGs. Production of bio-methane from paddy straw has the potential, the data revealed that 777.0 kWh electricity can be generated with one ton of paddy straw with output: input energy ratio of 5.5. On the other hand, one ton of paddy straw can generate electricity equivalent of 544.25 kWh with bio-ethanol formulation. Mulching with rice straw lower weed count, weed dry matter and increased weed control efficiency up to 52 per cent and reduced soil temperature up to 19.1°C. In wheat incorporation of rice stubbles increased organic carbon as compared to straw burning. Happy seeder is a foremost alternative to conventional sowing which saved 30 per cent of water and is also economically more workable with B:C ratio (3.87) as compared to conventional sowing (2.89) in wheat crop.

Future Thrust

In future, it can be expected that residue management will involve several technologies and government initiatives rather than rely solely on the burning of crop residues. By creating awareness and educating farmers about adverse implications of stubble burning on human and animal health through different media platforms including seminars, functions, rallies, wall paintings etc. Off-farm usages and *in-situ* management are the only concepts that can protect our environment to pollute.

References

Adjapong, A.O., Kwame, D.A., Faustina, A, and Henry OS. (2015). Maize Residue as a Viable Substrate for Farm Scale Cultivation of Oyster Mushroom *(Pleurotus ostreatus). Advances in Agriculture.*

Agarwal, S., Trivedi, R.C. and Sengupta, B. (2008). Air pollution due to burning of agricultural residue. *Indian Journal of Air Pollution Control* 8(1): 51-59.

Agarwal, S., Trivedi, R.C., and Sengupta, B. (2006). *Air pollution due to burning of residues. In Proceeding of the Workshop on Air Pollution Problems Due to Burning of Agricultural Residues,* held at PAU, Ludhiana Organized by the Indian Association for Air Pollution Control in collaboration with the Punjab State Pollution Control Board. New Delhi: Patiala and the Central Pollution Control.

Allen, A. and Voiland, A. (2017). *NASA Earth Observatory,* Haze Blankets Northern India. Available online: https://earthobservatory.nasa.gov/images/91240/haze-blankets-northern-india.

Anonymous. (2014, a). *Ministry of Statistics and Program Implementation (MOSPI).* http://www.mospi.gov.in/announcements/asi-2013-14-vol-i.

Anonymous. (2014, b). *National policy for management of crop residue.*Ministry of Agriculture Department of Agriculture & Cooperation (Natural Resource Management Division) Krishi Bhawan, New Delhi. Available online: http://agricoop.nic.in/sites/default/files/NPMCR_1.pdf (accessed on 6 March 2019).

Anonymous. (2016, a). *Annual Report,* Ministry of New and Renewable Energy, GOI, New Delhi, 2016, (http://mnre.gov.in).http://www.erewise.com/current-affairs/biomass-resources-inindia_art52cbbb9bcd5df. html#. Vd9atPmqqko.

Anonymous. (2017, a). *India's burning issues of crop burning takes a new turn.* Down to Earth.

Anonymous. (2017, b). *Renewable energy akshya Urja by Ministry of New and Renewable energy,* Government of India 10(6).

Anonymous. (2019, a). *Report of the committee on Review of the scheme "Promotion of Agricultural Mechanisation for in-situ Management of Crop Residue in States of Punjab, Haryana, Uttar Pradesh and NCT of Delhi".* Deptt. of Agriculture, Cooperation & Farmers Welfare, Ministry of Agriculture and Farmers Welfare, Government of India.

Anonymous. (2019, b). *Crop diversification, short duration paddy to help curb stubble burning: NRAA.* https://timesofindia.indiatimes.com/india/crop-diversification-short-duration-paddy-to-help-curb-stubble-burning-nraa/articleshow/71922666.cms.

Auffhammer, M., Ramanathan, V. and Vincent, J. (2006). Integrated model shows that atmospheric brown clouds and greenhouse gases have reduced rice harvests in India, *Proceedings of National Academy of Sciences* 130(52): 19668.

Badarinath, K.V.S., Kiranchand, T.R. and Krishna, P.V. (2006). Agriculture crop residue burning in ht Indo-Gangetic Plains- A study using IRS-P6 A WiFS satellite data. *Current Science* 91(8): 1085-1089.

Badarinath, K.V.S., Kumar, K.S. and Sharma, A. (2008). Long range transport of aerosols from agriculture crop residue burning in indo-Gangetic Plains- A study using LIDAR, ground measurements and satellite data. *Journal of Atmospheric and Solar-Terrestrial Physics* 59(3): 219-236.

Badve, V.C. (1991). Feeding systems and problems in the Indo-Ganges plain: Case study. In A. Speedy & R. Sansoucy (Eds.), Feeding dairy cows in the tropics. *Proceedings of the FAO Expert Consultation Held in Bangkok,* Thailand, July 7-11, 1989.

Bimbraw, A.S. (2019). Generation and Impact of Crop Residue and its Management. *Current Agriculture Research Journal* 7(3): 304-309.

Crutzen, P.J. and Andreae, M.O. (1990). Biomass burning In the tropics: impact on atmospheric chemistry and biogeochemical cycles. *Science* 250: 1669-1678.

Devi, S., Gupta, C., Jat, S.L. and Parmar, M.S. (2017). Crop residue recycling for economic and environmental sustainability: The case of India. *Open Agriculture* 2: 486-494.

Dhillon, G.S. (2016). Comparative evaluation of happy seeder technology versus normal sowing in wheat (*Triticum aestivum*) in adopted village Killi Nihal Singh of Bathinda district of Punjab. *Journal of Applied and Natural Science* 8: 2278-2282.

Gadde, B., Christoph, M.C. and Wassmann, R. (2009). Rice straw as a renewable energy source in India, Thailand, and the Philippines: Overall potential and limitations for energy contribution and greenhouse gas mitigation. *Biomass Bioenergy* 33: 1532-1546.

Garg, S.C. (2008). Traces gases emission from field burning of crop residues. *Indian Journal of Air Pollution Control* 8(1): 76–86.

Gupta, P.K. and Sahai, S. (2005). Residues open burning in rice-wheat cropping system in India: An agenda for conservation of environment and agricultural conservation. In I. P. Abrol, R. K. Gupta & R. K. Malik (Eds.), *Conservation Agriculture- Status and Prospects* (pp. 50–54). New Delhi: Centre for Advancement of Sustainable Agriculture, National Agriculture Science Centre.

Gupta, P.K., Sahai, S., Singh, N., Dixit, C.K., Singh, D.P. and Sharma, C. (2004). Residue burning in rice-wheat cropping system: Causes and implications. *Current Science* 87(12): 1713-1715.

Jain, N., Bhatia, A. and Pathak, H. (2014). Emission of air pollutants from crop residue burning in India. *Aerosol and Air Quality Research* 14: 422-430.

Koopmans, A. and Koppejan, J. (1997). Agricultural and forest residues-generation, utilization and availability'. *Paper presented at the Regional consultation on modern application of biomass energy* 6: 10.

Kumar, P., Kumar, S. and Joshi, L. (2015). The extend and management of crop residue stubbles. In *Socio economic and environmental implications of agricultural residue burning: A Case Study of Punjab, India;* Eds.; Springer briefs in environmental science: Berlin, Germany, p. 144, ISBN 978-81-322-2014-5.

Kumar, S., Kumar, P. (2016). *Economic impacts of air pollution from agricultural residue burning from human health.* 2016. Springer International Publishing Switzerland 2016 S. Nautical et al. (eds.), Climate Change Challenge (3C) and Social-Economic-Ecological Interface-Building, *Environmental Science,* DOI 10.1007/978-3-319-31014-5_17.

Lal, M.M. (2006). An overview to agricultural burning. In Proceeding of the Workshop on Air Pollution Problems Due to Burning of Agricultural Residues, held at PAU, *Ludhiana organized by the Indian Association for Air Pollution Control in collaboration with the Punjab State Pollution Control Board, Patiala and the Central Pollution Control Board,* New Delhi.

Lohan, S.K., Jat, H.S., Yadav, A.K., Sidhu, H.S., Jat, M.L., Chaudhary, M., Jyotsna, K.P. and Sharma, P.C. (2018). Burning issues of paddy residue management in north-west states of India. *Renewable and Sustainable energy Reviews* 81(1): 693-706.

Lopes, A.A., Viriyavipart, A. and Tasneem, D. (2020). The role of social influence in crop residue management: Evidence from Northern India. *Ecological Economics* 169: 106563.

Mandal, K.G., Misra, A.K., Hati, K.M., Bandyopadhyay, K.K., Ghosh, P.K. and Mohanty, M. (2004). Rice residue management options and effects on soil properties and crop productivity. *Food, Agriculture and Environment* 2: 224-231.

Mandal, T., Chandra, S. and Singh, G. (2018). Productivity and economics of rice-wheat cropping system under irrigation, nutrient and tillage practices in a silty clay loam soil. *International Journal of Current Microbiology and Applied Sciences* 7: 823-831.

Pathak, H., Bhatia, A., Jain, N. and Aggarwal, P.K. (2010). Greenhouse gas emission and mitigation in Indian agriculture - A review, In *ING Bulletins on Regional Assessment of Reactive Nitrogen*, Bulletin No. 19 (Ed. Bijay-Singh), SCON-ING, New Delhi. p. 34.

Ramanathan, V. and Carmichael, G. (2008). Global and regional climate changes due to black carbon. *Nature Geoscience* 1(4): 221-227.

Saini, D.M., Singh, V.K. and Kumar, A. (2018). *Stubble burning: either farmers to be punished or technology need to be improved.* Biomolecule reports, pp.1-8.

Sidhu, B.S. and Beri, V. (2005). Experience with managing rice residues in intensive rice-wheat cropping system in Punjab. In Abrol, I.P., Gupta, R.K. and Malik, R.K. (Eds.), *Conservation agriculture: Status and prospects* (pp. 55–63). New Delhi: Centre for Advancement of Sustainable Agriculture, National Agriculture Science Centre.

Singh, R.P., Dhaliwal, H.S., Sidhu, H.S., Singh, M. and Blackwell, J. (2008). Economic assessment of the happy seeder for rice-wheat systems in Punjab, India. *Conference Paper, AARES 52nd Annual conference,* Canberra. Australia: ACT.

Singh, C.P. and Panigrahy, S. (2011). Characterization of residue burning from agricultural system in India using space-based observations. *Journal of Indian Society of Remote Sensing* 39: 423-429.

Singh, J., Singhal, N., Singhal, S., Sharma, M., Agarwal, S. and Arora, S. (2018). *Environmental implications of rice and wheat stubble burning in North-Western states of India.* DOI: 10.1007/978-981-10-7122-5_6.

Singh, R.P., Chanduka, L. and Dhir, A. (2015). Impacts of stubble burning on ambient air quality of a critically polluted area – Mandi - Gobindgarh. *Journal of Pollution Effects and Control* 2: 1-6.

Singh, R.P., Dhaliwal, H.S., Sidhu, H.S., Singh, M. and Blackwell, J. (2008). Economic assessment of the happy seeder for rice-wheat systems in Punjab, India. *Conference Paper, AARES 52nd Annual conference,* Canberra. Australia: ACT.

Singh, Y., Singh, D. and Tripathi, R.P. (1996). Crop residue management in rice-wheat cropping system. Abstracts of poster sessions. *2nd International Crop Science Congress* (p. 43). New Delhi, India: National Academy of Agricultural Sciences.

Talhelm, T. (2019). *$PM_{2.5}$ vs. PM_{10}: The Difference in Particle Air Pollution.*https:// smartairfilters.com/en/blog/pm10-pm2-5-difference-particle-air-pollution/.

Tung, N.S., Cu, N.X. and Hai, N.X. (2004). Impact of rice straw burning methods on soil temperature and microorganism distribution in the paddy soil ecosystems. *Journal of Agricultural and Biological Science* 9: 157-160.

Venkataraman, C., Habib, G., Kadamba, D., Shrivastava, M., Leon, J., Crouzille, B., Boucher, O. and Streets, D. (2006).Emissions from open biomass burning in India: Integrating the inventory approach with high-resolution Moderate Resolution Imaging Spectro radiometer (MODIS) active-fire and land cover data, *Global biogeochemical cycles* 20(2): GB2013.

Zehra, R. (2017). *How clean is the air around you.* Available online: https://fit.thequint.com/health-news/clean-your-air-as-per-who-standards-2. (accessed on 10 June 2018).

Chapter 14

Fertilizer Recommendations by Using Different Geospatial Technologies in Precision Farming or Nanotechnology

B. R. Meena[1,*], H. S. Jatav[1], B. L. Dudwal[1], P. Kumawat[1], S. S. Meena[2], V. K. Singh[3], M. A. Khan[1] and Eetela Sathyanarayana[4]

[1]SKN Agriculture University, Jobner Rajasthan India
[2]GB Pant University of Agriculture and Technology, Pantnagar Uttarakhand India
[3]Department of Soil Science, Punjab Agriculture University Ludhiana, Punjab India
[4]Agricultural College-Palem, Nagarkurnool, PJTSAU, Hyderabad, Telangana, India

Abstract

The use of technology to improve the effectiveness and efficiency of farming practices has increased tremendously. Discoveries in the field of science and technology have enabled farmers to effectively use their input to maximize their yield. These advancements have been greatly assisted by the use of sophisticated machineries, planting practices, use of fertilizers, herbicides and pesticides and so on. In particular, dependence on more efficient use of synthetic fertilizers to increase crop yields poses technical, economic and environmental challenges. Emerging technologies, such as geospatial technologies (GIS, GPS), Internet of Things (IoT), Big Data analysis, and artificial intelligence (AI), could be utilized to make informed management decisions aimed to increase crop production. Precision agriculture (PA) entails the application of a suite of such technologies to optimize agricultural inputs to increase agricultural production and reduce input losses. These technologies and tools enable farmers to characterize spatial variability (e.g., soils) among farms and large crop fields that negatively affect crop growth and yields. These state-of-the-art technologies for the development and implementation of site-specific management are integral part of PA.Further, nano-technology has the potential to revolutionize agricultural systems. In the management aspects, efforts are made to increase the efficiency of applied fertilizer with the help of nano clays and zeolites and restoration of soil fertility by releasing fixed nutrients. Use of nanofertilizers causes an increase in nutrients use efficiency, reduces soil toxicity, minimizes the potential negative effects associated with over dosage and reduces the frequency of the application. Hence, precision farming andnanotechnology has a high potential for achieving sustainable agriculture, especially in developing countries like India.

---

* Corresponding Author's Email: brmeena.agro@sknau.ac.in.

In: Ecosystem Services
Editor: Hanuman Singh Jatav
ISBN: 978-1-68507-614-6

Keywords:fertilizer, geospatial, precision, GIS, GPS, nanotechnology

Introduction

Agriculture is the backbone of India's economy, which accounts for 18% of India's Gross Domestic Product (GDP) and hires about half of the countries workforce. More than 70% of rural families depend on agriculture for their livelihood. Housing about 17% of the total world population, India faces the challenge of fulfilling the demands of agricultural commodities for this ever-growing population. It's imperative to upgrade traditional farming practices and prepare for a tech-revolution to develop eco-friendly systems for enhancing crop productivity. Smart farming concepts like precision agriculture can be aptly deployed to achieve this goal. Precision farming is an integrated crop management system which uses remote sensing (RS), GPS, and geographical information system (GIS) to monitor the crop field at ground level. The disparities in crop or soil properties within a field are recorded and mapped. Then management decisions are taken as a result of continuous assessment of the spatial variability within that field. In the Indian agricultural scenario, it can be described as the precise utilization of agricultural inputs depending upon the crop, soil, and weather requirement to optimize the use of fertilizers, pesticide, and irrigation requirements for maximum productivity.

Precision agriculture (PA), satellite farming or site specific crop management (SSCM) is a farming management concept based on observing, measuring and responding to inter and intra-field variability in crops. The goal of precision agriculture research is to define a decision support system (DSS) for whole farm management with the goal of optimizing returns on inputs while preserving resources. In recent scenario the several problems occurs in conventional farming. The intensification of the conventional farming systems has led to extensive usage of agricultural inputs like machinery, high-demanding varieties and agro chemicals, resulting in negative environmental impacts such as groundwater pollution and atmospheric contamination that exacerbates the greenhouse effect. The environmental pressure has adverse effects not only on human health and natural resources but also on the sustainability of agricultural production itself. Precision farming is an approach where inputs are utilised in precise amounts to get increased average yields, compared to traditional cultivation techniques and its counter the all problems that occurs under in conventional farming. Sustainable PA is this century's most valuable innovation in farm management that is based on using Information and Communication Technologies (ICTs). This is the most recent innovation technology based on sustainable agriculture and healthy food production and it consists of profitability and increasing production, economic efficiency and the reduction of side effects on the environment.In India, one major problem is the small land holding. More than 58 per cent of operational holdings in the country have size less than one hectare (ha). In such area clubbing of small field in an unit is a better option to implementation of precision farming. Whereas in the states of Punjab, Rajasthan, Haryana and Gujarat, more than 20 per cent of agricultural lands have an operational holding size of about four ha so better scope of precision farming in those areas. Commercial as well as horticultural crops also show a wider scope for PA in the cooperative farms.

Need of Precision Farming

The conventional farming systems has led to extensive usage of agricultural inputs like machinery, pesticides, water, and other inputs resulting in negative environmental impacts such as pollution of environment by emission of green house gases. Research suggests educational and economic challenges as the two most important in the application of precision agriculture. PA and initial costs have more of an impact among the economic challenges compared to the other issues. Rather than this PF (Precision Farming) increase agriculture productivity with prevents soil degradation. PF reduces the use of chemical application in crop production and efficient use of water resources. It is also helpful in dissemination of modern farm practices to improve quality, quantity and reduced cost of production, developing favourable attitudes and changing the socio-economic status of farmers.

A farmer's expense sheet is often the thing of doom and dread. Precision farming aims to reduce a farmer's expenditure by minimising the need for things like fertiliser, pesticide and herbicide. Over a growing season, grower's are seeing significant reductions in the amount of money they are spending on all of the above where technology is using the components sparingly and only where needed. As an alternative to blanket spraying this has seen massive savings and allows farmers to better budget and keeps costs to a minimum.

More Time on their Hands

Accompanying the simplified processes, farmers are finding they are freeing up a lot more time to concentrate on the business operations of their farms rather than the nitty gritty jobs that would ordinarily be very time consuming. This means that they can focus their energy on making the farm more profitable and expansion planning rather than get bogged down completing menial tasks that technology can do instead.

Higher yields and More Profitability

Statistically, a precision farmer will make more money than a traditional farmer. This is for a number of reasons. Their yields are higher because they have improved growing practices and they are able to sell more produce at the end of the season as a result. They also have less man hours to expend as technology is filling gaps that workers would have previously filled. There is now a huge fruit picking technology industry which is minimisinglabourer costs across many large fruit farms. As well as these benefits, because farmers are reducing costs by employing technology they are hitting their profit margins a lot sooner than traditional farmers.

Better Quality Produce

Implementing better growing processes is providing produce that is of a higher quality. This is done in many ways such as actively monitoring the nutrients in soil, strip tilling and irrigating plants correctly and when irrigation is needed. Again; this not only boosts yields, but it also boosts profit margins further as when it comes to taking the crop to market, precision farmers are able to negotiate a higher price for the improved quality in the produce.

Less Waste

Lost crop is a nightmare for farmers, and over a growing season there is expectation that a certain number of plants will not see the season through. Making this worse is that historically storage practices inevitably resulted in further spoilage. With streamlined growing processes and healthier plants that crops have a much better chance of surviving and growing seasons are becoming more sustainable. Furthermore, improvements in storage by technological means have greatly reduced crop spoilage in the storage stage as well.

Less Debt

Unfortunately, a word that has become synonymous with farming is debt. Precision farmers are no more able to avoid debt as a traditional farmer however they are able to clear it quicker because they have more money to pay it down. This is as a result of the increased profitability and many precision farmers are now debt free or taking big steps towards becoming debt free.

Higher Quality of Life

Farming is a difficult and exhausting job. Farmer suicide rates are high and overall quality of life was typically non-existent. Precision farming is changing this rapidly, and as a result of the things we have mentioned on the list they are experiencing a better quality of life with a reduction in stress levels. With more time to spend with the family, more money to go on holidays with and less physical labour involved in the farming process, precision farmers are living healthier and more fulfilled lifestyles that should see those alarming suicide statistics fall as time goes by.

Nanotechnology

Nanotechnology can be defined as methods that create materials or structures with designed features in the 1–100 nm size range (Niemeyer, 2002) or in other words nanotechnology is the science of the nanoscale it means objects around a nano meter in size. Our capacity to construct large, intricate structures with nano meter precision is rapidly changing, and consists of top-down reductive approaches and bottom-up additive approaches. Alternatively, nature has perfected an array of biological machinery that functions at the nano scale structure which typically self-assembles driven by the molecular chemistry of subunit interactions. Nanotechnology has made enormous progress over the past few decades. Nanotechnology requires the measurement, prediction, and fabrication of matter on the scale of atoms and molecules. It is hoped that atomic-scale nanotechnology will have a revolutionary impact on the way we do, design, and produce things in the future. Nanotechnology is an atomic or molecular approach to building physically, chemically, and biologically stable structures one atom, or one molecule, at a time. To date, scientists and researchers can control atoms or molecules individually and can manipulate them as well use them to achieve phenomenal degrees of precision. Nano technology contributions to society may appear small at the moment but with continued manpower and funding it is hoped it will provide results and innovations. If nanotechnology continues to grow at its current rate it will touch the life of nearly everyone on the planet in the next few next years. Results of investigations and developments in nano technological fields are entering into all areas of our lives, such as agriculture, aerospace, energy, defence, materials science, environmental science, and medicine.

GeospatialTechnologies for FertilizerRecommendation

Geospatial technologyprovides better management of natural resources that leads to higher potential of increasing yields and economic returns in agricultural production. In this system, the agricultural management is tailored to the variability of conditions found in each field. Since fertilizer application is an important operation, it has to be carried out efficiently and judiciously to ensure optimum returns from its investment. Geospatial technologies for fertilizer recommendation are aimed to optimize the all available nutrient resources in combination for site specific crop management. Such types of techniques are employed to increase yield, reduce production costs, and minimize negative impacts to the environment. Using GIS analytical capabilities, variable parameters that can affect agricultural production can be evaluated. Some important geospatial technology used for fertilizer recomentdation is site specific nutrient. The main goal of this chapter to know comprehensive information's of advanced geospatial technologies, used in precision farming and nano-technology.

1. Site-Specific Crop Management

Site-specific nutrient management (SCNM) is a plant-based approach that provides principles that can be used everywhere. SCNM also gives guidelines for effective N, P, and K management, so that farmers give the right amount of essential nutrients to the crop. With SSCM, plant-essential nutrients are supplied as and when required to ensure the feeding of the crop to optimally meet its nutrient needs. SCNM enables the tailoring of nutrient management to field- and location-specific conditions. Site-specific crop management (SSCM) is one facet of precision agriculture which is helping in increasing production with minimal input. It has enhanced the cost-benefit scenario in crop production. Even though the SSCM is very widely used in row crop agriculture like corn, wheat, rice, soybean, etc. it has very little application in cash crops like fruit and nut. Guidelines developed by International Plant Nutrition Institute (IPNI) addresses specific issues related to site-specific soil and crop management some of which are discussed.

1. Site-Specific Use of the Environmental Phosphorus Index Concept
Phosphorus (P) loss to surface water can have negative impacts on the environment. The risk of such loss depends on both source (added fertilizer and manure, soil P) and transport factors (erosion and surface runoff). Fields at risk are those where areas of high P application or soil P coincide with zones of active surface runoff or erosion. A P index has been developed to rank field vulnerability to P loss so that high risk fields may be identified for site-specific management. The index provides a framework that can be regionally adapted to prevailing topography, geology, and climatic conditions and requires only readily available data.

2. Management Zone Concepts
Varying the application rates of plant nutrients and other crop inputs across variable fields makes good agronomic sense. For every input some reasonable strategy must be used to guide that application. Grid soil sampling for phosphorus (P) and potassium (K) has greatly improved the accuracy of fertilizer application, although even greater accuracy can be attained by considering additional site characteristics within sub-regions of fields. A "management zone" is a sub-region of a field that expresses a relatively homogeneous combination of yield-limiting factors for which a single rate of a specific crop input is appropriate. Spatial information that is most helpful in defining management zones should be quantitative (numerical), densely or continuously sampled, stable over time, and directly related to crop yield.

3. Profitability of Site-Specific Farming
Profitability results from field research studies show that high-value crops give the biggest payoff to VR fertilizer application. Many yield map benefits come from whole-field improvements such as drainage, land levelling, windbreaks, and fencing. Farmers and agribusinesses should remember that because SSF practices are site-specific, their profitability potential also will be site-specific.

4. Developing Management Zones to Target Nitrogen Applications
Whether the goal is to determine the level of soil nitrogen (N) or the soil yield potential, management zones for N fertilizer management can be constructed using a variety of tools, including topography, aerial photographs, satellite imagery, soil electrical conductivity sensors, yield maps, and intensive soil survey data. For the producer starting out, viewing satellite images and/or aerial photos that are relatively inexpensive to obtain and comparing them with landscape features would be a good place to start. Zones can be constructed and managed for N using a fraction of the number of soil samples required to reveal the same zones through grid sampling. Zone sampling results in lower sampling costs for variable-rate fertilizer application and allows precision farming to be much more practical for producers of commodity crops.

5. Global Positioning System Receivers
GPS and GIS receivers provide the means to determine position at locations anywhere on earth. Developed by the U.S. Department of Defence (DOD) and used for many civilian purposes, from fishing to flying, GPS has also made precision farming a reality. A typical configuration for on-farm agricultural applications includes a GPS receiver and antenna, a differential correction receiver and antenna, and cables to interface differentially-corrected (DGPS) data from the receiver to other electronic equipment such as a yield monitor or a variable rate

controller. Accurate, automated position tracking with GPS receivers allows farmers and agricultural service providers to record geo-referenced data and to apply variable rates of inputs to smaller areas within larger fields.

6. Variable Rate Equipment—Technology for Weed Control

Sprayer controllers have been developed by agricultural equipment vendors to minimize variation of applied rates of chemicals within fields. The control systems that allow these devices to compensate for changes in vehicle speeds now also provide the potential to apply variable rates of pesticides according to pre planned maps. The types of sprayer systems and controllers are capable of variable rate control.

7. Standardization and Precision Agriculture

'The Promised Land' Progress toward increased use of electronic systems for precision farming applications will be enhanced by the introduction of standards for electronic communications on agricultural equipment and translation of spatial data formats. The standard will provide a uniform approach to communications on tractors and implements.

8.The Pioneer Split-Planter Comparison Method

The Pioneer Split-Planter Comparison Method is a simple, low-cost technique for making treatment evaluations using a global positioning system (GPS) – equipped yield monitor in whole fields. Comparisons can be made between two hybrids, varieties, or agronomic treatments applied in alternating strips throughout a field. Pooling results from similar comparisons made at multiple locations is much preferred to relying on single-location results. The ability to bring a yield difference map into a geographic information system (GIS) and overlay it with other spatial data layers will greatly increase the value of the map as a crop management tool.

9. Earth Model—Calculating Field Size and Distances between Points Using GPS Coordinates

An ever-increasing number of farmers have global positioning system (GPS) receivers on their combines. When not harvesting, GPS receivers are useful for more than locating one's favourite spot. The distance between two sampling points and the area of a field can be found using GPS coordinates and knowledge of the Earth Terrestrial Coordinate System. The objective of this guideline is to provide a method to those farmers and agricultural practitioners can use to calculate distances between points and to calculate the size of a field using Excel, a commonly available spreadsheet.

10. Assessing Crop Nitrogen Needs with Chlorophyll Meters

One of the most difficult challenges faced by farmers is to determine the appropriate fertilizer nitrogen (N) application rate. Irrespective of sources, most N in soil is eventually transformed to the nitrate (NO3) form. To minimize NO_3 leaching, cropping systems and management practices must minimize excess NO_3 in the soil and the potential for percolation below the root zone. The problem is basically one of synchronizing soil N availability (from all N sources) with crop N needs. This task is complicated because it is difficult to accurately predict climatic variables that influence crop growth, soil microbial activity, and NO_3 leaching.

11. Identifying Good Candidates for Precision Phosphorus Management

In fields with high soil test P variability, precision management of P produced the greatest level of profitability when the composite soil test P level was in the high to very high soil test P categories. Average soil P test level and prior field histories can be used as a decision aid to reduce economic risks associated with adopting precision farming techniques. Appropriate P response models and yield goals must be used to accurately assess potential profitability associated with precision P management.

12.Variable-Rate Technology (VRT)

Developing accurate variable-rate technology (VRT) fertilizer application maps is critical in implementing precision farming management. Intensive grid soil sampling may be used to develop application maps. However, the cost and labour intensity associated with intensive grid sampling suggests other approaches may be more feasible. Management zone technology may provide a more economical method of developing VRT application maps.

13. Getting Specific with Site-Specific Nutrient Management

Over simplification of site-specific nutrient management can lead to reduced profits and production. Currently, site-specific nutrient management typically involves applying a definite set of recommendations to different areas in a field, based upon a few factors, such as soil test levels and yield goals. However, if these recommendations do not consider other site-specific factors that influence response to fertilizer application, substantial opportunities to increase profits and production may be lost. Proper evaluation of these yield-limiting factors and appropriate management changes based on readily available information could make site-specific nutrient management more profitable.

14. Grain Protein Sensing to Identify Nitrogen Management in Wheat

Protein concentration in grain is greatly influenced by the level of nitrogen (N) fertility. However, there is significant spatial variation in N fertility in a field. Conventional uniform application ignores this variability and leads to over fertilization in some areas and under-fertilization in others. Therefore, the question arises as to whether grain protein can be optimized on a site-specific basis by accounting for spatial variability of N fertility within individual field.

2. GIS, GPS and Remote Sensing for Fertilizer Recommendations

Geographic Information Systems (GIS) applications enable the storage, management, and analysis of large quantities of spatially distributed data. These data are associated with their respective geographic features. For example, water quality data would be associated with a sampling site, represented by a point. A GIS can manage different data types occupying the same geographic space. For example, a biological control agent and its prey may be distributed in different abundances across a variety of plant types in an experimental plot. Although predator, prey, and plants occupy the same geographic region, they can be mapped as distinct and separate features. The ability to depict different, spatially coincident features is not unique to a GIS, as various computer aided drafting (CAD) applications can achieve the same result.

Global Positioning System GPS technology has provided an indispensable tool for management of agricultural and natural resources. GPS is a satellite- and ground-based radio navigation and location system that enables the user to determine very accurate locations on the surface of the Earth. Although GPS is a complex and sophisticated technology, user interfaces have evolved to become very accessible to the non-technical user. Simple and inexpensive GPS units are available with accuracies of 10 to 20 meters, and more sophisticated precision agriculture systems can obtain centimetre level accuracies.

The uses of GIS, GPS, and RS technologies, either individually or in combination, span a broad range of applications and degrees of complexity. Simple applications might involve determining the location of sampling sites, plotting maps for use in the field, or examining the distribution of soil types in relation to yields and productivity. More complex applications take advantage of the analytical capabilities of GIS and RS software. These might include vegetation classification for predicting crop yield or environmental impacts, modelling of surface water drainage patterns, or tracking animal migration patterns.

Precision agriculture provides better management of natural resources that leads to higher potential of increasing yields and economic returns in agricultural production. In this system, the agricultural management is tailored to the variability of conditions found in each field. Since fertilizer application is an important operation, it has to be carried out efficiently and judiciously to ensure optimum returns from its investment. Precision Agriculture practices are aimed to optimize the use of soil resources to compensate the GIS and GPS technologies are used in combination for precision farming and site-specific crop management. Precision farming techniques are employed to increase yield, reduce production costs, and minimize negative impacts to the environment.

Using GIS analytical capabilities, variable parameters that can affect agricultural production can be evaluated. It can be used to create fertilizer variable rates map from crop yield data and foliar analysis. The information and fertilizer rates map created will guide the actual application of fertilizer in the field in such a way that will optimize fertilizer usage and maximize yield production.

Some Other Tools for Recommendation

1. Nutrient Expert

Fertilizer recommendation tool 'Nutrient Expert' for food and nutritional security challenges in South Asia. Yields of cereals in small scale farms in Asia are 40-65% of their potential largely because nutrient management does not consider the crop's dynamic response to the environment (Dobermann and Cassman, 2002). Data from several environments in India revealed gaps between potential and attainable yields of maize of up to 100% and between attainable and actual yields of up to 25-50%. Among the cereals, maize is rapidly emerging as a favourable option for the farmers of South Asia due to higher productivity and profitability of maize over rice or wheat. Maize extracts large amounts of mineral nutrients from the soil due to high grain and Stover yields. There is a possibility of nutrient mining from the soil if fertilizers are not added as per the requirements for high target yield of maize. Intensification of maize will therefore need nutrient management that produces high yields while preserving soil quality and the environment.

South Asian agriculture is characterized by small land holdings for cultivation with high variability in plant nutrient availability between fields. However, fertilizer recommendations for crops in this region are usually done over large geographic areas that often fail to meet the demand of high yielding cops like maize. Site-specific nutrient management (SSNM), on the other hand, integrates information from different scales to make field specific decisions on N, P and K management. SSNM, which was developed for rice in Asia, was later adapted for maize in Asia (Pasuquinet al. 2010). There is little acceptance of SSNM by farmers beyond the on-farm trials. Apart from shortcomings in the strategy to promote SSNM, and farmer's inability to buy fertilizers, extension agents lack confidence in using the methodology (Gabinete and Buresh, 2009). Many extension agents still perceive SSNM as complex, requiring an understanding of concepts and methods outside their experience. Such knowledge requirements have slowed the wide-scale promotion and adoption of SSNM by the farmers and extension agents (Buresh, 2008). The decision support system developed for South Asia in 2010-11 is an easy-to-use, interactive computer-based decision tool that can rapidly provide nutrient recommendations for an individual farmers' field in the presence or absence of soil testing data. It utilizes information provided by a farmer or a local expert to suggest a meaningful yield goal for his location and formulates a fertilizer management strategy required to attain that yield goal.

2. DRIS Approach

The Diagnosis and Recommendation Integrated System (DRIS) was developed by Beaufils in 1973, this method consist in dual relation between a pair of nutrients (N/P, P/N, N/K, K/N...) Instead of the use of sufficiency range or critical level that are called univariate methods, because only the individual concentration of the nutrients in leaf tissue is taken into consideration while no information about the nutritional balance is provided. DRIS enables the evaluation of the nutritional balance of a plant, ranking nutrient levels in relative order, from the most deficient to the most excessive. With the use of dual relation on DRIS, the problem with the effect of concentration or dilution on the nutrients in plants is solved, because, according to Beaufils (1973); Walworth & Sumner (1987) with the growth of leaf tissue, on one hand the concentration of nitrogen, phosphorus, potassium and sulphur decrease in older plants and the concentration of calcium and magnesium increase in older plants on the other hand. When it is used the DRIS method, where the dual ratio is used, the values remain constant, minimizing the effect of biomass accumulation, that is one of the major problem with sufficiency range and critical level method. It is feasible to find on literature some crops on which DRIS had already been used to assess the nutritional status of fruit plants and crops. According to Baldock& Schulte (1996), there are four advantages of DRIS; (1) the scale of interpretation is continuous numeric scale, and easy to use, (2) put the nutrients in order of the most deficiency to the most excessive, (3) identify cases where the yield of plant is been limited by into factor as nutritional status and (4) the Nutritional Balance Index (NBI) give a result of combined effects of nutrients. Nevertheless, the disadvantage of this methodology is that the DRIS index is not independent, because one nutrient concentration can have hard influence on the other DRIS index for one nutrient but this problem can be corrected in parts with a hard selection of the nutrient that will compound the DRIS norms.

3. Variable Rate Applicator

Variable rate application help farmer increase their yields by using optimal levels of fertilizer, seed, and chemical in the areas of the farm fields where the inputs are going to add the most value, while lowering the inputs in places that they have less impact on returns and yields. In precision agriculture, variable rate application is an area of invention that focuses on the automated precise application of materials to a given location. The application of these materials is based on data that is collected by technologies like sensors, GPS and maps. These materials are inputs like chemicals, fertilizers, and seeds. The key focus is optimizing crop's production. There are many technological tools that can be used in application of variable rate in precision agriculture. These include everything ranging from drones and artificial intelligence to satellites and hyperspectral imaging. Whichever variable rate application tool that you decide to use, it is critical that you understand the general way in which such technology is applied. Variable rate application focuses on different areas in farming activities like fertilization, lime application, weed control, irrigation and seeding. Some variable application can be used by the help of geographical positioning system (GPS) or even without GPS system. Two basic technologies of variable rate application are:

- Map based; and
- Sensor based.

Map-based variable rate application adjusts the relevant application rate based on a generated electronic map, also known as prescription map. The GPS receiver gives the field position and a prescription map at a desired rate. The applicator then applies the field position and prescription from a GPS as it moves in the field. On the other hand, the sensor-based variable rate application requires no positioning system or map. Sensors that are mounted on the applicator measures the soil properties or crop characteristics at the same time. The information is then streamed on real-time basis the control system calculates the quantity of inputs that are required based on the crop or soil needs. Because both sensor based and map based variable system have advantages and disadvantages, some manufacturers have made tools that take advantage of the benefits from both methods.

Geospatial Data

Geospatial data is information that describes objects, events or other features with a location on or near the surface of the earth. Geospatial data typically combines location information (usually coordinates on the earth) and attribute information (the characteristics of the object, event or phenomena concerned) with temporal information (the time or life span at which the location and attributes exist). The location provided may be static in the short term (for example, the location of a piece of equipment, an earthquake event, and children living in poverty) or dynamic (for example, a moving vehicle or pedestrian, the spread of an infectious disease). Geospatial data typically involves large sets of spatial data gleaned from many diverse sources in varying formats and can include information such as census data, satellite imagery, weather data, cell phone data, drawn images and social media data. Geospatial data is most useful when it can be discovered, shared, analyzed and used in combination with traditional business data.

Geospatial analytics is used to add timing and location to traditional types of data and to build data visualizations. These visualizations can include maps, graphs, statistics and cartograms that show historical changes and current shifts. This additional context allows for a more complete picture of events. Insights that might be overlooked in a massive spreadsheet are revealed in easy-to-recognize visual patterns and images. This can make predictions faster, easier and more accurate.

Types of Geospatial Data

Geospatial data is information recorded in conjunction with a geographic indicator of some type. There are two primary forms of geospatial data: vector data and raster data.

Vector data is data in which points, lines and polygons represent features such as properties, cities, roads, mountains and bodies of water. For example, a visual representation using vector data might include houses represented by points, roads represented by lines and entire towns represented by polygons.

Raster data is pixelated or gridded cells which are identified according to row and column. Raster data creates imagery that's substantially more complex, such as photographs and satellite images.

Examples of Geospatial Data

Examples of geospatial data include:

- *Vectors and attributes:* Descriptive information about a location such as points, lines and polygons
- *Point clouds:* A collection of co-located charted points that can be recontextured as 3D models
- *Raster and satellite imagery:* High-resolution images of our world, taken from above
- *Census data:* Released census data tied to specific geographic areas, for the study of community trends
- *Cell Phone data:* Calls routed by satellite, based on GPS location coordinates
- *Drawn images:* CAD images of buildings or other structures, delivering geographic information as well as architectural data
- *Social Media Data:* Social media posts that data scientists can study to identify emerging trends

Dealing with large geospatial data sets presents many challenges. For this reason, many organizations struggle to take full advantage of geospatial data.

First, there is the sheer volume of geospatial data. For example, it is estimated that 100 TB of weather-related data is generated daily. This alone presents considerable storage and access problems for most organizations. Geospatial data is also stored across many different files, which makes it difficult to find the files that contain the data needed to solve your specific problem. In addition, geospatial data is stored in many different formats and calibrated by different standards. Any effort to compare, combine or map data first requires a significant amount of data scrubbing and reformatting. Finally, working with raw geospatial data requires specialized knowledge and the application of advanced mathematics to conduct necessary tasks, such as geospatial alignment of data layers. Unless analysts are proficient and experienced at this work, they will not get value from the data or make progress toward their organization's business goals.

Geospatial Data Collection

Because the sheer volume of geospatial data routinely required by enterprises is prohibitively large, many organizations look to using a service to obtain curated geospatial data. Regardless of where you source your geospatial data, data quality must always be maintained. Poor data results in models of little or limited use. (The cautionary phrase "Bad data in — bad insights out" proves brutally true.) It seems self-evident that organizations can benefit significantly from having a solution in place that curates and checks data, so any "garbage" data gets properly accounted for.

Geospatial Data Management

With so much data now in abundance, managing it takes on considerable importance. Many organizations are finding themselves overrun with data and are turning to their in-house data scientists to help them manage it. It has been estimated that as much as 90% of data scientists' time is spent on data-curation activities, including organizing, "cleaning" and reformatting data. That leaves those data scientists with only 10% of their workday to devote to analyzing data trends and using those insights to help shape business policy.

Although geospatial analysis, as empowered by GIS, was originally used in connection with life sciences such as geology, ecology and epidemiology, its use has since become manifest throughout most industries. Its applications now touch industries as diverse as defense and social sciences. And the insights that geospatial analysis generates affect matters as critically important as natural resource management and national intelligence. Geospatial analysis lends itself to the study of many things at once, monitoring hundreds or even thousands of events and collecting pertinent data from them. This provides enterprises of all sizes the chance to leverage data to make more informed business decisions:

- Utilities providers can analyze the performance of hundreds of thousands of miles of power lines.

- Restaurant chains can use geospatial analysis techniques to track incoming shipments from food providers.
- Wind and solar installations can analyze environmental conditions by pinpointing local assets.

Efforts to analyze massive amounts of data have become more challenging in recent years due to a relative explosion within the Internet of Things (IoT). Objects and devices of all types and purposes are now being engineered to be able to transmit data relevant to that device's performance or protocols. That's good news for geospatial analysis, which involves a profusion of data in order to glean valuable insights.

Conclusion

The soil testing and fertilization recommendation technology can prevent improper fertilization and low utilization ratio of fertilizers. Precision farming is a good tool to increase the yield of farm without much hampering of soil fertility and other essential inputs. Geospatial technology is the main component of precision farming. Several geospatial technology like nutrient expert, VRT's, SSNM and other GIS and GPS based farming systems are commonly used for deciding the nutrient recommendations in fallow or in standing crop fields. They are useful for estimating the site-specific fertilizer rate requirements; and to use decision support system to create information from the foliar analysis to develop the foliar nutrient maps. All these technology are helpful to the farmer for increasing farmer income by the reducing of input quantity. Apart from this nanotechnology has a vast role in better fertilizer management and recommendations.

References

Abbas, A. & Khan, S. (2007).Using remote sensing techniques for appraisal of irrigated soil salinity, in: *International Congress on Modelling and Simulation (MODSIM)*, edited by: Oxley, L. and Kulasiri, D., Modelling and Simulation Society of Australia and New Zealand, Brighton, 2632–2638.

Akinola, A.A., Arega, D.A., Adeyemo, R., Sanogo, D., Olanrewaju, A.S., Nwoke, C., Nzigaheba, G. & Diels, J. (2008). Determinants of adoption and intensity of use of balanced nutrient management systems technologies in the northern Guinea savanna of Nigeria, No 52007, 2007 *Second International Conference, August 20-22, 2007, Accra, Ghana, African Association of Agricultural Economists (AAAE).*

Al Khaier, F. (2001). Soil Salinity Detection Using Satellite Remote Sensing, International Institute for Geo-Information Science and Earth Observation, Enschede, the Netherlands, 2003. AlaviPanah, S. K. and Goossens, R.: Relationship between the Landsat TM, MSS data and soil salinity, *J. Agric. Sci. Technol.*, *3*, 21–23.

Alam, M. M. (2005). Leaf color chart for managing nitrogen fertilizer in lowland rice in Bangladesh. *Agronomy Journal,97*, (3), 949-59.

Ali, K. O., Hashim, N., Rostam, K. &Jusoh, H. (2011).Spatial growth of the semi-squatter settlement in Tripoli, Libya, *International Journal of Research and Reviews in Applied Sciences*, *9*, 478– 485.

Amundson, R., Berhe, A. A., Hopmans, J. W., Olson, C., Sztein, A. E. & Sparks, D. L. (2015). Soil science, soil and human security in the 21st century, *Science*, *348*, 1261071, https://doi.org/10.1126/science.1261071.

Ananda, J. &Herath, G. (2003).Soil erosion in developing countries: a socio-economic appraisal, *J. Environ. Manage.*, *68*, 343–353.

Angrist, Joshua D., Guido W. Imbens. & Donald B. Rubin. (1996). Identification of causal effects using instrumental variables. *Journal of the American Statistical Association*, *91*, (434) (Jun.): 444-55.

Asio, V. B., Jahn, R., Perez, F. O., Navarrete, I. A. &Abit Jr., S. M. (2009). A review of soil degradation in the Philippines, *Annals of Tropical Research*, *31*, 69–94.

Bai, Z. G., Dent, D. L., Olsson, L. &Schaepman, M. E. (2008). Global assessment of land degradation and improvement, identification by remote sensing, Report No. 12, 2008/01, ISRIC, Wageningen, 70 pp.

Balasubramanian, V. (1999). Farmer adoption of improved nitrogen management technologies in rice farming: Technical constraints and opportunities for improvement. *Nutrient Cycling in Agroecosystems*,*53*, (1), 93-101.

Barbero-Sierra, C., Marques, M. J., Ruiz-Pérez, M., Escadafal, R. &Exbrayat, W. (2015). How is desertification research addressed in Spain? Land versus soil approaches, *Land Degrad. Dev.*, *26*, 423–432, https://doi.org/10.1002/ldr.2344.

Barrett-Lennard, E. G. &Hollington, P. A.: Development of a national program on saline

Batjes, N. H. (2014). Projected changes in soil organic carbon stocks upon adoption of recommended soil and water practices in the Upper Tana River Catchment, Kenia, *LandDegrad. Dev.*, *25*, 278–287, https://doi.org/10.1002/ldr.2141.

Baugher, T.,Schupp, J., Travis, J., Hull, L.,Ngugi, H.,Krawczyk, G., Harsh, M.,Reichard, K., Ellis, K.,Remcheck, J.,Crassweller, R., Marini, R., Harper, J.,Kime, L., Heinemann, P., Liu, J., Lewis, K.,Hoheisel, G., Jones, V., Glenn, M., Miller, S., Tabb, A., Park, J., Slaughter, D., Johnson, S., Landers, A.,Reichard, G., Singh, S.,Bergerman, M., Kantor, G.&Messner, W. (2009). Speciality Crop Innovations: Progress and Future Directions; Specialty Crop Innovations Progress Report; College of Agricultural Sciences, Penn State University: University Park, PA, USA, pp. 1-16.

Behmanesh, B., Barani, H., Sarvestani, A. A., Shahraki, M. R. &Sharafatmandrad, M. Rangeland degradation assessment: a new strategy based on the ecological knowledge of indigenous pas

Blair, N., Faulkner, R. D., Till, A. R., Korschens, M. & Schulz, E. (2006). Long term management impacts on soil C, N and physical fertility; Part II; Bad Lauchstadt static and extreme FYM experiments. *Soil Till Res.*,*91*, 39-47.

Bouman, B.A.M., Humphreys, E., Tuong, T.P.& Barker, R. (2006). Rice and water. *Advances in Agronomy*,*92*, 187 – 237.

Buresh, R.J., Witt, C., Ramanathan, S., Chandrasekaran, B.&Rajendran, R. (2005). Sitespecific nutrient management: managing N, P, and K for rice. *Fert. News*,*50* (3), 25– 28, 31–37. Buresh, RJ. 2006. Site-specific Nutrient Management. International Rice Research Institute, Los Baños, Philippines. Burrows,

Chattha, Zaman, Chang, Read, Schumann, Brewster.&Farooque. (2014). "Variable Rate Spreader for Real-time Spot-application of Granular Fertilizer in Wild Blueberry." *Computers and Electronics in Agriculture*, *100*, 70-78.

Cross, J.V.&Walklate, P.J. (2008). The UK PACE scheme for adjusting the dose to suit apple crops. Agr. Eng. Int.: *CIGR EJournal*, *6*, Manuscript ALNARP 08 003.

Dai, Do Dinh, Nguyen Thi Ha. & Vu Xuan Thanh. (2005). Food security and reasonable use of rice land in Vietnam. In "*Agricultural Technology and Science and Rural Development, 20 years of renovation*", Vol *3*. Vietnam National Political Publisher, pp 147-167.

Das, D.K. (2007). In: *Micronutrients: Their Behaviour in Soils and Plants*. Second revised Ed., Kalyani Publishers, Ludhiana/New Delhi. International Plant Nutrition Institute (IPNI). Website: www.ipni.net Site Specific Management Guidelines.

Das, D.K. (2011). In: *Introductory Soil Science*, Third revised Ed., Kalyani Publishers, Ludhiana/New Delhi.

Deering, D.W. (1978). Rangeland Reflectance Characteristics Measured by Aircraft and Spacecraft Sensors. Ph.D. Dissertation, Texas A&M University, College Station, TX, USA, p. 338.

Doberman, A. (2002). Site-specific nutrient management for intensive rice cropping systems in Asia. *Field Crops Research*,*74*, (1), 37-66.

Doran, J. W., Sarrantonio, M.& Liebig,M. A. (1996). Soil health and sustainability. *AdvAgron*,*56*, 1-54.

Doss, Cheryl R. (2006). Analyzing technology adoption using microstudies: Limitations, challenges, and opportunities for improvement.(author abstract). Vol. *34*, 31

Duflo, Esther, Michael Kremer, and Jonathan Robinson (2005). "Understanding Fertilizer Adoption: Evidence from Field Experiments," Mimeo, MIT.

Feder, G. (1985). Adoption of agricultural innovations in developing countries: A survey. *Economic Development and Cultural Change*,*33*, (2), 255-98. 1993.

Fox, Jennifer E., Jay Gulledge, Erika Engelhaupt, Matthew E. Burow.& John A. McLachlan. (2007). Pesticides reduce symbiotic efficiency of nitrogen-fixing rhizobia and host plants. *Proceedings of the National Academy of Sciences of the United States of America*,*104*, (24) (06/12).

Fung, T.&LeDrew, E. (1987). Application of principal component analysis to change detection. Photogramm. *Eng. Remote Sensing*, *53*, 1649-1658.

Grisso, Robert, Alley, Marcus, Thomason, Wade &Holshouser, D.& Roberson, O.T. (2011). Precision farming tools: Variable-rate application. *Precision, Geospatial, & Sensor Technologies*. 442-505.

Heckman, James J. & Edward Vytlacil. (2005). Structural equations, treatment effects, and econometric policy evaluation. *Econometrica*,*73*, (3) (May), 669-738.

Heffer, P. (2008). Assessment of Fertilizer Use by Crop at the Global Level. International Fertilizer Industry Association. Paris, France.

Hien, Bui Huy. (2005). Results on crop nutrition and efficient use of fertilizer during Renovation period and plan for 2006-2010. In "*Agricultural Technology and Science and Rural Development, 20 years of rennovation*". Vol *3*. Vietnam National Political Publisher, pp 245-264.

Imbens, G. (2004). Nonparametric estimation of average treatment effects under exogeneity: A review. *The Review of Economics and Statistics*, *86*, (1), 4-29.

Jain, A.K., Murty, M.N.& Flynn, P.J. (1999). Data clustering: A review. *ACM Comput. Survey.*, *31*, 264-323.

Jat, R.K., Sapkota, T.B., Singh, R.G., Jat, M.L., Kumar, M. & Gupta, R.K. (2014). Seven years of conservation agriculture in a rice-wheat rotation of Eastern Gangetic Plains of South Asia: yield trends and economic profitability. *Field Crops Research*, *164*, 199-210.

Khan, Fahad Sarwa. & Dalhousie University. (2012). Department of Engineering. Mapping Soil Properties and Water Table Depths Using Electromagnetic Induction Methods.

Khurana, H. S. (2007). Performance of site-specific nutrient management for irrigated, transplanted rice in northwest India. *Agronomy Journal*, *99*, (6), 1436-47. 32

Krewer, G.& Meyers, S. (1989). Commercial Blueberry Culture; Circular 713; University of Georgia, Cooperative Extension Service, The University of Georgia, Athens, GA, USA.

Krewer, G., NeSmith, D.S. & *Blueberry cultivars for Georgia*. (2006). Available online: http://www.smallfruits.org/Blueberries/production/06bbcvproc_Nov0206.pdf (Accessed on 26 February 2010).

Maiti, D. & Das, D.K. (2006). Management of nitrogen through the use of Leaf colour chart (LCC) and Soil Plant Analysis Development (SPAD) in wheat under irrigated ecosystem. *Archives of Agronomy and Soil Science*, *52*, 105–112.

Margaret, Achimugu., Ojone & Lajide, L. (2010). *Physico chemical characteristics of fortified compost and its effect on the growth of tomato*. Pub., By Federal University of Technology Akure, Theses and Dissertation (Chemistry).

Mendola, M. &Mendola. (2007). Agricultural technology adoption and poverty reduction: A propensity-score matching analysis for rural Bangladesh. *Food Policy*, *32*, (3), 372-93.

Ouedraogo, E., Mando, A., Brussard, L. &Stroosnijder, L., (2007). Tillage and fertility management effect on soil organic matter and sorghum yiled in semi-arid West Africa. *Soil Till Res.*, *94*, 64-74.

Ouma, James O., Festus M. Murithi, Wilfred Mwangi, Hugo Verkuijl, Macharia Gethi. & Hugo De Groote. (2002). Adoption of Maize Seed and Fertilizer Technologies in Embu District, Kenya. Mexico: *CIMMYT*.

Pampolino, M. F., Manguiat, I. J., Ramanathan, S., Gines, H. C., Tan, P. S., Chi, T. T. N., Rajendran, R. &Buresh, R. J.(2007). Environmental impact and economic benefits of site-specific nutrient management (SSNM) in irrigated rice systems. *Agricultural Systems*, *93*, (1-3) (3), 1-24.

Panda, S.S. (2003). *Data Mining for Precision Management of Crops (Paper 2).* Ph.D. dissertation, North Dakota State University, Fargo, ND, USA.

Panda, S.S. (2008). Rainfall Patterns. *Encyclopedia of Global Warming and Climate Change*; Philander, S.G., Ed.; Sage Publications: Los Angeles, CA, USA, pp. 846-847.

Panda, S.S. (2008). Spatial image mining. *Encyclopedia of Geographical Information Science*; Sekhar, S., Xiong, H., Eds; Springer: New York, NY, USA, pp. 475-479.

Panda, S.S., Hoogenboom, G. & Paz, J. (2010). Image classification algorithm efficiency comparison: A contextual study of blueberry orchard delineation. *Comput. Electron. Agr.*, In Review.

Paynter, B. (2008). Feeding the Masses: Data in, Crop Predictions Out. *Wired Magazine, 16* July.

Potera, Carol. (2007). Pesticides disrupt nitrogen fixation. *Environmental Health Perspectives, 115*, (12) (12), A579. Reetz, H.F. Jr. and P.E. Fixen. 2000. Strategic approach to site-specific systems.

Purakayastha, T. J., Rudrappa, L., Singh, D., Swarup, A. & Bhadraray, S. (2008). Long term impacts of fertilizers on soil organic carbon pools and sequestration rates in maze-wheat-cowpea cropping system. *Geoderma, 144*, 370-378.

Saleem, Shoaib Rashid. & Dalhousie University. (2012). *Department of Agricultural Engineering. Variable Rate Fertilization in Wild Blueberry Fields to Improve Crop Productivity and Reduce Environmental Impacts* By Shoaib Rashid Saleem.

Sapkota, T.B., Jat, R.K., Singh, R.G., Jat, M.L., Stirling, M.G., Jat, M. K., Bijarniya, D., Kumar, M., Singh, Y., Saharawat, Y. S. & Gupta, R. K. (2017).Soil organic carbon changes after seven years of conservation agriculture in a rice–wheat system of the eastern Indo-Gangetic Plains. *Soil Use and Management, 33*, 81-89.

Sawyer, J. E. (1994). Concepts of Variable Rate Technology with Considerations for Fertilizer Application. *J. Prod. Agric., 7*, 195-201.

Senay, G.B., Ward, A.D., Lyon, J.G., Fausey, N.R. & Nokes, S.E. (1998). Manipulation of high spatial resolution aircraft remote sensing data for use in site specific farming. *Trans. ASAE, 41*, 489-495.

Son, Tran Thuc. (2006). Combining nutrition management and IPM is the key to reduce fertilizer, pesticides, to increase crop yield, quality, and to improve rice farmers' income. Paper presented at the workshop.

Swain, G. (2008). Water Scarcity: Preventing Water Shortages. Trends in America February.

Thomas M. (1983). Pesticide demand and integrated pest management: A limited dependent variable analysis. *American Journal of Agricultural Economics, 65*, (4) (Nov.), 806-10. Clayton, Sophie. 2010. "50 years of rice science for a better world – and it's just the start!" Rice Today, IRRI.

Verma, S. & Sharma, P. K. (2007). Effect of long term manuring and fertilizers on carbon pools, soil structures and sustainability under different cropping systems in wet-temperate zone of North West Himalayas. *Biol Fertility Soil., 44*, 235-240.

VinhPhuc province. Son, Tran Thuc, Nguyen Van Chien, Vu Thi Kim Thoa, A. Dobermann, & Witt, C.(2004). Site-specific nutrient management in irrigated rice systems of the Red River Delta of Vietnam. *UNEP.* 2005. Integrated Assessment of the Impact of Trade Liberalization. A Country Study on the Viet Nam Rice Sector.

Wang, Shaohua, Yan Zhu, Haidong Jiang. &Weixing Cao. (2006). Positional differences in nitrogen and sugar concentrations of upper leaves relate to plant N status in rice under different N rates. *Field Crops Research, 96*, (2-3) (4/30), 224-34.

Williamson, J. & Lyrene, P. (2004). *Blueberry Gardener's Guide*; Circular 1192; Horticultural Sciences Department, Florida Cooperative Extension Service, Institute of Food and Agricultural Sciences, University of Florida: Gainesville, FL, USA.

Williamson, J.G. (1994). Dooryard blueberry production in Florida. *Proc. Fla. State Hort. Soc., 107*, 399-403.

Yang, W. H. (2003). Using leaf color charts to estimate leaf nitrogen status of rice. *Agronomy Journal*, 95, (1), 212-7.

Chapter 15

Use of Organic Soil Amendments for Improving Soil Ecosystem Health and Crop Productivity

Vijay Kant Singh[1], Gurdeep Singh Malhi[1,*], Manpreet Kaur[2], Gurpreet Singh[1] and Hanuman Singh Jatav[3]

[1]Punjab Agricultural University, Ludhiana, Punjab, India
[2]CCS Haryana Agricultural University, Hisar, Haryana, India
[3]Department of Soil Science and Agriculture Chemistry,
Sri Karan Narendra Agriculture University, Jobner, Jaipur, India

Abstract

Innovative agronomic practices are needed to increase food supply while minimizing tradeoffs to feed a rapidly growing human population. Any material applied to a soil to improve its physical properties, such as infiltration, aeration, water-holding capability, permeability, drainage, and compaction, while indirectly affecting the soil's chemical and biological properties, is referred to as a soil amendment. Sustainable soil amendments are essential for increasing agricultural yield and nutrient efficiency while reducing carbon efflux from the soil. A major challenge for agricultural sustainability is maintaining and enhancing agricultural product yield and nutrient quality while retaining soil quality, including the soil carbon reserve. Organic soil amendments (OSA) increase aggregate stability, bulk density, aeration, porosity, water-holding capacity, water flow, and nutrient cycling in soil. The use of suitable OSA, including green manure (GM), vermicompost (VM), farmyard manure (FYM), rice straw (RS), coir pith, municipal solid waste (MSW), and other materials as a novel technique for increasing crop production and soil fertility has been investigated by several scientists. Compost is a stabilized kind of organic matter (OM), which enhances soil physical characteristics such as nutrient and water reserve, aggregation, macro-pores and micro-pores resistance to erosion, temperature insulation, and apparent density tothe soil, chemical and biological characteristics such as microbial biomass population. Sewage sludge is high in organic matter and, in terms of holding soil fertile and stimulating plant growth, it works similarly to a slow-release organic fertilizer. As a result, using soil amendments to improve soil quality and crop sustainability could be a viable option.

Keywords: soil amendments, compost, soil properties, yield

[*] Corresponding Author's Email: 89gurdeep.malhi@gmail.com.

In: Ecosystem Services
Editor: Hanuman Singh Jatav
ISBN: 978-1-68507-614-6

1. Introduction

The quality and health of the soil are affected by fertilizers used in agricultural management (Singh and Ryan 2015; Doran and Zeiss 2000). Soil fertility refers to a soil's ability to meet a plant's physico-chemical and biological requirements for growth, reproduction, and quality, while taking into account plant and soil type, land use, and climatic conditions (Abbott and Murphy 2007). Soil health refers to a soil's ability to act as a living system, supporting plant and animal production, maintaining or improving water quality, and promoting crop productivity. Soil quality is widely defined as a soil's ability to interact with the ecosystem to preserve biological productivity and the efficiency of other environmental compartments, thereby promoting the health of plants and animals, including humans (Doran and Parkin1994). Soil is a dynamic natural environment at the crossroads of soil, air, water, and lives, and provides critical ecosystem services for the survival of humanity (Needelman 2013). The preservation of soil quality is one of the significant threats and opportunities we face in the 21st century.

The recent rise in environmental concerns related to the overuse of chemical fertilizers necessitates extensive research into potential methods for addressing such risks. Excessive agrochemical use, deep tillage, and luxury irrigation are popular agricultural practices that have degraded soils, poisoned water sources, and polluted the atmosphere. One of the major causes of the reduction of soil health and crop productivity is the lack of organic matter content in Indian soil, in addition to several other factors (FAO 2005). Food production would have to double in the next few decades to feed the world's rising population, according to projections. Since the beginning of the 1960s, the Green Revolution has resulted in sharp rises in food production from agricultural systems, thanks to the introduction of new crop varieties and livestock breeds, as well as the widespread use of agrochemicals like fertilizers, pesticides, irrigation, farm machinery. Intensification, not the expansion of agricultural space, was the driving force behind this global rise in food production (Pretty and Bharucha2014). Therefore, for the planning of soil fertility systems, a better understanding of soil quality control factors is essential.

Agricultural intensification was sustained through indiscriminate use of synthetic agrochemicals, overuse of water, and alteration of the soil ecosystem, all at a considerable cost to the climate, to increase crop output. Indeed, substituting external inputs for internal soil processes caused the soil's fundamental properties, such as its ability to self-regulate, to deteriorate over time (Bender et al., 2016).

Organic wastes can be used as soil amendments to increase soil fertility. They are important sources of nutrients for growing crops (Davies and Lennartsson 2005). In addition, organic amendments help to mitigate climate change by storing carbon in the soil, the size of which is determined by the type, strength, and frequency with which they are used (Diacono et al., 2010). According to different sources, the country has around 300 million tonnes of alternate sources of soil organic amendments (Indoria et al., 2018). It is widely acknowledged that using sewage sludge in agriculture is environmentally friendly because it is a cost-effective method of processing municipal waste (Ankush et al., 2020). Biosolids are generated by solid waste treatment plants and are helpful for sustainable agriculture because they provide essential plant nutrients while also reducing emissions (Solanki and Debnath, 2014). Fruit peel biochar has a wide range of applications. Biochar is an oxygen-restricted thermo-chemical conversion of low-cost biomass that is a carbon-rich fine-grained by-product. The benefits of using organic manure and straw to maintain soil quality are becoming more commonly recognized (Chander et al., 1997). When compared to chemical fertilizers, long-term use of organic amendments to the soil increases many metrics such as organic carbon, aggregate stability, and crop yield (Diacono and Montemurro2011; Zimmer et al., 2018; Cheng et al., 2019).

Soil amendments are both inorganic and organic compounds blended into the soil to strengthen the soil constitution in terms of plant production. Soil amendments aimed to improve the root proliferation and plant growth environment, which includes improving water holding capacity (WHC), soil structure and nutrient accessibility, and the living conditions for microbes that are essential for plant growth. Improved soil texture and root growth also help to avoid soil degradation during windy areas or heavy rains. The use of organic amendments also aids the nutrient cycle (e.g., manure). The form of alteration you choose is entirely dependent on how you alter the soil. By amending the soil, almost any type of soil can be made fertile. Sustained organic amendment activities may be a valuable way to preserve or raise OM content in agricultural soils, as well as to protect and boost soil fertility. According to Tilman et al., 2000, global nitrogen (N) and phosphorus fertilizer usage have risen by 7 times and 3.5 times in the last six decades, respectively, with both fertilizer products predicted to increase by three times by 2050.

Mulching, or the placement of materials on top of the surface, is not regarded as a soil amendment. Different substances are used with various soils and plants to optimize soil conditions. Because of its low production costs, OM, such as compost, is a common amendment. For thousands of years, OSA has improved soil quality and fertility. Several years ago, the Greeks, Romans, and Indians applied animal waste and human urine to the earth. Waste, manure from the plant, and other waste products were also being used to help in the growth of crops. OSA has been used to increase soil quality and fertility for thousands of years. Seashells, vegetable waste, FYM, and other waste products have all been used in the past to encourage crop growth. Fabric mulch was linked to higher soil moisture, organic matter, and other chemical and biological properties than paper mulch, which influenced soil properties. The fabric mulch helped in the early emergence and harvesting of bean and zucchini plants. With the application of compost amendments, the yield of bean and zucchini crop improved, increased soil pH, OM, and many nutrient concentrations, when compared to chemical fertilization alone, with compost amendments increasing soil pH, organic matter, and many nutrient concentrations, as well as crop emergence and yield (Larkin 2020). This chapter discusses several types of OSA and their role in improving soil physical, chemical, and biological properties, and crop yield as a substitute for chemical fertilizer in conjugation.

2. Organic Soil Amendments: An Alternative Source for Sustainability

Soil amendments application for achieving sustainability is a golden option, although several options are present. Several organic wastes can be used after treating as organic soil amendments (OSA) so that their better utilization can be done to achieve agricultural sustainability. OSA strengthens the physical properties of the soil, resulting in better water conservation, effective root growth, and soil ecosystems. Soil amendments aid in the aeration of the soil. Soil amendments, especially organic ones (such as compost), can be made locally and relatively inexpensive. They can also be used by almost anybody, almost anywhere.

OSA can be derived naturally from the organic plant product such as GM, crop residues, moss from peat bogs, by-product material from the processing units or mills (rice hulls, bark, bagasse, cedar chips, sawdust, etc.), or from waste disposal units (biosolids, sewage sludge, compost). OSA are mainly used to slacken off the soil and build large pores to maximize the amount of water absorbed. Compost and animal dung are the most essential organic soil additives globally. Compost and animal dung are often readily accessible and inexpensive. In intensive agriculture, peat moss is the most commonly used organic amendment. OSA includes grass clippings, straw, wood chips, compost (produced in large or small scale amenities), manure, treated sewage sludge (biosolids), sawdust, etc. Organic substances are frequently more readily available and less costly than inorganic substances. Using good crop rotations and picking the right crop improves soil amendment (Davis and Whiting 2000). Before using in ground-beds, most OSA, predominantlycedar chips, sawdust, bark, and bagasse, should be composted or aged, and untainted if possible because all of these OSA have a high C: N ratio. Use of non-compostable amendments with a high C: N ratio can deplete N in the soil, cause a salt or ammonia/ammonium burn, or cause heat buildup damage, after decomposition, amendments with a low C: N ratio release nitrogen, acting as an organic fertilizer. Organic matter (OM) with a low C: N ratio should also be composed to avoid the release of rapid NH_3/NH_4^+ gas and toxicity. OSA such as compost and animal manure are the most common nowadays, but peat moss, wood chips, straw, sewage sludge, and sawdust are still used. Some of the alternative sources of OSA available in the country are listed below.

2.1. Animal Manure

Manure is a mixture of faeces, urine, and animal bedding that has been stacked and keep turning until it has been composted to a certain extent. It's made from beef, dairy, pork, poultry, and turkey, and the composition varies depending on where it's made, how urine and faeces are excreted and mixed, and how long it sits before being added to soil. Manure not only feeds crops, but it also adds organic matter to the soil, making it more fertile (Goss et al., 2013). As solid manure is turned regularly and not closely packed, organic matter degrades under aerobic conditions. Under aerobic conditions, the pace of the process is faster than under anaerobic conditions, and more C is lost as CO_2. This is partly because aerobic breakdown produces more energy, which in turn supports approximately 5.5 times more microbial biomass per unit of organic substrate absorbed than anaerobic fermentation. Appropriate use of these manures as an amendment necessitates calculating the nutrient content, especially N, P, and K, concerning the requirements of crop and soil test values to attain sustainability that can reduce pollution on the surface and groundwater (Al-Kaisi et al., 1998a, 1998b). Intensive agriculture creates extra pressure to conserve soil, water, and air resources for sustaining crop and animal productivity. It necessitates the immediate requirement to develop and implement advanced manure management technologies and practices to minimize the harmful microbial population prior to land application (Millner et al., 2009). Manure composting is an effective way to utilize animal manure as soil amendments, made by controlled aerobic, thermophilic (self-heating) decomposition of organic matter by the aerobic microbes for nutrient stabilization and pathogen reduction.

2.2. Cover Crops

Cover crops include plants that are planted for positive turning of physical, chemical, or biological properties, minimizing erosion or increasing soil fertility i.e., for in-between periods when the principal cash crops are produced (where the soil would otherwise be left blank) other than harvest for direct commercial value. Cover crops may also be used to fill in short periods of non-cultivation in crop rotations to protect soils, prepare the soil for a perennial crop, or provide animal feed. Cereal crops supply straw that remains after harvest. At the same time, legumes like soybeans, cowpeas, and clover are mostly favoured because they can fix nitrogen from the atmosphere using bacteria at the root stage. The key advantage of a cover crop is that it supplies additional nutrients and OM to the soil and increases microbial activity and water retention. Cover crops may also be used to fill in gaps in crop rotations when there is just a short period of non-cultivation, as well as for soil conservation, perennial cropland preparation, and animal feed. Cereal crops that supply straw left after harvesting do not fix nitrogen. In contrast, legumes such as soybeans, cowpeas, and clover are preferred because they can fix nitrogen from the environment at the root level using bacteria.

2.3. Compost

Composting is a long-term method of converting organic wastes into OSA, useful as potting media or soil conditioners (Barthod et al., 2018). Composting is an aerobic method that can be done on a large scale in windrows or piles (Hobson et al., 2005) or on a smaller scale in a home composter (Andersen et al., 2011). Composting is a method of treating manures, green wastes, industrial solid wastes, and their fractions (Goyal et al., 2005). Manure, the most commonly used OSA, is formed when organic waste decomposes. Composting (Pérez-Piqueres et al., 2006) is both a fascinating agricultural practice and a waste recycling process. Moreover, compost as a fertilizer has many benefits, including increased organic C content and microbial activity (Scotti et al., 2015), higher concentrations of plant nutrients such as N, P, K, and Mg, and root reinforcement (Scotti et al., 2015; Donn et al., 2014). In addition, compost can affect soil microflora by killing certain soil borne pathogens and diseases such as Pythium, Phytophthora, and Fusariumspp (Szczech and Smoliska2001; Borrero et al., 2004). Composting on-farm may be a dependable, cost-effective, and environmentally friendly biological solution for recycling residual agricultural biomasses (Pane et al., 2015).

2.4. Crop Residues

Only one-third of all crop residues are available for direct recycling, (compost-making), while two-thirds are used as animal feed (FAO 2005). Crop residues are generated in large quantities during the processing of agricultural crops at harvesting. These wastes are used as soil mulch, fertilizer, domestic and industrial oils, and thatching material, among other things. It is an important source of OM for bettering soil quality and crop productivity. However, their handling and collection are difficult due to their bulky nature; it is also a time-consuming and labour-intensive operation. Owing to a lack of appropriate equipment, shredding and mixing crop residues are difficult.

There is a common misconception among farmers that residues burning improve soil fertility and aids in the control of weeds, disease, and pests. India's agriculture produces crop residues annually approximately 500–550 mt (NAAS 2012). However, a significant segment of these crop residues (roughly 90-140 mt) is burned on-farm each year, mainly to clear the fields to sow the next crop (Pathak et al., 2010). In a long-term experiment (2005-2012), Sharma et al. (2017) originate that exterior application of sorghum residue @ 6 t ha^{-1} and 4 t ha^{-1} increased grain yields of sorghum by 21% and 16%, respectively, over no residue, while the corresponding increase in cowpea yield was 50% and 60%, respectively. In all procedures, concentrations of different fractions of soil organic carbon (SOC) in the 0-5 cm depth were substantially higher than lower layers (5-15 cm depth). When sorghum stover was applied at a rate of 6 t ha^{-1}, the labile fraction of carbon in the 0-5 cm soil layer was significantly higher than in the other treatments.

2.5. Aquatic Weed Biomass

Aquatic weeds such as *Eichhorniacrassipes* (water hyacinth), *Salviniamolesta*, Chara spp., Nitella spp., and algal scum can be found in large areas of India's northern region, including Kerla (Varshney et al., 2008). It covered approximately 2 lakh hectares of water surface in India's seasonal rivers, lakes, and other water bodies (Murugesanet al., 2005). As green manure, it is a potent source of organic soil amendment. It generates an adequate amount of organic matter and vital plant nutrients. A fresh plant contains 95.5 percent moisture, 0.04 percent nitrogen, 0.06 percent P_2O_5, 0.20 percent potassium, and 3.5 percent organic matter (Vidya and Lakshmi2014). According to DWSR (2014), *Eichhornia crassipes* can be used as mulch material in potato crops. The cost of transportation is high due to the distance from the farmers' fields, and the majority of the water sources are located near towns. One of the constraints is the lack of appropriate technology to turn the immense aquatic weed into nutrient-rich compost or liquid fertilizers.

2.6. Organic Solid Waste from Agro-Industries

Agro-industrial wastes include sugarcane bagasse and press mud, paddy husk, waste from vegetables, agricultural ingredients, tea, oil refining, jute fibre, groundnut shell, wooden mill waste, coconut husk, and cotton stalk. Solid-state fermentation (SSF) produces animal feed, biofuels, enzymes, and other chemicals from agro-industrial wastes, which are rich in nutrient composition and bioactive compounds (Sadh et al., 2018). It should be treated as "raw material" rather than "waste" in other manufacturing processes because the presence of specific nutrients residues provides ideal conditions for microbe proliferation. Soil health and crop growth improve as the microbial population grows. The use of agricultural and agro-based industry wastes as raw materials will help reduce production costs while also contributing to waste recycling and making the environment more environmentally friendly. Using agro-industrial wastes as raw materials will lower production costs while also lowering emission levels in the atmosphere. Many studies have shown that sugarcane agro-industry waste (bagasse and press mud) contains a high concentration of plant nutrients and can increase soil properties and sugarcane yield (Kumar and Verma 2002). The plant samples treated with biofertilizer made from watermelon, banana wastes, papaya and had promising soil physical characteristics, according to Lim et al., 2015. However, the watermelon biofertilizer had the highest pH value (5.15). The potassium content of the banana biofertilizer was the highest; at 3.932 g K/L. Farmers' key barriers to using agro-industrial waste are shipping costs and other environmental concerns.

2.7. Biosolids (Municipal Solid Waste)

Solid waste includes garbage, refuse, and discarded products from mining, farming activities, manufacturing, community events, and consolidated materials from a wastewater and water treatment plant (Singh et al., 2021). Since biosolids contain nutrients and OM, they can be applied to farming soils, but only under strict regulatory guidelines that control heavy metals, weeds, and human and plant pathogens. As an efficient means of handling solid waste, biosolids, also known as municipal solid waste (MSW), is one of the most capable and profitable sources of OSA. According to Srivastava et al. (2016), using compost made from MSW improves the physico-chemical characteristics of the soil and improves the biological population of the cultivated property. MSW compost improves soil nutrient profile, microbial activity, soil texture, and buffering capability while increasing plant yield (Hargreaves et al., 2008a, b, c; Carbonell et al., 2011; Weber et al., 2014; Bouzaiane et al., 2014). India's cities generate about 48 million tonnes of MSW per year (Pappu et al., 2007). The amount of waste produced per capita per day in India varies between 273 and 657 g/day/capita, and the amount of waste generated per capita per day is expected to be rising at a rate of 1-1.5 percent per year. The agriculturist's limitation in using biosolids is that they contain various types of pollutants and harmful pathogens; as a result, their composition varies depending on location, city population, and industry form (Indoria et al., 2006). As a result, it is essential to treat them with appropriate methods to not degrade the soil or human health once they reach the food chain before being applied to cropland. Radiation therapy, which is rich in plant nutrients and OM, does not contain radioactivity, and destroys dangerous bacteria, is one of the most efficient methods for treating sewage sludge (Zhongming et al., 2018). The cost of transportation and a lack of understanding among farmers about the benefits of using biosolids are also significant obstacles.

2.8. Biochar

The pH, surface area, nutrient concentration, porosity, and metal-binding potential of biochars differ significantly due to the variety of feedstock materials and thermal conversion conditions under which it is produced. Biochar's diverse chemical and physical properties have led to its use as an amendment to rebuild soil health, improve crop yields, increase soil water storage, and restore mining-impacted soils/spoils (Novak et al., 2016). Biochar is an important vital soil amendment because its many uses for enhancing soil health and quality. Because of its porous structure and capacity to increase pH, maintain water in its pore space, and bind nutrients and metals on its functional groups, biochar has the potential to improve soil health characteristics

(Ippolito et al., 2015). The chemical composition of biochar materials varies, and it is dependent on the biochar made from various feed materials and using various methods. The labile components in most biochars are small compared to the stable components. Biochar has been shown in many studies to increase the availability of micronutrients, thus lowering Al toxicity by increasing the pH of the soil (Alling et al., 2014; Srinivasarao et al., 2013). Effective biochar as an amendment can be created by customizing its physical and chemical properties. It also reduces the movement of microbes through the soil, thereby decreasing human health risks. Since biochars are costly to produce and cannot be purged from the soil after use, they should be aimed for particular uses in the agricultural and environmental sectors (Novak et al., 2016). The production of suitable biochar is hampered by a knowledge gap and a lack of appropriate technology at the farmer level.

2.9. Coir Pith

Coir pith is a by-product of the coir industry, consisting of short fibres and dust left over after the industrially applicable long coir fibre has been removed from the coconut husk. It's a lingo-cellulosic substance made from coir fibre extracted from coconut husk that's found as waste near coir processing plants (Thomas et al., 2013). However, because of the complexity in chemical and structural terms of the lignin-cellulose complex, which has a high content of lignin (30%) and cellulose (26%) and a low pentosan-lignin ratio (> 0.5), its decomposition rate in the soil is prolonged (Ramalingam et al., 2005). Coir-processing units in India are expected to produce 0.5 million tonnes of coir pith per year, with 10 million metric tonnes in southern India (Ghosh et al., 2007). It has numerous advantages and can be used as a potentially useful resource in the agricultural sector if properly utilized (Prabhu and Thomas 2002). Coir pith has a cation exchange capacity (CEC) of 38.9 to 60 p^{+}kg/1000 g (Mapa and Kumaru 1995) and a moisture retention capacity of 500-600% (Evans et al., 1996), allowing it to maintain a significant amount of nitrogen as well as exchangeable potassium, sodium, and magnesium (Mapa and Kumaru1995).

2.10. Vermicompost

Composting or breaking down organic materials supplemented with earthworms produces vermicompost. Vermicomposting is an environmentally sustainable and efficient method of transforming agricultural waste (such as crop residues, cow dung, tree leaves biomass, fruit and vegetable waste, kitchen waste, weeds, and so on) into nutrient-rich produce with the help of earthworms (Nedgwa and Thompson 2001). It is high in essential plant nutrients and has the potential to improve the soil's condition and quality (Mahmud et al., 2018). Furthermore, since it is homogeneous, has attractive aesthetics, and has a low level of contaminants, vermicompost is an outstanding natural commodity with the potential to retain more nutrients for a longer time frame, an increased soil enzyme, and microbial population (Tajbakhsh et al., 2011). In sandy soil, vermicomposting is more beneficial because it helps to increase soil organic matter (SOM), which improves soil aeration, maintains good soil aggregation, increases macro and micronutrient availability, and protects against soil erosion (Roy et al., 2006; Baligar et al., 2001).

2.11. Green Manuring

Green manuring (GM) is a method of enhancing the physical, chemical, and biological properties of soil by turning undecomposed plant material beneath the soil, which provides organic material and nutrients to the soil and an alternate source of soil amendments (Fageria 2007). GM has long been recognized as having a positive impact on crop production. Because of the high cost of chemical fertilizers, the increased risk of environmental contamination, and the need for sustainable cropping systems, the importance of this soil ameliorating activity has grown in recent years. GM can increase crop yields by improving the physical, chemical, and biological properties of the soil. Furthermore, reduced chance of nitrate (NO_3^-) leaching and lower fertilizer N needs for effective crops are possible benefits of GM (Fageria 2007). *Some of the most common GM crops areSesbania aculeate, Sesbaniarostrata, Crotalaria juncea, Tephrosiaperpurea, Sesbaniaspeciosa, Indigoferatinctoria, Vigna radiata, Vigna mungo, and Vigna unguiculata.*Green leaf manure is also made from tree species, including Glyricidiamaculeata, Pongamiaglabra, Calatropisgigantecum, Azadirachtaindica, and Calatropisgigantca. GM crops are rich in plant nutrients and OM. Even though GM crops have a high N_2-fixing capacity, lower nitrate (NO_3^-) leaching risk, lower fertilizer N requirements for succeeding crops, and positive effects on soil physical and chemical parameters, as well as crop yields, the area under GM crops has not increased in India over the last few decades. GM use is also limited by the unreliability of its output, the scarcity of seeds, and labour-intensive operations. GM application increased OM content, and microorganism behaviour enhanced soil physico-chemical properties, and increased crop yields in other studies (Choi et al., 2014; Duyar, 2014; Krobel et al., 2014; Patil et al., 2014; Ye et al., 2014; Alam et al., 2016; Pinto et al., 2017). GM has also been shown to increase soil structure, reduce nematode populations, and reduce certain diseases and disease-causing factors (Larkin 2013; Thakur 2014; Azimzadeh et al. 2014; Gardiano et al. 2014). However, there is inadequate knowledge on how the use of GM, integrated at various plant growth stages and in various planting systems for crops, affects soil microbial biomass and yields in organic production systems.

2.12. Tank Silt

Tank silt is a mixture of sand and clay particles, or degraded soil, accumulated or collected in the village's tanks or lakes, as well as in delta regions. It is primarily used to improve soil and moisture conservation, increase water-holding ability, and improve the soil's aeration, porosity, and nutrient status to promote healthy crop growth. Desilting dams, water holding tanks, or reservoirs are a financially feasible operation among farmers for increasing water storage capability and returning silt to the fields as a supply of organic matter and other plant nutrients. As a result, successful use of silt collected in tanks by transporting it to agricultural lands is critical. It will aid in the restoration of tank storage capacity and the productivity of farmlands treated with tank silt. Tank silt has been shown in several studies to aid in releasing nutrients such as total nitrogen and to be beneficial to sandy and alkaline soils. It is used for crop production as a source of manure. It's also a natural amendment to increase water quality and productivity. Organic carbon, microbial biomass carbon, residual pesticides, and other substances are contained in the tank silt. From tank to tank, the sediment composition and depth of deposition differ. It is recommended that silt samples taken from tanks in vulnerable areas be tested and analyzed for toxicity in soil testing laboratories to make reasonable recommendations for use in the fields. Patil et al. (2017) discovered that the treatment Tank silt @15 Mg ha^{-1} + RDF had the highest soil moisture. Organic carbon (OC), nitrogen (N), phosphorus (P), potassium (K), and DTPA extractable micronutrient and yield of soybean crop were increased in treatment T_7 (Tank silt @ 5 Mg ha^{-1} + FYM @ 2.5 Mg ha^{-1} + RDF), which was significantly superior to application of tank silt, FYM, and RDF alone.

3. Organic Matter as a Substrate for Soil Amendments

OM is important in the soil environment because it provides substrates for microbes that decompose, (which provide nutrients to plants), boost soil structure and water retention ability (Abiven et al., 2009), improve environmental resistance to soil-borne pathogens (Bonanomi et al., 2010), and reduce the toxicity of heavy metals (Bonanomi et al., 2010; Park et al., 2011). Organic additives, such as manure, have been shown to improve soil structure and chemicals (Scotti et al., 2013) and biological fertility (Ros et al., 2003), as well as suppress soil-borne pathogens (Zaccardelli et al., 2013). Humic compounds are the primary organic constituent of organic soil amendments. Plants were once thought to use humus directly, but it was later discovered that inorganic compounds are needed for plant growth. When the organic matter was broken down, its constituent nutrient components were released into inorganic forms, which increased soil fertility. Humus affects soil fertility by influencing the retention capacity of water, as OM’s spongy nature of structure can bind water and specific inorganic molecules that act as micro or macronutrients. As a result, humic acids reduce the rate at which water evaporates from soils. This is particularly true in soils that lack or have a low concentration of clay, as well as in arid areas and sandy soils that can't retain water. Another benefit of organic amendments is that they feed soil bacteria, which release nutrients into the soil, increasing soil fertility.

4. Effects of Soil Amendments on Soil Properties

In agricultural soils, maintaining an adequate organic C stock is essential for nutrient contribution (Tian et al., 1992), soil structure improvement (Abiven et al., 2009), effective microorganism activity, and pathogen suppressiveness (Mader et al., 2002; Bonanomi et al., 2010). When the amount of OM inputs exceeds the amount of organic C outflows, primarily governed by temperature and water supply, soil organic C depletion can occur. The effects of OSA in agriculture systems on soil quality (physical, chemical, and biological fertility) and crop yields will be addressed in this section.OM affects crop growth and yields directly by supplying nutrients or indirectly by changing soil physical properties such as aggregate stability, porosity, and available water content, which can improve the root condition and promote plant growth.

4.1. Effects of Soil Amendments on Physical Properties

SOM is essential for soil structural stability, air and water infiltration, water conservation, and erosion control (Liu et al., 2016). OSA, including cow manure, sheep manure, rice husk, reeds, and wheat straw improved soil aggregate stability while lowering bulk density (Karami et al., 2012). OM affects crop growth and yield either directly by supplying nutrients or indirectly by altering the physical properties of the soil including aggregate stability, porosity, and available water volume, all of which can enhance the root ecosystem and promote plant growth (Darwish et al., 1995). Since OSA tend to increase soil organic C while decreasing soil bulk density, the latter soil property is inextricably linked to the former (Bauer and Black 1994). OM has been shown to improve soil structure and water retention capability, whether crop residues or field manure (Bhagat and Verma1991). Rasoulzadeh and Yaghoubi (2010) found that applying 3 and 6 t ha^{-1} of cattle manure significantly decreased soil bulk density and particle density ($P < 0.05$). Increased organic matter in the soil and aggregate stability are the main reasons for the decrease in bulk density. In the 3 and 6 t ha^{-1} treatments, the aggregates' mean weight diameter (MWD) was significantly ($P < 0.05$) greater than in the control treatment. When 3 and 6 t ha^{-1} cattle manure was applied, saturated hydraulic conductivity improved slightly, but there was no significant difference between the two ($P < 0.05$). The use of cattle manure increased both field capacity (FC) and permanent wilting point (PWP) significantly ($P < 0.05$). Cattle manure increased the water content at the FC and PWP. As a result, the difference in water content between FC and PWP was approximately the same. When OC concentrations were increased by adding cattle manure, increases in FC and PWP were almost identical. Elwell et al. (1986) discovered a similar relationship between OC and water-stable aggregates in some ferrallitic clay soils in Zimbabwe. Increasing the OC content of the soil improved its aggregation status, water infiltration rate, and bulk density (Singh et al., 2020). When compared to inorganic fertilizer alone, OSA significantly increased SOM content. Mondal et al. (2015) found that MBS had a positive impact on various soil parameters, including bulk density, which decreased by about 21% in the surface layer, mean weight diameter (MWD), and porosity.

4.2. Effects of Soil Amendments on Chemical Properties

Organic carbon in the soil is an essential component of soil fertility and growth potential. According to Ankush et al., 2021, adding sewage sludge (SS) significantly increased soil organic carbon content (SOC), available nitrogen (N), available phosphorus (P), and available potassium (K). However, SS (5 t ha^{-1} + 75 percent RDF) F_4 had the highest soil biomass, N, P, and K content, which was comparable to (RDF) F_3 and SS (5 Mg ha^{-1}) F_2) and higher than (RDF) F_5 and (no fertilizers) F_1. According to several studies involving long-term fertility trials, OM applications increased the organic carbon stock and, as a result, increased the cation exchange capacity (CEC). This effect was caused by the strong negative charge of organic matter. Habteselassie et al. (2006) discovered that the C pool increased by 115 percent in dairy-waste compost treated soil over five years. Total OC increased by 24.0 and 43.2 percent for cocksfoot and alfalfa plots, respectively, after three years of municipal solid waste (MSW) compost and olive pomace compost application, relative to the unfertilized control, showing that these amendments positively affected the organic matter (Montemurro et al., 2006). Humic acid in MSW compost improves CEC and buffering capacity of the soil (Garcia-Gil et al., 2004).

4.3. Effects of Soil Amendments on Biological Properties

Application of OSA control the majority of soil properties, OC and its fractions (labile and non-labile), such as particulate organic carbon (POC) and its components (coarse POC and fine POC), light fraction organic carbon (LFOC), readily oxidizable organic carbon (VOC), and dissolved organic carbon (DOC), are essential for the long-term viability of any agricultural production system (Singh et al., 2019). In three subtropical paddy soils, Hao et al. (2008) discovered that microbial biomass was significantly higher in plots receiving FYM+NPK fertilizer than in plots receiving only NPK fertilizer. MBC and MBN levels rose when straw was combined with inorganic fertilizer. When high-rate compost treatments were used, MBC increased by up to 100%, and enzymatic activity increased by 30% when sludge was applied. The amount of biomass in the soil at any given time cannot be used to decide if the soil's organic matter consistency is improving or deteriorating. To assist in resolving this conundrum, the MBC/SOC ratio provides an indicator of organic matter dynamics (Carter1991). The lowest ratios were found in the treatment absolute N or control soils, and the highest in the NP+FKM treatment, according to Lie et al. (2010). In a similar study, Masto et al. (2006) discovered a ratio of 2.88 in control plots and 3.87 in farmyard manure plus inorganic fertilizers. Kaur et al. (2005) found that after seven years of continuous use of poultry manure had the highest MBC: SOC ratio (6.3), while soils receiving chemical fertilizers had the lowest (2.6-2.9).

OSA had a significant effect on basal respiration. According to several studies, qCO2 is higher in soil treated with OSA than in soil treated with inorganic fertilizer (Heinze et al., 2010). Respiratory activity was higher in the long-term FYM applied map due to higher OM levels built up with continuous FYM additions. Since dehydrogenase activity is found only in living cells, it is thought to reflect the full range of oxidative activity in soil microflora. As a result, it may be a good indicator of microbial activity (Nannipieri et al., 2017). The more readily decomposable components of FYM on soil microorganism metabolism may justify the stronger effects of FYM on dehydrogenase activity. Microbial biomass and enzyme activities were higher in OSA than in mineral fertilizer treatments because OSA can provide abundant OM for microbial growth and activity (Gong et al., 2009). The combined application of biochar with manure effectively increases microbial biomass properties. It has been shown that applying biochar to the soil will either increase or decrease soil activity; however, when mixed with manure, it also encourages microbial abundance and activity (Brtnicky et al., 2019). In comparison to RDF and control treatments, Ankush et al. (2021) found that combining SS (5 Mg ha-1) and 75 percent RDF resulted in a substantial increase in soil microbial biomass carbon and alkaline phosphatase production. The control treatment had the lowest operation in which no fertilizer was applied. In both wheat and pearl millet, the direct and residual effects of SS application resulted in a substantial increase in urease activity at each crop growth rate. The use of SS and mineral fertilizers together may have resulted in a significant increase in organic matter in the soil (Sharma and Dhaliwal 2019). Furthermore, Fan et al. (2019) studied straw decomposition and its priming effect (PE) on native soil organic carbon (SOC) decomposition in soils with 23 years of chemical fertilizer (NPK) application and partial replacement of chemical fertilizer with organic manure (NPKM) and discovered that NPKM substantially increased straw decomposition. Increased straw decomposition was related to increases in total bacteria, fungi, and cellulose-degrading enzyme activity. The use of NPK had no impact on the amount of PE caused by straw, while the use of NPKM substantially reduced PE.

4.4. Effects of Soil Amendments on Plant Growth, Productivity and Economic Returns

With low lignin content, abundant labile C fraction, and a low C/N ratio, the organic C biochemical stability of the reduced fraction of plant debris integrated into the soil (e.g., root exudates, death roots, and a minor fraction of leaf litter) is high. There was a significant increase in the percentage of germination, fresh weight, dry weight, biomass, root, and shoot length of wheat plants in E. crassipes-treated compost as compared to control plots; higher soil moisture and organic matter were also found in the treated plots (Sharda and Lakshmi, 2012). Indoria et al. (2013) and Mondal et al. (2015) found a 3 percent (on an oven-dry basis) increase in the yield of some oilseed crops (*Brassica juncea*, *Brassica napus*, and *Eruca sativa*) amended with sewage sludge (municipal biosolids) compared to regulation. They also found that soil amended with biosolids (sewage sludge) had higher levels of heavy metals (Cd and Ni) in various plant parts (stem, leaf, and seed) than soil that had not been amended. When rice straw, sugarcane trash, and water hyacinth vermin-compost were used instead of the 100 percent recommended dose of fertilizer (RDF), (N, P, K), yield increases of 17.17 percent, 30.29 percent, and 47.31 percent were observed, respectively, in rice crops (Sudhakar, 2000). Crop growth and yields can be improved with the combined application of inorganic fertilizer and OSA. It may be possible that the slower breakdown rate of FYM added or straw-added plots were due to a slower breakdown rate (less and constant mineralization rate).

The use of vermin-compost in conjunction with RDF, as well as the use of vermin-compost alone, was found to increase or sustain similar yields of fruit crops, pulses, cereals, and vegetables (Nagavallemmaet al., 2004; Ramesh et al., 2005). Using a combination of compost @ 4 Mg ha^{-1} + 2 Mg ha^{-1} gliricidia lopping during the sorghum crop growing season and 2 Mg compost and 1 Mg ha^{-1} gliricidia lopping during the mungbean crop growing season could save 50% of the nitrogen needed by sorghum and mungbean, while also improving soil properties like pH, nitrogen, phosphorus, potassium, sulphur, labile carbon and microbial biomass carbon.

The economic returns at farm level are the most important determinants of the adoption of technology. The application of soil organic amedments viz. FYM and organic fertilizers has been reported to increase the wheat yield and consequently the net returns by 22 percent and 24 percent, respectively. The household size, machine ownership, dependency ratio and non-paid labours determine the adoption of OSA (Zheng et al., 2020). The yield of maize crop is also reported to increase on the application of soil amendments viz. rice husk biochar, compost, bentonite clay, clay-manure compost and rice husk biochar compost. The incremental yield of 0.77-3.79 t/ha at Naphok and 1.21-5.14 t/ha at Veunkham was obtained which led to increased per hectare net revenue of -\$794 to \$841 and -\$331 to \$1391 in Naphok and Veunkham, respectively. The negative economic revenue is observed in case of high cost soil amnedments such as bentonite clay which requires atleast five seasons to be economically beneficial. On the other hand, low cost soil amendments such as compost, rice husk biochar compost, rice husk biochar and clay manure was found to be economically viable from first year itself (Mekuria et al., 2013). The use of gypsum as soil amendment is found to be quite profitable and results in partial benefit cost ratio of 1.68. These benefits of using gypsum as a soil amendment are not instantaneous and even increased over time (Batte and Forster, 2015). However, the use of biochar as soil amendment is economically unviable because of variable rates of application, initial soil state and uncertain feedstock effects, which builds up a wide range of costs for increased yield from biochar application (Filiberto and Gaunt, 2013).

Conclusion

Soil amendments provide an alternate source of nutrients while improving soil fertility and crop sustainability. Further, OSA improved the soil quality and functionality as nutrient dynamics in the soil are controlled by various physical, chemical, and biological properties of soil. As a result of the changed soil climate scenario, the alternate sources of plant nutrients, either soil organic amendments alone or in combination with the recommended dose of fertilizers, resulted in considerably higher yields in different crops grown in varying agro-climatic conditions. The most efficient methods for transforming biomass from alternative organic amendment sources into plant-nutrient-rich products are composting and vermicomposting. Therefore, compost and organic matter additions have been emphasized as an essential source for improving soil quality needed for improved crop production and yield.

References

Abbott, L. K., Murphy, D. V. (2007). What is soil biological fertility?. In *Soil biological fertility* (pp. 1-15). Springer, Dordrecht.

Abiven, S., Menasseri, S., Chenu, C. (2009).The effects of organic inputs over time on soil aggregate stability–A literature analysis. *Soil Biology and Biochemistry*, 41(1), 1-12.

Alam, M. Z., Lynch, D. H., Sharifi, M., Burton, D. L., Hammermeister, A. M. (2016). The effect of green manure and organic amendments on potato yield, nitrogen uptake and soil mineral nitrogen. *Biological Agriculture & Horticulture*. 32(4), 221-236.

Al-Kaisi, Mahdi M., Jessica Gwyn Davis, R. Waskom. "Liquid manure application methods." PhD diss., Colorado State University. Libraries, 1998.

Alling, V., Hale, S. E., Martinsen, V., Mulder, J., Smebye, A., Breedveld, G. D., Cornelissen, G. (2014).The role of biochar in retaining nutrients in amended tropical soils. *Journal of Plant Nutrition and Soil Science*. 177(5), 671-680.

Andersen, J. K., Boldrin, A., Christensen, T. H., &Scheutz, C. (2011). Mass balances and life cycle inventory of home composting of organic waste. *Waste management*. 31(9-10), 1934-1942.

Ankush, A., Prakash, R., Kumar, R., Singh, V., Harender, H., Singh, V. K. (2020). Soil microbial and nutrient dynamics influenced by irrigation-induced salinity and sewage sludge incorporation in sandy–loam textured soil. *International Agrophysics*. 34(4), 451-462.

Azimzadeh, Y., Shirvani, M., Shariatmadari, H. (2014). Green manure and overlapped rhizosphere effects on Pb chemical forms in soil and plant uptake in maize/canola intercrop systems: a rhizobox study. *Soil and Sediment Contamination: An International Journal*. 23(6), 677-690.

Baligar, V. C., Fageria, N. K., He, Z. L. (2001).Nutrient use efficiency in plants. *Communications in soil science and plant analysis*. 32(7-8), 921-950.

Barthod, J., Rumpel, C., Dignac, M. F. (2018). Composting with additives to improve organic amendments.A review. *Agronomy for Sustainable Development*. 38(2), 1-23.

Batte, M. T., Forster, D. L. (2015). Old is new again: The economics of agricultural gypsum use. *Journal of ASFMRA*. 1-9.

Bauer, A., Black, A. L. (1994). Quantification of the effect of soil organic matter content on soil productivity. *Soil Science Society of America Journal*. 58(1), 185-193.

Bhagat, R. M., Verma, T. S. (1991). Impact of rice straw management on soil physical properties and wheat yield. *Soil Science*. 152(2), 108-115.

Bonanomi, G., Antignani, V., Capodilupo, M., Scala, F. (2010).Identifying the characteristics of organic soil amendments that suppress soilborne plant diseases. *Soil Biology and Biochemistry*. 42(2), 136-144.

Borrero, C., Trillas, M. I., Ordovás, J., Tello, J. C., Avilés, M. (2004). Predictive factors for the suppression of Fusarium wilt of tomato in plant growth media. *Phytopathology*. 94(10), 1094-1101.

Bouzaiane, O., Jedidi, N., Hassen, A. (2014). Microbial biomass improvement following municipal solid waste compost application in agricultural soil. *Composting for sustainable agriculture*.199-208.

Brtnicky, M., Dokulilova, T., Holatko, J., Pecina, V., Kintl, A., Latal, O., Datta, R. (2019).Long-term effects of biochar-based organic amendments on soil microbial parameters. *Agronomy*. 9(11), 747.

Carbonell, G., de Imperial, R. M., Torrijos, M., Delgado, M., Rodriguez, J. A. (2011).Effects of municipal solid waste compost and mineral fertilizer amendments on soil properties and heavy metals distribution in maize plants (Zea mays L.). *Chemosphere*. 85(10), 1614-1623.

Carter, M. R. (1991). Ninhydrin-reactive N released by the fumigation-extraction method as a measure of microbial biomass under field conditions. *Soil Biology and Biochemistry*. 23(2), 139-143.

Chander, K., Goyal, S., Mundra, M. C., Kapoor, K. K. (1997). Organic matter, microbial biomass and enzyme activity of soils under different crop rotations in the tropics. *Biology and fertility of soils*. 24(3), 306-310.

Choi, B., Lim, J. E., Sung, J. K., Jeon, W. T., Lee, S. S., Oh, S. E., Ok, Y. S. (2014). Effect of rapeseed green manure amendment on soil properties and rice productivity. *Communications in soil science and plant analysis*. 45(6), 751-764.

Darwish, O. H., Persaud, N., Martens, D. C. (1995). Effect of long-term application of animal manure on physical properties of three soils. *Plant and soil*. 176(2), 289-295.

Davies, G., Lennartsson, M. (2005).Organic vegetable production. *A complete guide. A complete guide*. Henry Doubleday Research Association, London.

Davis, J. G., Whiting, D. (2000). *Choosing a soil amendment* (Doctoral dissertation, Colorado State University. Libraries).

Diacono, M., Montemurro, F. (2011). Long-term effects of organic amendments on soil fertility. *Sustainable agriculture volume*.2, 761-786.

Donn, S., Wheatley, R. E., McKenzie, B. M., Loades, K. W., Hallett, P. D. (2014). Improved soil fertility from compost amendment increases root growth and reinforcement of surface soil on slopes. *Ecological engineering*. 71, 458-465.

Doran, J. W., Parkin, T. B. (1994). Defining and assessing soil quality. *Defining soil quality for a sustainable environment*. 35, 1-21.

Doran, J. W., Zeiss, M. R. (2000). Soil health and sustainability: managing the biotic component of soil quality. *Applied soil ecology*. 15(1), 3-11.

Duyar, H. (2014). Effect of green manure and poultry manure on yield and fruit quality in organic greenhouse tomato production. *TarimBilimleriDergisi*. 20(1), 10-18.

DWSR (2014). Marching Ahead, Directorate of Weed Science Research, Jabalpur, 1-52.

Elwell, H. A. (1986). Determination of erodibility of a subtropical clay soil: a laboratory rainfall simulator experiment. *Journal of Soil Science*. 37(2), 345-350.

Evans, M. R., Konduru, S., Stamps, R. H. (1996).Source variation in physical and chemical properties of coconut coir dust. *Hort Science*. 31(6), 965-967.

Fageria, N. K. (2007). Green manuring in crop production. *Journal of plant nutrition*. 30(5), 691-719.

Fan, F., Yu, B., Wang, B., George, T. S., Yin, H., Xu, D., Song, A. (2019). Microbial mechanisms of the contrast residue decomposition and priming effect in soils with different organic and chemical fertilization histories. *Soil Biology and Biochemistry*. 135, 213-221.

FAO (2005) Fertilizer use by crop in India, Land and Plant Nutrition Management Service, Land and Water Development Division, Food and Agriculture Organization of the United Nations, Rome, 1-45.

Filiberto, D. M., Gaunt, J. L. (2013).Practicality of biochar additions to enhance soil and crop productivity. *Agriculture*. 3(4), 715-725.

Garcıa-Gil, J. C., Ceppi, S. B., Velasco, M. I., Polo, A., Senesi, N. (2004). Long-term effects of amendment with municipal solid waste compost on the elemental and acidic functional group composition and pH-buffer capacity of soil humic acids. *Geoderma*. 121(1-2), 135-142.

Gardiano, C. G., Krzyzanowski, A. A., Abi Saab, O. J. G. (2014). Efficiency of green manure species on the population of reniform nematode. *Semina: CiênciasAgrárias*. 35(2), 719-726.

Ghosh, P. K., Sarma, U. S., Ravindranath, A. D., Radhakrishnan, S., Ghosh, P. (2007). A novel method for accelerated composting of coir pith. *Energy & fuels*. 21(2), 822-827.

Gong, W., Yan, X., Wang, J., Hu, T., Gong, Y. (2009). Long-term manure and fertilizer effects on soil organic matter fractions and microbes under a wheat–maize cropping system in northern China. *Geoderma*. 149(3-4), 318-324.

Goss, M. J., Tubeileh, A., Goorahoo, D. (2013). A review of the use of organic amendments and the risk to human health. *Advances in agronomy*. 120, 275-379.

Goyal, S., Dhull, S. K., Kapoor, K. K. (2005).Chemical and biological changes during composting of different organic wastes and assessment of compost maturity. *Bioresource technology*. 96(14), 1584-1591.

Habteselassie, M. Y., Miller, B. E., Thacker, S. G., Stark, J. M., Norton, J. M. (2006). Soil nitrogen and nutrient dynamics after repeated application of treated dairy-waste. *Soil Science Society of America Journal*. 70(4), 1328-1337.

Hao, X. H., Liu, S. L., Wu, J. S., Hu, R. G., Tong, C. L., Su, Y. Y. (2008). Effect of long-term application of inorganic fertilizer and organic amendments on soil organic matter and microbial biomass in three subtropical paddy soils. *Nutrient Cycling in Agroecosystems*. 81(1), 17-24.

Hargreaves, J. C., Adl, M. S., Warman, P. R. (2008). A review of the use of composted municipal solid waste in agriculture. *Agriculture, Ecosystems & Environment*. 123(1-3), 1-14.

Hargreaves, J. C., Adl, M. S., Warman, P. R., Rupasinghe, H. V. (2008). The effects of organic and conventional nutrient amendments on strawberry cultivation: Fruit yield and quality. *Journal of the Science of Food and Agriculture*. 88(15), 2669-2675.

Hargreaves, J., Adl, M. S., Warman, P. R., Rupasinghe, H. V. (2008). The effects of organic amendments on mineral element uptake and fruit quality of raspberries. *Plant and Soil*. 308(1), 213-226.

Heinze, S., Raupp, J., Joergensen, R. G. (2010). Effects of fertilizer and spatial heterogeneity in soil pH on microbial biomass indices in a long-term field trial of organic agriculture. *Plant and Soil*. 328(1), 203-215.

Hobson, A. M., Frederickson, J., Dise, N. B. (2005). CH_4 and N_2O from mechanically turned windrow and vermicomposting systems following in-vessel pre-treatment. *Waste Management*. 25(4), 345-352.

Indoria, A. K., Mehta, S. C., Poonia, S. R., Sharma, M. K., Panwar, B. S. (2006). Effect of sewage sludge and farmyard manure on the Ni sorption in a sandy loam soil. *Annals of Agri Bio Research*. 11(1), 15-20.

Indoria, A. K., Poonia, S. R., Sharma, K. L. (2013). Phytoextractability of Cd from soil by some oilseed species as affected by sewage sludge and farmyard manure. *Communications in soil science and plant analysis*. 44(22), 3444-3455.

Indoria, A. K., Sharma, K. L., Reddy, K. S., Srinivasarao, C., Srinivas, K., Balloli, S. S., Raju, N. S. (2018). Alternative sources of soil organic amendments for sustaining soil health and crop productivity in India–impacts, potential availability, constraints and future strategies. *Current Science*. 115, 2052-2062.

Ippolito, J. A., Spokas, K. A., Novak, J. M., Lentz, R. D., Cantrell, K. B. (2015). Biochar elemental composition and factors influencing nutrient retention. *Biochar for environmental management: Science, technology and implementation*. 139-163.

Karami, A., Homaee, M., Afzalinia, S., Ruhipour, H., Basirat, S. (2012). Organic resource management: impacts on soil aggregate stability and other soil physico-chemical properties. *Agriculture, Ecosystems & Environment*. *148*, 22-28.

Kaur, K., Kapoor, K. K., Gupta, A. P. (2005). Impact of organic manures with and without mineral fertilizers on soil chemical and biological properties under tropical conditions. *Journal of Plant Nutrition and Soil Science*. 168(1), 117-122.

Kröbel, R., Lemke, R., Campbell, C. A., Zentner, R., McConkey, B., Steppuhn, H., Wang, H. (2014). Water use efficiency of spring wheat in the semi-arid Canadian prairies: Effect of legume green manure, type of spring wheat, and cropping frequency. *Canadian Journal of Soil Science*. 94(2), 223-235.

Kumar, V., Verma, K. S. (2002). Influence of use of organic manure in combination with inorganic fertilizers on sugarcane and soil fertility. *Indian Sugar*. 52(3), 177-181.

Larkin, R. P. (2013).Green manures and plant disease management. *CAB Reviews*. 8(37), 1-10.

Larkin, R. P. (2020).Effects of Selected Soil Amendments and Mulch Type on Soil Properties and Productivity in Organic Vegetable Production. *Agronomy*. 10(6), 795.

Lim, S. F., Matu, S. U. (2015). Utilization of agro-wastes to produce biofertilizer. *International Journal of Energy and Environmental Engineering*. 6(1), 31-35.

Liu, E., Yan, C., Mei, X., He, W., Bing, S. H., Ding, L., Fan, T. (2010). Long-term effect of chemical fertilizer, straw, and manure on soil chemical and biological properties in northwest China. *Geoderma*. 158(3-4), 173-180.

Liu, T., Chen, X., Hu, F., Ran, W., Shen, Q., Li, H., Whalen, J. K. (2016). Carbon-rich organic fertilizers to increase soil biodiversity: Evidence from a meta-analysis of nematode communities. *Agriculture, ecosystems & environment. 232*, 199-207.

Mäder, P., Fliessbach, A., Dubois, D., Gunst, L., Fried, P., Niggli, U. (2002). Soil fertility and biodiversity in organic farming. *Science*. 296(5573), 1694-1697.

Mahmud, M., Abdullah, R., Yaacob, J. S. (2018).Effect of vermicompost amendment on nutritional status of sandy loam soil, growth performance, and yield of pineapple (Ananascomosus var. MD2) under field conditions. *Agronomy*. 8(9), 183.

Mapa, R. B., Kumaru, G. K. K. P. (1995). Potential of coir dust for agricultural use. *Sri Lankan J. Agric. Sci*, 32, 161-164.

Masto, R. E., Chhonkar, P. K., Singh, D., Patra, A. K. (2006). Changes in soil biological and biochemical characteristics in a long-term field trial on a sub-tropical inceptisol. *Soil biology and Biochemistry*. 38(7), 1577-1582.

Mekuria, W., Getnet, K., Noble, A. D., Hoanh, C. T., Mccartney, M., Langan, S. (2013).Economic valuation of organic and clay-based soil amendments in small-scale agriculture in Lao PDR.*Field Crops Research*.149, 379-389.

Millner, P. D. (2009). Manure management. In *The produce contamination problem* (pp. 79-104).Academic Press.

Mondal, S., Singh, R. D., Patra, A. K., Dwivedi, B. S. (2015). Changes in soil quality in response to short-term application of municipal sewage sludge in a typichaplustept under cowpea-wheat cropping system. *Environmental nanotechnology, monitoring & management*. 4, 37-41.

Montemurro, F., Maiorana, M., Convertini, G., Ferri, D. (2006). Compost organic amendments in fodder crops: effects on yield, nitrogen utilization and soil characteristics. *Compost Science & Utilization*. 14(2), 114-123.

Murugesan, A. G., Ruby, J., Paulraj, A. G., Sukumaran, N. (2005). Impact of different densities and temperature regimes on the feeding behaviour of water hyacinth weevils, Neochetinabruchiand Neochetinaeichhorniae on Eichhorniacrassipes. *Asian Journal of Microbiology, Biotechnology and Environmental Sciences*. 7(1), 73-76.

NAAS. (2012). Management of crop residues in the context of conservation agriculture. Policy Paper No. 58, National Academy of Agricultural Sciences, New Delhi., 1-12.

Nagavallemma, K. P., Wani, S. P., Lacroix, S., Padmaja, V. V., Vineela, C., Rao, M. B., Sahrawat, K. L. (2004). *Vermicomposting: Recycling wastes into valuable organic fertilizer*. Global Theme on Agroecosystems Report no. 8.

Nannipieri, P., Greco, S., Ceccanti, B. (2017). Ecological significance of the biological activity in soil. *Soil biochemistry*. 293-356.

Ndegwa, P. M., Thompson, S. A. (2001). Integrating composting and vermicomposting in the treatment and bioconversion of biosolids. *Bioresource technology*. 76(2), 107-112.

Needelman, B. A. (2013). What are Soils?. *Nature* [online] 4. Disponível:< http.

Novak, J. M., Ippolito, J. A., Lentz, R. D., Spokas, K. A., Bolster, C. H., Sistani, K., Johnson, M. G. (2016). Soil health, crop productivity, microbial transport, and mine spoil response to biochars. *BioEnergy Research*. 9(2), 454-464.

Pane, C., Celano, G., Piccolo, A., Villecco, D., Spaccini, R., Palese, A. M., Zaccardelli, M. (2015). Effects of on-farm composted tomato residues on soil biological activity and yields in a tomato cropping system. *Chemical and Biological Technologies in Agriculture*. 2(1), 1-13.

Pappu, A., Saxena, M., Asolekar, S. R. (2007).Solid wastes generation in India and their recycling potential in building materials. *Building and environment*. 42(6), 2311-2320.

Park, J. H., Lamb, D., Paneerselvam, P., Choppala, G., Bolan, N., Chung, J. W. (2011). Role of organic amendments on enhanced bioremediation of heavy metal (loid) contaminated soils. *Journal of hazardous materials*. 185(2-3), 549-574.

Pathak, H., Bhatia, A., Jain, N., Aggarwal, P. K. (2010). Greenhouse gas emission and mitigation in Indian agriculture–A review. *ING bulletins on regional assessment of reactive nitrogen, bulletin*. 19, 1-34.

Patil, G. D., Kharade, M. S., Sawale, D. D. (2014). Effect of green manures on physico-chemical properties of soil. *Journal of Agriculture Research and Technology*. 39(1), 160-163.

Patil, P. D. *Effect tank silt and fym application on soil quality and yield of soybean (glycine max l.) under inceptisol* (Doctoral dissertation, Vasantrao Naik Marathwada Krishi Vidyapeeth, Parbhani).

Pérez-Piqueres, A., Edel-Hermann, V., Alabouvette, C., Steinberg, C. (2006).Response of soil microbial communities to compost amendments. *Soil Biology and Biochemistry.* 38(3), 460-470.

Pinto, R., Brito, L. M., Coutinho, J. (2017).Organic production of horticultural crops with green manure, composted farmyard manure and organic fertiliser. *Biological Agriculture & Horticulture.* 33(4), 269-284.

Prabhu, S. R., Thomas, G. V. (2002). Biological conversion of coir pith into a value-added organic resource and its application in Agri-Horticulture: Current status, prospects and perspective. *Journal of Plantation Crops*. 30(1), 1-17.

Pretty, J., Bharucha, Z. P. (2014).Sustainable intensification in agricultural systems. *Annals of botany*. 114(8), 1571-1596.

Ramesh, P., Singh, M., Singh, A. B. (2005).Performance of macaroni (Triticum durum) and bread wheat (Triticumaestivum) varieties with organic and inorganic source of nutrients under limited irrigated conditions of vertisols. *Indian journal of agricultural science*. 75(12), 823-825.

Rasoulzadeh, A., Yaghoubi, A. (2010). Effect of cattle manure on soil physical properties on a sandy clay loam soil in North-West Iran. *Journal of Food, Agriculture & Environment*. 8(2), 976-979.

Ros, M., Hernandez, M. T., García, C. (2003).Soil microbial activity after restoration of a semiarid soil by organic amendments. *Soil Biology and Biochemistry*. 35(3), 463-469.

Roy, R. N., Finck, A., Blair, G. J., Tandon, H. L. S. (2006). Plant nutrition for food security. *A guide for integrated nutrient management.FAO Fertilizer and Plant Nutrition Bulletin*. 16, 368.

Sadh, P. K., Duhan, S., Duhan, J. S. (2018). Agro-industrial wastes and their utilization using solid state fermentation: a review. *Bioresources and Bioprocessing*. 5(1), 1-15.

Scotti, R., Bonanomi, G., Scelza, R., Zoina, A., Rao, M. A. (2015). Organic amendments as sustainable tool to recovery fertility in intensive agricultural systems. *Journal of soil science and plant nutrition*. 15(2), 333-352.

Scotti, R., Conte, P., Berns, A. E., Alonzo, G., Rao, M. A. (2013). Effect of organic amendments on the evolution of soil organic matter in soils stressed by intensive agricultural practices. *Current organic chemistry*. 17(24), 2998-3005.

Sharma, S., Dhaliwal, S. S. (2019).Effect of sewage sludge and rice straw compost on yield, micronutrient availability and soil quality under rice-wheat system. *Communications in Soil Science and Plant Analysis*. 50(16), 1943-1954.

Singh, B., Ryan, J. (2015). Managing fertilizers to enhance soil health. *International Fertilizer Industry Association (IFA), Paris, France*.

Singh, O., Shahi, U. P., Singh, V. K., Shivangi, Kunwar, S. S., Abhineet and Naresh, R. K. (2019).Effect of minimal soil disturbance and residue retention on soil carbon stocks and microbial biomass in different land use practices of *TypicUstochrept*soil: A review, *International Journal of Chemical Studies*. 6: 886-894.

Singh, V. K., Solanki, P., Ghosh, A., Pal, A. (2021). Solid Waste Management and Policies Toward Sustainable Agriculture. *Handbook of Solid Waste Management,* Springer Nature Singapore Pte Ltd. 1-22. (2021). https://doi.org/10.1007/978-981-15-7525-9_27-1.

Solanki, P., Debnath, P. Role of biosolids in sustainable development. *Agrotechnology*. 2(4), 220.(2014) https://doi.org/10.4172/2168-9881.S1.008.

Srinivasarao, C., Gopinath, K. A., Venkatesh, G., Dubey, A. K., Wakudkar, H., Purakayastha, T. J., Sikka, A. K. (2013). Use of biochar for soil health management and greenhouse gas mitigation in India: Potential and constraints. *Central Research Institute for Dryland Agriculture. Hyderabad, Andhra Pradesh*.

Srivastava, V., De Araujo, A. S. F., Vaish, B., Bartelt-Hunt, S., Singh, P., Singh, R. P. (2016). Biological response of using municipal solid waste compost in agriculture as fertilizer supplement. *Reviews in Environmental Science and Bio/Technology*. 15(4), 677-696.

Sudhakar, G. (2000). Investigation to Identify Crop Wastes/Low Land Weeds as Alternate Source to Organics to Sustain the Productivity of Rice Based System (Doctoral dissertation, Tamil Nadu Agricultural University; Coimbatore).

Szczech, M., Smolińska, U. (2001). Comparison of suppressiveness of vermicomposts produced from animal manures and sewage sludge against Phytophthoranicotianae Breda de Haan var. nicotianae. *Journal of Phytopathology*. 149(2), 77-82.

Tajbakhsh, J., Goltapeh, E. M., Varma, A. (2011). Vermicompost as a biological soil amendment.In *Biology of Earthworms* (pp 215-228).Springer, Berlin, Heidelberg.

Tian, G., Kang, B. T., Brussaard, L. (1992).Biological effects of plant residues with contrasting chemical compositions under humid tropical conditions—decomposition and nutrient release. *Soil biology and Biochemistry*. 24(10), 1051-1060.

Tilman, D., Cassman, K. G., Matson, P. A., Naylor, R., Polasky, S. (2002). Agricultural sustainability and intensive production practices. *Nature*. 418(6898), 671-677.

Varshney, J. G., Sushilkumar, M. J. (2008). Current status of aquatic weeds and their management in India. In *Proceedings of Taal 2007: the 12th world lake conference* (pp. 1039-1045).

Weber, J., Kocowicz, A., Bekier, J., Jamroz, E., Tyszka, R., Debicka, M., Kordas, L. (2014).The effect of a sandy soil amendment with municipal solid waste (MSW) compost on nitrogen uptake efficiency by plants. *European journal of Agronomy*. 54, 54-60.

Ye, X., Liu, H., Li, Z., Wang, Y., Wang, Y., Wang, H., Liu, G. (2014).Effects of green manure continuous application on soil microbial biomass and enzyme activity. *Journal of Plant Nutrition*. 37(4), 498-508.

Zaccardelli, M., De Nicola, F., Villecco, D., Scotti, R. (2013).The development and suppressive activity of soil microbial communities under compost amendment. *Journal of soil science and plant nutrition*. 13(3), 730-742.

Zheng, H., Ma, W., Li, G. (2020). Adoption of organic soil amendments and its impact on farm performance: evidence from wheat farmers in China. *Australian Journal of Agricultural and Resource Economics*. 65(2), 367-390.

Zimmer, D., Kruse, J., Siebers, N., Panten, K., Oelschläger, C., Warkentin, M., Leinweber, P. (2018). Bone char vs. S-enriched bone char: Multi-method characterization of bone chars and their transformation in soil. *Science of the Total Environment*. 643, 145-156.

Chapter 16

Fluorescent Pseudomonades: Diversity, Abundance and Potential Biotechnological Applications in Agriculture

Mudasser Ahmed Khan[1,*], Samiran Gangopadhyay[2], Hanuman Singh Jatav[1], Ravi Kumar Meena[3], Sanjay Kumar Attar[1], Suresh Kumar[5], Ramu Meena[4], Kailash Chandra[1], Subhita Kumawat[1], Mujahid Khan[4], Champa Lal Khatik[4], Kailash Chand Verma[4], Subhash Chandra Mahala[4] and Mukesh Nitharwal[1]

[1]College of Agriculture, Fatehpur Shekhawati (SKNAU, Jobner) Rajasthan, India
[2]Swami Keshvanand Rajathan Agricultural University, Bikaner, India
[3]College of Agriculture, Baseri (SKNAU, Jobner), Rajasthan, India
[4]Agriculture Research Station, Fatehpur Shekhawati (SKNAU, Jobner), Rajasthan, India
[5]Agriculture Research Station, Navgaon, Alwar (SKNAU, Jobner), Rajasthan, India

Abstract

Fluorescent pseudomonades are considered important bioagents in agriculture due to their rhizospheric competence, biocontrol abilities and biofertilizing properties. These bacteria exhibit profound diversity due to their significant genotypic, phenotypic, physiological, biological and functional variability. Due to these diversities, they are capable of degrading a number of organic substances *viz.* aromatic, halogenated compounds and many organic industrial effluents. Further, they can produce a variety of antagonists, plant hormones and other biochemical metabolites. Phenotypically, they have been divided into biotypes and biovars which is reflected in their genotypic structures including their chromosomes and other genetic materials. They are the most abundant bacteria which inhabit most of the niches on the earth *viz.* terrestrial, marine, freshwater and glaciers.

Recently, biotechnological interventions are being increasingly used in the development of new strains of these bacteria and to understand their interaction in rhizosphere. With the advent of these technologies, bacterial strains with enhanced plant growth promoting abilities and disease resistance inducing capabilities are being developed. These novel strains are having increased antimicrobial production capabilities with higher rhizospheric competence and also more compatible to pesticides used in agriculture. Some of the genes of these bacteria are also used for development of disease

*Corresponding Author's Email: makhan.ppath@sknau.ac.in.

In: Ecosystem Services
Editor: Hanuman Singh Jatav
ISBN: 978-1-68507-614-6

resistant GMO crops. The molecular technologies like rRNA–DNA hybridization based on 16S rRNA, REP, ERIC and BOX-PCR help in the diversity study and classification of this bacterium. These biotechnological applications help in the transformation of these fluorescent pseudomonades into better bioagents which are more assessable and reliable to the farmers as tool in sustainable agriculture.

Keywords: bioagent, biotechnology, diversity, fluorescent pseudomonas

1. Introduction

The antagonistic microorganisms are gaining foremost importance in plant disease management playing important role in agricultural systems. These bioagents show multifunctional plant growth promoting (PGP) characters like production of antagonistic substances, growth hormones, siderophores, inducing disease resistance and making micronutrients available to the plants. Among these microbes, fluorescent pseudomonas is the most important group of bacteria in agriculture as it aggressively colonizes the plant roots due to its higher rhizosphere competence and biocontrol abilities.

Pseudomonas was first described by Migula in late 19th century as rod-shaped, Gram-negative and polar-flagellated bacteria. Pseudomonas word is derived from the Greek words pseudēs (Greek: - false) and monas (Greek: - a single unit) and word fluorescens is given due to the fluorescent pigment called pyoverdin (siderophore) secreted by its cell. It is a ubiquitous bacteria having vast number of species(Cox and Adams 1985; Naik et al. 2008). These fluorescent bacteria produce antagonistic substances, induce disease systemic resistance (O'Sullivan and O'Gara 1992, Bano and Musarrat 2004, Naik and Sakthivel 2006; Ramamoorthy et al. 2001, Renwick et al. 1991, Zahnder et al. 2001). Some pseudomonades are also able to degrade soil pollutants (Ravindra Naik and Sakthivel 2006, Bano and Musarrat 2003) and manage weeds (Charudattan R 1991, Gurusiddaiah and Gealy 1994, Kremer et al. 1990). These bacteria are also capable to produce plant growth hormones and antifungal enzymes (O'Sullivan and O'Gara F 1992, Ramamoorthy et al. 2001, Cattelan et al. 1999, Glick et al. 1995, Sunish Kumar et al. 2005). These abilities make fluorescent pseudomonas an important bioagent in agriculture.

The importance of this microbe increased in today's era due to indiscriminate use of pesticides in modern agriculture which is very hazardous to the environment and lethal towards the humans as well as to the beneficial microorganisms. This illegitimate use of these hazardous chemicals also develops resistance in the target pest or pathogens. These pseudomonades have very appealing potential and natural abilities to replace these pesticides in agriculture (Bloemberg and Lugtenburg 2001, deLorenzo et al. 2001, Haas et al. 2001, Morrissey et al. 2002, Vance 2001, Walsh et al. 2001). But there are some limitations in utilizing these bacteria like rhizospheric competence, variation in metabolite production, different inoculants delivery systems and variable performance in field. These factors obstruct the utilization of these fluorescent bioagents in place of conventional chemical pesticides. With the advancement of biotechnological applications, new strategies has been developed utilizing genetic modifications to overcome these obstructions. These latest technologies can help to develop strains with increased resistance towards pathogens, herbicides and stress and make them ecologically more adaptable with higher antagonistic activity, vaccine production and plant molecular farming ability (Mark et al. 2006, Mark et al. 2002).

Considering the multiple utility of fluorescent pseudomonades, this chapter covers its diversity, abundance and potential applicability in biotechnological techniques. It will be useful in designing environmental and farmer's friendly strategies for sustainable and organic agriculture.

2. Diversity

The *Pseudomonas* genus is the most diverse bacterial group on the earth and its genetic and physiological variability make it ubiquitous in its distribution (Spiers et al. 2000). Migula in 1894, first described the genus *Pseudomonas* based on morphological characters of its members (Migula, 1900). This genus belongs to family Pseudomonadaceae and they are having significant genetic, physical, biological, functional and other diversities (Gomila et. al. 2015).

Due to these diversities, Pseudomonades developed many qualities like they are resistant to higher and lower temperatures, antibiotics, become adaptable to oxygen, moisture, nutrient and oxygen stress etc. These bacteria can act as plant growth and plant health promoters as well as a great bioagent which is able to control many soil-borne diseases. These bacteria are capable of utilizing different organic compounds as a carbon and energy source and also having ability to degrade many toxic compounds (Novik et al. 2015).The diversity of this bacterium will be discussed as follows.

2.1. Species Level Diversity

The Gram-negative Pseudomonas classified in the phylum-Proteobacteria, class-Gamma proteobacteria, order-Pseudomonadales, family-Pseudomonadaceae. These bacteria are divided into three major groups: (1) phytopathogenic cytochrome *c* oxidase-positive species, *viz. Pseudomonas tolaasii, P. cichorii* and *P. marginalis*; (2) non-necrogenic, non-phytopathogenic strains, *viz. P. fluorescens, P. aeruginosa, P. aureofaciens, P. chlororaphis* and *P. putida* type species and (3) necrogenic, phytopathogenic fluorescent *Pseudomonas* spp. without cytochrome *c* oxidase, *viz. P. viridiflava* and *P. syringae* (Palleroni, 1993). *Pseudomonas* genus contains the largest number of species of bacteria found on earth. As per the approved lists of bacterial names, at present 216 species with 18 sub species have beenreported and the species are continuously increasing (Tindall et al. 2006; Özen and Ussery 2012; Sarker et al. 2014; Novik et al. 2015; Gomila et al. 2015).

2.2. Functional Diversity

The florescent pseudomonades have variety of functions which affect its biocontrol efficacy and other plant growth promoting activities. These functions vary with strains which are studied by many researchers. Khan and Gangopadhyay (2012) tested 4 isolates of *Pseudomonas fluorescens in vitro* as well as *in vivo* against *F. oxysporum* f.sp. *ciceri* and found that isolate *P. fluorescens* (Pf-I) is most effective in controlling the fungus amongst the four strains (Plate 1). Naik et al. (2008) reported that out of 80 fluorescent pseudomonad strains, 33 strains displayed antagonistic effect against the pathogenic fungus and inhibited the pathogen mycelium which ranges from 10-36 mm diameter in 50 mm Petriplate.

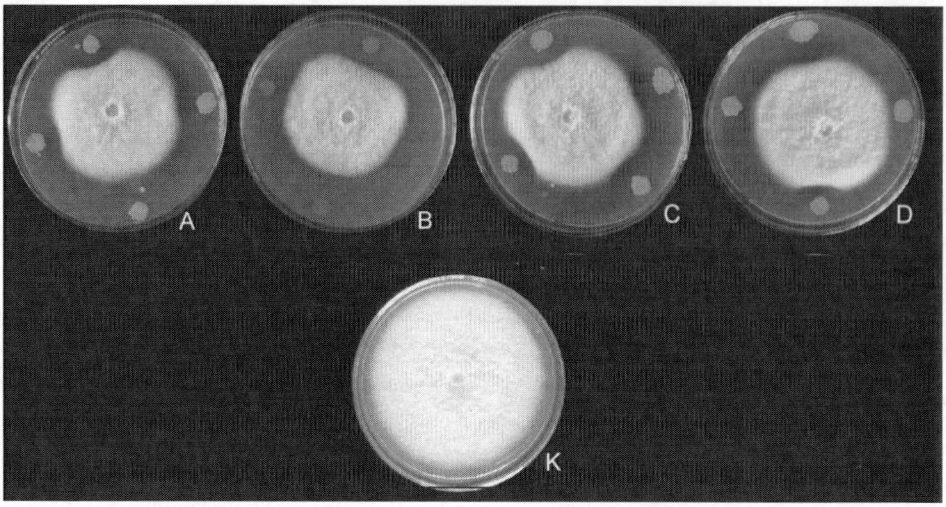

Plate 1..*In vitro* testing of *P. fluorescens* isolates against FOC mycelium (Khan 2012).

In another experiment, which included 60 fluorescent pseudomonades, three levels of diversities were observed. Under *in vitro* fungal growth inhibition assay, only 26% isolates of bacteria from bulk soil inhibited the phytopathogenic fungus. In contrast to soil zone isolates, all rhizospheric isolates inhibited the fungus. In ectomycorrhizal fungal assay test, the bacterial isolates selectively stimulate the ectomycorrhiza based on the fungus involved and it also depends on the ecological niche of the bacteria. There inhibition pattern on effect is also different on Douglas fir–*L. bicolor* mycorrhizal establishment. These tests suggested that fluorescent pseudomonades isolated from different soil zone are very diverse in their response towards pathogenic and mycorrhizal fungi (Frey-Klett et al. 2005).

Earlier, it was thought that *Pseudomonas* species (sensu stricto) cannot fix nitrogen but now it has been reported that some strains are having nitrogen fixation abilities and their nitrogen fixing genes closely resemble that of *A. vinelandii* (Rediers et al. 2004; Yan et al. 2008; Young and Park 2007).In another experiments regarding influence of different rhizosphere on these bacteria, it was observed that different rhizosphere have variable effect on these bacteria. It was observed that plants affect community of these bacteria selectively and the intensity of selection depends on plants. The analysis also suggested the major population in flax is *Pseudomonas putida*bv. A and *P. fluorescens* bv. II in tomato. Some experiments also showed that the *Pseudomonas* population in the rhizosphere is more affected by young plants as compared to mature plants (Lemanceauet al.1995; Gomeset al.2001).

2.3. Physiological Diversity

The group fluorescent pseudomonades has wide range of physiological diversity which has been proved from early biochemical studies of many scientists. Due to these diversities, it can utilize aromatic derivatives, halogenated compounds andhas remarkable capacity to degrade a wide range of substrates including aromatic compounds, halogenated derivatives and intractable residues. Further, these bacteria are also capable to produce great range of antagonistic compounds, plant hormones and other biochemical metabolites. These physiological diversities make them adaptable to survive in adverse environments.These bacteria grow most favourably at 28°C temperature, but some species were able to grow at 4°C or 41°C. It cannot grow at below pH 4.5 (Palleroni 2005). The physiological diversity is mainly described in two sections i.e., nutritional diversity and metabolic diversity:

2.3.1. Nutritional Diversity

Fluorescent pseudomonades have elevated metabolic versatility due to which it can degrade a variety of organic compounds. These bacteria are mostly aerobic, but some of them like *P. aeruginosa, P. stutzeri* and some biovars of *P. fluorescens* can also utilize nitrate as alternate electron acceptor to carry out denitrification process. *P. chloritidismutans* also uses chlorate (ClO_3-) as an electron acceptor (Wolterink et al. 2002). They prefer organic acids over other complex organic compounds. Most of the species lacks 6-phosphofructokinase and carry out incomplete glycolysis. Therefore, they use Entner-Doudoroff pathway for dissimilation of sugars and organic acids (Moore et al. 2006; Palleroni 2005; Lessie and Phibbs 1984).

Pseudomonas can be grown with only carbon source and without nitrogen source like ammonium or nitrate on minimal media as they do not need any specific growth elements for their growth. Some phytopathogenic strains like P. syringae grow very slow as compared to saprophytes, but through addition of additional factors (yeast extract, peptones) the growth of these pathogenic bacteria can be enhanced. The Pseudomonas has characteristic physiology as it cannot accumulate polyhydroxy butyrate but while growing on gluconate or alkanes, it will accumulate polyhydroxy alkanoates (Palleroni 2005).

It has been found that *Pseudomonas* has total 11 central pathways for degradation of substrates in which many sub pathways converge. These sub pathways convert complex compounds into a few simple intermediates compounds like dihydroxy aromatic acids or dihydroxy benzenes.These compounds further degrad into TCA (tricarboxylic acid cycle) by its eleven main pathways. These different pathways have strain specific ability which enable a strain to degrade a specific compound. Availability of different pathways in pseudomonades for nutrients, make them able to degrade a variety of organic compounds which may be natural derivatives (phenol, naphthalene, toluene and styrene) or industrial derivatives (polychlorobiphenyls, dioxins and nitrotoluenes) which are toxic to other organisms. The availability of different mechanism conforms that this bacterium can inhabit in a variety of niches. (Jiménez et al. 2010).

Naik et al. (2008) tested 80 strains of fluorescent pseudomonades for their ability to utilize different organic compounds. These strains showed higher degree of variability while utilizing, L-arabinose, lactose, xylose, ribose, fructose, D-arabinose, sorbose, trehalose, malonate, glycerol, α-methyl-D-mannoside, esculin, xylitol, glucosamine, sorbitol, mannitol, sorbitol, mannitol and adonitol. They also found that none of the strain utilized sucrose, maltose, inositol, dulcitol, xylitol, α-methyl-D-gluconate, rhamnose, cellobiose, melazitose, salicin and ONPG. On the basis of this study these strains were divided into three major phenons *viz.* Phenons I, II and III. Liao and Wells (1978) also found that fluorescent *Pseudomonas* of the same physiological groups could be divided into several subgroups according to their ability to use organic compounds.

2.3.2. Metabolic Diversity

These bacteria are able to produce a whole spectrum of bioactive molecules i.e., antibiotics, enzymes, siderophores, volatile compounds and growth hormones. Siderophores are iron chelating agent which binds irons with high affinity under iron stress conditions e.g., fluorescein (green-yellow pyoverdine pigment). *Pseudomonas* produces a variety of pigments which are important in its diagnosis *Pseudomonas.* They bind Fe^{3+} ions very tight and these ferripyoverdine transported to bacterial cells and creat iron starvation (Meyer JM and Abdallah 1978; Meyer and Hornsperger 1978; Bultreys et al. 2001). Each genomic group produces specific pyoverdine. Other siderophore pigments are chlororaphin (*P. chlororaphis*, green color) pyomelanin (*P. aeruginosa*, brown/black color), pyorubin (*P. aeruginosa*, red color), carotenoid (*P. mendocina*) and pyocyanin (*P. aeruginosa*, blue color) (Palleroni 2005; Blondel-Hill et al. 2007). During a diversity study of 80 isolates of fluorescent pseudomonades, it was reported that all the stains produced siderophores. Among these 39 strains produced IAA growth promoting hormone and 13 strains produced ACC deaminase. Out of 80 isolates, 22 isolates having proteolytic ability, seven isolates produced chitinase, 33 isolates showed pectinase activity and 25 showed cellulose activity. This experiment conform the variability in metabolite productions by these bacteria (Naik et al. 2008).

In another experiment on 60 isolates of *P. fluorescens*, it was found that majority of isolates from mycosphere solubilize tri-calcium phosphate but of the most bulk soil isolates did not solubilize it. There siderophore production ability also depends on their origin. Mycospheric isolates have higher frequency in siderophore production than bulk soil isolates. Further, their ability to produce IAA was not significantly different on the basis of their origin. But again, their HCN production ability depends on isolates origin as most of the bulk soil isolates can synthesize HCN but most of the mycorrhizal isolates were not able synthesize it (Frey-Klett et al. 2005; Khan and Zaidi 2002).

These bacteria also show diversity in production of salicylic acid, pyoluteorin (PLT), phenazine-1-carboxylic acid (PCA), 2, 4-diacetylphloroglucinol (DAPG), HCN and activity of ACC deaminase and chitinase activity (Sarma et al. 2014). Pseudomonades strains which are having denitrification ability and pyverdine production genes (*narG* and *nosZ*) are having better survival ability in rhizosphere (Delormeet al. 2003).

Subramanian and Satyan (2014) conducted an intensive study on characterization of 144 fluorescent pseudomonades strains isolated from rhizospheric soil. These strains were tested for their ability for production of urease, amylase, oxidase, gelatinase, caseinase, catalase and their efficacy to reduce sulfates and glucose. They observed a vast diversity regarding production of these metabolites among these strains. These strains were also evaluated for different PGP traits *viz.* production of phytohormones, siderphores, ammonia and hydrogen cyanide, biofilm formation (motility), phosphate solubilization, promotion of root augmentation and germination (Figure 1). It was found that these strains show diversity regarding these traits (Subramanian and Satyan 2014).

Genus Pseudomonas showed different metabolic pathways for aromatic compounds: homogentisate (phenylalanine, tyrosine), gentisate (salicylate, 3-hydroxybenzoate, m-cresol), phenylacetyl-CoA (phenylethylamine, nicotinate (nicotinic acid), phenylethanol, styrene, tropate) and homoprotocatechuate (4-hydroxyphenylacetate) etc. (Jiménez et al. 2010).

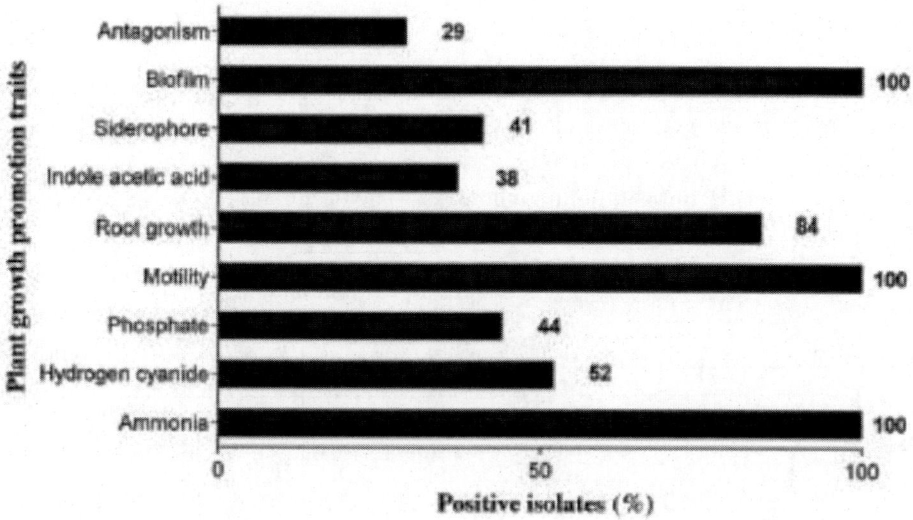

Figure 1. Percentage of different isolates percentage in plant growth promoting traits (Frey-Klett et al. 2005).

2.4. Phenotypic Diversity

Pseudomonads are non spore former, straight or curved rod shaped having single or many polar flagella for motility. Like other diversities, *Pseudomonas* also has varied phenotypes. Several species like *P. stutzeri* and *P. mendocina* also having lateral flagella, in addition to their polar flagella. These flagella also have taxonomic importance. Some bacteria like *P. aeruginosa*, *P. alcaligenes* and *P. mendocina* were characterized on the basis of their single flagellum (Shinoda and Okamoto 1977; Moore et al.2006). Species like *P. alcaligenes, P. syringae* and *P. aeruginosa* also form sex pili which are important in formation of biofilms on substrate which ultimately help to survive under unfavourable conditions (Novik et al. 2015; Woods et al. 1980; Doig et al. 1988; Bradley 1980; O'Toole and Kolter 1998; Ghafoor 2011).

Fluorescent pseudomonas has been phenotypically characterized and subdivided into biotypes, biovars A and B. *P. fluorescens* is also further divided into 7 biotypes *viz*., A, B, C, D and F also called biovars (bv. I to V). Biotypes D and E were characterized were as *P. chlororophis* and *P. aureofaciens*, respectively which later on clustered as the *P. chlororaphis* (Jonhson and Palleroni, 1989). *P. fluorescens* bv. V contains strains that cannot be further classified due to lack in differential characters. This shows that these bacteria have highly phenotypic heterogenecity (Sorensen et al. 1991; Janse et al. 1992; Lemanceau et al. 1995; Grimont et al. 1996; Latour et al. 1996). Mulyukin et al. 2008 studied Cyst like resting cells (CRC) of the *Pseudomonas* bacteria which are dormant forms and resistant to adverse conditions and the result showed that there is interspecies diversity in its dormant forms which also add to its ability to survive in adverse conditions.

2.5. Genotypic Diversity

Intraspecific diversity is not limited to phenotypical traits but all diversities of an organism are governed mainly by its genome architecture (of both chromosomes and accessory genetic elements). Many researchers found diversity in restriction fragment polymorphism between the strains of a species and even within the same strain. The bacterial genome size varies from 3.7 Mbp to 7.1 Mbp (Rainey et al. 1994; Ginard et al. 1997; Grothues et al. 1988).

Table 1. Genome size of different *Pseudomonas* species

S. No.	Bacteria	Genome size (Mbp)	Reference
1	*Pseudomonas stutzeri*	3.7	Ginard et al. 1997
2	*Pseudomonas aeruginosa*	7.1	Schmidt et al., 1996
3	*Pseudomonas fluorescens SBW25*	6.7	Rainey & Bailey, 1996
4	*P. syringae pv. phaseolicola*	5.6	De Ita et al., 1998
5	*P. stutzeri*	3.7 to 4.7	Ginard et al. 1997
6	*P. aeruginosa*	5.2 to 7.1	Schmidt et al., 1996

Adapted from Spiers et al. 2000.

Naik et al. 2008 also analysed same 80 strains of fluorescent pseudomonad bacteria for their genetic diversity using bacterial repetitive BOX element polymerase chain reaction (BOX-PCR). BOX PCR analysis divided these strains into three genomic clusters and 26 distinct BOX profiles at 80% similarity strains. The 16S rRNA gene amplification, sequencing and phylogenetic tree analysis generated three 3 major clusters. Among these strains, cluster I had 58 strains, cluster II had 8 strains and cluster III had 14 strains. Their results revealed a high degree of genetic diversity among these bacteria.

Frey-Klett et al. 2005 studied the genotypic fingerprinting of 60 *P. fluorescens* isolates from bulk soil and mycorrhiza which belongs to five 16S rDNA genotypic groups (Laguerre et al. 1994; Frey et al., 1997). They found two genotypic groups 22 and 23 in bulk soil isolates but in mycorrhiza isolates only one additional group was detected.

Naik et al. 2008 carried out genotypic analysis by BOX of 95 isolates from banana rhizospheric soil. This study divided these bacteria into three clusters based on genetic diversity at similarity level of 50% and also 62 BOX profiles. On the basis of 16S rRNA analysis and subsequent analysis of phylogenetic tree these islolates were identified as *P. plecoglossicida, P. monteilii, P. fulva, P. fluorescens, P. alcaligenes, P. pseudoalcaligenes, P. mosselii* and *P. aeruginosa.*

On the basis of molecular phylogenetic analysis of 120 strains of fluorescent pseudomonads, Sarma et al. 2014 found 35 isolates closest (GGRJ86 to GGRJ120). All these strains were found closely related to *P. aeruginosa*. Rest of 85 strains were analysed through 16S rRNA sequence and compared against NCBI nucleotide database which confirmed presence of 23 species of pseudomonades. Kwon et al. 2005 compared the diversity of 43 fluorescent pseudomonades strains along with 53 culturable fluorescent pseudomonades from soils using 16S ribosomal DNA sequences. Ten different clusters that included two phylogenetic groups appeared in this study which were not represented by previously described type strains and the authors confirmed that soil management practices are responsible for their variability. Moore et al. 1996 grouped *Pseudomonas* into 2 distinct intrageneric divisions i.e., "*P. aeruginosa* intrageneric cluster" and the "*P. fluorescens* intrageneric cluster" including four (*P. mendocina, P. flavescens, P. resinovorans, P. aeruginosa*) and five (*P. fluorescens,*

P. syringae, P. cichorii, P. putida, P. agarici) lineages, respectively, on the basis of 16S rRNA sequencing.

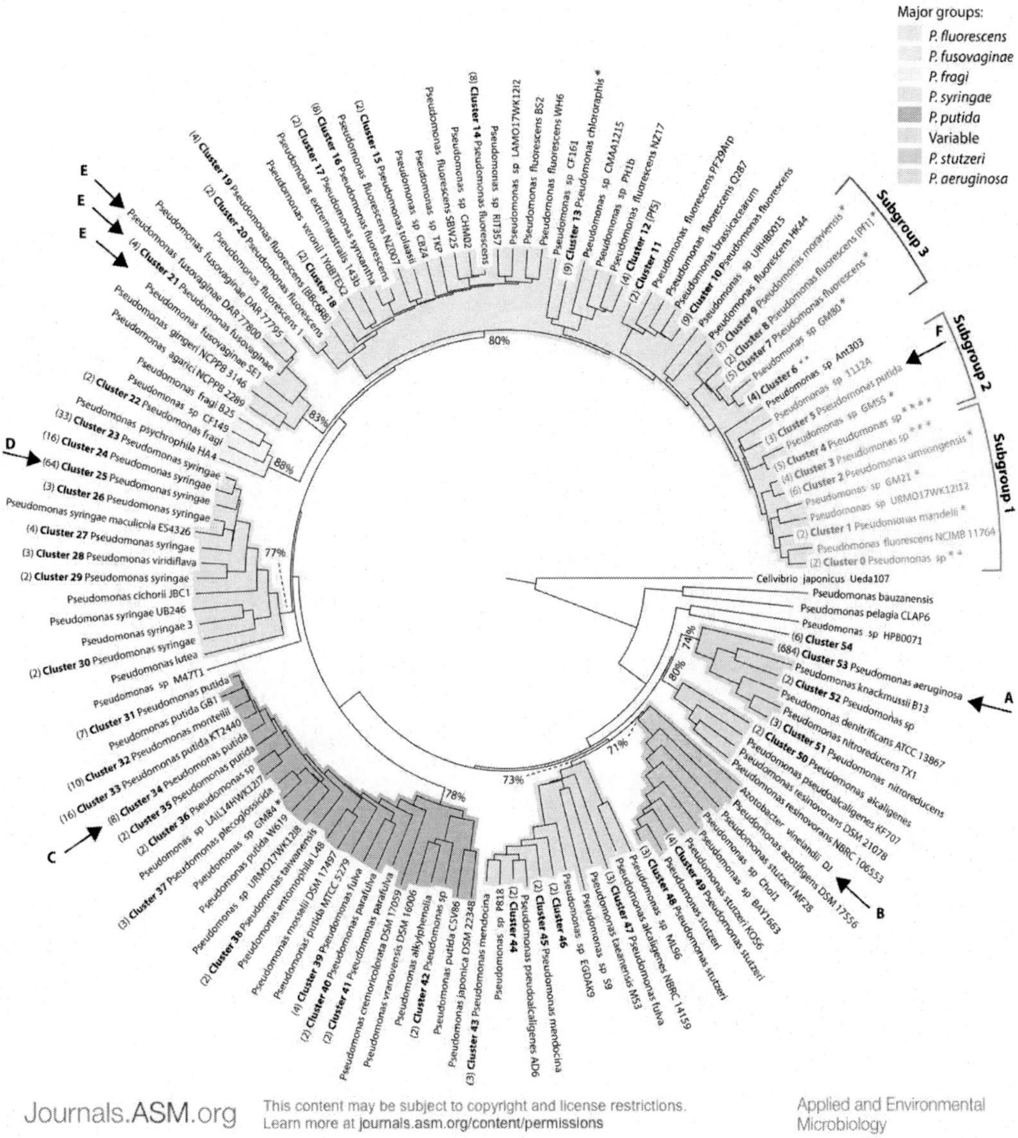

Figure 2. AAI-based tree of 1,073 *Pseudomonas* genomes with collapsed branches for more than one genome per genomic cluster. (Jun et al. 2016).

3. Abundance

The fluorescent pseudomonades inhabit various ecological niches exhibiting great adaptation abilities in different fluctuating environments with versatile metabolic activity. Due to their diverse genetic and physiological adaptability they are abundantly found in all environmental conditions *viz.*, terrestrial, freshwater and marine (Silby et al. 2011 and Spiers et al. 2000).The elevated resistance of these bacteria make them ubiquitous distributed in water and soil, as well as PGPR (plant growth-promoting rhizobacteria), plant and animal pathogens (Figure 3). These bacteria can degrade various organic compounds hence they are able to colonize habitats with limited nutrition (Novik et al. 2015). It has been reported that *Pseudomonas syringae* is found in the rain, snow, lakes, plants as well as in clouds as an ice nucleus in the water cycle. Owing to this abundance, the *Pseudomonas* genus was first characterized hundred years ago and with the passage of time, its taxonomy has been revised by various prominent workers (Morris et al. 2008; Schleifer 2009; Özen and Ussery 2012).

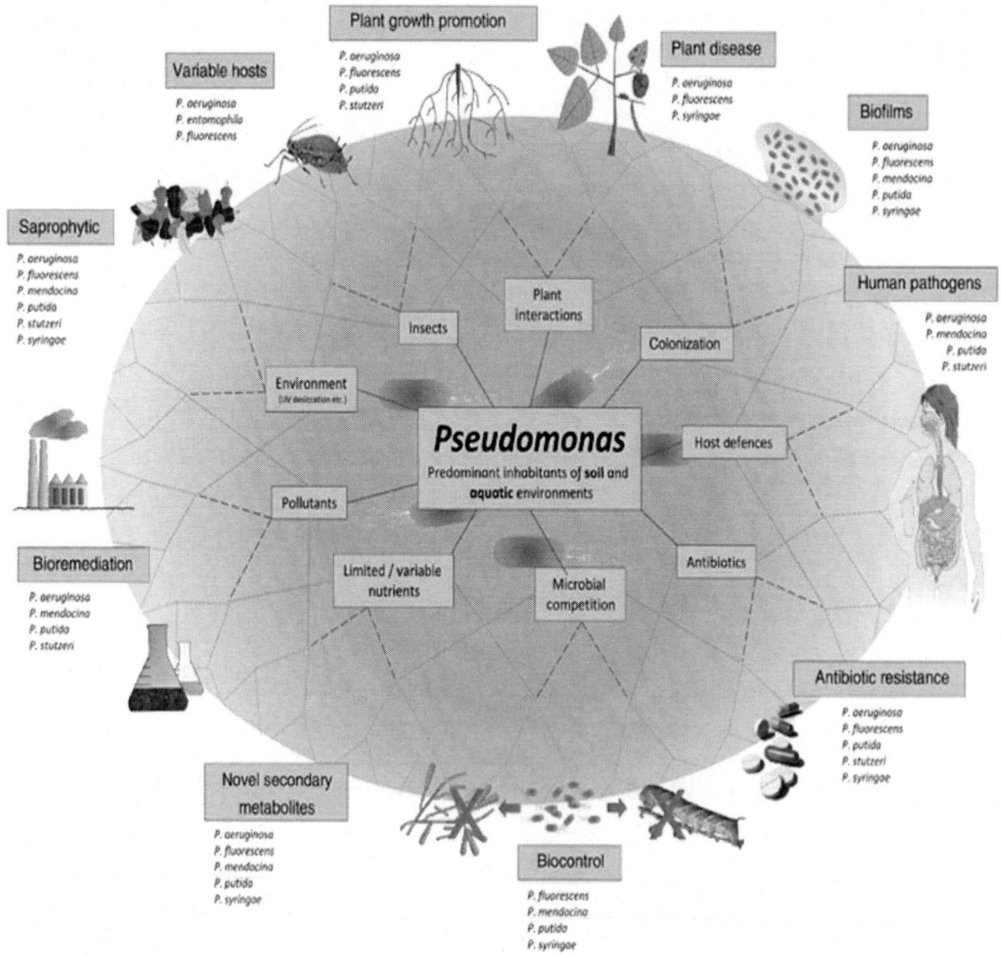

Figure 3. The functional and environmental range of *Pseudomonas* spp. (Silby et al. 2011). Unless provided in the caption above, the following copyright applies to the content of this slide: © 2011 Federation of European Microbiological Societies.

In the recent years, many studies were conducted to decipher the abundance and diversity of fluorescent pseudomonades and the main reasons for its abundance are: [1] it has higher rhizosphere competence as it can aggressively colonize endo and ectorhizopshere (Hofte 2010; Weller 1988); [2] it can produce antagonistic secondary metabolites which inhibit other microbes (Haas and Defago 2005; Salman 2010) and [3] and fluorescent pseudomonades produce different phenolics, polyketides, phenazines, pyrrole-type compounds and siderophore which act synergistically against the pathogenic microbes (Pfender et al.1993, Pierson and Thomashow 1992).

Yadav et al. (2017) also mentioned that many *Pseudomonas* spp. are psychrophilic as well as psychrotolerant e.g., *Pseudomonas aeruginosa*, *P. baetica*, *P. extremaustralis*, *P. antarctica*, *P. cedrina*, *P. costantinii*, *P. corrugate*, *P. deceptionensis*, *P. gessardii*, *P. azotoformans*, *P. extremorientalis*, *P. fluorescens*, *P. fragi*, *P. frederiksbergensis*, *P. geniculata*, *P. graminis*, *P. kilonensis* and *P. jessani*. In another rhizospheric study, it was found that P. flourescens is most abundant in soil. Khan (2012) reported that while isolating various soil microbes viz., *Pseudomonas fluorescens*, *Trichoderma* spp., *Bacillus subtilis*, actinomycetes, Sterptomyces spp. and Fusarium oxysporum f. sp. ciceri the dilution factor was 10^{14},10^{6}, 10^{6}, 10^{3}, 10^{6} and 10^{5} respectively. The dilution factor 10^{14} shows the many fold abundance of this bacterium in rhizosphere as compared to other soil microbes.

4. Potential Biotechnological Applications in Agriculture

Currently, the fluorescent pseudomonas has become the attractive object for biotechnological approach in agriculture. During last few decades, there has been significant advancement in biotechnological applications. These technologies are applied in fluorescent pseudomonas for the development of new strains with enhanced biocontrol and biofertilizing abilities which will help in sustainable agriculture. Recent genomic level studies on biocontrol mechanism and rhizospheric interaction have developed new tools for efficient utilization of these microbes which will help to exploit their plant-beneficial properties (Wiyono et al. 2008; Couillerot et al. 2009). The major biotechnological applications used for the development of fluorescent pseudomonas are discussed below.

4.1. Application in Diversity Study

The genus *Pseudomonas* is the most diverse group which has approx. 216 species with 18 subspecies and their number is constantly increasing (Annonymous, 2015). Earlier these classifications were based on phenotypic characters. From 1960s, the diversity studies were started on nucleic acid similarity like DNA-DNA hybridization. These studies showed great genetic diversity in its species. rRNA–DNA hybridization divided the *Pseudomonas* into five groups (Palleroni et al. 1978). With the advancement in molecular techniques, the use of 16S rRNA sequencing is mostly utilized for the classification and identification of Pseudomonas. Moore et al. 1996 reported that 6% sequence of 16S rRNA govern around 44% of the total genetic variability of *Pseudomonas* species and on the basis of this sequence grouped *Pseudomonas* into 2 distinct intrageneric divisions i.e., "*P. aeruginosa* intrageneric cluster" and the "*P. fluorescens* intrageneric cluster" which include fourand five lineages, respectively.

The 16S rRNA sequence having 1492 nucleotide positions containing 148 variable positions and out of which 65 nucleotide positions are in three hypervariable (hv) regions which contribute about 44% of the total genetic variability of these bacteria. The hv1 regions of 16S rRNA have highest nucleotide sequence variability in 16S rRNAs of bacteria across the phylogenetic spectrum. These sequences are useful for differentiation of *Pseudomonas* and its intrageneric lineages. These hypervariable regions of 16S rRNAs are also utilized to recognize new strains and analysis of natural occurring indigenous isolates (Moore et al., 1996; Gutell et al., 1985; Woese, 1987). Godfrey and Marshall (2002) used primers on hv regions of 16S rRNAs to study the diversity of Pseudomonades isolates from different parts of the world. They found that by this method they were able to identify different *Pseudomonas* species from the soil from all over the world. But there are some limitation in this technique as it cannot separate different gene on the basis of gene size because size of the genes are constant (García-Martínez et al., 1999). Through amplifying inter specific regions of 16S and (ITS1) rRNA gene, these limitations can be overcome as these regions are variable in size (Gürtler and Stanisich, 1996). Now with development of new primers, both ITS1 and major part of 16S rDNA can amplified simultaneously (Locatelli et al. 2002).

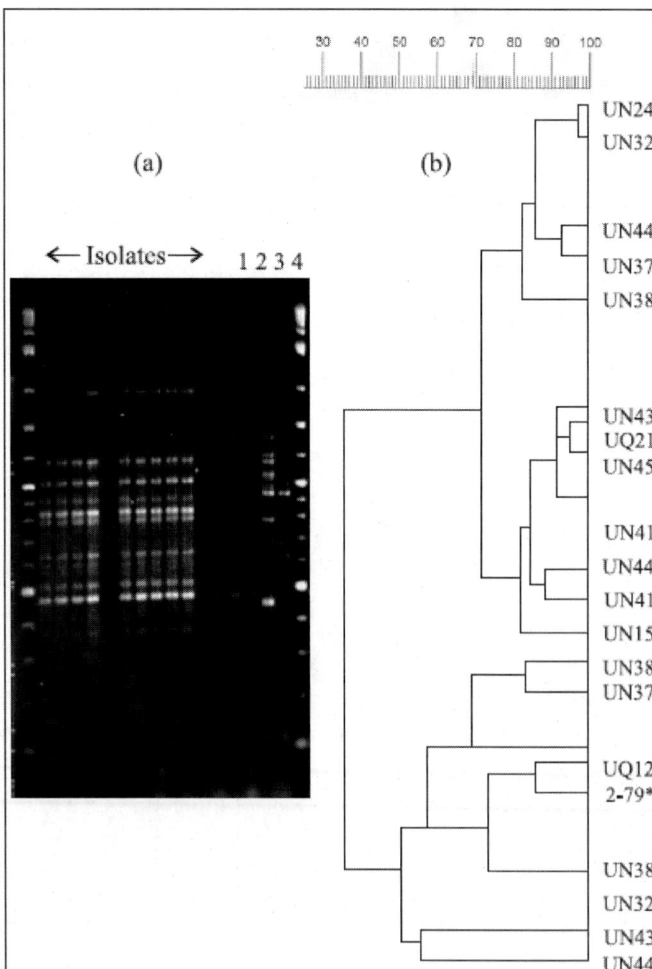

Figure 4. Genetic characterization of fluorescent pseudomonades using BOX –PCR (Botelho 2001).

Louws et al. (1994) described that35 - 40 bp repetitive extragenic palindromic sequence (REP), 124-127 bp, enterobacterial repetitive intergenic consensus sequence (ERIC) and bacterial repetitive three subunits boxA, boxB and boxC (BOX-PCR), collectively known as rep-PCR, can be utilized for rapid molecular characterization of the *Pseudomonas* (Figure 4). Other studies demonstrated that the diversity of membrane-bound nitrate reductase (narG) and nitrous oxide reductase (nosZ) genes, led to the identification of two different groups of strains (Delorme et al. 2003).

Some protein-encoding genes like rpoD (σ70 factor) and gyrB (DNA gyrase B subunit) which evolve much faster than rRNA genes were used for detection of two new major intrageneric clusters (Yamamoto and Harayama 1998; Yamamoto 2000). Some other neucleotide sequences like atpD, rpoB, OprI, carA, oprF and recA can also serve as alternative phylogenetic markers for taxonomical study of *Pseudomonas* (De Vos 1998; Ait Tayeb 2005; Hilario 2004; Bodilis and Barray 2006).

Recently introduced Multilocus sequence typing/analysis (MLST/MLSA) technique is a molecular typing method consisting of 400-600 bp long sequencing of housekeeping genes (most commonly present bacterial genes). This technique is better than 16S rRNA sequencing as these genes have higher variability and increased length of nucleotide which reduces the horizontal gene transfer risk as it uses more than one gene (Vinatzer and Bull 2009). This technique subdivided the already grouped two lineages into nine groups (Mulet 2010).

In addition to genes sequencing, some other methods like PCR-RCBH (polymerase chain reaction-reverse cross-blot hybridization) (Jaturapahu et al. 2005) and RFLP (restriction fragment length polymorphism) (Guasp et al. 2000) could be used. PFGE (Pulsed-field gel electrophoresis) regarded as the "gold standard" in molecular typing methods which can be used for identification and differentiation of single strains (Geider 1997; Spencker et al. 2000). The genetic diversity can be studied by several molecular techniques but gene sequencing is considered as most sophisticated and advance technique. Study of vast diversity in fluorescent pseudomonas requires further advancement in these methodologies (Novik et al. 2015).

4.2. Application in Development of Novel Strains for Enhanced Biocontrol Efficacy

The whole genome sequences are now available for many fluorescent pseudomonades which are used as bioagents like *P. fluorescens* lineage (strain SBW25) as well as for the non bioagents species like *P. fluorescens* Pf0-1. The novel methods of gene array-based expression studies, comparative genomics and integrated in situ molecular analyses of microbe-host interactions have provided detailed information on plant protection properties and rhizosphere competence of these bioagents (Barret et al. 2009; Loper et al. 2007; Kiely et al. 2006). New bioagents developed either by reprogramming the regulatory mechanism of available bioagents traits *via* functional genomics (Mark et al. 2006) or addition of new mechanisms like ACC deaminase activity (Wang et al. 2000) or degradation of pathogen quorum-sensing molecules (Molina et al. 2003). These development strategies are summarized in following heads:

4.3. Development of Strains for Over Production of Anti-Microbial Compounds

The production of antimicrobial compounds by fluorescent pseudomonades makes them ideal candidates for utilization as bioagents in agriculture. Duffy and Defago (1999) studied the biosynthetic regulation of antimicrobial compounds *viz.* siderophores, salicylic acid, pyoluteorin, pyrrolnitrin, phychelin and 2, 4-Diacetylphloroglucinol (PHL) by *P. fluorescens* CHAO strain isolated from Switzerland's disease-suppressive soil. The recombinant DNA technology established direct link between antimicrobial production and disease biocontrol efficacy of *Pseudomonas* such as 2, 4-Diacetylphloroglucinol (PHL) production is linked with its biocontrol potential (Fenton et al., 1992; Keel et al., 1992). Through genomic reprogramming of biocontrol traits regulating systems can be developed which govern antifungal metabolites production in fluorescent pseudomonades and it is possible to make new improved bioagents (Couillerot et al. 2009; Mark et al. 2002).

Walsh et al. (2001) developed strategies through schematically illustrated design which can be applied for the development of advance biocontrol strains of *Pseudomonas*. One method is the introduction of additional genes to develop genetically modified (GM) inoculants in these bacteria which will enhance their rhizosphere competence. 2, 4-diacetylphloroglucinol (PHL) production is the basic mechanism which is utilized by *P. fluorescens* F113 in biocontrol. The PHL production is encoded by PHL biosynthetic gene cluster (Figure 5) in *P. fluorescens* F113 (Abbas et al. 2002). The PHL production is enhanced *in vitro* in *P. fluorescens* F113 by overexpression of PHL biosynthetic gene and mutation in *phlF* gene which increased the production of PHL in this bioagent. This over production of PHL enhance the biocontrol ability of this GM *P. fluorescens* F113 strain against *Rhizoctonia* and *Pythium* infection in soil-based microcosm (Delany et al. 2000; Resca et al. 2001). These GM bacteria had no significant influence on indigenous microbes.

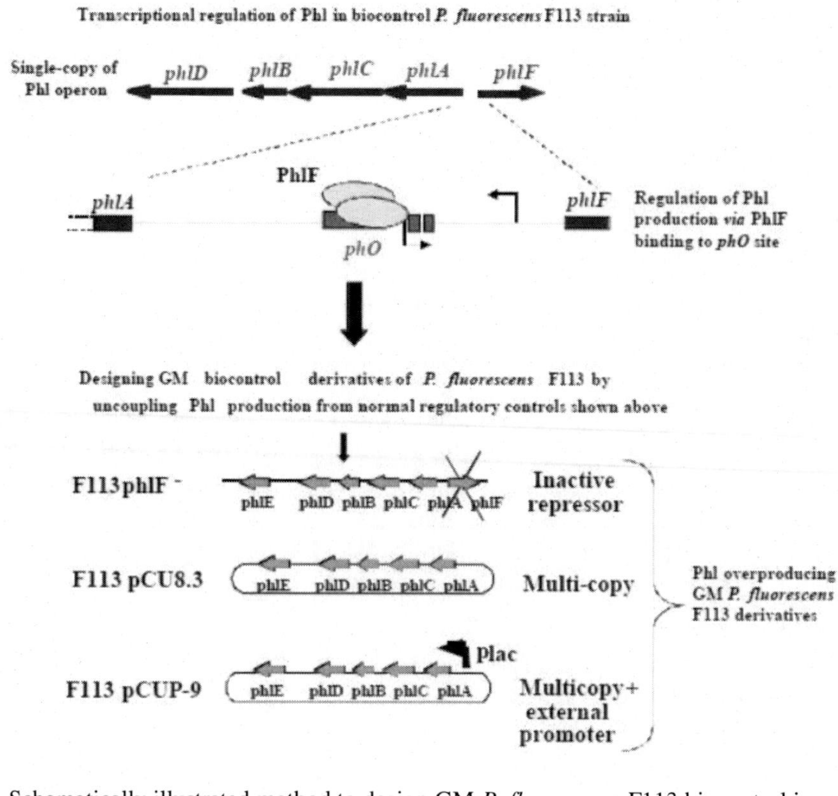

Figure 5. Schematically illustrated method to design GM *P. fluorescens* F113 biocontrol inoculants (adapted from Abbas et al. 2002; Delany et al. 2000; Delany et al. 2001) Source Mark et al. 2002.

In wild type *Pseudomonas,* the introduction of additional copies of *sss* colonisation genes increased the colonization of these bacteria. In *P. fluorescens* a new gene *espR* is identified, which influence its rhizospheric fitness which can be utilized to increase its bioagents efficiency (Dekkers et al. 2000).

Table 2. List of *Pseudomonas* wild strains with traits of overproducing secondary metabolites

Pseudomonas WT strain	GM trait (GM derivative)	Reference
P. putida WCS358	Produce Phz and Phl (WCS358r::phz) (WCS358r::phl)	Glandorf et al. (2001), Bakker et al. (2002), Viebahn et al. (2003)
P. fluorescens F113	Overproduce Phl	Resca et al. (2001)
P. fluorescens F113	Overproduce Phl	Barea et al. (1998)
P. fluorescens CHA0	Overproduce Phl and Plt (CHA0[pME3424])	Shaukat & Siddiqui (2003)
P. fluorescens CHA0	Overproduce Phl and Plt (CHA0 -Rif[pME342])	Girlanda et al. (2001)
P. fluorescens CHA0	Overproduce IAA (CHA0[pME3468])	Beyler et al. (1999)
P. fluorescens CHA0	Overproduce Phl and Plt (CHA0-Rif[pME3424])	Natsch et al. (1998)
P. fluorescens CHA0	Overproduce Phl and Plt (CHA0 [pME3424])	Niemann et al. (1997)

Adapted from Mark et al. 2006.

Table 3. List of secondary metabolites produced by fluorescent pseudomonas with their role in biocontrol and their genetic constituents

S. No.	Secondary metabolites	Mechanism of action	Genetic constituents	Reference
1.	2, 4-di-acetylphloro glucinol (DAPG or Phl)	It is a broad-spectrum antibiotic, Phl is a phenolic molecule produced by many fluorescent pseudomonads and exhibits antifungal, antibacterial, antihelmenthic and phytotoxic activities	A total of five complete open reading frames (ORFs) and one partial ORF, within the 6.8 kb segment of DNA, is responsible for biosynthesis of DAPG. The genes *phlA, phlC, phlB and phlD* are contained in a large transcriptional unit transcribed in the same direction.	Abbas et al., 2002; Delany et al. 2000.
2.	Phenazines	It is broad spectrum antibiotic activity against bacteria and fungi and they exhibit it by inhibiting the electron transport system of the pathogens.	Seven genes, *phzABCDEFG*, are involved in the synthesis of PCA which are localized within a 6.8 kb fragment in *P. fluorescens* 2-79.	Mavrodi et al. 1998
3.	Pyrrolnitrin	It is a broad-spectrum antifungal metabolite which inhibit the fungal respiratory chain	*prnABCD*spans 5.8 kb DNA in which four ORFs, *prnA, prnB, prnC* and *prnD* are involved.	Arima et al.1964; Gaffeny et al.,1994.
4	Pyoluteorin	It is an aromatic polyketide antibiotic	Ten genes, *pltLABCDEFG* are involved in the biosynthesis of Plt. They span a 24 kb genomic region in *P. fluorescens* Pf-5.	Kitten et al. 1998; Nowak et al., 1999; Pierson et al., 1998

Sarker et al. 2014.

In UK, during 1993-94 first free living GMO *Pseudomonas fluorescens* SBW 25 was released on spring wheat and sugar beet. This bacterium readily colonise the rhizosphere and phylloplane of wheat and sugar beet (Rainey and Bailey 1996; Lynch 2002).

Studies have found that in *Pseudomonas* spp., two component system comprising of the sensor kinase GacS and response regulator GacA exists which control the production of antifungal secondary metabolites. GacS is a group of genes which encodes functionally homologous cognate sensor kinases like *pheN, lemA*, *apdA* and *repA* (Kitten et al. 1998). The expression of some biocontrol metabolites *viz.* pyoluteorin, Phl, HCN are strictly governed by the GacS system in *P. fluorescens*CHAO (Haas and Défago 2005; Haas and Keel 2003). The GacA genes encode RsmA and RsmE proteins and these proteins act as post-transcriptional repressors of typical biocontrol genes as they interfere in ribosome binding site's functions (Haas and Défago 2005; Reimmann 2005). These component systems can be utilized for the development of strains with over production potential of anti-microbial compounds.

4.4. Development of Highly Rhizosphere Competent Strains

Superior colonization and effective functioning in rhizosphere are important traits required in a bioagents for disease management as well as for bioremediation. Fluorescent pseudomonades are aggressive root colonizer and are ideal carriers for PCB (polychlorinated biphenyl) degradative genes. To manage contaminated biosystem, new rhizoremediation derivatives of *Pseudomonas fluorescens* F113 are developed in which PCB degradative genes are regulated by specific root-exudates that allow controlled degradation of contaminants (Couillerot et al. 2009).

Some scientists have reported that fluorescent pseudomonades having ability to produce pyoverdine and carry out denitrification process which makes them more competent to survive in rhizosphere (Mirleauet al.98). Some *Pseudomonas* strains having both genes (narGandnosZ) for denitrification process are having better surviving ability in soil. Those fluorescent pseudomonades which have total denitrification ability show the best rhizosphere competence (Delormeet al. 2003).

The fluorescent pseudomonades survive under adverse conditions with the help of alternative sigma factors RpoS (σs) and RpoE ($\sigma 22$; also known as AlgU or AlgT). The sigma factor RpoS (σs) controls the susceptibility of bacterial cells towards adverse conditions. There are reports on rpoS mutant *P. aeruginosa* PAOl in which RpoS sigma factor are encoded by rpoS gene. The study revealed that with the RpoS sigma factor, death rate of bacterial cell in stationary-phase is two to three fold high when exposed to heat, high osmolarity, low pH, ethanol and hydrogen peroxide. However, RpoS-negative cells are more resistant in stationary-phase as compared to exponentially growing RpoS-positive cells of *P. aeruginosa* (Jørgensen et al. 1999).

The RpoE (AlgU) sigma factor add to tolerance against oxidative, osmotic and heat stresses in *P. syringae* and *P. aeruginosa* which are encoded by mucABCD genes. When bacteria are exposed to adverse environment, the MucB-MucA-AlgU complex destabilize which leads to release of AlgU sigma factor into the cytosol which become active (Martin et al. 1994; Keith and Bender 1999; Schurr and Deretic 1997; Schurr et al. 1995; Yu 1995; Govan and Deretic 1996; Mathee 1997). ATP and ppGpp levels are enhanced by Polyhydroxyalkanoates (PHAs) availability and expression of RpoS gene are induced by ppGpp (Gentry et al. 1993).

Gram-negative bacteria like *Pseudomonas* possess modified cell envelope with membrane vesicles having adhered solvent molecules (Sardessai et al. 2002). These modifications provide ability to form biofilm on substrate which enhances resistance to unfavourable conditions like exposure to antibiotic, heavy metals. These extracellular polymeric substances (EPS) check the diffusion of heavy metals in biofilm by binding them which protect bacterial cells under stress conditions (Teitzel et al. 2003). These mechanisms can be utilized to develop GM fluorescent pseudomonades with enhanced rhizosphere colonizing ability. These genetically modified bacteria can be monitored in rhizosphere for their survival and activity by *in vivo* expression technology. Genomic based new designs can be used for improvement of these novel strains like a novel gene *espR* found in *P. fluorescens* which influence its rhizosphere competence.

4.5. Identification and Development of Plant Growth Promoting Strains

The PGPR promote plant growth by several mechanisms and the general traits include (Alizadeh et al. 2013):

1. They fix atmospheric nitrogen which is supplied to the plants.
2. They synthesize phytohormones like IAA, auxins and cytokinins
3. They solubilize minerals such as phosphorus
4. They lower the production of stress ethylene (inhibitor of root growth) in roots through ACC deaminase enzyme (Penrose et al. 2001).

The new biotechnological applications are utilized to identify the genes/strains having PGP abilities which are then utilized to develop new strains. Choi et al. (2008) analysed DNA sequencing to identify eleven putative open reading frames consisting of seven known and four unknown pyrroloquinoline quinone (PQQ) biosynthetic genes. These PQQ genes express in nutrient limiting conditions and PQQ also act as an antioxidant in plants. Saber et al. (2015) used16S rRNA gene sequencing technique for selection of *Pseudomonas otitidis* isolates having plant growth promoting characters like production of cyanide, IAA, their antagonistic effect and their increased ability to solubilize phosphorus and potassium. Studies were done for the identification of new biosynthetic genes and metabolites and their function in biocontrol mechanism and plant growth promoting activities in two strains of *P. fluorescens* strains i.e., SBW25 andSS101. For this purpose different novel biotechnological application *viz.* microarray-based expression profiling, bioinformatics analysis of the genome sequences, bioactivity assays, mass spectrometric image analysis (MALDI-IMS), screening of genomic libraries and GC/MSMS were done to complete this study. These tests concluded that GacS/GacA two-component system of these bacteria play a pivotal role in induced systemic resistance, plant growth promotion and also in antagonistic activity. These bacterial strains promote plant growth by influencing steroid biosynthesis, carbohydrate metabolism, sulfur assimilation and auxin biosynthesis and transport (Cheng 2016).

4.6. Development of Strains Which Enhance Disease Resistance in Plants

Plant activates its defense mechanism against wide range of pests and pathogens after infection of an inducing agent which is called induced systemic resistance (ISR) and if it acquires resistance by any other agencies then it is called systemic acquired resistance (SAR) (Alizadeh et al. 2013; Hammerschmidt and Kuc 1995). A variety of enzymes which play role in plant protection are known as PR proteins. Fluorescent pseudomonades induce various PR proteins *viz.* peroxidase (PO), polyphenol oxidase (PPO), β-1,3-glucanase, chitinase, phenylalanine ammonia lyase (PAL), catalase (CAT), lipoxygenase (LOX), superoxide dismutase (SOD), ascorate peroxidase (APX) and proteinase inhibitors in plants after their inoculation through ISR (Van Loon, 1997, Khan 2012). Many workers reported that *P. fluorescens* increase resistance in plant by inducing accumulation of defense enzymes (PR proteins) in plants against phytopathogens through production of salicylic acid (Van Loon et al., 1998; Srivastava et. al. 2001; Karthikeyan et al. 2006).Metraux et al. (1990) reported that *pmsB* gene encodes pyruvate lyase enzyme in *P. fluorescence* which converts isochorismic acid into salicylic acid.

Maurhofer et al. (1998) utilized salicylic acid encoding genes *pchA* and *pchB* of *P. aeruginosa* for the development of two new strains of root-colonizing *P. fluorescens.* These genes were cloned into expression vector and then introduced into two non salicylic producing *P. fluorescens* strains. By introduction of *pchBA* genes, these strains gained production salicylic acid ability and significantly induced systemic resistance in tobacco against tobacco necrosis virus.

Further, PHL (2,4-diacetylphloroglucinol) have important role in inducing systemic disease resistance in plants through bioagents (Van Loon et al., 1998). *P. fluorescens* F113 were genetically modified by uncoupling transcriptional regulatory controls on PHL. This modification gives bacteria the ability to overproduce PHL which increased its disease resistance inducing ability (Fenton et al., 1992; Delany et al., 2001).

4.7. Development of Strains Fungicides/Antibiotics Compatible Strains

Khan and Gangopadhyay (2008) tested two strains of *Pseudomonas fluorescens* for their compatibility with fungicides. They found that strain *P. fluorescens* strain PFBC-25 is less sensitive to fungicide as compared to strains PFBC-26. Burr et al. (1978) also reported that *P. fluorescens* and *P. putida* are not affected by fungicides like benomyl, captan, PCNB used on potato seed pieces.

Moreover, it has been reported that *P. aeruginosa* develops antibiotic resistance by utilizing several mechanisms like cell wall mutations (reduction in porin channels), β-lactamases (degrade β-lactams antibiotic) and carbapenemases production (imipenem resistance), production of aminoglycoside nucleotidyl transferases (ANT) and aminoglycoside acetyltransferases (AAC) enzymes (modify aminoglycoside antibiotics) and multidrug efflux pumps upregulation (remove antibiotic from membrane) which leads to a less permeability of cell membrane for antibiotics and degradation of antibiotics (Lambert 2002).Genes like aac(3) and aac(6′) encodes AAC(6′) and AAC(3) enzymes are also linked with transposons or integrons which carry genes for extended-spectrum metallo-β-lactamases and β-lactamases (Poole 2005). Mutations in gyrA and parC genes which encode topoisomerases II and IV enzymes, respectively also enhance antibiotic resistance and generally mutations in parC gene are closely related with highly resistant gyrA mutants (Yonezawa et al. 1995; Higgins et al. 2003; Salma 2013).

Fluorescent Pseudomonades have many biochemical mechanisms which enable them to degrade many compounds including toxic chemicals (Spiers et al. 2000; Kononova and Nesmeyanova 2002). Thereby, utilizing these genes and other factors, new pesticide resistant fluorescent pseudomonas inoculants can be developed which can be applied with other pesticides during seed treatment or by any other application method.

4.8. Application in Genetic Modified Crops

Some genes of fluorescent pseudomonas are incorporated in genetically modified plants to develop disease resistance. *Pseudomonas fluorescence* and *Escherichia coli* produce salicylic acid that activates SAR (systemic acquired resistance) through induction of pathogenesis-related (PR) proteins (Metraux et al. 1990). It is encoded through *pmsB* gene in *P. fluorescens* and through *entC* gene in *E. coli.* Both of these genes *entC* and *pmsB* are incorporated in tobacco plants through biolistic methods. These genes are constitutively expressed in its chloroplast. These plants showed a 1000 times higher accumulation of salicylic acid than wild-type tobacco. They showed increased level of disease resistance against the fungus *Oidium lycopersicon* without any adverse effect on plant. This strategy can be used to develop disease resistance GM varieties in other commercially important crops (Verberne et al. 2000).

4.9. Application in Detection of Target Bacteria

Biotechnological techniques also make it possible to detect the targeted fluorescent *Pseudomonas* strain. In this method direct staining techniques with advanced immunofluorescence microscopy are used with fluorescentin situ hybridization, fluorescent antibodies and marker gene technology which provide the detailed image of single-cell of these bacteria in root colonization (Kragelund and Nybroe 1996; Troxler et al. 1997). The new fluorescent in situ hybridization (FISH) techniques which target rRNA enable us to possibly detect target bacteria in situ and without extracting them from rhizosphere. In this method fluorescence-labelled oligonucleotide probes are used which are having specific sequence of 16S rRNA. These probes bind to specific 16S rRNA sites of the fixed and permeabilized targeted bacterial cell (Aman et al. 1995). But it is possible in very few strains of *Pseudomonas* (Boye et al. 1995). However, use of immunofluorescence labeled dual staining technique detects specific strains and FISH technique is more suitable to differentiate different growth levels of *Pseudomonas* populations in rhizosphere. The fluorescens signals from fluorescent pseudomonas can be masked by soil particles by autofluorescence which can obstruct the accurate detection of bacteria.

To overcome this issue, confocal laser scanning microscopy (CLSM) is to be used with these immunochemical methods and/or FISH techniques (Lübeck et al. 2000). Marker genes are common method for detection of specific *Pseudomonas* which commonly encodes enzymes (gusA, luxAB, xylE, luc and lacZ,) which catalyze bioluminescence reactions for identification (Götz et al. 2006; Sørensen et al. 2001). Further, in UK two marker genes (*lacZY* and *kanrxylE*) have been used for detection of GMO bacterium *Pseudomonas fluorescens* SBW 25 in rhizosphere and phylloplane of wheat and sugar beet. They were identified by culture methods and positioned 1 Mb apart on the 6.5 Mb chromosome to ensure phenotypic and genotypic accuracy, as well as to detect any gene exchange between host crop and its microbial populations.

4.10. Application in Study of Rhizospheric Interaction

The performance of *Pseudomonas* in soil may differ from one field to another or from year to year. This variation may occur due to its difference in its efficacy to colonise roots of different crops or variation in its biocontrol traits (Mark et al. 2006; Weller 2007).Therefore, better understanding of molecular basis of these traits of fluorescent pseudomonas and their signaling process which regulates its ecology will help in its better rhizospheric colonization and expression of its traits (Mark et al. 2005; Mavrodi et al. 2006; Kiely et al. 2006; Dubuis et al. 2007; Barret et al. 2009).

The bacteria communicate in rhizosphere through many signal molecules called quorum-sensing (QS) and it links gene expression with cell density (Sharma et al. 2003; Keller and Surette 2006). These signal molecules (autoinducers) like acyl-homoserine lactones are diffused by the bacteria in environment which after reaching at a threshold level induce gene expression. At this cell density level, bioagents express their traits i.e., inhibition the pathogens by producing antagonistic metabolites or nitrogen fixation in soil. The pseudomonades also use quorum-sensing system for their persistence and survival on plant surface (Loh et al. 2002, Fonseca et al. 2004). Advance research in quorum-sensing can increase the efficiency of these bioagents in soil health improvement, plant productivity, bioremediation (Sharma et al. 2003). In *P. aureofaciens* strain 30-84 it is observed that two QS systems regulate its biocontrol efficacy traits *viz.*, exoproteases activity, cell surface features and production of phenazine compounds and secondary metabolites against take all disease (Zhang and Pierson 2001; Chin-A-Woeng 2001). The phenazine production depends on two genes i.e., *phzR* and *phzI* which are members of *luxI/luxR* gene group (Wood and Pierson 1996; Mavrodi et al. 1998).

The *in vivo* expression technology (IVET) of gene fusion can help in study of gene expression of bacteria in rhizosphere (Rainey 1999; Rediers et al. 2005). With the help of this technique 20 genes were identified with elevated expression in rhizosphere inoculated with *P. fluorescens* SBW25. Further study of these expressed 20 genes, revealed that 14 genes showed homology to *Pseudomonas* genes which are responsible for stress response, nutrient acquisition and the type-three secretion system (Preston et al. 2001). This technique is also used to identify survival genes in another bioagent i.e., P. fluorescens strain Pf0-1 (Silby and Levy 2004). But this technology has one limitation that it cannot prolify global gene expression. For this purpose new technologies like 'Omics' are developed which analyse the global gene expression in rhizosphere such as transcriptomics. These transcriptomics will be helpful to understand the interaction of fluorescent pseudomonades with pathogens and host plants in rhizosphere. Now studies are being conducted on quorum sensing generated global gene expression in *Pseudomonas aeruginosa* (Schuster et al. 2003).

Mark et al. (2006) developed a model through application of system biology approach by adopting'omics' technology to identify those bacterial genes which play role under stress conditions in rhizosphere (Figure 6). They studied the effect of rhizosphere on entire genome (transcriptome) of *Pseudomonas* model strain (*P. aeruginosa* PA01Affymetrix Gene-Chip*s*) by the application of Gene-Chip technology which is a good root colonizer. They identified the genes which are involved in rhizosphere interaction by their model and also profiled the effect of root exudates of sugarbeet on this bacterium (Walker et al., 2004).

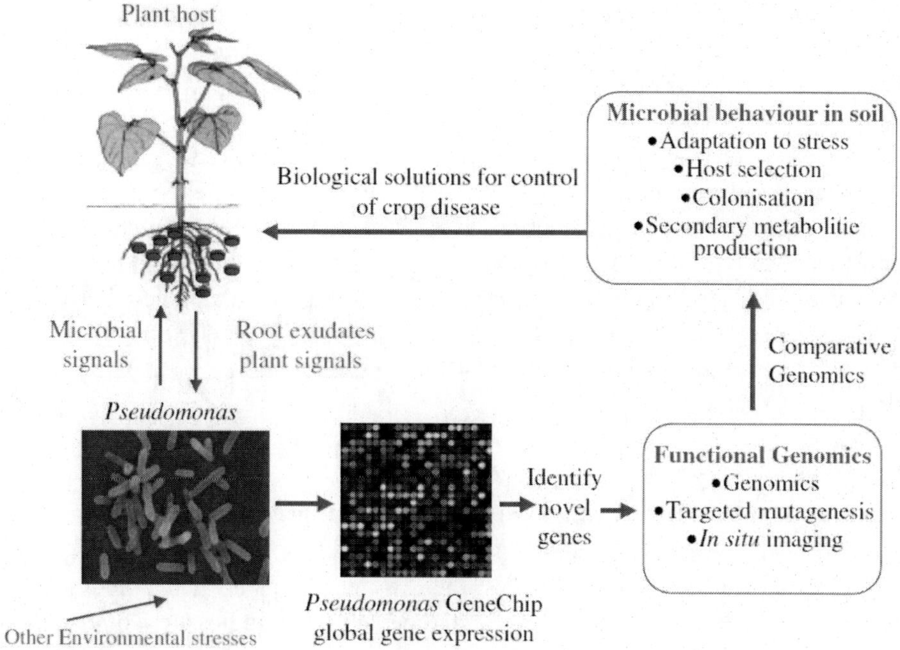

Figure 16. A systemic molecular signaling based (GeneChip) technique to study *Pseudomonas*–plant relationship (Mark et al. 2006).

By utilizing proteomic, functional genomics and *in situ* imaging techniques we are able to understand the role of genes in the host plant -*Pseudomonas* interaction in rhizosphere. Further advancement in commercial microarray technology along with the availability of complete sequencing of fluorescent pseudomonas genomes will help in profiling of global gene expression in bacteria (Pühler et al. 2004; Morrissey et al. 2004). The understanding of quoram sensing and its rhizospheric functions will play a major role in overcoming existing limitations and development of new fluorescent pseudomonades which will help in sustainable agriculture.

5. Major Challenges in Biotechnological Interventions

Biotechnology offers great potential in exploitation of fluorescent pseudomonades in sustainable agriculture but also warns about the risks as described by Tebbe (2002) with the release of GMO bacteria which are as follows: (i) There may be changes in fitness of genetically modified strains and their persistence in soil, (ii) the level and effect of expression of recombinant genes in soil, (iii) there may be replacement of indigenous microbes by these engineered bacteria in rhizosphere (iv) the indigenous bacteria may show resilience to remove GMOs from soil, and (v) there may be gene transfer in indigenous from non-indigenous bacteria.

Moreover, other than these risks the genus Pseudomonas is very adaptive in nature and it can easily change its phenotypic and genotypic characters. It is known to cause many toxic and pathogenic disorders in plants, animal as well as in humans as the infections caused by *Pseudomonas* are generally opportunistic. The persons who are suffering from cystic fibrosis, major trauma or burns and with low immunity are at risk of infection. Some pseudomonades like *Pseudomonas aeruginosa* also attack a variety of non vertebrate hosts like *Galleria mellonella* (greater wax moth) *Caenorhabditis elegans* (nematode) (Anonymous 2006; Choi et al. 2002).

Hence, before releasing these bacteria in field, we should first release these non-hazardous marker gene tagged soil bacteria in small scale controlled plot experiments to test their interaction in the rhizosphere microclimate and also their future prospects in adherence to biosafety measures (Novik et al. 2015). The novel inoculants of fluorescent pseudomonas may lose cell survivability during their field inoculation as well as in storage (Haas and De´fago 2005). However, recent advances show that storage viability of these formulations can be enhanced for longer period (Guo et al. 2004) along with retaining their biocontrol potential (Wiyono et al. 2008).

Conclusion and Future Prospects of Biotechnological Applications

Despite advancement in new technologies, there is dearth of knowledge about the complex rhizospheric interaction between plant and microbes. Further studies are required for in depth knowledge and to decipher how these bacteria interact with plants as well as with other microbes and how these relationships are affected by biotic and abiotic factors prevailing in environment. Advance technologies such as phylo-chips detection of microbes composition in the environment, metabolic profiling and metatranscriptomics of root exudates offer great promise. Some other advance technologies, such as confocal laser scanning microscopy, immunofluorescence microscopy and reporter genes have provided in depth knowledge on behavior of Pseudomonas inoculants in the environment. The recent techniques available for understanding inter and intraspecies communication through quoram sensing, signaling, help to decipher the effect of other microbes on gene expression of bioagents.

The recent technological interventions have laid focus on research related to beneficial actions of *Pseudomonas* and its molecular mechanism and in this surge genomics and proteomics has started to yield considerable results. These methods will also provide insight on other naturally-occurring biocontrol microbes in the rhizosphere like nonfluorescent *Pseudomonas* reported to produce a strain specific iron repressed nonfluorescent peptide siderophore. With the advancement in these technologies, we can better understand the interaction and efficacy of these bacteria in rhizosphere and provide solutions to their current bottlenecks in commercial use.

The application of these technologies in the development of fluorescent pseudomonades as a bioagent provides significant avenues to make them more assessable and reliable to the farmers. However, there is need for continuous exploration of new ways to enhance the field efficacy and delivery methods of these bacteria as bioagents so that they can be used sensibly for disease management and for the bigger goal of sustainable agriculture.

Acknowledgment

The authors are grateful to Dr. S. R. Dhaka, Dean, College of Agriculture, Fatehpur Shekhawati, SKNAU Jobner for their constant encouragement and providing assistance and necessary facilities in this venture.

References

Abbas A, Morrissey PJ, Marquez PC, Sheehan MM, Delany IR, O'Gara F (2002) Characterization of interactions between the transcriptional repressor PhlF and its binding site at the phlA promoter in *Pseudomonas fluorescens* F113. *J Bacteriol* 184:3008-3016.

Ait Tayeb L, Ageron E, Grimont F, Grimont PAD (2005) Molecular phylogeny of the genus Pseudomonas based on rpoB sequences and application for the identification of isolates. *Research in Microbiology* 156: 763-773.

Alizadeh O, Azarpanah A, Ariana L (2013) Induction and modulation of resistance in crop plants against disease by bioagent fungi (arbuscular mycorrhiza) and hormonal elicitors and Plant Growth Promoting Bacteria. *Intl J Farm & Alli Sci* 2: 982-998.

Amann RI, Ludwig W, Schleifer K (1995) Phylogenetic identification and in situ detection of individual microbial cells without cultivation. *Microbiol Rev* 59: 143-169.

Annomyous (2006) *Pseudomonas*. In: *Safety Assessment of Transgenic Organisms,* Volume 2: OECD Consensus Documents. OECD Publishing, pp 312-393. DOI: 10.1787/9789264095403-10-en.

Annonymous (2015) List of prokaryotic names with standing in nomenclature. http://www.bacterio.net (accessed 30 January 2015).

Arima K, Imanaka H, Kausaka M, Fukuda A, Tameera C (1964) Pyrrolnitrin a new antibiotic substance produced by *Pseudomonas. Agric Biol Chem*28:575-576.

Bakker PA, Glandorf DC, Viebahn M, Ouwens TW, Smit E, Leeflang PK et al. (2002) Effects of *Pseudomonas putida* modified to produce phenazine-1-carboxylic acid and 2,4-diacetylphloroglucinol on the microflora of field grown wheat. *Antonie Van Leeuwenhoek* 81: 617–624.

Bano N, Musarrat J (2003) Characterization of a new *Pseudomonas aeruginosa* strain NJ-15 as a potential biocontrol agent. *Curr Microbiol* 46:324-328.

Bano N, Musarrat J (2004) Characterization of a novel carbofuran degrading *Pseudomonas* sp. with collateral biocontrol and plant growth promoting potential. *FEMS Microbiol Lett* 231:13-17.

Barea JM, Andrade G, Bianciotto V, Dowling D, Lohrke S, Bonfante P et al. (1998) Impact on arbuscular mycorrhiza formation of *Pseudomonas* strains used as inoculants for biocontrol of soil-borne fungal plant pathogens. 64: 2304–2307.

Barret M, Frey-Klett P, Boutin M, Guillerm-Erckelboudt AY, Martin F, Guillot L et al. (2009) The plant pathogenic fungus *Gaeumannomyces graminis* var. *tritici* improves bacterial growth and triggers early gene regulations in the biocontrol strain Pseudomonas fluorescens Pf29Arp. *New Phytol* 181: 435–447.

Beyler M, Keel C, Michaux P, Haas D (1999) Enhanced production of indole-3-acetic acid by a genetically modified strain of *Pseudomonas fluorescens* CHA0 affects root growth of cucumber, but does not improve protection of the plant against *Pythium* root rot. *FEMS Microbiol Ecol* 28: 225–233.

Bloemberg GV and Lugtenberg BJ (2001) Molecular basis of plant growth promotion and biocontrol by rhizobacteria. *Curr Opin Plant Biol.* 4: 343-350.

Blondel-Hill E, Henry DA, David P (2007) *Pseudomonas.* In: Murray PR, Baron EJ, Jorgensen JH, Landry ML, Pfaller MA (eds.) *Manual of Clinical Microbiology.* 9th ed. Washington, DC: ASM Press, pp 734-748.

Bodilis J, Barray S (2006) Molecular evolution of the major outer-membrane protein gene (oprF) of *Pseudomonas*. *Microbiology* 152:1075-1088.

Botelho GR(2001) Seleção*de Pseudomonas fluorescentes para Controle Biológico da Podridão Vermelha da Raiz causada por Fusarium solani.*Rio de Janeiro, 2001, 108p. (Dr. Thesis, Biotecnologia Vegetal, UFRJ).

Botelho GR, Mendonça-Hagler LC (2006) Fluorescent Pseudomonades associated with the rhizosphere of crops - an overview Brazilian Journal of Microbiology. *Braz. J. Microbiol.* 37 Print version ISSN 1517-8382 On-line version ISSN 1678-4405.

Boye M, Ahl T, Molin S (1995) Application of a strain-specific rRNA oligonucleotide probe targeting *Pseudomonas fluorescens*Ag1 in a mesocosm study of bacterial release into the environment. *Appl Environ Microbiol* 61:1384-1390.

Bradley DE(1980) A function of *Pseudomonas aeruginosa* PAO polar pili: twitching motility. *Can Journal of Bacteriology* 26: 146-154.

Bultreys A, Gheysen I, Maraite H, De Hoffmann E (2001) Characterization of Fluorescent and Nonfluorescent Peptide Siderophores Produced by *Pseudomonas syringae* Strains and Their Potential Use in Strain Identification. *Applied and Environmental Microbiology* 67: 1718–1727.

Burr TJ, Schroth MN, Suslow T (1978) Increased potato yield on treatment of seed pieces with specific strains of *Pseudomonas fluorescens* and *P. putida*. *Phytopathology* 68: 1377-1383.

Cattelan AJ, Hartel PG, Furhmann FF (1999) Screening for plant growth promoting rhizobacteria to promote early soybean growth. *Soil Sci Soc Am J,* 63:1670-1680.

Charudattan R (1991) The mycoherbicide approach with plant pathogens. In TeBeast DO (eds.) *Microbial control of weeds.* Chapman and Hall, New York, 24-57.

Cheng X (2016) Plant growth promotion by *Pseudomonas fluorescens*: mechanisms, genes and regulation. Thesis, Wageningen University.ISBN 9789462578753 – 192.

Chin-A-Woeng TFC, Bloemberg GV, Lugtenberg JJ (2003) Phenazines and their role in biocontrol by *Pseudomonas* bacteria. *New phytologist* 157: 503-523.

Choi JY, Sifri CD, Goumnerov BC, Rahme LG, Ausubel FM, Calderwood SB (2002) Identification of virulence genes in a pathogenic strain of *Pseudomonas aeruginosa* by representational difference analysis. *Journal of Bacteriology* 184: 952-961.

Choi O, Kim J, Kim JG, Jeong Y, Moon JS, Park CS et al. (2008) Pyrroloquinoline Quinone Is a Plant Growth Promotion Factor Produced by *Pseudomonas fluorescens* B161. *Plant Physiology* 146: 657–668.

Couillerot O, Prigent-Combaret C, Caballero-Mellado J, Moënne-Loccoz Y (2009) *Pseudomonas fluorescens* and closely-related fluorescent pseudomonads as biocontrol agents of soil-borne phytopathogens. *Letters in Applied Microbiology* 48:505–512.

Cox CD, Adams P (1985) Siderophore Activity of Pyoverdin for *Pseudomonas aeruginosa. Infection and Immunity* 48: 130–138.

De Ita ME, Marsch-Moreno R, Guzmán P Alvarez-Morales A (1998) Physical map of the chromosome of the phytopathogenic bacterium *Pseudomonas syringae* pv. *phaseolicola. Microbiology*144: 493-501.

De Vos D, Bouton C, Sarniguet A, De Vos P, Vauterin M, Cornelis P (1998) Sequence diversity of the oprI gene, coding for major outer membrane lipoprotein I, among rRNA group I pseudomonads. *Journal of Bacteriology* 180: 6551-6556.

Dekkers LC, Mulders IHM,Phoelich CC, Chin-A-Woeng TFC, Wijfjes AHM, Lugtenberg BJJ (2000) The *sss* colonisation gene of the tomato-*Fusarium oxysporum* f.sp. *radicis lycopersici* biocontrol strain *Pseudomonas fluorescens* WCS365 can improve root colonisation of other wild-type *Pseudomonas* spp. bacteria. *Mol Plant Microbe Interact*13: 1177-1183.

Delany IR, Walsh UF, Ross I, Fenton AM, Corkery DM, O'Gara F (2001) Enhancing the biocontrol efficacy of *Pseudomonas fluorescens* F113 by altering the regulation and production of 2,4-diacetylphloroglucinol. *Plant Soil* 232: 195–205.

Delany IR, Sheehan MM, Fenton A, Bardin S, Aarons S, O'Gara F(2000) Regulation of production of the antifungal metabolite 2,4-diacetylphlorglucinol in *Pseudomonas fluorescens* F113: genetic analysis of *phlF* as a transcriptional repressor. *Microbiol* 146: 537-546.

Delany IR., Walsh UF, Ross I, Fenton AM, Corkery DM, O'Gara F (2001)Enhancing the biocontrol efficacy of *Pseudomonas fluorescens* F113 by altering the regulation and production of 2,4-diacetylphlorglucinol. *Plant Soil*232: 195-205.

deLorenzo V (2001) The potential for genetically modified bacteria to break down toxic pollutants in the environment. *EMBO reports* 2: 357-359.

Delorme S, Philippo L, Edel-Hermann V, Deulvot C, Mougel C, Lemanceau P (2003) Comparative genetic diversity of *narG*, *narZ*, and 16S rRNA genes in fluorescent Pseudomonads. *Appl Environ Microbiol* 69:1004-1012.

Doig P, Todd T, Sastry PA, Lee KK, Hodges RS, Paranchych W, Irvin RT (1988) Role of pili in adhesion of *Pseudomonas aeruginosa* to human respiratory epithelial-cells. *Infection and Immunity*. 56: 1641-1646.

Dubuis C, Keel C, Haas D (2007) Dialogues of root colonizing biocontrol pseudomonads. *Eur J Plant Pathol* 119, 311–328.

Duffy BK, Defago G (1999) Environmental factors modulating antibiotic and siderophore biosynthesis by *Pseudomonas fluorescens* biocontrol strains. *Appl Environ Microbiol* 65: 2429 - 2438.

Fenton AM, Stephens PM, Crowley J, O'Callaghan M, O'Gara F (1992) Exploitation of gene(s) involved in 2,4- diacetylphloroglucinol biosynthesis to confer a new biocontrol capability to a *Pseudomonas* strain. *Appl Environ Microbiol* 58: 3873–3878.

Fonseca MCC, Rumjanek NG, Xavier GR (2004) "Quorum sensing": uma nova interpetação da incidência e controle de doenças de plantas. *Seropédica: Embrapa Agrobiologia*,37p. (Embrapa Agrobiologia. Documentos, 170).

Frey P, Frey-Klett P, Garbaye J, Berge O, Heulin T (1997) Metabolic and genotypic fingerprinting of fluorescent pseudomonads associated with the Douglas-fir–*Laccaria bicolor* mycorrhizosphere. *Applied and Environmental Microbiology*63: 1852–1860.

Frey-Klett P, Chavatte M, Clausse ML, Courrier S, Le Roux C, Raaijmakers J et al. (2005) Ectomycorrhizal symbiosis affects functional diversity of rhizosphere fluorescent pseudomonads. *New Phytologist* 165: 317–328.

Gaffney TD, Lam ST, Ligon J, Gates K, Frazelle A, Di Maio J et al. (1994) Global regulation of antifungal factors by a *Pseudomonas fluorescens* biological control strain. *Mol Plant Microb. Interact.* 7:455-463.

García-Martínez J, Acinas SG, Antón AI, Rodríguez-Valera F (1999) Use of 16-23S ribosomal genes spacer region in studies of prokaryotic diversity. *J. Microbiol Meth* 36:55-64.

Geider K (1997) Differentiation and identification of *Pseudomonas syringae* pathovars by PCR- and PFGE-Analyses. In: Rudolph K, Burr TJ, Mansfield JW, Stead D, Vivian A, von Kietzell J. (eds.) *Developments in Plant Pathology,* Volume 9: Pseudomonas syringae Pathovars and Related Pathogens. Springer Netherlands, pp 459-464. DOI: 10.1007/978-94-011-5472-7_82.

Gentry DR, Hernandez VJ, Nguyen LH, Jensen DB, Cashel M (1993) Synthesis of the stationary-phase sigma factor σ^s is positively regulated by ppGpp. *Journal of Bacteriology* 175: 7982-7989.

Ghafoor A, Hay ID, Rehm BH (2011) Role of exopolysaccharides in *Pseudomonas aeruginosa* biofilm formation and architecture. *Applied and Environmental Microbiology* 77: 5238-5246.

Ginard M, Lalucat J, Tummler B, Romling U (1997) Genome organization of *Pseudomonas stutzeri* and resulting taxonomic and evolutionary considerations. *Int J Syst Bacteriol* 47: 132-143.

Girlanda M, Perotto S, Moenne-Loccoz Y, Bergero R, Lazzari A, D´efago G et al. (2001) Impact of biocontrol *Pseudomonas fluorescens* CHA0 and a genetically modified derivative on the diversity of culturable fungi in the cucumber rhizosphere. *Appl Environ Microbiol* 67: 1851–1864.

Glandorf DC, Verheggen P, Jansen T, Jorritsma JW, Smitt E, Leeflang P et al. (2001) Effect of genetically modified *Pseudomonas putida* WCS358r on the fungal rhizosphere microflora of field-grown wheat. *Appl Environ Microbiol* 67: 3371–3378.

Glick BR, Karaturovic DM, Newell PC (1995) A novel procedure for rapid isolation of plant growth promoting pseudomonas. *Can J Microbiol* 41:533-536.

Godfrey SAC, Marshall, JW (2002) Soil on imported shipping containers provides a source of new Pseudomonad biodiversity into New Zealand. *New Zealand J Crop Hortic Sci* 30:19-27.

Gomes, NCM, Heuer H, Schönfeld J, Costa R, Mendonça-Hagler L, Smalla K (2001) Bacterial diversity of the rhizosphere of maize (*Zea mays*) grown in tropical soil studied by temperature gradient gel electrophoresis. *Plant and Soil*232: 167-180.

Gomila M, Peña A, Mulet M, Lalucat J, García-Valdés E (2015) Phylogenomics and systematics in *Pseudomonas. Front Microbiol* 6:1-214.

Götz M, Gomes NCM, Dratwinski A, Costa R, Berg G, Peixoto R et al. (2006) Survival of*gfp*-tagged antagonistic bacteria in the rhizosphere of tomato plants and their effects on the indigenous bacterial community. *FEMS Microbiol Ecol*56: 207-218.

Govan J, Deretic V (1996) Microbial pathogenesis in cystic fibrosis: mucoid *Pseudomonas aeruginosa* and *Burkholderia cepacia*. *Microbiological Reviews* 60: 539-574.

Grimont PAD, Vancanneyt M, Lefèvre M, Vandemeulebroecke K, Vauterin L, Brosch R et al. (1996) Ability of Biolog and Biotype 100 systems to reveal the taxonomic diversity of the *Pseudomonads*. *Syst Appl Microbiol* 19:510-527.

Grothues D, Koopmann U, von der Hardt H, Tummler B (1988) Genome fingerprinting of *Pseudomonas aeruginosa* indicates colonization of cystic fibrosis siblings with closely related strains. *J Clin Microbiol*26: 1973-1977.

Guasp C, Moore ER, Lalucat J, Bennasar A (2000) Utility of internally transcribed 16S-23S rDNA spacer regions for the definition of *Pseudomonas stutzeri* genomovars and other Pseudomonas species. *International Journal of Systematic and Evolutionary Microbiology* 50: 1629-1639.

Guo JH, Qi HY, Guo YH, Ge HL, Gong LY, Zhang LX et al. (2004) Biocontrol of tomato wilt by plant growth-promoting rhizobacteria. *Biol Control* 29:66-72.

Gürtler V, Stanisich VA (1996) New approaches to typing and identification of bacteria using the 16S - 23S rDNA spacer. *Microbiol*142:3-16.

Gurusiddaiah S, Gealy DG (1994) Isolation and characterization of metabolites from *Pseudomonas fluorescens*-D7 for control of downy brome (*Bromus techtorum*). *Weed Sci* 42:492-501.

Gutell RR, Weiser B, Woese CR, Noller HF (1985) Compartive anatomy of 16S-like ribosomal RNA. *Prog Nucleic Acid Res Mol Biol*32:155-216.

Haas D, Blumer C, and Keel C (2001) Biocontrol ability of fluorescent pseudomonads genetically dissected: importance of positive feedback regulation. *Curr Opin Biotechnol* 11: 290-297.

Haas D, Defago G (2005) Biological control of soil-borne pathogens by fluorescent pseudomonads. *Nat Rev Microbiol* 3: 307–319.

Haas D, Keel C (2003) Regulation of antibiotic production in root-colonizing *Pseudomonas* spp. and relevance for biological control of plant disease. *Annu Rev Phytopathol* 41: 117-153.

Hammerschmidt R, Kuc J (1995) *Induced Resistance to Disease in Plants.* Kluwer Academic Publishers, Dordrecht, The Netherlands, p.182.

Higgins PG, Fluit AC, Milatovic D, Verhoef J, Schmitz FJ (2003) Mutations in GyrA, ParC, MexR and NfxB in clinical isolates of Pseudomonas aeruginosa. *The International Journal of Antimicrobial Agents* 21: 409-413.

Hilario E, Buckley TR, Young JM (2004) Improved resolution on the phylogenetic relationships among Pseudomonas by the combined analysis of atp D, car A, rec A and 16S rDNA. *Antonie Van Leeuwenhoek* 86: 51-64.

Hofte M, Altier N (2010) Fluorescent pseudomonads as biocontrol agents for sustainable agricultural systems. *Res Microbiol* 61: 464–471.

Janse JD, Derks JHJ, Spit BE, Van DTWR (1992) Classification of fluorescent soft rot *Pseudomonas* bacteria including *P marginalis* strains using whole cell fatty acid analysis. *Syst Appl Microbiol* 15:538-553.

Jaturapahu T, Puttinaowarat S, Somsiri T (2005) Detection and identification of Pseudomonas Spp. by polymerase chain reaction-reverse cross-blot hybridization (PCRRCBH) with 16S-23S ribosomal RNA intergenic spacer probes. In: Walker P, Lester R, Bondad-Reantaso MG. (eds.) *Diseases in Asian Aquaculture:* Proceedings of the Fifth symposium on Diseases in Asian Aquaculture, 24-28 November 2002, Queensland, Australia. Fish Health Section, Asian Fisheries Society, Manila, Philippines, pp 447-456.

Jiménez JI, Nogales J, García JL, Díaz E (2010) A genomic view of the catabolism of aromatic compounds in *Pseudomonas*. In: Timmis KN. (ed.) *Handbook of Hydrocarbon and Lipid Microbiology.* Springer Berlin Heidelberg;. Pp 1297-1325. doi; 10.1007/978-3-540-77587-4_91.

Jonhson J, Palleroni NJ (1989) Deoxyribonucleic acid similarities among *Pseudomonas* species. *Int J Syst Bacteriol* 39:230-235.

Jørgensen F, Bally M, Chapon-Herve V, Michel G, Lazdunski A, Williams P et al. (1999) RpoS-dependent stress tolerance in Pseudomonas aeruginosa. *Microbiology* 145: 835-844.

Jun S, Wassenaar TM, Nookaew I, Hauser L, Wanchai V, Land M et al. (2016) Diversity of Pseudomonas Genomes, Including Populus-Associated Isolates, as Revealed by Comparative Genome Analysis Applied and Environmental Microbiology82: 375-383.

Karthikeyan M, Radhika K, Mathiyazhagan R, Bhaskaran R, Samiyappan R, Velazhahan R (2006)Induction of phenolics and defense related enzymes in coconut (*Cocos nucifera* L.) roots treated with biocontrol agents. *Braz. J. Plant Physiol*. 18: 367-377.

Keel C, Schnider U, Maurhofer M, Viossard C, Laville J, Burger U et al. (1992) Suppression of root diseases by *Pseudomonas fluorescens* CHA0: importance of the bacterial secondary metabolite 2, 4-diacetylphloroglucinol. *Mol Plant Microbe Interact* 5: 4–13.

Keith LMW, Bender CL (1999) AlgT (σ^{22}) controls alginate production and tolerance to environmental stress in Pseudomonas syringae. *Journal of Bacteriology* 181: 7176-7184.

Keller L, Surette MG (2006) Communication in bacteria: an ecological and evolutionary perspective. *Nature Rev Microbiol* 4: 249-258.

Khan MA (2012) Studies on variability and management of *Fusarium oxysporum* f. sp. *ciceri* (Padwick) Synd. and Hans. incitant of wilt in chickpea. Ph. D. thesis (Plant Pathology).Rajasthan Agricultural University, Bikaner.

Khan MA, Gangopadhyay S (2008) Efficacy of *Pseudomonas fluorescens* in Controlling Root Rot of Chickpea Caused by *Macrophomina phaseolina. J Mycol Pl Pathol* 38:580-587.

Khan MA, Gangopadhyay S (2012) Effect of soil inhabitiaing antagonistic microflora against *Fusarium oxysporum* f.sp. *ciceri* incitant of wilt of chickpea. *Journal of Mycology and Plant Pathology* 42: 341-346.

Khan MS and Zaidi A (2002) Plant growth promoting rhizobacteria from rhizospheres of wheat and chickpea. *Ann Pl Prot Sci* 10: 265-271.

Kiely PD, Haynes JM, Higgins CH, Franks A, Mark GL, Morrissey JP O'Gara F (2006) Exploiting new systems-based strategies to elucidate plant-bacterial interactions in the rhizosphere. *Microb Ecol* 51: 257–266.

Kitten T, Kinscherf T, McEvoy G, Willis DK (1998). A newly identified regulator is required for virulence and toxin production in *Pseudomonas syringae. Mol. Microbiol.* 28:917-929.

Kononova SK, Nesmeyanova MA (2002) Phosphonates and their degradation by microorganisms.*Biochemistry*(Moscow), 67: 184-195.

Kragelund L, Nybroe O (1996) Competition between *Pseudomonas fluorescens*Ag1 and*Alcaligenes eutrophus*JMP134 (pJP4) during colonization of barley roots.*FEMS Microbiol Ecol* 20: 41-51.

Kremer RJ, Begonia MFT, Lynn S, Lanham ET (1990) Characterization of rhizobacteria associated with weed seedlings. *Appl Environ Microbio* 56:1646-1655.

Kwon SW, Kim JS, Crowley DE, Lim CK (2005) Phylogenetic diversity of fluorescent pseudomonads in agricultural soils from Korea. *Letters in Applied Microbiology* 41: 417-423.

Laguerre G, Rigottier-Gois L, Lemanceau P (1994). Fluorescent *Pseudomonas* species categorized by using polymerase chain reaction (PCR) /restriction fragment analysis of 16S rDNA. *Molecular Ecology* 3: 479–487.

Lambert PA (2002) Mechanisms of antibiotic resistance in *Pseudomonas aeruginosa. Journal of the Royal Society of Medicine* 95: 22-26.

Latour X, Corberand T, Laguerre G, Allard F, Lemanceau P (1996). The composition of fluorescent Pseudomonad populations associated with roots is influenced by plant and soil type. *Appl Environ Microbiol* 62:2449-2456.

Lemanceau P, Corberand T, Gardan L, Latour X, Laguerre G, Boeufgras J.M, Alabouvette C (1995) Effect of two plant species, flax (*Linum usitatissinum* L.) and tomato (*Lycopersicon esculentum* Mill.), on the diversity of soilborne populations of fluorescent pseudomonads. *App Environ Microbiology*61: 1004-1012.

Lessie TG, Phibbs PV Jr (1984) Alternative pathways of carbohydrate utilization in pseudomonads. *Annual Review of Microbiology* 38: 359-388.

Liao CH, Wells JM (1978) Diversity of Pectolytic, Fluorescent Pseudomonads Causing Soft Rots of Fresh Vegetables at Produce Markets. *Postharvest Pathology and Mycotoxins* 77: 673-677.

Locatelli L, Tarnawski S, Hamelin J, Rossi P, Aragno M, Fromin N (2002) Specific PCR amplification for the genus *Pseudomonas* targeting the 3' half of 16S rDNA and the whole 16S-23S rDNA spacer. *Syst. Appl. Microbiol.* 25:220-227.

Loh J, Pierson E, Pierson III LS, Stacey G, Chatterjee A (2002) Quorum sensing in plant-associated bacteria.*Curr Op Plant Biol* 5: 285-290.

Loper JE, Kobayashi DY, Paulsen IT (2007) The genomic sequence of *Pseudomonas fluorescens* Pf-5: insights into biological control. *Phytopathology* 97: 233–238.

Louws FJ, Fullbright DW, Stephens CT, de Bruijn FJ (1994) Specific genomic fingerprintings of phytopathogenic *Xantomonas* and *Pseudomonas* pathovars and strains generated with repetitive sequences and PCR. *Appl Environ Microbiol* 60: 2286-2295.

Lübeck PS, Hansen M, Sørensen, J. (2000) Simultaneous detection of the establishment of seed-inoculated *Pseudomonas fluorescens* strain DR54 and native soil bacteria on sugar beet root surfaces using fluorescence antibody and in situ hybridization technique. *FEMS Microbiol Ecol* 33: 11-19.

Lynch JM (2002) Impact of marker and functional genes on soils and plants.In: Lelley T, Balázs E, Tepfer M (eds.) *Ecological impact of GMO dissemination in Agro-Ecosystem.* OECD Cooperative Research programme workshop on "Ecological impact of GMO Dissemination in Agro-ecosystems," Grossrussbach, Austria, pp25-30 ISBN 3-85076-631-4.

Mark GL, Dow JM, Kiely PD, Higgins H, Haynes J, Baysse C et al. (2005) Transcriptome profiling of bacterial responses to root exudates identifies genes involved in microbe-plant interactions. *Proc Natl Acad Sci, USA* 102: 17454–17459.

Mark GL, Morrissey J P and O'GARA F (2002) Designing improved GM bacteria for application in environmental biotechnology. In: Lelley T, Balázs E and Tepfer M (eds.) *Ecological impact of GMO dissemination in Agro-Ecosystems.* International OECD Workshop held in Grossrussbach, Austria pp 8-45.

Mark GL, Morrissey JP, Higgins P and O'Gara F (2006) Molecular-based strategies to exploit Pseudomonas biocontrol strains for environmental biotechnology applications. *FEMS Microbiol Ecol* 56:167–177.

Mark GL, Morrissey JP, O'Gara F (2002) Designing improved GM bacteria for application in environmental biotechnology.In: Lelley T, Balázs E, Tepfer M (eds.) *Ecological impact of GMO dissemination in Agro-Ecosystem.* OECD Cooperative Research programme workshop on "Ecological impact of GMO Dissemination in Agro-ecosystems," Grossrussbach, Austria, pp11-24 ISBN 3-85076-631-4.

Martin DW, Schurr MJ, Yu H, Deretic V (1994) Analysis of promoters controlled by the putative sigma factor AlgU regulating conversion to mucoidy in Pseudomonas aeruginosa: relationship to σ^E and stress response. *Journal of Bacteriology* 176: 6688-6696.

Mathee K, McPherson CJ, Ohman DE (1997) Posttranslational control of the algT (algU)-encoded sigma22 for expression of the alginate regulon in Pseudomonas aeruginosa and localization of its antagonist proteins MucA and MucB (AlgN). *Journal of Bacteriology* 179: 3711-3720.

Maurhofer M, Reimmann C, Schmidli-Sacherer P, Heeb S, Haas D, Défago G (1998) Salicylic acid biosynthetic genes expressed in *Pseudomonas fluorescens* strain P3 improve the induction of systemic resistance in tobacco against tobacco necrosis virus. *Phytopathology* 88: 678-684.

Mavrodi DV, Ksenzenko VN, Bonsall RF, Cook RJ, Boronin AM, Thomashaw LS (1998) A seven gene locus for synthesis of phenazine-1-carboxylic acid by *Pseudomonas fluorescens* 2-79. *J Bacteriol* 180:2541-2548.

Mavrodi OV, Mavrodi DV, Weller DM, Thomashow LS (2006) Role of ptsP, orfT, and sss recombinase genes in root colonization by *Pseudomonas fluorescens* Q8r1-96. *Appl Environ Microbiol* 72: 7111–7122.

Metraux JP, Signer H, Ryals J, Ward E, Wyss-Benz M, Gaudin J, Raschdorf K, Schmid E, Blum W, Inverardi B (1990) Increase in salicylic acid at the onset of systemic acquired resistance in cucumber. *Science* 250: 1004-1006.

Meyer JM, Abdallah MA (1978) The fluorescent pigment of *Pseudomonas fluorescens*: biosynthesis, purification and physicochemical properties. *Journal of General Microbiology* 107: 319-328.

Meyer JM, Hornsperger JM (1978) Role of pyoverdine Pf, the iron binding fluorescent pigmentof *Pseudomonas fluorescens* in iron transport. *Journal of General Microbiology* 107: 329-331.

Migula W (1900)*System der Bakterien. Handbuck der Morphologie, Entwickelung-Geschichte und Systematik der Bakterien*, In Verlag von Gustav Fischer (eds.) Jena 2:.doi: 10.1099/00207713-37-4-463.

Mirleau P, Delorme S, Philippot L, Meyer JM, Mazurier S, Lemanceau P (2001) Involvement of nitrate reductase and pyoverdine in competitiveness of *Pseudomonas fluorescens* strain C7R12 in soil. *Appl Environ Microbiol*67: 2627-2635.

Molina L, Constantinescu F, Michel L, Reimmann C, Duffy B, De´fago G (2003) Degradation of pathogen quorum-sensing molecules by soil bacteria: a preventive and curative biological control mechanism. *FEMS Microbiol Ecol* 45: 71–81.

Moore ERB, Mau M, Arnscheidt A, Böttger EC, Hutson RA, Collins MD et al. (1996) The determination and comparison of the 16S rRNA gene sequences of species of the genus *Pseudomonas* (sensu stricto) and estimation of the natural intrageneric relationships. *Systematic and Applied Microbiology* 19: 478-492.

Moore ERB, Tindall BJ, Martins Dos Santos VAP, Pieper DH, Ramos JL, Palleroni NJ. (2006) Nonmedical: *Pseudomonas*. In: Dworkin M, Falkow S, Rosenberg E., Schleifer KH, Stackebrandt E. (eds.) *The Prokaryotes* Volume 6: Proteobacteria: Gamma Subclass. Springer, New York, pp 646-703. DOI: 10.1007/0-387-30746-X_21.

Morris CE, Sands DC, Vinatzer BA, Glaux C, Guilbaud C, Buffière A et al. (2008) The life history of the plant pathogen *Pseudomonas syringae*is linked to the water cycle. *ISME J* 2:321–334.

Morrissey JP, Dow JM, Mark GL & O'Gara F (2004) Are microbes at the root of a solution to world food production? *EMBO Rep* 5: 922–926.

Morrissey JP, Walsh UF, O'Donnell A, Moënne-Loccoz Y, O'Gara F (2002) Exploitation of genetically modified inoculants for industrial ecology applications. *Antonie van Leeuwenhoek* 81: 599-606.

Mulet M, Lalucat J, García-Valdés E (2010) DNA sequence-based analysis of the Pseudomonas species. *Environmental Microbiology* 12: 1513-1530.

Mulyukin AL, Suzina NE, Duda VI, El'-Registan GI (2008) Structural and physiological diversity among cystlike resting cells of bacteria of the genus *Pseudomonas. Microbiology* 77:455–465.

Naik PR, Raman G, Narayanan KB, Sakthivel N (2008) Assessment of genetic and functional diversity of phosphate solubilizing fluorescent pseudomonads isolated from rhizospheric soil. *BMC Microbiology* 8:230.

Naik PR, Sahoo N, Goswami D, Ayyadurai N, Sakthivel N (2008) Genetic and functional diversity among fluorescent pseudomonads isolated from the rhizosphere of banana. *Microb Ecol.* 56:492-504.

Naik PR, Sakthivel N (2006) Functional characterization of a novel hydrocarbonoclastic *Pseudomonas* sp. strain PUP6 with plant-growth-promoting traits and antifungal potential. *Res Microbiol* 157:538-546.

Natsch A, Keel C, Hebecker N, Laasik E, Defago G (1998) Impact of *Pseudomonas fluorescens* strain CHAO and a derivative with improved biocontrol activity on the culturable resident bacterial community on cucumber roots. *FEMS Microbiol Ecol* 27: 365–380.

Niemann S, Keel C, Puhler A, Selbitschka W (1997) Biocontrol strain *Pseudomonas fluorescens* CHA0 and its genetically modified derivative with enhanced biocontrol capability exert comparable effects on the structure of a *Sinorhizobium meliloti* population in gnotobiotic systems. *Biol Fertil Soils* 25: 240–244.

Novik G, Savich V, Kiseleva E (2015) An Insight Into Beneficial Pseudomonas bacteria. In: Shah MM (eds.) *Microbiology in Agriculture and Human Health.* IntechOpen, Chapter 5, DOI: 10.5772/60502.

Nowak-Thompsan B, Chancey N, Wing JS, Gould SJ, Loper JE (1999) Characterization of pyoluteorin biosynthetic gene cluster of *Pseudomonas fluorescens* Pf-5. *J Bacteriol* 181:2166-2174.

O'Sullivan DJ, O'Gara F (1992): Traits of fluorescent *Pseudomonas* spp. involved in suppression of plant root pathogens. *Microbiol Rev* 56:662-676.

O'Toole GA, Kolter R (1998) Flagellar and twitching motility are necessary for *Pseudomonas aeruginosa* biofilm development. *Molecular Microbiology* 30: 295-304.

Özen AI and Ussery DW (2012) Defining the *Pseudomonas* Genus: Where Do We Draw the Line with *Azotobacter*? *Microb Ecol* 63: 239–248.

Palleroni NJ (1993) *Pseudomonas* classificassion: a new case history in the taxonomy of Gram- negative bacteria. Antonie van Leeuwenhoek. 64:231-251.

Palleroni NJ (2005) Pseudomonas. In: Brenner DJ, Krieg NR, Staley JT, Garrity GM. (eds.) *Bergey's Manual of Systematic Bacteriology,* Volume Two the Proteobacteria Part B the Gammaproteobacteria. 2nd ed. Springer, US, pp 323-379. doi:10.1007/0-387-28022-7.

Palleroni NJ, Kunisawa R, Contopoulou R, Doudoroff M (1973) Nucleic acid homologies in the genus Pseudomonas. *International Journal of Systematic Bacteriology* 23: 333-339.

Patel JB (2001) 16S rRNA gene sequencing for bacterial pathogen identification in the clinical laboratory. *Molecular Diagnosis* 6: 313-321.

Paulsen IT, Press CM, Ravel J, Kobayashi DY, Myers GSA, Mavrodi DV (2005) et al. Complete genome sequence of the plant commensal *Pseudomonas fluorescens*Pf-5.*Nat Biotechnol* 23:873–878.

Penrose DM, Mofatt BA, Glick BR (2001) Determination of 1-aminocyclopropane-1- carboxylic acid (ACC) to assess the effects of ACC deaminase-containing bacteria on roots of canola seedlings. *Canadian J Microbiol* 47: 77-80.

Pfender WF, Kraus J, Loper JE (1993) A genomic region from *Pseudomonas fluorescens* Pf-5 required for pyrrolnitrin production and inhibition of *Pyrenophora tritici-repentis* in wheat straw. *Phytopathology* 83: 1223–1228.

Pierson LS, Thomashow LS (1992) Cloning and heterologous expression of the phenazine biosynthetic locus from *Pseudomonas aureofaciens* 30–84. *Mol Plant- Microbe Interact* 5: 330–339.

Pierson LS, Wood DW, Pierson EA, Chancey ST (1998) *N*-acyl homoserine lactone-mediated gene regulation in biological control by fluorescent pseudomonads: current knowledge and future work. *Eur J Plant Pathol*104:1-9.

Poole K (2005) Aminoglycoside resistance in *Pseudomonas aeruginosa. Antimicrobial Agents and Chemotherapy* 49: 479-487.

Preston GM, Bertrand N, Rainey PB (2001) Type III secretion in plant growth-promoting Pseudomonas fluorescens SBW25. *Mol Microbiol* 41: 999–1014.

Pühler A, Arlat M, Becker A, G¨ottfert M, Morrissey JP, O'Gara F (2004) What can bacterial genome research teach us about bacteria–plant interactions? *Curr Opin Plant Biol* 7: 137–147.

Raaijmakers JM, Vlami M & de Souza JT (2002) Antibiotic production by bacterial biocontrol agents. *Antonie Van Leeuwenhoek* 81: 537–547.

Rainey PB (1999) Adaptation of *Pseudomonas fluorescens* to the plant rhizosphere. *Environ Microbiol* 1: 243–57.

Rainey PB, Bailey MJ (1996) Physical and genetic map of the *Pseudomonas fluorescens* SBW25 chromosome. *Mol Microbiol*19: 521-533.

Rainey PB, Bailey MJ, Thompson IP (1994). Phenotypic and genotypic diversity of fluorescent pseudomonads isolated from field-grown sugar beet. *Microbiology*140: 2315-2331.

Ramamoorthy V, Viswanathan R, Raguchander T, Prakasam V, Samiyappan R (2001) Induction of systemic resistance by plant growth promoting rhizobacteria in crop plants against pests and diseases. *Crop Protection* 20:1-11.

Rediers H, Rainey PB, Vanderleyden J, De Mot R (2005) Unraveling the secret lives of bacteria: use of in vivo expression technology and differential fluorescence induction promoter traps as tools for exploring niche-specific gene expression. *Microbiol Mol Biol Rev* 69: 217–261.

Rediers H, Vanderleyden J, Mot R (2004) *Azotobacter vinelandii*: a *Pseudomonas*in disguise? *Microbiology* 150:1117–1119.

Reimmann C, Valverde C, Kay E, Haas D (2005) Posttranscriptional repression of GacS/GacA-controlled genes by the RNA-binding protein RsmE acting together with RsmA in the biocontrol strain *Pseudomonas fluorescens* CHAO. *J Bacteriol* 187:276-285.

Renwick A, Campbell R, Coe S (1991) Assessment of *in vivo* screening systems for potential biocontrol agents of *Gaeumannomyces graminis*. *Plant Pathol* 40:524-532.

Resca R, Basaglia M, Poggiolini S, Vian P, Bardin S, Walsh UF, Barreiros CME, O'Gara F, Nuti MP, Casella S, Peruch U (2001) An integrated approach for the evaluation of biological control of the complex *Polymyxa betae*/Beet Necrotic Yellow Vein Virus, by means of seed inoculants. *Plant and Soil* 232: 215–226.

Saber FMA, Abdelhafez AA, Hassan EA, Ramadan EM (2015) Characterization of fluorescent pseudomonads isolates and their efficiency on the growth promotion of tomato plant. *Annals of Agricultural Science* 60: 131–140.

Salma R, Dabboussi F, Kassaa I, Khudary R, Hamze M (2013) gyrA and parC mutations in quinolone-resistant clinical isolates of *Pseudomonas aeruginosa* from Nini Hospital in north Lebanon. *Journal of Infection and Chemotherapy* 19: 77-81.

Salman M (2010) Determination of antibiotic activity on plasmids from fluorescent pseudomonad isolates CW2, WB15 and WB52 against pre emergence damping-off caused by *Pythium ultimum* and *Rhizoctonia solani* in cucumber. *Biol Control* 53: 161–167.

Sardessai Y, Bhosle S (2003) Tolerance of bacteria to organic solvents. *Research in Microbiology* 153: 263-268.

Sarker K, DuttaS and Mohapatra PKD (2014) Fluorescent pseudomonads: Milestones achieved in the last two decades. *Afr J Microbiol Res* 8:1544-1561.

Sarma RK, Gogoi A, Dehury B, Debnath R, Bora T C, Saikia R (2014) Community Profiling of Culturable Fluorescent Pseudomonads in the Rhizosphere of Green Gram (*Vigna radiata* L.). *PLoS ONE* 9: e108378. doi:10.1371/journal.pone.0108378.

Schleifer KH (2009) Classification of *Bacteria* and *Archaea*: past, present and future.*Syst Appl Microbiol*32:533–542.

Schmidt KD, Tummler B, Romling U (1996) Comparative genome mapping of *Pseudomonas aeruginosa* PAO with *P. aeruginosa* C, which belongs to a major clone in cystic fibrosis patients and aquatic habitats. *J Bacteriol*178: 85-93.

Schurr MJ, Deretic V (1997) Microbial pathogenesis in cystic fibrosis: co-ordinate regulation of heat-shock response and conversion to mucoidy in Pseudomonas aeruginosa. *Molecular Microbiology* 24: 411-420.

Schurr MJ, Yu H, Boucher JC, Hibler NS, Deretic V (1995) Multiple promoters and induction by heat shock of the gene encoding the alternative sigma factor AlgU (σE) which controls mucoidy in cystic fibrosis isolates of Pseudomonas aeruginosa. *Journal of Bacteriology* 177: 5670-5679.

Schuster M, Lostroh CP, Ogi T, Greenberg EP (2003) Identification, timing, and signal specificity of Pseudomonas aeruginosa quorum-controlled genes: a transcriptome analysis. *J Bacteriol* 185: 2066–2079.

Sharma A, Sahgal M, Johri BN (2003) Microbial communication in the rhizosphere: Operation of quorum sensing.*Curr Sci* 85: 1164-1172.

Shaukat SS, Siddiqui IA (2003) Impact of biocontrol agents *Pseudomonas fluorescens* CHA0 and its genetically modified derivatives on the diversity of culturable fungi in the rhizosphere of mungbean. *J Appl Microbiol* 95: 1039–1048.

Shinoda S, Okamoto K (1977) Formation and function of *Vibrio parahemolyticus* lateral flagella. *Journal of Bacteriology* 129: 1266-1271.

Silby MW, Levy SB (2004) Use of in vivo expression technology to identify genes important in growth and survival of *Pseudomonas fluorescens* Pf0-1 in soil: discovery of expressed sequences with novel genetic organization. *J Bacteriol* 186: 7411–7419.

Silby MW, Winstanley C, Godfrey SAC, Levy SB, Jackson RW (2011) *Pseudomonas* genomes: diverse and adaptable. *FEMS Microbiology Reviews* 35: 652–680.

Sørensen J, Jensen LE, Nybroe O (2001) Soil and rhizosphere as habitats for *Pseudomonas* inoculants: new knowledge on distribution, activity and physiological state derived from micro-scale and single-cell studies.*Plant Soil* 232: 97-108.

Sørensen J, Skouv J, Nybroe O (1991) Rapid identification of environmental isolates *Pseudomonas aeruginosa, P. fluorescens* and *P. putida* by SDS PAGE analysis of whole- cell protein patterns. *FEMS Microbiol Ecol* 101: 41-50.

Spencker FB, Haupt S, Claros MC, Walter S, Lietz T, Schille R et al. (2000) Epidemiologic characterization of Pseudomonas aeruginosa in patients with cystic fibrosis. *Clinical Microbiology and Infection* 2000;6(11) 600-607. doi: 10.1046/j.1469-0691.2000.00171.x.

Spiers AJ, Buckling A, Rainey PB (2000) The causes of *Pseudomonas* diversity. *Microbiology* 146: 2345–2350.

Srivastava AK, Singh T, Jana TK, Arora DK, Singh TSO (2001) Induced resistance and control of charcoal rot in *Cicer arietinum* (chickpea) by *Pseudomonas fluorescens*. *Can J Bot* 79: 787-795.

Subramanian J, Satyan K (2014) Isolation and selection of fluorescent pseudomonads based on multiple plant growth promotion traits and siderotyping. *Chilean J Agric Res*74: http://dx.doi.org/10.4067/S0718-58392014000300010.

Sunish Kumar R, Ayyadurai N, Pandiaraja P, Reddy AV, Venkateswarlu Y, Prakash O, Sakthivel N (2005) Characterization of antifungal metabolite produced by a new strain *Pseudomonas aeruginosa* PUPa3 that exhibits broad-spectrum antifungal activity and biofertilizing traits. *J Appl Microbiol* 98:145-154.

Tebbe CC (2002) Dissemination of genetically engineered microorganisms in terrestrial ecosystems –case studies for identifying risk potentials. In: Lelley T, Balázs E, Tepfer M (eds.) *Ecological impact of GMO dissemination in Agro-Ecosystem.* OECD Cooperative Research programme workshop on "Ecological impact of GMO Dissemination in Agro-ecosystems," Grossrussbach, Austria, pp 31-44 ISBN 3-85076-631-4.

Teitzel GM, Parsek MR (2003) Heavy metal resistance of biofilm and planktonic Pseudomonas aeruginosa. *Applied and Environmental Microbiology* 69: 2313-2320.

Tindall BJ, Kampfer P, Euzeby JP, Oren A (2006) Valid publication of names of prokaryotes according to the rules of nomenclature: past history and current practice. *Int J Syst Evol Microbiol*56:2715–2720.

Troxler J, Zala M, Natsch A, Moënne-Loccoz Y, Défago G (1997) Autecology of the biocontrol strain *Pseudomonas fluorescens* CHA0 in the rhizosphere and inside roots at later stages of plant development. *FEMS Microbiol Ecol* 23: 119-130.

Van Loon LC (1997) Induced resistance in plants and the role of pathogenesis-related proteins. *European J Pl Pathol* 103: 753-765.

Van Loon LC, Bakker PAHM, Pieterse CMJ (1998) Systemic resistance induced by rhizosphere bacteria. *Annu Rev Phytopathol* 36: 453-483.

Vance CP (2001) Symbiotic nitrogen fixation and phosphorus acquisition. Plant nutrition in a world of declining renewable resources. *Plant Physiol* 127: 390-7.

Verberne MC, Verpoorte R, Bol JF, Mercado-Blanco J, Linthorst HJM (2000) Overproduction of salicylic acid in plants by bacterial transgenes enhances pathogen resistance.*Nature Biotechnology* 18: 779-783.

Viebahn M, Glandorf DC, Ouwens TW, Smit E, Leeflang P, Wernars K et al. (2003) Repeated introduction of genetically modified *Pseudomonas putida* WCS358r without intensified effects on the indigenous microflora of field-grown wheat. *Appl Environ Microbiol* 69: 3110–3118.

Vinatzer BA, Bull CT (2009) The impact of genomic approaches on our understanding of diversity and taxonomy of plant pathogenic bacteria. In: Jackson RW. (ed.) *Plant Pathogenic Bacteria: Genomics and Molecular Biology.* Horizon Scientific Press, pp 37-61.

Walker TS, Bais HP, Deziel E, Schweizer HP, Rahme LG et al. (2004) *Pseudomonas aeruginosa*–plant root interactions. Pathogenicity, biofilm formation, and root exudation. *Plant Physiol* 134: 320–331.

Walsh UF, Morrissey JP, O'Gara F (2001) Pseudomonas for biocontrol of phytopathogens: from functional genomics to commercial exploitation. *Curr Opin Biotechnol* 12: 289-295.

Wang C, Knill E, Glick BR, De´fago G (2000) Effect of transferring 1-aminocyclopropane-1-carboxylic acid (ACC) deaminase genes into Pseudomonas fluorescens strain CHA0 and its derivative CHA96 on their growth-promoting and disease-suppressive capacities. *Can J Microbiol* 46: 1–10.

Weller DM (1988) Biological control of soil-borne plant pathogens in the rhizosphere with bacteria. *Annu Rev Phytopathol* 26: 379–407.

Weller DM (2007) *Pseudomonas* biocontrol agents of soilborne pathogens: looking back over 30 years. *Phytopathology* 97: 250–256.

Wiyono S, Schulz DF, Wolf GA (2008) Improvement of the formulation and antagonistic activity of *Pseudomonas fluorescens* B5 through selective additives in the pelleting process. *Biol Control* 46: 348–357.

Woese CR (1987). Bacterial evolution. *Microbiol Rev*51: 221-271.

Wolterink AFWM, Jonker AB, Kengen SWM, Stams AJM (2002) *Pseudomonas chloritidismutans* sp. *nov*., a non-denitrifying, chlorate-reducing bacterium. *International Journal of Systematic and Evolutionary Microbiology* 52: 2183-2190.

Wood DW, Pierson III LS (1996) The *PhI* gene of *Pseudomonas aureofaciens*30-84 is responsible for the production of a diffusible signal required for phenazine antibiotic production. *Gene* 168: 49-53.

Woods DE, Straus DC, Johanson WG, Berry VK, Bass JA (1980) Role of pili in adherence of Pseudomonas aeruginosa to mammalian buccal epithelial-cells. *Infection and Immunity.* 29: 1146-1151.

Yadav AN, Kumar R, Kumar S, Kumar V, Sugitha TCK, Singh B et al. (2017) Beneficial microbiomes: Biodiversity and potential biotechnological applications for sustainable agriculture and human health. *Journal of Applied Biology & Biotechnology* 5:45-57.

Yadav AN, Verma P, Sachan SG and Saxena AK (2017) Biodiversity and biotechnological applications of psychrotrophic microbes isolated from Indian Himalayan regions. *EC Microbiology* ECO 01: 48-54.

Yamamoto S, Harayama S (1998) Phylogenetic relationships of Pseudomonas putida strains deduced from the nucleotide sequences of gyrB, rpoD and 16S rRNA genes. International Journal of Systematic Bacteriology 48: 813-819.

Yamamoto S, Kasai H, Arnold DL, Jackson RW, Vivian A, Harayama S (2000) Phylogeny of the genus Pseudomonas: intrageneric structure reconstructed from the nucleotide sequences of gyrB and rpoD genes. *Microbiology* 146:2385-2394.

Yan Y, Yang J, Dou Y, Chen M, Ping S, Peng J (2008) Nitrogen fixation island and rhizosphere competence traits in the genome of root-associated *Pseudomonas stutzeri*A1501.*Proc Natl Acad Sci USA* 105:7564–7569.

Yonezawa M, Takahata M, Matsubara N, Watanabe Y, Narita H (1995) DNA gyrase gyrA mutations in quinolone-resistant clinical isolates of *Pseudomonas aeruginosa. Antimicrobial Agents and Chemotherapy* 39: 1970-1972.

Young JM, Park DC (2007) Probable synonymy of the nitrogen-fixing genus *Azotobacter* and the genus *Pseudomonas. Int J Syst Evol Microbiol.*57:2894–2901.

Yu H, Schurr MJ, Deretic V (1995) Functional equivalence of Escherichia coli σE and *Pseudomonas aeruginosa* AlgU: E. coli rpoE restores mucoidy and reduces sensitivity to reactive oxygen intermediates in algU mutants of *P. aeruginosa. Journal of Bacteriology* 177: 3259-3268.

Zahnder GW, Murphy JF, Sikora EJ, Kloepper JW (2001) Application of rhizobacteria for induced resistance. Eur J Plant Pathol 107:39-50. Morris CE, Sands DC, Vinatzer BA, Glaux C, Guilbaud C, Buffière A, Yan S, Dominguez H, Thompson BM (2008) The life history of the plant pathogen *Pseudomonas syringae* is linked to the water cycle. *ISME J* 2:321–334.

Zhang Z, Pierson III LS (2001) A second quorum-sensing system regulates cell surface properties but not phenazine antibiotic production in *Pseudomonas aureofaciens. Appl Environ Microbiol* 67: 4305-4315.

Chapter 17

Soil Fertility Management: Issues and Challenges in Tropical Areas of Nigeria

Mukhtar Iderawumi Abdulraheem[1,2,3,4,*],
Syed Muhammad Zaigham Abbas Naqvi[1,2,3], Linze Li[1,2,3], Shakeel Ahmed[1,2,3], Wentao Wei[1,2,3], Su Rui[1,2,3], Baolei Li[1,2,3], Can Li[1,2,3], Yanyan Zhang[1,2,3], Yuan Feng[1,2,3] and Jiandong Hu[1,2,3†]

[1]Department of Electrical Engineering,
Henan Agricultural University, Zhengzhou, China
[2]Henan International Joint Laboratory of Laser Technology in Agriculture Science, Zhengzhou, China
[3]State Key Laboratory of Wheat and Maize Crop Science, Zhengzhou, China
[4]Department of Agricultural Science, Oyo State College of Education, Lanlate, Nigeria

Abstract

Soil fertility management has two requirements: establishing and maintaining the soil pH and essential plant nutrient elemental content within their desired ranges for that soil type and crop or cropping sequence, with its associated cultural management practices. It is obvious that none of the soil fertility management systems will meet all these requirements. However, there are basic principles that do apply, requiring moderate modification to suit the specifications of soil type, crop species, and climatic/weather characteristics. Management requirements for achieving a moderate yield and average product quality, require fewer inputs and skill requirements than those required for achieving maximum yield and highest product quality, the latter not allowing for errors in procedural practices. For most cropping situations, the maximum biological yield potential based on the combination of soil and plant parameters is unknown. It is also not possible to advance quickly from a moderate soil fertility/plant nutrition status to one that results in high yield or quality product achievement. Those management practices applied to one set of soil, plant, or climatic conditions are not applicable to all ranges of conditions. Hence, this chapter tends to discuss the problems of soil fertility and productivity, nutrients and mining and imbalance, and inadequate plant nutrients supply in Nigeria's agriculture, as well as the problems affecting the use of chemical fertilizers, the types, and role of organic sources of plant nutrients. Finally, a case will be made for integrated nutrient supply that combines chemical fertilizers with biological and organic sources.

* Corresponding Author's Email: m.iderawumi@gmail.com.
† Corresponding Author's Email: jdhu@henau.edu.cn.

In: Ecosystem Services
Editor: Hanuman Singh Jatav
ISBN: 978-1-68507-614-6

Keywords: soil fertility, organic fertilizer, soil nutrients, integrated soil nutrition, fertilizer use

1. Introduction

The activities of research and development in producing any agricultural technology for the farmer should be seen to be improving or incorporating farmers' indigenous knowledge and not producing modern technology that seems to have no input from them. The adoption of any technology developed in a participatory manner among the farmer, the researcher, and the extension agent *seems* to be faster than when indigenous knowledge *of* the farmers is ignored. This situation calls for in-depth studies on traditional practices and the strategies farmers employ to cope with the declining soil fertility status of the savanna region.

Tropical soils are characterized by high spatial variability on both macro and/or micro-scales due to the combined effects of intrinsic (e.g., biophysical and chemical processes), and extrinsic factors (e.g., crop management, fertilizer and tillage, among others) operating at different intensities and on different spatiotemporal scales (Tripathi et al. 2015). Spatial variability in soils and the corresponding variability of yield response of various crops to nutrient application have already been observed across Sub-Sahara Africa (Kihara et al. 2016; Kihara et al. 2017; Shehu et al. 2018; Akinseye et al. 2020).

Fertilizers play a critical role in enhancing productivity of crops along with other inputs. Food grains meet the basic food demand of the majority population in the country. Therefore, it is important to examine food grain production behavior and the relationship between the production of food grains and the consumption of fertilizers. The food grain production of the country has increased nearly 5.5 times, from 51.99 million tonnes in the year 1951-52 to 285.21 million tonnes in the year 2018-19. During the green revolution fertilizer responsive high-yielding varieties were introduced along with assured irrigation facilities, which immensely increased production of food grains.

Soils in Nigeria are not as fertile as people believe. The soils require development and implementation of special management approaches beyond appropriate fertilizer application, some of which have been developed to support food production on a continuous, sustainable basis. There is a need to establish a National Soil Research Institute to handle all issues relating to sustainable soil use and management (Fasina et al. 2015).

A discussion and appraisal of sources of nutrients for crop plants in Nigeria is a worthwhile exercise since the inadequate availability of plant nutrients and the associated degradation of agricultural lands is a major problem in Nigeria agriculture. One of the major factors responsible for food insecurity in Nigeria is poor crop yield, due mainly to unfavorable soil conditions (Fasina et al. 2015). A large proportion (70%) of the soils in Nigeria is made up of low activity clay (LAC) soils, which cannot naturally sustain crop production on a continuous basis (Ogunkunle, 2009). There is a need to give a current appraisal of soil fertility management practices and plant nutrient sources in order to provide information on the role of known and less known sources, their applicability, influence and suggestion to combat the problem of inadequate nutrient supply. Application of wastes generated from domestics, agricultural and industrial activities on agricultural land is a means of their disposal. The organic wastes could increase the presence of plant nutrients in the soil (Ojeniyi et al. 2017).

Research studies in Nigeria have focused mainly on the effects and use of synthetic fertilizers, and government efforts have been concentrated on supply and manufacture of chemical fertilizers. Recently, it has been revealed that dependence on chemical fertilizers as the main supply of plant nutrients is not sustainable at the level of majority of peasant farmers who produce 90% of the food needs in Nigeria. It therefore becomes essential to examine the role of traditional organic sources of plant nutrients in sustaining the dominant smallholding crop production system in Nigeria. This is more so because the average farm size continues to decrease in Nigeria and farmers will continue to depend largely on the local organic sources of plant nutrient supply.

It is therefore not surprising that low soil fertility has always been mentioned by farmers as one of the constraints affecting cereal production in Northern Ghana (RELC, 2005). This is confirmed by the fact that low grain yields of cereals attributed to poor soils for the past decade have been ranked first among the constraints collated from all the districts at the various regional planning sessions.

Soil degradation is widespread in Nigeria and continues unabated due to the absence of a 'National Soil Research Institute' with the mandate to oversee sustainable management of the nations' nonrenewable natural resource (soil). Integrated soil health/quality management approach should be adopted in agricultural soil use to ensure sustainable agricultural productivity, food security and environmental conservation (Chude and Odunze, 2015).

2. Loss of Soil Fertility and Productivity

As in most parts of tropical Africa, the traditional method of maintaining soil fertility and productivity in Nigeria has, hitherto, been the bush-fallow system whereby arable land is allowed to revert to fallow after 3 - 4 years of continuous cultivation. The system evolved out of natural exigencies, and the degree of regeneration of soil fertility is generally dependent on the length of the fallow period, which, in turn, is related to the availability of land. In view of the growing human population, and other socio-economic pressures, attempts were made to shorten the fallow period from about 7 - 10 years to 2 - 3 years by planting leguminous and grass fallows. Although it was clearly demonstrated that soil fertility could be effectively maintained in this way, its acceptance by the peasant farmer was hampered by the tedium of land preparation with native hoes and the economics of the practice; it was not an easy task to convince a farmer to adopt a system that included unproductive fallows.

Despite the high natural fertility of the soils, intensive cultivation without adequate replenishment of soil nutrients through organic or inorganic fertilizers has resulted in the decline of soil fertility, particularly, in the dryland areas, where the only source of nutrients for maize crops comes from intercropped beans. The soil organic matter and carbon content is low ($<2\%$) due to lack of soil cover, removal of crop residue from farmlands, and high rate of organic matter decomposition (Baptista, 2015).

The low soil fertility in this part of the country is therefore blamed on the bush burning which usually occur annually during the dry season commencing from October to April the following year (SARI, 1995). This situation renders the soil bare exposing it to both wind and water erosion in the dry and rainy seasons respectively, thereby depleting the macro-nutrients such as Nitrogen, Phosphorus, Potassium (NPK), and organic matter from the soil. Initially, farmers used to replenish the soil with its nutrients by practicing shifting cultivation or land rotation. However, with the increase in population which has put pressure on land use, this practice is not being sustained and this therefore calls for other measures to maintain soil fertility for sustainable crop production in the savanna zone.

Shifting cultivation and the fallow system have been the reliable strategy for maintaining soil fertility in Nigeria. This traditional system functions perfectly as long as population pressure does not intervene. However, because of the increasing population and decreasing available land per head for food production, the bush fallow is no more efficient to maintain soil productivity. The fallow period is either too short or non-existent. The series of studies conducted by IITA in Kaduna, Sokoto, and Bauchi States showed that the fallow period had disappeared in 59% of the villages and declined in 26% others. The fallow period is now less than 3 years in most of southwest and no more existing in most of the southeast Nigeria. Whereas, the minimum period of 6 years in the south and 10 years in the North is required for soil fertility restoration.

Several studies in Sub-Sahara Africa (Kihara et al. 2016; Kihara et al. 2017; Shehu et al. 2018; Tittonell et al. 2008) have recommended that nutrient management and fertilizer recommendation strategies should be tailored towards field-, site-, or soil-specific conditions to achieve balanced and effective fertilizer use with close nutrients related yield gaps. Matching the right fertilizer with the right recommendation is therefore regarded as critical in optimizing and sustaining crop yield and economic returns (Kamaluddin, et al. 2020).

In Nigeria, soil fertility and crop yields decline with years of cultivation even when chemical fertilizers are used. This is attributed to degradation in soil physical, biological and chemical qualities, negative nutrients balance, loss of organic matter, nutrient removal as a result of crop harvesting, poor nutrient cycling process, soil acidity, and other problems associated with the use of chemical fertilizers, bush burning, erosion, nutrients leaching, etc. The common factor associated with degradation in soil physical, chemical, and biological conditions is the associated fall in soil organic matter contents with cultivation. For example, studies conducted in the Department of Soil Science, Obafemi Awolowo University, indicated 75% decline in soil organic matter content during 18 years of cultivation. Also within a year of opening a fallow land for cultivation, Nigerian soil undergoes about 30% loss in organic matter and 27% loss in organic phosphorus. The characteristic low organic matter content of Nigerian soils, coupled with its rapid dissipation under cultivation, suggests that Nigerian soils may be poorly adapted to continuous cultivation unless a reasonable amount of organic matter can be maintained in the soil. Several studies have demonstrated that a combined application of organic materials and synthetic fertilizer to soil benefits both the crop and soil tremendously.

The recent move towards market agriculture and intensive production led to a negative effect on the supply of plant nutrients. When agricultural products are sold, plant nutrients leave the soil. The replacement of these nutrients with synthetic fertilizers is often inadequate and unbalanced.

The problem of negative nutrient balance and soil fertility loss shall become more serious in the future with the decrease in agricultural land area per capital. In 1997, the amount of cultivable land ha per head was just 0.85. This will progressively diminish to 0.77 in 2000, 0.41 in 2020, and 0.15 in 2050. Agricultural land use will become more intense in order to feed the increasing population. Although the annual food production increased by an average of 3.1% between 1989 and 1993, this food production growth has failed to meet up with the 3.3% annual population growth rate of the same period. Nutrient deficiencies and imbalances are the main constraints to crop production. A negative nutrient balance means the natural capital of the farm is being mined; yield and protective plant biomass cover will gradually decrease. Hence, sufficient organic and chemical fertilizers should be used, and nutrient cycling should be ensured.

3. Fertilizers Use in Nigeria

Agricultural production techniques have remained rudimentary. The use of inputs such as fertilizers and improved seeds is low. The current average application of nutrients per ha of arable land is estimated at 6 kg. The average yields of millet and sorghum, which together account for almost one-third of the harvested area, remain virtually unchanged at about 1000 kg per ha. From a food-deficit situation during the late 1970s and early 1980s, Nigeria's agricultural production grew throughout the 1990s to attain self-sufficiency in major staples such as maize, sorghum, millet, as well as cassava, yams, and cocoyam. Rice production also increased. Over the past 20 years, value-added per capita in agriculture has risen by less than 1 percent per year. Food production gains have not kept pace have population growth, resulting in rising food imports and declining levels of national food nutrients (World Bank, 2005).

Sustainable management of soil is of prime importance for increasing and sustaining crop yields globally and, more importantly, in sub-Saharan Africa (SSA), where food production is not at pace with the high food demand that resulted from the rapidly growing population (Vlek et al. 2008; Dewitte et al. 2013). Substantial area with high production potential in the SSA region has been degraded due to continuous cropping with poor nutrient management (Bationo et al. 2015).

For continuous crop production and sustained yields, inorganic fertilizers are used to supplement the native and organically derived nutrients in the soil. Although there is renewed interest in the world in the use of organic manures and biofertilizers rather than chemical fertilizers, the reasons are economical and ecological and not due to unsatisfactory performance. According to the FAO, chemical fertilizers have been responsible for 50 percent of the yield increase in food crops In developing countries since the 1960s.

Fertilizers are agro-chemicals. When applied appropriately and in the correct dosage, fertilizers, enhances vegetative growth in plants, keep the environment healthy. The leaf canopy protects the soil from the direct impact of rain drops which could result in soil erosion. The life of microorganisms living beneath the soil and animals is protected. During eutrophication, fertilizer elements are eroded into the rivers and lakes for the nutrition of aquatic life. The ecosystem is maintained and the biodiversity is rejuvenated.

In urban and rural areas, organic wastes, consisting of human excrement animal dungs, bird guano sawdust and bush burning, domestic and industrial wastes, and others, abound, if not adequately managed can constitute potential environmental hazard. However, when converted technologically or biologically into organic fertilizers and used as feeds for animals, fishes in the ponds and rivers, biodegradation of the organic matter helps in improving the environment and making it safe from pollution.

The soil management (or soil fertility management) zoning approach is a popular and feasible approach to developing and implementing site-specific nutrient management strategies to manage spatial variability on macro- and microscales (Tripathi et al. 2015). The tropical world, particularly Africa, is poor in its level of fertilizer application to crops. Whereas, the world average consumption of NPK fertilizers per ha in 1986 was 90.4 kg, it was 19.3 kg for Africa and 9.4 kg for Nigeria. She needs 2.5million tonnes of fertilizer yearly, but has rarely attained one million tonnes yearly, both from local production and importation in the last few years. Up to 70% of the fertilizers are consumed in the Northern states. In Nigeria 87 kg NPK/ha/yr is mined, the value being higher than those recorded for Zimbabwe, Coted'voire, Benin and Ethiopia. Spatial variation of pH and macronutrients (N, P, and K) was largely influenced by indigenous or inherent bio-physical factors, while variation of most micronutrients was significantly influence by both indigenous and artificially human-induced soil management factors. This emphasizes the need for the development of site- or area-specific soil management advises. (Kamaluddin et al. 2015)

India, China, and the UK were 17, 44, 57, 75, 273 and 343 kg/ha respectively. Fertilizer consumption in Nigeria dropped drastically between 1993/94 and 1996/97. Compared with the world average consumption of 89.4 kg/ha nutrients for 1996, the value for Nigeria was 4.1. Fertilizer use in Nigeria in 1996/97 amounted to 1.17 kg/persons compared with 18.4 kg/person for South Africa.

The high cost of chemical fertilizers has greatly contributed to their low use by Nigerian farmers. As at 2000, the subsidy on fertilizer was reduced to zero, and prices per 50 kg bag of fertilizer increased from N150 in 1996 to N7,800 – N10,000 in 2020. Fertilizer requests rose from 298,000 tonnes in 1977 to 7.4 million tonnes in 2020, whereas the highest supply was less than 0.6 million tonnes. Local fertilizer production does not meet domestic requirements, even at the low rate of fertilizer application that currently prevails.

Some researchers in Nigeria reported that wood ash increased nutrients availability (Owolabi et al., 2003; Ojeniyi et al., 2002) and crop performance in production of crops such as Soybeans (Akinola et al., 2013, Soretire and Olayinka, 2013a, 2013b), Sorghum (Abdulraheem et. al., 2012), Okra (Abdulraheem, 2018; Abdulraheem and Ojeniyi, 2015), cassava (Babadele and Ojeniyi, 2013a, 2013b, Ojeniyi et al., 2015), 217 tomato and pepper (Ewulo et al., 2009, Odedina et al., 2003), amaranthus (Abdulraheem et al. 2021; Ojeniyi and Adejobi, 2002), groundnut (Awodun and Ojeniyi, 2005, Obiefuna et al., 2003) and vegetables (Ojeniyi et al., 2002).

The use of chemical fertilizers by the majority of small holding crop farmers is not common. Apart from the high cost, farmers do not have access to fertilizers. Also, a high percentage of the peasant farmers are illiterate and the problem of misuse of fertilizers occur, even though, the extension workers may help in disseminating information on the use of fertilizers. Most farmers apply fertilizers without Soil testing. Deficiency of secondary and micronutrients occurs in soil and crop with continuous use of the common NPK fertilizers. Soil acidity may also occur. Continuous cropping with the aid of chemical fertilizers may be accompanied by yield declined attributable to loss of soil organic matter, nutrient imbalance and degradation of soil physical properties. Farmers in the Northern parts of Oyo State of Nigeria discontinued the use of certain fertilizers for the following reasons;

1. that fertilizers promote weed;
2. that maize and yam grown with fertilizers had poorer taste and shorter storage life;
3. that fertilizers usually lead to greater vegetative growth rather than fruit/tuber formation.

Fertilizers are chemicals often applied to crops to enhance growth in situations where mineral elements required by the crops are in short supply. Fertilizers are more frequently used in developing nations, principally to sustain food sufficiency, irrespective of the environmental problems such use could present. Inhabitants in developing nations need food. They need to improve on their traditional agricultural practices and apply fertilizers.

However, such application of fertilizers should be done with care and through adequate study of the needs of the soil and plants. Recent soil fertility concepts are pointing to the "ecologizing" of agriculture by emphasizing ecological, organic, or biodynamic systems of farming both in rural and urban locations. This concept appears to de-emphasize the use of synthetic fertilizers.

In temperate agriculture, fertilizer has been used to maintain continuous arable crop production, but not so in Nigeria. The efficiency of fertilizer use in Nigeria agriculture is low due to these reasons:

a) lack of soil testing programme;
b) prolonged use of fertilizer has led to soil acidification;
c) leaching of applied fertilizers under open clean cultivation
d) use of incomplete fertilizers,
e) low light intensity;
f) water stress;
g) increasing weed problem,
h) fertilized yam turns black after cooking and does not store well.

One outstanding and general hindrance to effective fertilizers use in Nigeria is limited use of soil tests for making fertilizers recommendations. Soil acidity, lack of soil testing, poor handling of fertilizer, lack of quality control, illiteracy on part of farmers, negative impressions on the effect of fertilizers and ineffective extension service are also problems militating against effective and efficient use of fertilizers in Nigeria.

The growth of fertilizer use has been constrained by non-adoption of research recommendation on management of fertilizers leading to soil degradation and declining crop yields, high costs and low profitability of fertilizers, scarcity and relegation of soil testing as a sound basis of fertilizers recommendation, and absence of effective quality control mechanism. Most small scale farmers apply about 33% of the recommended fertilizers rate due to high cost. Whereas large scale individual and corporate farmers apply high use of N-fertilizers leading to soil acidity and declining crop yields. Indiscriminate use of fertilizers and wrong methods of application also hinder effectiveness of fertilizers. Increased aggregate food production has not necessarily translated into improved household food security and nutrition. During the pilot phase of the Nigeria Program for Food Security (NPFS), it was found that out of a total of 1,718 households in four Local Government Councils of Kano State, 80 percent were food insecure; almost half suffered from temporary food insecurity while 34 percent suffered from chronic food insecurity (FAO, 1996).

It appears that the soil cannot be rendered productive by synthetic fertilizers alone, but plus appropriate soil management and farming techniques that build up soil organic matter. Nearly all attempts to maintain continuous crop production with chemical fertilizers alone have failed in the tropics. Therefore, agricultural and organic wastes will continue to be important source of nutrients for intensive crop production, especially among the majority smallholders.

4. Organic Fertilizers

Meeting the food demand of the people would be a big challenge with the burgeoning population of Nigeria and limited land and water resources. The only option to confront this rise in demand is to increase the productivity of agriculture in a sustainable and cost-effective manner.

Crop production in Nigeria is sustained mainly by nutrients derived from organic sources and bush fallows. However, the maintenance of organic matter in soil is a major problem, especially in semiarid and arid zones, where there are high competing demand for organic and agricultural wastes. Large quantities of human, agricultural, forestry, and agro-industrial waste are produced annually, which are not being effectively utilized. However, because of the increasing cost of chemical fertilizers coupled with disposal problems posed by these wastes, their use as a means of maintaining organic matter levels and boosting agricultural productivity has become an economic proposition.

The use of poultry dung, cow dung and household refuse increases the efficiency of inorganic fertilizers by providing secondary and micronutrients not present in the inorganic fertilizers. Long term studies in northern Nigeria (Lombin and Abdullahi 1977), showed that 5t/ha of dung annually combined with 100kg N and 50kg P205 will maintain yield under continuous cultivation.

In an Ultisols at Agege in South Western Nigeria, maize yields were doubled from 1.2 to about 2.5ton/ha for the duration of the study (1962 - 1967) by addition of 5t/ha of dung annually. From the analysis of the dung and also other treatments included in the study, the dung provided lime effect, Mg and micronutrients deficiency in the soil.

Studies by the author showed that plant wastes such as wood ash, ground cocoa pod husk, brewery waste (spent grain), rice bran, oil palm bunch ash, and sawdust ash are effective as fertilizers for okra and amaranthus. The effects of these agro wastes were enhanced by mixing them with poultry, pigs, goats, and excreted materials. The residues increased soil organic matter, N, P, K, Ca, Mg and pH. Sawdust and rice barn with higher C:N ratios gave the lowest values for these soil parameters. Compared with NPK fertilizers, the residues increased soil organic matter, Ca, Mg, and pH. Application of the residues sole or amended with types of animal manure increased soil N, P, Ca and Mg contents, growth and leaf yield of amaranthus. The chemical analysis of the plant residues showed that they contain nutrients such as N, P, K, Ca, Mg, Fe, Mn, Cu and Zn. Studies carried out by the author also found that wood ash applied as fertilizers increased soil pH, nutrients content, growth and yield of maize and okra and other vegetables in Southwest Nigeria, and that goat dung manure increased soil P and K and yield of okra and amaranthus. Ash derived from plant wastes also served as liming material.

However, with agriculture becoming more and more intensive, coupled with the introduction of higher-yielding and more nutrient-demanding crop varieties, it became obvious that FYM could not be obtained in sufficient quantities to meet farmers' demand. Even where available, transportation problems and labour costs (unavoidably) limit its use on a routine basis. In these circumstances, attention was turned to mineral fertilizers as the alternative.

Weeds and weed residues are important sources of plant nutrients and could be used to sustain soil fertility and productivity. This is especially so in humid Southern Nigeria where large volumes of weed residues are generated during manual clearing after fallow. Chromolaena odorata (Siam weed) is particularly recognized by farmers for its soil improving capability. If the soil is fallowed heavily with chromolaena, it is very likely that the soil will be fertile enough to sustain good crops for at least two years. In the past 30 years, the dominant fallow species in Southwest Nigeria is Chromolaena which provides substantial biomass. On decomposing chromolaena increased availability of soil organic matter, N, P, K, Ca and Mg and yield of crops such as coffee, okra and maize. Also, residues of chromolaena and other weeds such as ageratum, elephant grass, guinea grass and aspilia were investigated as to their effects as sources of nutrients to maize. The weed residues applied as surface mulch to maize at 10 and 20Ot/ha increased growth and yield of maize and leaf contents of N, Ca, Mg and K. The increases in grain yield and leaf N were significant. Mulching of tomato plants with residues of axenopus, cynodon and eleusine at 12t/ha increased growth, flowering, fruit setting and yield in Southwest Nigeria. Apart from serving as mulch, the weed residues supply nutrients and improve yield. The residues are more effective if used as a mulch material rather than being burnt into ash or incorporated. Recently, Mexican sunflower (Tithonia diversifolia) has become a dominant aggressive weed for soil improvement in Nigeria. Mexican sunflower and siam weed also have the ability to improve soil physical properties.

An increase in the price of chemical fertilizers and the withdrawal of government intervention have stimulated Nigerian farmers to search for local resources and biological means of maintaining soil fertility. The addition of major quantities of organic matter to soil has proven to be the most important means of sustaining soil productivity by the majority of peasant farmers. For example, continuous homestead farming is expanding in the Eastern States of Nigeria as a result of the use of organic fertilizers. Advance booking is made for poultry dropping. For the planting of pepper at household level, the use of poultry droppings, cow dung, wood ash and plant residues for improving the fertility of nutrient-depleted soil has been adopted. In the Northern States, domestic and agricultural wastes are deposited on farmland and incorporated at the beginning of the cropping season. This enables continuous cropping of the same land. The wastes such as animal excreta, ash from crop wastes such as sugar cane, guinea corn and cowpea are composted and returned to farm land. The organic wastes supply required plant nutrients in balanced or nearly balanced proportions, and in addition stabilizing soil structural and physical conditions; improving biotic life, moderating soil temperature, and improving water availability and conservation as opposed to synthetic fertilizers. These attributes cumulate in improved growth, yield, and nutritional status of crops while production is made more sustainable.

Recycling of crop residues will replace 40 to 50% of N exported by a crop, 25 to 40% of the P and 70% of the K. In the Nigerian Savanna, about 31.4 million tonnes of crop residues are produced annually. Substantial quantities are generated from sorghum, millet, cotton, maize, groundnut, and cowpea. Residues contain substantial amounts of macro and micronutrients which are rich in K, low in P and poor to medium rich in N. About 45% N, 40% P, 86% K, 92% Ca, 72% Mg and 59% S removed from the soil by crops are contained in crop residues.

4.1. Cases for Integrated Soil Fertility Management Practices

In recent years, the farming population has aged, reflecting the migration of youth to the rapidly expanding urban economy where services are better than in rural areas. Smallholder or subsistence agriculture is unattractive to young people. The steady out-migration of younger workers to urban centers has deprived the agricultural sector of labor, creating the paradox of rural labor shortages in a labor -surplus economy (World Bank, 2005). The high cost, scarcity, and low efficiency make fertilizer use unprofitable for farmers. Also, without an adequate supply of organic matter, continuous use of NPK fertilizers leads to soil acidification, nutrient imbalances, and degradation in soil physical quality. Other problems associated with the use of synthetic fertilizers have been mentioned earlier. However, it is often difficult to obtain an adequate quantities of organic fertilizers for use, especially, in large scale arable cropping. On average, 20 kgN, 4 kg P, and 20 kgK are absorbed in the production of 1 tonne grain of some of the major cereals in West Africa, and it is difficult to supply these nutrients as manure alone.

The concept of bio-nutrition, whereby plants are induced to effectively utilize the natural nutritional potentials of the soil, deserves further study. It is possible that plants growing in their natural habitat may not require added fertilizers until such a time that they begin to exhibit obvious nutrient stress symptoms. In addition, the screening of crop/plant varieties for tolerance or adaptation to the mineral nutrient content of the soil would result in effective and economic use of fertilizer.

It is more efficient to combine organic and inorganic nutrient sources than to use either separately. In terms of residual effect, the effect of chemical fertilizers may not extend to the following cereal crop, especially in terms of yield, whereas the effects of organic manure, such as poultry manure, extend to following one or two crops in terms of increased yields, plant nutrients content and soil fertility. Organic products and animal wastes can be converted and recycled to feed the plant and improve soil fertility. Further research is needed in the conversion of organic waste to fertilizers and the use of the biogas so generated for domestic needs.

Combining the application of organic and inorganic sources of plant nutrients also reduces the quantities of either source required. Unlike in the case of NPK fertilizers alone, application of the poultry manure in combination with NPK fertilizers also increased Ca and Mg status of soil and plant and reduced soil acidity. Other advantages of involving poultry manure are improved soil physical condition, increased water holding capacity, and enhanced soil organic matter status.

Organic manures are excellent sources of organic matter but are relatively low in nutrients. Therefore, to obtain optimum yields from continuous cropping, additional nutrients from inorganic fertilizers will be needed to complement organic fertilizer.

However, it should be noted that fertilizer-use, though highly desirable for increased crop yield, can be subjected to abuse or serious misuse. Misuse of fertilizers can emanate from the poor knowledge or awareness of the users of fertilizers, the distribution channels, the soil fertility or chemical status, crop-specific needs and farming systems adopted by the farmers. The practice of providing fertilizer recommendations that are not based on adequate soil and sometimes plant tissue tests, the use of generalized (blanket) application of fertilizer rates for all crops and errors in calculation of crops, needs and mode of application constitute misuse and should be avoided. Fertilizer users should, therefore, be adequately informed and advised by extension agents so as to derive maximum benefits from the use of fertilizers.

Recent developments in soil fertility management and plant nutrient supply focus on integrated nutrient management systems that fully utilize nutrients supplied from both organic and inorganic sources. This is emphasized in the FAO's approach to soil fertility and productivity maintenance in the past decade. One of the objectives of FAO'S IPNS (Integrated plant Nutrients System) is to maintain or enhance soil productivity through a balance use of mineral fertilizers combined with organic sources including green manuring, mulch, animal manure, use of legumes for biological fixation, and use of crop residues and domestic waste. The concept of organic recycling to make nutrients available for soil fertility maintenance is sound and needs encouragement, particularly in the dominant smallholder production systems of the tropics.

Improving the phytoavailability of nutrients in soil is one of the primary goals of plant nutrition research. NP application can influence nutrient availability in soil by modifying the adsorption-desorption mechanism. Diffusional transportation and diffusion coefficients lie within low ranges (10–12 to 10–15 cm2 s−1) in soils that ensure an adequate phosphorus supply to plants (Lynch et al. 2012). Phosphorus supply was reported to improve in *Brassica napus* by using P-loaded Al2O3 NP in a hydroponic system (Santner et al. 2012).

Furthermore, the precise application of NPs may improve the use of soil pore water occurring in lower soil horizons, resulting in better dissolution of NP complexes, thereby improving the diffusion of bound phosphorus into the rhizosphere. Various approaches could be used to improve the phytoavailability of phosphorus in soils, such as changing the rhizosphere soil pH or improving the root exudation.

The future requires that farmers farm the same piece of land continuously and they will be faced with the challenge of increasing food production on existing land that is already in cultivation. This has to be done in a way that does not lead to diminished soil productivity. Combined the use of chemical and organic fertilizers is required for sustainable soil productivity under intensive continuous cultivation. There is a need to encourage the development of organic-based fertilizers in their pure form or fortified with chemical fertilizers. There should be a prioritized research programme focusing on organic matter and inorganic fertilizers for building soil and the development of indigenous materials for recapitalization and maintenance of soil fertility.

It is commendable that some state governments are setting up organic fertilizer plants and Some agro-based industries are known to be processing poultry manure for sale as fertilizer. Subsistence farming based on ecologically sound and organically based soil management practices seem to be the sustainable basis for livelihood within the reach of the majority small holdings farmers.

Conclusion

Fertilizers play a critical role in enhancing crop productivity, farmers income, and the sustainability of production systems. Consumption of fertilizers has been increasing over time, and currently, Nigeria stands among the top users of nitrogen (N), phosphorus (P), and potash (K) consumption. The demand projections showed that, at the current rates, we would need an additional 2.57 to 2.97 million tonnes of fertilizer by 2023-24. However, the major concern were the imbalanced use of N, P and K fertilizers, which has been arising mainly because of limited purchasing power of farmers, and the use of urea because of its comparatively low prices. The nutrient-based subsidy has addressed this problem to an extent, but more efforts are needed not only to promote balanced use of fertilizers but also organic and inorganic sources of fertilizers, which have more impact on crop productivity besides the positive impact on microbiome and environment.

Small holder farmers' decisions to adopt fertilizer and their subsequent decisions about application rates are influenced by agroecology, type of crops cultivated, age of farmer, educational status, access to credit, access/ distance to fertilizer market, and gender. Soil fertilizer extension issues generally include:

- Generalized/ blanket fertilizer recommendations, regardless of the complexity and diversity of the context in which farmers are operating. This often leads to an imbalance in nutrients and soil degradation where the nutrients are under-applied.
- Low adaptation of fertilizer recommendations to farmers' socioeconomic conditions/ relative annual changes in economic conditions which imply changes in optimal/ most profitable rates of fertilizer for farmers to use. Fertilizer recommendations as currently made available are not dynamic/ responsive to these changes that take place within the farming system.
- Lack of technical knowledge by farmers on appropriate fertilizer use.
- Limited or no access to soil test results/ soil testing laboratories.
- Limited access to extension services.
- Other demand and supply issues

The raw materials for the production of fertilizers are concentrated in few countries. This has also led to a limited number of countries from which both raw materials and finished products are imported. The dependence on imports has resulted in greater exposure to international markets. The major problems faced are price fluctuations, policy uncertainty in the exporting country, and the viability of the overseas investment. These problems directly impact the availability and amount of subsidies paid for fertilizers by the central exchequer. The acquisition policy of fertilizers and their feedstock should depend upon the global outlook of these products. The available evidence indicates the global outlook for fertilizer looks optimistic in terms of the increasing availability of all fertilizers. The estimates of the International Fertilizers Association indicate that there is already a surplus production of N, P & K fertilizers globally and that it is likely to increase in the future with the creation of additional production capacity.

In the continuing efforts at improving fertilizer use efficiency in Nigeria, therefore, the next logical step is to be able to provide specific recommendations for groups of soil having fairly similar properties and fertilizer response potentials. To achieve this, the following areas of research must now be pursued vigorously.

1. Calibration studies to determine the fertilizer nutrient response characteristics of benchmark soils across the country. In this connection, the initial concerns should be N, P, K, S, Zn, and B.
2. Determination of the critical soil nutrient fertility values in the different soil groups.
3. Use of modeling to improve decision making on Fertilizer Management.

Considering farming in a broad general sense, one notes that farming is changing constantly. Therefore, it is recommended that farmers, in addition to making proper use of fertilizers, should also aim at gaining knowledge of all the underlying principles and processes, to enable them to cope with new situations or new and different problems. Farmers are forced to change their farming system or management practices when technical, economic, or social conditions change.

Acknowledgments

The authors are thankful to the National Natural Science Foundation of China (grant number 32071890) and Supported by Henan Center for Outstanding Overseas Scientists.

References

Abdulraheem, M. I. (2018): Growth and Yield Responses of Okra *(Abelmoscus esculentum L.)* As Influenced By Sawdust Ash and Ammonium Nitrate. *Sumerianz Journal of Agricultural and Veterinary*. Vol. 1 (1): 8 – 13.

Abdulraheem, M. I. and Ojeniyi, S. O. (2015) Combined application of urea and sawdust ash in okra production effects on yield and nutrients availability. *Nigerian Journal of Soil Science* 25, 146-154.

Abdulraheem, M. I., Ojeniyi, S. O. and Charles, E. F. (2012). Integrated application of urea and sawdust ash effect on soil chemical properties plant nutrients and soghum performance. *IOSR Journal of Agriculture and Veterinary Science* 1(4), 38-41. DOI: 10.9790/2380-014384.

Abdulraheem, M. I., Charles, E. F. and Fudzagbo, J. (2021): Amaranthus Performance as Influence by Integrated Application of Goat Manure and Urea Fertilizer. Journal of Soil Science and Plant Physiology. 3(1):1 - 4. DOI: https://doi.org/10.36266/JSSPP/134.

Akinola, M. O., Ojeniyi, S. O. and Amusan, A. O. (2013). Soil chemical composition and soybean yield as influenced by tillage and ash. *Nigerian Journal of Soil Science* 22(1), 24- 28.

Akinseye, F. M., Hakeem, A. A., Kamara, A. Y., Elijah, A. A., Anthony, M. W. Understanding the response of sorghum cultivars to nitrogen applications in the semi-arid Nigeria using the agricultural production systems simulator. *J. Plant. Nutr.* 2020, 43, 834–850.

Awodun, M. A. and Ojeniyi, S. O. (2005). Response of groundnut yield and nutrient status to tillage and wood ash. *Nigerian Agricultural Journal* 36, 80-87.

Babadele, D. O., Ojeniyi, S. O. and Agbede, O. O. (2013a). Periodic nutrient content of cassava as influenced by sawdust ash and siam weed residue. *Proceedings 37th Annual Conference of Soil Science Society of Nigeria*, Lafia. P 133-140.

Babadele, D. O., Ojeniyi, S. O. and Agbede, O. O. (2013b). Periodic evaluation of growth and root count of cassava and effect of sawdust ash and Siam weed residue. *Proceedings 37th Annual Conference of Soil Science Society of Nigeria*, Lafia. P 141-150.

Baptista I. (2015): Strategies towards sustainable soil management in Cabo Verde: Environemntal and livelihood challenges. In a Book "Sustainable Soil Management: Key to Food Security and Nutrition in Africa." *Nature and Fauna*. 30(1): 26-29.

Bationo, A., Lamers, J., Lehmann, J. Recent achievement of sustainable soil management in Sub-Saharan Africa. *Nutr. Cycl. Agroecosyst.* 2015, 102, 1–3.

Chude V. O. and Odunze A. C. (2015): Priopities for Sustainable Soil Management in Nigeria. In a Book "Sustainable Soil Management: Key to Food Security and Nutrition in Africa." *Nature and Fauna*. 30(1): 17-21.

Dewitte, O., Arwyn, J., Otto, S., Henrik, B. M., Michel, B., Almami, D., Jozef, D., Tahar, G., Stephen, H., Robert, J. Harmonization of the soil map of Africa at the continental scale. *Geoderma* 2013, 211, 138–153.

Ewulo, B. S., Babadele, D. O. and Ojeniyi, S. O. (2009). Sawdust ash and effects on soil and plant nutrient contents and yield of tomato, *American Eurasian Journal of Sustainable Agriculture*3(1), 88-92.

Fasina A. S., Oluwadare, D. A., Omoju, O. J., Oluleye A. K, Ogbonnaya U. O and Ogunleye K. S. (2015): Can Nigeria Soils Sustain Crop Production? The dilemma of a Soil Scientist. In a Book "Sustainable Soil Management: Key to Food Security and Nutrition in Africa." *Nature and Fauna*. 30(1): 30-38.

Kamaluddin T. A., Alpha Y. K., Jibrin M. J., Jeroen E. H., Bello M. S., Julius B. A., Ibrahim B. M., Reuben S., Adam M. A. and Ayuba M. S. (2020): Delineation of Soil Fertility Management Zones for Site-specific Nutrient Management in the Maize Belt Region of Nigeria. *Sustainability* 2020, 12, 9010, doi:10.3390/su12219010.

Kihara, J., Nziguheba, G., Zingore, S., Coulibaly, A., Esilaba, A., Kabambe, V., Njoroge, S., Palm, C., Huising, J. Understanding variability in crop response to fertilizer and amendments in sub-Saharan Africa. *Agric. Ecosys. Env*. 2016, 229, 1–12.

Kihara, J., Weldesemayat, G., Nziguheba, G., Kinyua, M., Zingore, S., Sommer, R. Application of secondary nutrients and micronutrients increases crop yields in sub-Saharan Africa. *Agron. Sustain. Dev*. 2017, 37, 25.

Lynch J, Marschner P, Rengel Z (2012) *Effect of internal and external factors on root growth and development. In: Marschner's mineral nutrition of higher plants*. Academic, Amsterdam/Boston, pp 331–346.

Obiefuna, J. C., Ibeawuchi, I., Ekong, E. A., Okoli, N. A. and Alagba, R. A. (2012). Liming efficiency of organic and inorganic limes for acid soil amelioration and production of groundnut in Tropical rainforest - ultisols of Southeastern Nigeria. *Proceedings of the 8th Annual Conference on Organic Agriculture University of Calabar* P 18-22.

Odedina, S. A., Odedina, J. N., Ayeni, S. O., Arowojolu, S. A., Adeyeye, S. D. and Ojeniyi, S. O. (2003). Effects of types of ash on soil fertility nutrient availability and yield of tomato and pepper. *Nigerian Journal of Soil Science* 13, 61-67.

Ogunkunle, A. O. (2009). Management of Nigeria soil Resources for sustainable Agricultural productivity and food rd security, *Proc. of the 33 Annual conference of soil science society of Nigeria held at university of Ado – Ekiti* (March 9 –13, 2009) 9 – 24pp.

Ojeniyi S. O., Oso, O. P. and Arotolu, A. A. (2002). Response of vegetables to wood ash fertilizer. *Proceedings of the 35th Annual Conference of Agricultural Society of Nigeria, UNAAB*, P 39-43.

Ojeniyi S. O., Oyatoye, A. and Abdulraheem, M.I (2017): Use of Ash for Soil Fertility Improvement: Effect on Cowpea. *Nigeria Journal of Soil Science*. 27: 216 – 221.

Ojeniyi, S. O. and Adejobi, K. B. (2002). Effects of ash and goat manure on leaf nutrient composition growth and yield of amaranths. *The Nigeria Agricultural Journal* 33, 46-57.

Ojeniyi, S. O., Adejuwon, A. and Abdulraheem, M. I. (2015). Soil Physico-chemical properties and cassava yield in alfisol given tillage ash treatments. *Nigerian Journal of Soil Science (Special)* 25,71-82.

RELC, NR. (2005). Outcomes of Northern Region RELC Planning Session for 2005. Tamale.

Santner J, Smolders E, Wenzel WW, Degryse F (2012) First observation of diffusion-limited plant root phosphorus uptake from nutrient solution. *Plant Cell Environ* 35(9):1558–1566.

SARI (1995). *Savanna Agricultural Research Institute (SARI) annual Report for 1995,* Nyakpala, Ghana. 133pp.

Shehu, B. M., Merckx, R., Jibrin, J. M., Kamara, A. Y., Rurinda, J. Quantifying variability in maize yield response to nutrient applications in the Northern Nigerian Savanna. *Agron. J.* 2018, 8, 18.

Soretire, A. A. and Olayinka, A. (2013b). Soybean nodulation and N2 fixation as influenced by cowdung wood ash and rhizobial inoculation in tropical acid soils (Typic Dystrudept). *Nigerian Journal of Soil Science* 23(2), 114-125.

Soretire, A. A., and Olayinka A. (2013a). Response of soybean (*Glycine max*. L.) to cowdung and wood ash application in tropical acid soils of Southwestern Nigeria. *Nigeria Journal of Soil Science* 23(2), 103-113.

Tittonell, P., Vanlauwe, B., Corbeels, M., Giller, K. E. Yield gaps, nutrient use efficiencies and response to fertilisers by maize across heterogeneous smallholder farms of western Kenya. *Plant. Soil* 2008, 313, 19–37.

Tripathi, R., Nayak, A. K., Mohammad, S., Lal, B., Priyanka, G., Raja, R., Mohanty, S., Anjani, K., Panda, B. B., Sahoo, R. N. Delineation of soil management zones for a rice cultivated area in eastern India using fuzzy clustering. *Catena* 2015, 133, 128–136.

Vlek, P., Le, Q. B., Tamene, L. *Land Decline in Land-Rich Africa—A Creeping Disaster in the Making,* CGIAR Science Council Secretariat: Rome, Italy, 2008, p. 55.

Chapter 18

Ecosystem Management of Degraded Red and Lateritic Soil through Agroforestry

P. Das[1,*] and S. Panda[1,2]

[1]Department of Soil and Water Conservation,
Bidhan Chandra Krishi Viswavidyalaya, Mohanpur, Nadia, India
[2]AICRP on Agroforestry, Regional Research Station, Jhargram, India

Abstract

Agroforestry is now recognized as an important approach for ensuring food security and maintaining stable rural ecosystem. In India various types of agroforestry systems (AFS) exist across different ecozones ranging from humid tropical low land to high altitude temperate areas and rainforest zone to dryland situations. Agroforestry is also successfully practised in highly erodible flood prone and environmentally sensitive land, because it can offer both economically and ecologically viable options to farmers and rural ecosystems to a large scale for its efficiency to supply supplemental fuel, fodder, fruits and fibres and as well as environmental amelioration. In this regard, agroforestry systems play major role in protecting environment and forests and integrating farming components in a lesser space and time to produce crop per drop of water and per plot of land, and, thus, it can make better use of environmental or natural resources. In red and lateritic soil, crop production under rainfed condition with low and erratic rainfall with poor management of marginal land accelerates land degradation. The soils of the red and laterite zone are less fertile with poor rooting conditions due to shallow depth with poor structure and consistency of soil. Arable crops hardly sustain on such poor marginal lands during dry season. Monocropping is not sufficient to meet family expenses. Out of the various types of AFS, fruit-based agroforestry may be considered as an alternative land-use system for those areas. From long term studies on red and lateritic soil under the humid and sub-humid zone, it has been found that fruit-based agroforestry systems can give good production of agricultural crops with additional fruits and wood from the same piece of land through fast-growing tree species. Such AFS systems are also capable of improving soil health and, thereby, increase crop production on a sustained basis. That effect of AFS is evidenced from various field trials on red and lateritic soils from time to time. Experimental results have also revealed that such land use through AFS can improve residual soil moisture content for crops to be raised after rainy season. In such practice of land use, apart from crop harvests, a single piece of land can also give harvests of fruits and wood. In those ways fruit-based agroforestry systems are not only capable of increasing crop production but also has a sound effect on development and improvement of soil health in less fertile degraded soil through increasing status of soil organic carbon, pH and available major plant nutrients.

---

[*] Corresponding Author's Email: pinakeshpsb17@gmail.com.

In: Ecosystem Services
Editor: Hanuman Singh Jatav
ISBN: 978-1-68507-614-6

Keywords: agroforestry, soil health, red and lateritic soil, crop productivity

1. Introduction

Agroforestry is an age-old practice of intimately combining trees with the cultivation of agricultural crops. Farmers are used to follow this crop growing system throughout the globe as well as in India. In the context of following intensive cropping and seeking more arable lands to satisfy the 'Grow More Food' and ultimately successful operationalisation of the 'Green Revolution' towards overcoming horrible hunger during thirties to mid-fifties in the last century, role of association of trees within the crop fields might be overlooked. Thus, the practice of agroforestry as system of crop cultivation went to the back foot of agricultural practices. With the continuously following modern intensive cropping, steady increased harvest faced a setback mainly due to obtaining diminishing effect of fertiliser application unified with the decreasing fertility of soil year after year with hardening of soil due to low soil moisture holding capacity and other unfavourable physical, chemical and physico-chemical conditions of soils under cultivation. That hard times of the advent of modern agriculture, also faced decline in forest covers and the trees outside forest (TOF). So, the aberrant weather conditions, followed by unusual climate in combination with decline in reaping of forest products. Such unwanted conditions compelled both farming community and conscious section of the society to look back for returning of the trees within cultivable lands called as the agroforestry system (AFS) since first half of eighties in the last century (Nair 1993; Annon 1951; Bansil 1960; Gollin et al. 2018; Panda et al. 2018).

From the time immemorial farmers are used to practice cultivating trees with agricultural crops in intimate combination with one another throughout the world. Developing countries, including India, are all-time leaders in such type of cultivation, i.e., agroforestry (Nair 1993). The world of agroforestry system (AFS) spans across different ecozones ranging from humid tropical low land to high altitude and temperate biomes and rainforest zone to dryland. This system of cultivation plays important role in the economy by the way of providing various intangible and tangible benefits for the farmers and the society. Basically, agroforestry fulfils three important objectives of protecting and stabilizing the ecosystems, producing a high level of output of economic goods, and improving income (Dhyani 2019). In this way AFS plays a key role in maintaining the natural resource and increasing overall productivity in the rainfed areas in general and specifically in the arid and semi-arid regions. Improving employment generation, carbon sequestration, and optimization of farm productivity are now being focused through agroforestry system. In India, AFS is the only alternative to meet the target of increasing present level of forest cover of less than 25 percent to 33 percent (Sharma et al. 2017). A given land-use system or practice must permit significant economic and ecological interactions between the woody and non-woody components to qualify as agroforestry. Agroforestry is practised in India for thousands of years on agricultural lands for different purposes like food, fodder, fruit, firewood, medicinal trees, bio-fertilizer, non-timber forest products (NTFP), shelterbelt, etc. (Pathak et al. 2000; Prasad and Dhyani 2010; Kumar et al. 2012). AFS is now recognized as an important approach for ensuring food security and maintaining a stable rural ecosystem through diversifying farm activities and making better use of natural resources. Agroforestry also provides nutritional security because of its diverse production systems, which may include fruit, vegetables, legumes, pulses, oilseeds, medicinal and aromatic plants in addition to normal food crops grown by the farmers. Different agroforestry systems can provide the desired diversification options to increase food security and act as insurance against low production during drought and other stress conditions (Reddy and Shanker 2018). Agroforestry is commonly defined as "a collective name for land use system in which woody perennials (tree, shrubs, etc.) are grown in association with herbaceous plants (crops, pastures) or livestock, in spatial arrangement, a rotation or both; there are usually both ecological and economic interactions between the trees and other components of the system" (Lundgren 1982). This can be either in some form of spatial arrangements or in a time sequence. In other words, it is a collective name for a sustainable land-use system for obtaining social, economic and environmental benefits (Sanchez 1995). Griffith (2000) considers agroforestry as an alternative ecologically sustainable land-use option to the prevalent nourishment of farming patterns for conservation and development, particularly in the tropics. According to a report of the World Bank (2004), around 1.2 billion rural people in the world are used to practice agroforestry; with the inclusion of more than 200 plant species (Rao et al. 2000).

Red and lateritic soil under rainfed conditions of aberrant monsoon, represents one of the low and unstable productive soils. Poor management of marginal lands results in land degradation and poor contents of soil nutrients. In most cases, monocropping is not sufficient to meet the family expenses of rural people (Dhara et al. 2016). Out of the various types of AFS, fruit-based agroforestry, silvopastoral system, high-density plantation of silvi tree species may be considered as an alternative land-use system for those areas by integrating cultivation of arable crops, fruit trees, and silvi components. Under such system, fruit trees can be successfully grown with legumes like moth bean (*Vigna aconitifolius*), guar (*Cyamopsis tetragonoloba*), and cowpea (*Vigna unguiculata*) as bonus crops which can serve as an economical alternative system for marginal and sub-marginal lands (Singh 1996). The present review is aimed at elaborating potentialities of different agroforestry systems for improving livelihoods, maintaining soil health and managing red and lateritic degraded lands.

2. Role of Agroforestry in Soil Conservation

Soil and water conservation has attracted global attention in this millennium due to increased soil erosion, reduced water availability as well as deterioration in soil health and water quality. Two anthropogenic activities i.e., increased land use and fossil fuel burning are the main causes of soil and water degradation, which are very crucial issues in the modern era, and, thereby, affecting the availability of major inputs for agricultural production. Therefore, to resolve those problems, sustainable soil and water conservation option should be given due emphasis with agroforestry. In the agroforestry system, the only crop is not enough to lower the velocity of runoff; where trees with deep root system improve groundwater quality through trapping of nutrient metals which are deposited in the surface and subsurface soils (Dury, 1991). In the red and lateritic zone, soil is highly degradable and porous in nature of which nearly 50 percent of lands under rainfed cultivation and are facing some kinds of land degradation. Therefore, some alternate land-use systems are to be developed for those types of lands (Dhara and Sharma 2015). Fruit-based agroforestry system, integrated with cultivation of arable crops including pulses, fruit trees, and silvi components, maintain the stability of farming environment through conserving nutrients and control soil erosion hazards in marginal and sub-marginal lands (Singh 1996). Afforestation with suitable tree species like *Acacia nilotica* (Gum), *Azadirachta indica* (Neem), *Butea monosperma* (Palash), *Prosopis juliflora* (Velvet mesquite), *Dalbergia sissoo* (Sissoo), *Tectona grandis* (Teak), *Bambusa* spp. (Bamboo) and *Dendrocalamus strictus* (Male Bamboo) and other adaptable species such as grasses like *Dichanthium annulatum* (Shade Grass), *Bothriochloa pertusa* (Hurricane Grass), *Cynodon dactylon* (Scutch grass) and *Sehima nervosum* (White Grass) will help in soil conservation in the Red and Laterite zone critically affected by wind and water erosions (Tewari 2008). Agroforestry systems have potentiality to conserve moisture, improving water use efficiency by reducing run-off, soil evaporation and drainage (Murmu et al. 2016). Trees play a key role in controlling soil erosion. The effect of perennial vegetation and alley cropping (hedgerow intercropping) has often been emphasized for erosion control (Young 1989; Lai 1989). Rudramurthy et al. (2008) have found that arecanut *(Areca catechu* L.*)* and mixed forest land-use system compared to fallow and sugarcane *(Saccharum officinarum)* land-use system showed higher value of soil moisture holding capacity, field capacity with lowering of permanent wilting point in Karnataka, India. According to the Planning Commission of India (Sharma 2017), 33 percent of forest cover can be achieved only through implementation of agroforestry systems. The inclusion of the agroforestry system in Indian agriculture has many advantages such as improving soil conservation, reducing erosion, soil fertility improvement, soil structure, etc. Thus, agroforestry systems conserve natural resources just like conservation agriculture in rainfed areas (Sharma 2018). In the Eastern Ghat region of India, multitier agroforestry system is potential to conserve soil and water and pay back improved productivity in a sustainable way (Jakhar et al. 2017). Ber (*Ziziphus mauritiana* Lam.) with cluster bean (*Cyamopsis tetragonolobus*), cowpea *(Vigna unguiculata),* horse gram *(Macrotyloma uniflorum),* or other legumes are widely adapted in the red and lateritic tracts of Karnataka and Andhra Pradesh in India to control wind erosion (Narain 2000). Hombegowda et al. (2020) has reported that soil-water uptake from the topsoil increased with increasing diameter, within and across shade tree species in coffee (*Coffea arabica*) agroforestry system. Eucalyptus trees can alter soil water availability (SWA) in a silvopastoral system when compared to full sun pasture, sometimes varied; besides the soil-water recharge tends to be higher close to the tree, with faster withdrawal of soil-water below the trees with their vigorous root system (Bosi et al. 2020). De Carvalho et al. (2021) have observed higher water loss in the unshaded area (338 L ha^{-1}) compared to the shaded area (150 L ha^{-1}), and with lower soil temperature and higher moisture under shaded conditions, thus, improving microclimatic condition and drainage system under shaded coffee *(Coffea arabica)* based AFS. Roy et al. (1989) has observed the ability of the agroforestry system to conserve soil moisture at different depths under trees on degraded red and lateritic soil even in the dry season.

3. Trees Outside Forests – Role in Improving Farming

Trees outside the forest (TOF) consist of a heterogeneous and locally quite different natural resource called a tree resource outside the forest (TROF), which is defined by FAO as "trees on land not defined as forest or other wooded land." Due to increasing demands and scarcity of forest produce, TOF has assumed a greater importance in national economy. TOF, present in natural and cultivated landscapes, serves several ecological, and economic functions (Kleinn 2005) and play a key role in securing rural livelihoods (Tewari 2009). For the last two decades, TOF has significantly buffered the pressure on our declining forest resources and catered to the burgeoning need for domestic biomass and escalating demands for industrial wood. According to the report of FAO, 93 percent of industrial wood demand is met from TOF (MoEF & CC, GOI, 2018). In the Indian scenario, the Ministry of Environment, Forest, and Climate Change (MoEF & CC, GOI, 2018) has reported on the huge potentiality for developing TOF through expansion of agroforestry to focus on climate-resilient farming in a big way in the face of uncertainty of monsoon rains and meeting the goal of 33 percent of forest and tree cover (FTC) for achieving Nationally Determined Contribution (NDC) target of additional CO_2 sink, reducing land degradation, improving the farmers' income and reducing the trade deficit in wood. After 1980, with the introduction of social forestry, huge numbers of trees of different species have been planted, of which 35-40 percent on private, communal, or village lands outside the forest domain. The Government has also sponsored tree planting along roads, railways, canals, and around ponds for the purpose of stabilizing dunes, checking erosion along coastlines and rivers, reclaiming gully-eroded areas, and establishing windbreaks and providing shade and shelterbelts along highway, in parks, and gardens for aesthetic purpose (Pandey 2000). Over 70 percent of those trees were grown outside forest areas (FSI 2000b). In India, tree planting is an important activity, and being a natural habit, no priority has been attached to it. Under the Van Mahotsav (the tree planting festival), the local government/ forest department are used to distribute seedlings to local people for about a week to a month every year during the rainy season, which has been celebrated since 1950. However, those rituals have made little impact on the development of viable tree resources outside the forest (Panday 2008). Agriculture and forest both are connected culturally with human societies since the time immemorial. With the expansion of agriculture and civilization, community people are still continuing to their indigenous close association with various species of plants and trees in their homestead garden, around the temples, houses, and community lands for their various rituals and beauties around their habitations. So, this association is deeply rooted within human feelings (Shiva and Mathur 1996; Blatter and Millared 1954; Talbot 1984). Trees in homestead gardens are helpful for supply of timber and wood for their own use or cash to meet up family expenses even in meeting up emergency needs. So, that type of agroforestry system needs more attention for increasing TOF areas (Panda et al. 2018). Farmers mostly prefer fruit-based agroforestry system successfully, along with silvi trees as boundary plantation in the red and lateritic zone under humid and sub humid climate in West Bengal, India (Dhara 2016). The aonla (*Phyllanthus emblica*) based agri-horticulture system has been proven to be one of the most profitable agroforestry systems in block plantation on poor and marginal degraded lands of drought-prone Bundelkhand region. It has helped farmers to support even in conditions of crop failure due to vagaries of weather (Newaj et al. 2006). Panday (2008) has observed that 95 percent of TOF volume occurred in rural areas and 5 percent in urban areas of India with a huge contribution of block and boundary plantation in rural areas with several species of mango (*Mangifera indica*) with a maximum contribution of 11.18 percent of the total volume, followed by coconut (*Cocos nucifera*), jamun (*Syzygium cumini*), neem (*Azadirachta indica*), and mahuya (*Madhuca latifolia*), with contributions of 4.94, 4.20, 3.91, and 3.72 percent respectively. From the review of several articles and research papers, it has become indicative that the huge potential of TOF in developing countries through the adaptation of agroforestry, homestead garden block, and boundary plantation are very much useful for creating a stable farming environment.

4. Role of Agroforestry in Protecting Environment

Many developing countries in the world, including India, are facing increasing pressure on natural resources due to increase in population in geometric progressions as well as increase in built-up areas, resulting in environmental instability. Such anthropogenic activities in combination with depletion of forest cover have attributed towards warranting conditions of global warming, causing a lot of unwanted environmental phenomena such as soil erosion, land degradation, flooding, depletion in soil fertility, loss of biodiversity as well as erratic behaviour of seasonal weather, and thereby, affecting rural and urban livelihoods and economies of the countries throughout the world (Sobola et al. 2015). To overcome this complex situation, agroforestry plays a major role, as part of a multifunctional working landscape, and can be a viable land-use option for reducing poverty and increasing several ecosystem services and environmental benefits viz. biodiversity conservation, soil enrichment, air and water qualities improvement and carbon sequestration (Jose 2009). Now-a-days many developing countries suffer from climate change due to increase in concentrations of atmospheric greenhouse gases (GHG). Therefore, mitigation strategy is the only way to check the effect of climate change (Verchot et al. 2007). The agroforestry system has played an influential role to reduce CO_2 emission and enhance carbon sinks, and thus, is able to provide great opportunities to create synergies between adaptation and mitigation activities. The potentialities of carbon sequestrations through various AFSs can be enhanced by stabilizing soil organic carbon through possible mechanisms of including biochemical recalcitrance, physical protection and reducing C losses (Dhyani et al. 2016). According to Alam et al. (2014) and Dev et al. (2019), agroforestry system is able in conserving the biodiversity, sustainable use of natural resources, and providing various ecosystem services such as flood reduction, water purification, micro-climatic improvement, nutrient cycling, and biodiversity conservation. The agroforestry systems play a vital role in the adaptation to climate change through changes in the microclimate, protection through provision of permanent cover, improving the efficiency of use of soil, water, and climatic resources and reducing carbon emissions (Rao et al. 2007). Agroforestry can conserve natural resources by protecting and stabilizing the ecosystems under the different agro-climatic regions in India by controlling soil erosion and acting as an air purifier by removing carbon dioxide and other pollutants from the atmosphere. Trees also provide shade and windbreak to the protected local climate under different seasons (Dev et al. 2019). As per Kumar et al. (2006) increase in soil carbon through plantations may also act as an important carbon sink. Bio-drainage tree including eucalyptus played a major role in combating the water logging conditions. According to Handa et al. (2016), trees in agroforestry systems not only provide direct benefit but also reduce soil erosion, improve soil fertility, filter atmospheric pollutants, and maintain carbon balance. Amadi et al. (2018) has reported that changes in precipitation patterns and soil moisture regime due to climate change have affected soil-atmosphere exchange of available N, C and soil moisture dependent greenhouse gas (GHG) emissions in shelter belts. Compared to the horticulture-based system, farm forestry has a significant positive effect on rhizospheric soil properties with higher microbial activity and nutrient availability due to increase in soil organic carbon (SOC) after the post rainy season (Neha et al. 2020) at Shivalik foothills of the Punjab, India. Iwasaki et al. (2020) has found that during the early growth stage of maize, the mean soil temperature with an increase in 0.4–0.8°C with wind speed reduction at a distance of 3 – 4 times of height of the windbreak and a decrease in 0.9–1.0°C by shade from windbreaks at a distance of 0.5 H. From various literatures and research papers, it is found that agroforestry system has many ecological and environmental benefits to protect the environment from adverse climatic conditions along with the following significant roles on ecosystem services:

- Trees can capture nutrient leached from topsoil and they can return to top soil by litter.
- Activity of soil biota is increased with the addition of litter in soil surface layer by trees, and this can improve water infiltration into the soil and reduce surface runoff.
- In agroforestry systems, tree species have potential to conserve soil moisture and improve fertility status of the soil, to reduce competition for nutrients and moisture with crops by uptake of those from deeper layers of soil.
- Agroforestry helps in improvement of microclimate by trees by significantly lowering air temperature through shade under tree canopies, as compared to open land, and thus, efficiently reducing heat stress for plants and animals during the hottest hours of the day.
- Reduction in wind speed directedly beneficial to physical damage to crops and other trees.
- Trees increase the rainfall interception to control soil erosion and allow conservation of soil and water.
- Trees improve ground water quality by taking up excess nutrients that have been leached below the rooting zone of agricultural crops.
- Agroforestry helps in reducing biodiversity loss by providing a protective tree cover along agricultural fields and by providing habitat for a diversity of flora and fauna.
- SOC and available nutrients substantially increased compared to growing only trees or only agricultural crops to maintain soil fertility status.

5. Degraded Red and Lateritic Soils – Their Characteristics

Red soil is a recognised category in soil classification system because of its large areal coverage throughout the world and its great possibilities for agricultural productions. In Soil Taxonomy (Soil Survey Staff 1999), red soils are present in many soil Orders, such as Alfisol, Oxisol, Ultisol and occasionally Mollisol and even in Inceptisol. Red soils are predominantly found in tropical and sub-tropical areas of South America, Central Africa, South and Southeast Asia, China, India, Japan and Australia (Soil Science Society of America 1996), covering about 13 percent of the land area world over (Baligar et al. 2004). In the Red and Laterite agroclimatic zone most (55 percent) of the cultivated lands are rainfed. The soils of red and laterite zone are coarse textured and with moderately acidic pH (varying from 5.0 to 5.5), highly porous in nature, rich in sesquioxides of iron and aluminium. The Red and Lateritic soils are poor in organic matter, low in water holding capacity and poor in available nitrogen, phosphorus and potassium contents and are highly susceptible to erosion hazards (Naugraiya and Shaw 2014;

Dhara et al. 2015; Dhara et al. 2016). Crop production on red and lateritic soils under rainfed condition is low and unstable because of prevalent aberrant monsoon behaviour and high levels of Al and Mn and nutrient deficiencies of P, K, Ca, Mg and Zn (Sehgal 1998). Poor management of marginal lands results in land degradation as well as soil. The lateritic zone is an undulating tract of high ridges and low valleys. The fertility status in the low land situation is rich, whereas under the higher land situation, it is very deficient in available plant nutrients. The soil is of poor rooting conditions due to shallow depth with poor structure and consistency, and it contains gravels, stones, and morrum, etc. at the ploughing depth (Saha et al. 2014).

5.1. Role of Alley Crop Cultivation on Improving Soil Health of Red and Lateritic Soils

Alley cropping deals with growing of crops on the same piece of land within the interspaces under the trees. In the alley cropping system, trees or shrubs and agricultural crops are grown in alternate rows, whereas trees or shrubs are planted on two or more sets of single or multiple rows at wide spacings. The hedges are pruned periodically for providing enough spaces for crop growth and thus, providing biomass to soil as well as enhancing nutrient status and physical properties of soil. One of the most important properties of alley cropping is that the addition of organic mulch, especially nutrient-rich mulch, has a favourable effect on the physical and chemical properties of soil and on crop productivity (Nair 1993). On red and lateritic soil, crop production is low and unstable due to soil and water erosion and nutrient losses leading to severe land degradation (Dhara et al. 2016). The losses of surface soil and nutrients not only substantially decrease nutrient status and utilization efficiency but also has a great impact on the degradation of the environment. Other important reasons for poor crop productivity include the occurrence of light-textured red and lateritic soils with high erodibility and low water holding capacity (Chaudhary et al. 1999). Poor management of this type of land on the upper part of undulating topography has resulted in high land degradation. Soil erosion is the most important constraint to agricultural productivity in those regions. So, putting those lands to an applied land-use system like agroforestry can restore productivity through reduced soil and nutrient loss, increased soil moisture and soil organic carbon (SOC) storage, and consequent improvement in soil health (Lenka et al. 2012). Enhancing and maintaining soil fertility is vital for food security, reducing poverty, preserving the environment, and sustainability. Traditional agroforestry land-use systems are efficient ways of restoring soil organic matter and carbon sequestration. Tree components of any agroforestry system perform one primary function in controlling erosion; the trees may act as barriers to wind and water flows or as on soil surface.

5.2. Effect of SOC through AFS in Red and Lateritic Soils

From a study on the effect of soil organic matter (SOM) as well as soil organic carbon (SOC) on two contrasting leguminous (*Calliandra calothyrsus* Meissner) and non-leguminous (*Vernonia amygdalina* Del.) fodder shrubs, it has been found that fodder shrubs planted on a low-fertility Ferrelasol can improve soil fertility through increased SOC by up to 12 percent over two years with enhanced soil aggregation (Fungo et al. 2020). SOC was found significantly higher in farms with shiny-leaf buckthorn (*Rhamnus prinoides*) intercropping than farms without trees/shrubs; and high-density plantation of *R. prinoides* helped to increase SOC (Gebremeskel et al. 2021). Aggregate stability being a very important function of soil organic carbon (SOC), the carbon concentration and stock both play key role in determining the stability of soil aggregates (Saha et al. 2011; Kukal et al. 2007). In bulk soil SOC concentration was found to be higher in dry stable aggregates (DSA) than water stable aggregates (WSA). SOC concentration was reported to be higher under tree-based cropping system compared to sole crop plots whereas SOC status under agroforestry system resulted in higher SOC both within the DSA and WSA (Dhaliwal et al. 2017). Khalee et al. (2019) has observed that tree-based systems were highly potential to increase SOC stocks especially at deeper depths as well as beneath the trees. SOC stocks within the 1.25 m depth of soil were observed to be 16 percent higher compared to the adjacent field. Pomegranate–lemongrass-based agroforestry system stored SOC as high as 0.51 to 0.68 percent in surface layer and 0.42 to 0.64 percent in sub-surface layer (Kumar et al. 2021). SOC stock in the top and the middle layers of soil decreased significantly through land use change from forest to agroforestry. On the other hand, the conversion from agriculture to agroforestry and pasture/ grassland to agroforestry significantly stored SOC at various soil horizons and depths (De Stefano et al. 2018). Van Den Berge et al. (2021) found that SOC stocks under hedgerows at a depth of 23 cm stored a considerably higher amount of soil carbon (81.7 ± 28.8 Mg C ha^{-1}) compared to grass strips (56.6 ± 14.5 Mg C ha^{-1}), besides the fields adjacent to hedgerows, SOC stocks were only slightly increased compared to stocks in fields with grass strips (56.4 ± 6.3 *vs.* 55.6 ± 5.0 Mg C ha^{-1}) with exponential decay up to 30 m from the margin. Soursop (*Anonna muricata*), cola (*Cola nitida*) and kassod (*Senna siamea*) showed high nutrient contents and decomposition rate with high response for soil nutrient addition in an agroforestry system (Akinyele and Donald-Amaeshi 2021). On soil with low organic carbon content, SOC and CEC have significantly increased with positive effect on yield of cocoa and maintaining soil health (Adeniyi et al. 2018).

5.3. Carbon Sequestration through AFS on Red and Lateritic Soils

Presently, the world over challenging situation is managing the loss of carbon (C) stocks in the terrestrial ecosystems and controlling the increased levels of greenhouse gases in the atmosphere. Agroforestry system (AFS) is highly potential in sequestering aboveground and belowground (soil) C and helps in mitigating the greenhouse gas effects by reducing C emissions. (Dixon et al. 1994; Pandey 2002; Albrecht and Kandji 2003; Lai 2004). Trees can store C both ex *situ* (products) as well as in *situ* (biomass and soil) and are considered as effective C sinks (Montagnini and Nair 2004). Tree-based land-use system has been recognized as an effective strategy across all ecological regions and farming situations and as the most effective farmer-friendly technology to control greenhouse gas emission (Doddabasawa et al. 2019). In Londrina, southern Brazil Zaro et al. (2019) has found that the coffee and rubber-based agroforestry system is an excellent alternative to increase carbon stocks as tree biomass and soil organic C compared to the traditional system of cultivation fully exposed to sunlight. According to Birhane et al. (2020), above ground and below-ground biomass C and SOC stock also varied distinctly along the elevation gradient. Upper elevation land category resulted in higher biomass C, and SOC than lower elevation and the carbon stock is not enhanced or significantly changed to woody species diversity in tree – Ensete banana dominated agroforestry practices. The eucalyptus stands at the age of 20+ stored higher amount of carbon sinks that could store eight times the CO_2 emissions than from a degraded savannah (Noumi et al. 2018). In Southwest British Columbia temperate zone, non-production perennial vegetation (NPPV) stored 3.34 Mt C in aboveground biomass, and the conversion of NPPV to cropland was likely to be most vulnerable, resulting in 0.98–1.86 Mt of C lost to the atmosphere from biomass C alone (Rallings et al. 2020). Ramos et al. (2018) have found that in the eastern Amazonia oil palm and cacao-based AFS stored higher C (116.7 ± 1.5 Mg C ha^{-1}) than in oil palm-based AFS (99.1 ± 3.1 Mg C ha^{-1}), besides total higher litter-C stock in oil palm and cacao-based AFS (3.27 ± 0.01 Mg C ha^{-1}) than in oil palm-based AFS (2.26 ± 0.06 Mg C ha^{-1}). Between the agroforestry systems total root and soil C stocks did not differ significantly at 0-30 cm soil depth. Sarto et al. (2020) have observed that soil C stock increased in a 6-year-old eucalyptus (C_3) based integrated crop-livestock systems (ICLS) system as well as monoculture palisade grass (*Urochloa brizantha*) pasture, but the contribution of C of eucalyptus was deeper in the soil profile whereas the grass pasture stored higher C in the upper 0.20 m soil layers. Goncalves et al. (2021) opined that alternative land use from traditional agriculture like coffee-based agroforestry systems (CAS) could sequester large quantities of carbon as comparable even to forest patches. However, the economic impact of carbon sequestration could not encourage the adoption of coffee agroforestry systems.

5.4. Changes in Available Nutrients in Red and Lateritic Soils through AFS

The alley cropping with leguminous trees such as subabul (*Leucaena leucocephala*) was effective for controlling soil erosion and nutrient recycling on degraded sloping land (Dhyani 2019). Eastern Ghat region of India was cultivated with different multitier agroforestry systems, in which plantation of drumstick (*Moringa oleifera*) with *Gliricidia sepium* hedgerow and ginger (*Zingiber officinale*) along with pigeonpea (*Cajanus cajan*) intercropping saved 74 percent more soil organic carbon, 64 percent more phosphorus and 66 percent more potassium respectively than broadcasted finger millet cultivation. That system also recorded minimum mean runoff (8.26 percent) and soil loss of 3.45 Mg ha^{-1} (Jakhar et al. 2017). Cultivation of shisham (*Dalbergia sissoo)* with intercrops viz. wheat, mustard, barley, lentil in *rabi* (winter) season and green gram, blackgram in *kharif* (rainy) season; shisham with napier (*Pennisetum polystachion*) and other suitable grasses showed a positive impact on restoration of soil organic matter, and thereby, maintaining the soil fertility and controlling erosion (Gupta et al. 2019). In teak (*Tectona grandis*) based agri-silviculture system, cultivation of intercrops of sorghum and groundnut along with papaya *(Carica papaya)* system improved soil fertility over the years as well as increased bulk density, pH, organic carbon, potassium and phosphorus due to incorporation of leaf litterfall/ decomposition in soil (Mutanal et al. 2018). Consequently, it is to be noted that maintenance of soil fertility plays a vital role in global food security and environmental sustainability. In the agroforestry system, the most important trees are legumes for their nitrogen-fixing abilities. Apart from legumes, fruit trees, vegetable plantation, pulse crops can also be included in this system to maintain soil fertility. In the red and lateritic soil, particularly in West Bengal, India fruit-based agroforestry system cultivated with pulses played a major role in maintaining soil fertility and controlling soil erosion (Murmu et al. 2016).

After two years of experiment Zhang et al. (2018) have found that jujube trees and wheat in the intercropping system could uptake high amounts of nitrogen, with higher percentage of N derived from fertilizer (NDFF) and higher fertilizer-use efficiency (UFN) in intercropping system than sole crop plots. Partial shade significantly increased shoot N percentage except pasture and excess of legume-based swards (Sanna et al. 2018). The current market scenario in the Pennsylvania Nutrient Credit Trading Program (NCTP), nitrogen credit trading alone did not provide an adequate incentive for riparian buffer establishment. The cost of nitrogen credit production for all riparian buffers was significantly higher than the estimated market value of nitrogen credits (Boleman and Jacobson 2021). Relationship between soil factors and foliar nutrient concentrations showed variations between 60 and 97 percent in foliar nutrient concentrations during the wet period. Soil factors like soil nutrient content, texture, and pH significantly influenced during the wet period through foliar application of P, Fe, Mn and Zn compared to those applied in the dry season. In the dry season, rainfall had a significant effect on Fe concentrations besides the total annual rainfall had an antagonistic relationship during the dry winter period, and those showed the importance of soil and climatic factors on soil-plant nutrient dynamics (Ramachela 2018). In comparing 35 and 55-year-old cocoa-based agroforestry systems, it was estimated that 35-year-old cocoa-based AFS produced 23.1% higher litter throughout the year and the N–P–K–Ca–Mg contents in a litter of 35 and 55-year old cacao-AFS were with comparatively higher presence (1.2–0.4–1.2–1.7–0.4 percent) than the other (1.1–0.6–1.2–1.4–0.4 percent) respectively and those were enough to replace nutrients extracted by the cacao for crops in the 35 and 55-year-old cacao-AFS (Pe´rez-Flores et al. 2017). In multistrata AFS, the younger age system, produced a higher amount of standing litter and added nutrients to the soil-plant system compared to the natural regeneration areas in the Brazilian Atlantic Rainforest Biome (Froufe et al. 2019). Raveendra et al. (2021) have found that in coconut (*Cocos nucifera*) based Gliricidia (*Gliricidia sepium*) systems, Gliricidia differed in their soil chemical attributes and added higher levels of soil nutrients when compared to coconut monocrop at younger ages and that system was highly potential to improve poor soil characteristics particularly in dryer agroecosystems. Integrated tree species improved soil fertility and crop yield whereas combined tree species and fertilizers performed far better than that of the control areas compared to the neighbourhood and treeless outside zones and an organic practice (Diallo et al. 2021). *Gliricidia sepium* legume tree contributed to more rapid decomposition of senescent leaves and release of nutrients to the soil in the short term, increasing nutrient cycling due to higher N concentration and lesser C:N ratio and sabia *(Sabia limoniacea)* forming more stable organic matter, thus, improving stable ecosystem services by silvopastoral systems (SPS) (Herrer et al. 2020). Soil fertility status improved under (*Faidherbia albida*) tree canopy due to the addition of litter and N release from *F. albida* tree litters contrary to the inherent soil organic matter able to supply required N for a maize crop for maintaining food security (Yengwe et al. 2017). Application of all plant materials significantly increased the biological yield, and N uptake of maize with Gliricidia *(Gliricidia sepium)* and Giant Mexican sunflower *(Tithonia diversifolia)* showing the most significant impact, especially in the primary rainy season and in comparison with the other species *G. sepium* and *T. diversifolia* could offer the most suitable option with limited inorganic fertilizer resources (Partey et al. 2016). High density plantation of *(Acacia mangium*) offered opportunities for high fine root productivity on an area basis (5.78 Mg ha^{-1}) and a successively high release of organic carbon and mineral nutrients (C: 1.83 Mg ha^{-1}; N: 78 kg ha^{-1}; P: 2.54 kg ha^{-1} and K: 25.5 kg ha^{-1}) and its potential for enriching soil fertility (Rocha et al. 2021).

According to Beuschel et al. (2020), conversion of short-rotation coppices (SRC) to alley cropping agroforestry systems (ACS), resulted in the decrease in SOC content, microbial biomass and activity in upper topsoil (0–5 and 5–20 cm) within the arable alleyway irrespective of tree distance. *Inga edulis* ice cream-bean could be intercropped with forage crops in silvo-pastoral systems which increased forage yield and improved soil fertility status including their symbiotic relationship with nodule forming bacteria present in soil, and thereby, improving the growth of the plants and maintaining stable N content of the soil (De Freitas et al. 2020).

6. Increasing Farming Income through Agroforestry in Red and Lateritic Zone

Few decades ago, agroforestry system was recognised as an important integrated farming practice to fulfil domestic need of food, fodder, fuelwood, fibre and timber along with aesthetic and environmental services for the human habitats. Due to the increased demand for food grain on a limited land resource, woody perennials have been reduced gradually in the entire world. This situation is challenging to the future livelihoods of smallholder farmers worldwide and especially in the tropics (Kang and Akinnifesi 2000). Agroforestry systems could provide huge potential to improve food security and smallholder-farmer household income (Garrity 2004; Dubiez et al. 2018). Adaptation of agroforestry in any specific region are decided based on two factors viz. relative profitability compared to other existing crops, and the price volatility of wood. On the basis of market signal, farmers allocate their resources in the production of various commodities, and for that reason farmers in many regions of India are growing trees in farm as a more profitable option than field crops or monocropping system (Singh et al. 2017). So, agroforestry can improve rural livelihoods due to its capacity to offer multiple alternatives and opportunities to smallholders to increase farm production and incomes, and thereby, protecting the environment. Agroforestry increased farmers' incomes since cash crops can be cultivated simultaneously with forest trees, and thus, improving the biomass of trees. Those cultivation practices have increased the standard of living, economic growth and development (Julius et al. 2011). In the red and lateritic zone, where the arable crop is not able to sustain particularly during the drought years and facing some kinds of land degradation, integration of economic trees into agricultural landscapes on a massive scale under red and lateritic rainfed upland would create an effective income generation and sustainable crop production year after year, and thereby, that would help also to adapt to climate change (Dhanushkodi et al. 2019). Out of the different types of agroforestry systems, a fruit-based agroforestry system is considered as an effective strategy for improving productivity, employment opportunities, economic and nutritional security. In this system, fruit crop can provide additional income to the farmer in drought condition; besides, the silvi tree plantation would check the drift of sand, provide forage, fuel and timber and would help in creating favourable micro-climatic conditions that also help crop production (Panda et al. 2018; Dhara 2016). Honnayya et al. (2020) has opined that different growth factors, consequent crop performance and nutrient uptake are dependent on the direction, orientation and distance from the tree line. To get the information about adaptation of agroforestry system and its influence on income Dhanushkodi et al. (2019) conducted farmers meeting at farm-level. *Ailanthus excelsa* and *Melia dubia* tree species could be grown with drought tolerant agricultural crops such as sorghum, pearl millet, moth bean and red gram as an agroforestry system. Compared to agroforestry and pure agriculture, a net income increased by Rs. 37000/-, 88000/- and 185000/- during 7th, 14th and 21st year respectively and benefit-cost ratio increased by 2.8, 3.7 and 4.8 during those years of growth of agroforestry systems respectively. Cultivation of Aonla *(Emblica officinalis)* with field crops of green gram, black gram in rainy season and mustard or chickpea in winter for several years showed the recorded highest B:C ratio of 1.84. However, it increased to 3.28 at the age of 13 years of the system (Newaj and Ram 2019). Banerjee and Dhara (2010) have observed that fruit trees of sweet orange and guava were cultivated with silvi species of *Gmelina arborea* and arable crops of cowpea, groundnut, blackgram and bottle gourd. In that trial *G. arborea* + sweet orange + bottle gourd model gave maximum total return (Rs. 89,504) closely followed by *G. arborea* + guava + bottle gourd model (Rs. 87,446). According to Parthiban (2019), *Melia dubia* based agroforestry system significantly improved productivity level of silvi component from 10 m^3/ha/annum to over 25 m^3/ha/annum of wood and increased the profit of 1 lakh/acre/annum compared to less than Rs.50,000/annum from the traditional agricultural practices. That AFS gained significant attention across the state of Tamil Nadu, India due to fast-growing characters of Melia. Mohapatra et al. (2019) has found that after 5 years of plantation, mango based agri-horticulture system, gave production of fruits up to 8 t/ha per annum with a B:C ratio of 2.85 or more. Naugraiya and Puri (2001) conducted a study to find out the productivity of seven multipurpose tree species, viz. *Albizia procera, Azadirachta indica, Dalbergia sissoo, Eucalyptus tereticornis, Hardwickia binata, Leucaena leucocephala* and *Terminalia arjuna* under silvi-pastoral systems. The total aboveground biomass was maximum in *L. leucocephala* (38.0 kg/tree) followed by *E. tereticornis* (26.1 kg/tree) and *Albizia procera* (24.8 kg/tree) and the production of fuelwood, foliage and fruits of energy plantation system were maximum for *L. leucocephala* (32.12 q/ha) and minimum for *E. tereticornis* (10.82 q/ha). The increasing trend of biomass production helped to increase growth of tree that provide additional income to the farmer. Tree shading was primarily responsible for reductions in productivity, and that provided higher beneficial effect from maize than soybean grain yield (Figyantika et al. 2020). In the Anthropic Ferralsol, *Acacia angustissima* and *Leucaena pallida* performed better than *Mimosa scabrella*, showing higher shoot biomass from *A. angustissima* by cutting *L. pallida* at cutting height of 0.3 m compared to of 1.0 m. It was found that cutting height did not affect crude protein (CP), mean neutral detergent fibre (NDF), acid detergent fibre (ADF) it was also observed that total polyphenol (TP) being tended to be higher at 1.0 m cutting height for which low cutting height was recommended for *A. angustissima* and *L. pallida* for higher-quality fodder and higher shoot biomass production (Mukangango et al. 2020). *Acacia decurrens* based small-scale agroforestry (SSA) was taken as trial on land-use system for providing better income for small-holder farmers and in changing from the exploitative land use to productive land use in northwest Ethiopian highlands (Mekonnen et al. 2021). In central India *Dendrocalamus strictus*-based agroforestry system produced leaf biomass, bamboo stock as well as sustained crop production over the years; bamboo-based AFS, particularly (bamboo + sesame + chickpea), could be one of the best viable livelihood options for small and marginal farmers (Dev et al. 2020).

7. Role of Fruit-Based Agroforestry in Improving Degraded Red and Lateritic Soils – Concluding Remarks

Fruit-based agroforestry system has been developed and widely followed throughout different regions of India based on either planting of trees, pastures, arable crops etc. in existing fruit orchards or *vice versa* (Kumar and Chaturvedi 2017). The developed systems are agri-horti, agri-horti-silvi, horti-silvi-pastoral and horti-pastoral systems. Population growth and local economics are both driving factors for the extensification on marginal lands towards the intensification of food production, where the risks of crop failure and land degradation are high. Horticulture based production system is the most effective strategy to provide food, nutrition and income security to the people. The horticulture-based production system plays a key role in improving the productivity of the land, improving economic conditions and generating employment of small and marginal farmers through the growth of entrepreneurs and enhancing export. Adopting the agri-horti-silvi system or horti-silvi-pastoral system can ameliorate those lands in the long-term owing to soil conservation and improvement in soil fertility levels. In those systems, horticultural crops, silvi species and fodder crops are being grown together on the same piece of land, where several drought resistant and deep-rooted fruit tree species could survive and generated income under severe drought conditions. Besides, the silviculture plantations could check the removal of surface soil, and provided forage, fuel, timber, and for creating a favourable microclimate beneficial for alley crops (Kumar et al. 2019). Plantation of trees on boundary line leading to better soil moisture retention and reduced transpiration will be an added advantage apart from windbreak. Shelterbelts also enhanced soil organic carbon and reduced daily air temperature under the tree canopy. Among different types of agroforestry systems, the fruit-based AFS has helped to improve soils in various ways and have been found to conserve higher amount of organic matter and maintain the nutrient cycle (Manna et al. 2003). Intercropping, mixed cropping and mixed arable-livestock systems can sustainably increase productivity. Integration of fruit trees along with timber trees in agricultural production systems can store a higher amount of C to fight against the hazards due to climate change. In Mediterranean conditions, the no-tillage (NT) management of chestnut (*Castanea sativa* Mill.) plantations seemed to be beneficial and improved soil carbon sequestration and maintaining soil quality at the soil surface layer (Borges et al. 2017). Horti-pasture system was found to be an economically viable agroforestry system for class V and VI soil types, and it could supply protective food (like fruit) and fodder, and thus, helped in bridging the wide gap between demand and supply. (Singh 1996; Sharma 2004). Integration of *Bael* (*Aegle marmelos*) with perennial pasture and moisture conservation treatments have the potential to enhance nutrient content in the soil, improved soil moisture content and production under the semi-arid ecosystem of India (Shukla et al. 2018). Qaiser et al. (2016) also reported that apple-based horti-pasture system improved the soil characteristics. Trenches developed in aonla based horti-pasture system that could conserve higher soil moisture content and improved soil health. Ahmed et al. (2014) has reported in details long term studies of various field trials throughout the world. It has been found that fruit-based agroforestry system could give good production of arable crops, wood and additionally the production of fruits in a same piece of land to generate extra income for farmers in severe climatic condition that helped rural people to meet their family expenses and created congenial microenvironment. Under the fruit-based agroforestry system, productivity of alley crops could be maintained at a certain level year after year, while wood volume was recorded higher under tree + crop field than the sole silvi/ fruit tree plots, and such stable rate of production indicated the requirements of sustainable agriculture and that might be possible due to conservation of soil fertility under agroforestry. Besides, soil pH of acid soil could be improved to some extent by following the fruit-based AFS under red and lateritic soils, and by that way the status of soil organic carbon, available N, P, K could be increased significantly. Fruit-based AFS could also maintain soil moisture at different soil depths under trees even in the dry season. From the review of research results it is revealed that such kind of alternate land use through fruit-based agroforestry system was highly efficient in improving red and lateritic degraded soils.

References

Adeniyi, S. A., W. P. de Clercq, and A. van Niekerk. 2018. "Assessing the Relationship between Soil Quality Parameters of Nigerian Alfisols and Cocoa Yield." *Agroforestry Systems* 93: 1235–1250. doi:10.1007/s10457-018-0238-2.

Ahmed, A., Sunil Kumar, and A. K. Shukla. 2014. "Effect of different soil moisture conservation techniques for establishment of Aonla based Hortipasture systems in Bundelkhand region." In Proceedings of *National Symposium on Climate Resilient Forage Production and its Utilization* 13-14, West Bengal: BCKV.

Akinyele, Adejoke O., and Ugonma Donald-Amaeshi. 2021. "Leaf Litter Decomposition and Nutrient Release of Three Selected Agroforestry Tree Species." *Agroforestry Systems* 95: 559–570. doi:10.1007/s10457-021-00606-w.

Alam, Mahbubul, Alain Olivier, Alain Paquette, Jérôme Dupras, Jean-Pierre Revéret, and Christian Messier. 2014. "A General Framework for the Quantification and Valuation of Ecosystem Services of Tree-Based Intercropping Systems." *Agroforestry Systems* 88: 679–691. doi:10.1007/s10457-014-9681-x.

Albrecht, Alain, and Serigne T. Kandji. 2003. "Carbon Sequestration in Tropical Agroforestry Systems." *Agriculture, Ecosystems & Environment* 99: 15–27. doi:10.1016/s0167-8809(03)00138-5.

Amadi, Chukwudi C., Richard E. Farrell, and Ken C. J. Van Rees. 2018. "Dynamics of soil-derived greenhouse gas emissions from shelterbelts under elevated soil moisture conditions in a semi-arid prairie environment." *Agroforestry systems* 92: 321-334. DOI: 10.1007/s10457-017-0112-7.

Anonymous. 1951. "Mechanisation of Agriculture to Speed up "Grow More Food "Campaign." *The Economic Weekly* 3, no. 6, February 10.

Baligar, V. C., N. K. Fageria, H. Eswaran, M. J. Wilson, and Zhenli He. 2004. "Nature and Properties of Red Soils of the World." In *The Red Soils of China: Their Nature, Management Utilization*, edited by M. J. Wilson, Zhenli He, and Xiaoe Yang, 7–27. Dordrecht: Kluwer Academic.

Banerjee, H., and P. K. Dhara. 2010. "Evaluation of fruit-based agroforestry model for rainfed upland under red & lateritic tract of West Bengal." *Crop Research (Hisar)* 39: 200-206. https://www.cabdirect.org/cabdirect/abstract/20103243463.

Bansil, P. C. 1960. "Food Resources and Population of India a Historical Study." *Sankhyā: The Indian Journal of Statistics* (1933-1960) 22(1/2):183-208. http://www.jstor.org/stable/25048450.

Beuschel, René, Hans-Peter Piepho, Rainer Georg Joergensen, and Christine Wachendorf. 2020. "Effects of converting a temperate short-rotation coppice to a silvo-arable alley cropping agroforestry system on soil quality indicators." *Agroforestry Systems* 94: 389-400. doi:10.1007/s10457-019-00407-2.

Birhane, Emiru, Said Ahmed, Mengsteab Hailemariam, Mesele Negash, Meley Mekonen Rannestad, and Lindsey Norgrove. 2020. "Carbon Stock and Woody Species Diversity in Homegarden Agroforestry along an Elevation Gradient in Southern Ethiopia." *Agroforestry Systems* 94: 1099–1110. doi:10.1007/s10457-019-00475-4.

Blatter, E., and S. Millard. 1954. "Some beautiful Indian trees 2nd ed." *Revised by TS William (Bombay: Bombay Nat. Hist. Soc.).*

Boleman, Patrick, and Michael Jacobson. 2021. "Nitrogen Credit Trading as an Incentive for Riparian Buffer Establishment on Pennsylvania Farmland." *Agroforestry Systems* 1 – 13. doi:10.1007/s10457-021-00595-w.

Borges, Olga, Fernando Raimundo, João Coutinho, Berta Gonçalves, Ivo Oliveira, Afonso Martins, and Manuel Madeira. 2017. "Carbon Fractions as Indicators of Organic Matter Dynamics in Chestnut Orchards under Different Soil Management Practices." *Agroforestry Systems* 93: 301 - 310 doi: 10:1007/s10457-017-0088-3.

Bosi, Cristiam, José Ricardo Pezzopane, and Paulo Cesar Sentelhas. 2020. "Soil Water Availability in a Full Sun Pasture and in a Silvopastoral System with Eucalyptus." *Agroforestry Systems* 94(2): 429–40. doi:10.1007/s10457-019-00402-7.

Chaudhary, R. S., H. Gadekar, and U. S. Patnaik. 1999. "Erodibility under different land uses in soils of Eastern Ghat High Land zone." *Indian J Soil Cons* 27(2): 118-121.

de Carvalho, Anôr Fiorini, Elpídio Inácio Fernandes-Filho, Mayara Daher, Lucas de Gomes, Irene Maria Cardoso, Raphael Bragança Fernandes, and Carlos E. Schaefer. 2021. "Microclimate and Soil and Water Loss in Shaded and Unshaded Agroforestry Coffee Systems." *Agroforestry Systems* 95(1): 119–134. doi:10.1007/s10457-020-00567-6.

de Freitas, Aroldo Felipe, Joana Junqueira Carneiro, Nelson Venturin, Fatima Maria de Souza Moreira, Ana Isabella Guimarães Ferreira, Gil Pedro de Oliveira Lara, Rafael de Moura Miranda, Luiza Figueira de Siqueira, and Irene Maria Cardoso. 2020. "Inga edulis Mart. intercropped with pasture improves soil quality without compromising forage yields." *Agroforestry Systems* 94: 2355-2366. doi:10.1007/s10457-020-00555-w.

De Stefano, Andrea, and Michael G. Jacobson. 2018. "Soil carbon sequestration in agroforestry systems: a meta-analysis." *Agroforestry Systems* 92:285-299. doi:10.1007/s10457-017-0147-9. 297 285

Dev, Inder, Asha Ram, S. Bhaskar, and O. P. Chaturvedi. 2019. "Role of Agroforestry in Current Scenario." In *Agroforestry for Climate Resilience and Rural Livelihood*, edited by Inder Dev, Asha Ram, Naresh Kumar, Ramesh Sing, Dhiraj Kumar, A. R. Uthappa, A. K. Handa, and O. P. Chaturvedi, 1–10. Jodhpur, India: Scientific Publishers.

Dev, Inder, Asha Ram, Sudhir Pal Ahlawat, Dana Ram Palsaniya, Ramesh Singh, Shiv Kumar Dhyani, Naresh Kumar, Rama Tewari, Mahendra Sing Sridhar K B, Ram Newaj, Ragunandan Dwivedi, Ram Kumar, Ram Yadav, Lal Chand, Dhiraj Kumar, Jasti Prasad. 2020. "Bamboo-Based Agroforestry System (Dendrocalamus Strictus + Sesame–Chickpea) for Enhancing Productivity in Semi-Arid Tropics of Central India." *Agroforestry Systems* 94: 1725–1739. doi:10.1007/s10457-020-00492-8.

Dhaliwal, Jeevanjot, S. S. Kukal, and Sandeep Sharma. 2017. "Soil Organic Carbon Stock in Relation to Aggregate Size and Stability under Tree-Based Cropping Systems in Typic Ustochrepts." *Agroforestry Systems* 95:615–624. doi:10.1007/s10457-017-0103-8.

Dhanushkodi, V., S. Easwaran, and G. Amuthaselvi. 2019. "Agroforestry: A novel techniques to enhance farmers' income under changing climate." *Journal of Pharmacognosy and Phytochemistry;* SP2: 204-206. https://www.phytojournal.com/archives/2019/vol 8issue2S/PartF/Sp-8-2-55-476.pdf.

Dhara P K (2016). "Humid and Sub Humid Region – BCKV, Jhargram (West Bengal)." In "*Agroforestry Technologies for Different Agro-climatic Zones of the Country*." edited by O. P. Chaturvedi., A. K. Sikka, A. K. Handa, and C. K. Bajpai, 27-28. All India Coordinated Research project on Agroforestry, Jhansi: ICAR-Central Agroforestry Research Institute.

Dhara, P. K., and Babloo Sharma. 2015. "Evaluation of mango-based agroforestry is an ideal model for sustainable agriculture in red & laterite soil." *Journal of Pure and Applied Microbiology* 9(2): 265-272. https://www.researchgate.net/profile/Babloo-Sharma-2/publication/299283879_Evaluation_of_Mango_Based_Agroforestry_is_an_Ideal_Model_for_Sustainable_Agriculture_in_Red_Laterite_Soil/links/5741d3b308aea45ee84a33c2/Evaluation-of-Mango-Based-Agroforestry-is-an-Ideal-Model-for-Sustainable-Agriculture-in-Red-Laterite-Soil.pdf.

Dhara, P. K., S. Panda, S. Sarkar, and N. C. Das. 2016. "Agroforestry systems practiced in new alluvial zone of West Bengal." In *Proc Intl. Conf. Agriculture, Food Sci., Nat. Resource Management and Environmental Dynamics: The Technology, People and Sustainable Development*, 162-164. Kalyani, India, August 13 – 14.

Dhyani, S. K. 2019. "Agroforestry in Indian Perspective." *In Agroforestry Opportunities for Enhancing Resilience to Climate Change in Rainfed Areas*, edited by G. R. Rao., M. Prabhakar, G. Venkatesh, I. Srinivas and K. S. Reddy, 2:12 – 18. Hyderabad: ICAR - Central Research Institute for Dryland Agriculture.

Dhyani, S. K., Ashra Ram, and Inder Dev. 2016. "Potential of agroforestry systems in carbon sequestration in India." *Indian J Agric Sci* 86: 1103-1112. https://www.researchgate.net/profile/Asha-Ram2/publication/317687438_Potential_of_agroforestry_systems_in_carbon_sequestration_in_India/links/59490.

Diallo, Mariama B., P. B. Irénikatché Akponikpè, Tougiani Abasse, Dougbédji Fatondji, and Euloge K. Agbossou. 2021. "Combined effect of trees and soil fertility management practices on millet yields in the Sahel." *Agroforestry Systems* 95: 717-730. doi:10.1007/s10457-021-00624-8.

Dixon, Robert K., Jack K. Winjum, Kenneth J. Andrasko, Jeffrey J. Lee, and Paul E. Schroeder. 1994. "Integrated Land-Use Systems: Assessment of Promising Agroforest and Alternative Land-Use Practices to Enhance Carbon Conservation and Sequestration." *Climatic Change* 27: 71–92. doi:10.1007/bf01098474.

Doddabasawa, B. M. Chittapur, and M. Mahadeva Murthy. 2019. "Comparison of Carbon Footprint of Traditional Agroforestry Systems under Rainfed and Irrigated Ecosystems." *Agroforestry Systems* 94: 465–475. doi:10.1007/s10457-019-00413-4.

Dubiez, Emilien, Vincent Freycon, Jean-Noël Marien, Régis Peltier, and Jean-Michel Harmand. 2018. "Long Term Impact of Acacia Auriculiformis Woodlots Growing in Rotation with Cassava and Maize on the Carbon and Nutrient Contents of Savannah Sandy Soils in the Humid Tropics (Democratic Republic of Congo)." *Agroforestry Systems* 93: 1167–1178. doi:10.1007/s10457-018-0222-x.

Dury, S.J. 1991. "Agroforestry for Soil Conservation." *Agricultural Systems* 35: 472–73. doi:10.1016/0308-521x(91)90121-p.

Figyantika, Arom, Daniel S. Mendham, Marcus A. Hardie, Eko B. Hardiyanto, and Mark A. Hunt. 2020. "Productivity Benefits from Integrating Acacia Auriculiformis and Agricultural Cropping in Java, Indonesia." *Agroforestry Systems* 94: 2109–23. doi:10.1007/s10457-020-00534-1.

Froufe, Luís Cláudio, Daniel Kramer Schwiderke, Amanda Costa Castilhano, Raul Matias Cezar, Walter Steenbock, Carlos Eduardo Seoane, Itamar Antônio Bognola, and Fabiane Machado Vezzani. 2019. "Nutrient Cycling from Leaf Litter in Multistrata Successional Agroforestry Systems and Natural Regeneration at Brazilian Atlantic Rainforest Biome." *Agroforestry Systems* 94: 159–171. doi:10.1007/s10457-019-00377-5.

FSI. (Forest Survey of India). 2000b. "Trees Outside Forest Resource of Haryana." *Forest Survey of India*: Northern Zone, Shimla.

Fungo, Bernard, Joel Buyinza, Jude Sekatuba, Susan Nansereko, Gerald Ongodia, Phiona Kwaga, Scovia Mudondo, et al. 2019. "Forage Biomass and Soil Aggregate Carbon under Fodder Banks with Contrasting Management Regimes." *Agroforestry Systems* 94(3):1023 – 1035.doi: 10.1007/s10457-019-00473-6.

Garrity, D. P. 2004. "Agroforestry and the Achievement of the Millennium Development Goals." *Agroforestry Systems* 61: 5–17. doi.10.1023/b:agfo.0000028986.37502.7c.

Gebremeskel, Destaalem, Emiru Birhane, Meley Mekonen Rannestad, Semae Gebre, and Girmay Tesfay. 2021. "Biomass and Soil Carbon Stocks of Rhamnus Prinoides Based Agroforestry Practice with Varied Density in the Drylands of Northern Ethiopia." *Agroforestry Systems* 1- 19. doi: 10.1007/s10457-021-00608-8.

Gollin, Douglas, Casper Worm Hansen, and Asger Wingender. 2018. "Two Blades of Grass: The Impact of the Green Revolution." Working Paper 24744. Http://www.nber.org/papers/w24744ed. Vol. Working Paper 24744. Cambridge, Massachusetts: National Bureau of Economic Research Working Paper Series.

Goncalves, N., Andrade, D., Batista, A., Cullen, L., Souza, A., Gomes, H., Uezu, A. 2021. "Potential economic impact of carbon sequestration in coffee agroforestry systems". *Agroforest Syst* 95, 419–430. doi:10.1007/s10457-020-00569-4.

Griffith, Daniel M. 2000. "Agroforestry: a Refuge for Tropical Biodiversity after Fire." *Conservation Biology* 14 (1): 325–26. doi:10.1046/j.1523-1739.2000.99101.x.

Gupta V. K., S. K. Dhyani, N. Kumar.2019. "Agroforestry Model: Shisham based Agrisilvicultural and Silvi-pastoral System." In *Successful Agroforestry Models for Different Agro-Ecological Regions in India*, edited by Handa, A. K., Inder Dev, R. H. Rizvi, Naresh Kumar, Asha Ram, Dheeraj Kumar, Anil Kumar,

S. Bhaskar, S. K. Dhyani and Javed Rizvi, 26 – 31. Jhansi, Uttar Pradesh: Central Agroforestry Research Institute (CAFRI).

Handa, A. K., O. P. Toky, S. B. Chavan. 2016. "Innovative agroforestry for environmental security in India." *World Agriculture* 1-23. "http://www.world-agriculture.net/article/innovative-agroforestry-for-environmental-security-in-india.

Herrera, Ana María, Alexandre Carneiro Leão de Mello, Valéria Xavier de Oliveira Apolinário, José Carlos Batista Dubeux Júnior, Valdson José da Silva, Mércia Virginia Ferreira dos Santos, and Márcio Vieira da Cunha. 2020. "Decomposition of senescent leaves of signalgrass (Urochloa decumbens Stapf. R. Webster) and arboreal legumes in silvopastoral systems." *Agroforestry Systems* 94: 2213-2224. doi:10.1007/s10457-020-00542-1.

Hombegowda, H. C., Michael Köhler, Alexander Röll, and Dirk Hölscher. 2019. "Tree Species and Size Influence Soil Water Partitioning in Coffee Agroforestry." *Agroforestry Systems* 94 (1): 137–149. doi:10.1007/s10457-019-00375-7.

Honnayya, Chittapur B. M., and Doddabasawa. 2020. "Productivity of Pigeonpea (Cajanus Cajan L. Millsp.) in Neem (Azadirachta Indica A. Juss.) Based Agroforestry System on Alfisols in Semi-Arid Tropics." *Agroforestry Systems* 94: 1879–89. doi:10.1007/s10457-020-00507-4.

Iwasaki, Kenta, Hiroyuki Torita, and Tomoyuki Abe. 2020. "A Simple Process-Based Model for Estimating Windbreak Effects on Soil Temperature during Early Crop Growth Stage." *Agroforestry Systems* 94: 2401–2415. doi:10.1007/s10457-020-00561-y.

Jakhar, Praveen, Anchal Dass, Partha Pratim Adhikary, S. Sudhishri, B. S. Naik, H. C. Hombegowda, M. Madhu, N. K. Lenka, P. R. Chaudhary, and R. K. Panda. 2017. "Multitier Agroforestry System for Integrated Resource Conservation on Uplands of Eastern Ghats Region in India." *Agroforestry Systems* 91: 697–712. doi:10.1007/s10457-016-9976-1.

Jose, Shibu. 2009. "Agroforestry for Ecosystem Services and Environmental Benefits: An Overview." *Agroforestry Systems* 76: 1–10. doi:10.1007/s10457-009-9229-7.

Julius, A. J., O. N. Iluyomade, G. O. Akinyemi, N. S. Oniroko, Q. A. Onilude, and A. M. Aderemi. 2011. "Assessment of the Impact of Agroforestry Education on Biodiversity Conservation and Community Development." *International Journal for Cross-Disciplinary Subjects in Education (IJCDSE)* 1: 577–582. doi:10.20533/ijcdse.2042.6364.2011.0080.

Kang, B. T., and F. K. Akinnifesi. 2000. "Agroforestry as Alternative Land-Use Production Systems for the Tropics." *Natural Resources Forum* 24: 137–51. doi:10.1111/j.1477-8947.2000.tb00938.x.

Khaleel, Ala' A., Thomas J. Sauer, and John C. Tyndall. 2019. "Changes in Deep Soil Organic Carbon and Soil Properties beneath Tree Windbreak Plantings in the U.S. Great Plains." *Agroforestry Systems* 94: 565–581. doi:10.1007/s10457-019-00425-0.

Kleinn, C., and D. Moralos. 2005. "Obtaining Resource Information on Trees Outside the Forests: Experiences with Inventory Approaches in Central America." In *Multipurpose Trees in the Tropics: Management & Improvement Strategies*, edited by V. P. Tewari and R. Srivastava, 612–618. Jodhpur: Published by Scientific Publishers (India) on behalf of Arid Forest Research Institute.

Kukal, S. S., Manmeet Kaur, S. S. Bawa, and Naveen Gupta. 2007. "Water-Drop Stability of PVA-Treated Natural Soil Aggregates from Different Land Uses." *CATENA* 70: 475–479. doi: 10.1016/j.catena.2006.11.013.

Kumar, B. Mohan, Anil Kumar Singh, and S. K. Dhyani. 2012. "South Asian agroforestry: traditions, transformations, and prospects." In *Agroforestry-The Future of Global Land Use*, edited by P. K. R. Nair and D. Garrity, 359-389. Dordrecht: Springer.

Kumar, Rajeew, Sharad Pandey, and Apurv Pandey. 2006. "Plant roots and carbon sequestration." *Current Science* 91: 885-890. 885 https://www.jstor.org/stable/24094284?seq=1#metadata_info_tab_contents.

Kumar, S., and O. P. Chaturvedi. 2017. "Agroforestry Interventions in Fruit Orchards for Improving Productivity, Profitability and Sustainability." *In Doubling Farmers Income through Horticulture*, edited by S K Sing, P Kalia, W S Dhilon, T K Bahera, and J Prakash, 569 - 576. New Delhi: Daya Publishing House, a division of Astral International Pvt. Ltd.

Kumar, Sudhir, Rajendra Prasad, Anil Kumar, and S. K. Dhyani. 2019. "Integration of fruit trees in agroforestry for sustainability and profitability of farming systems in arid and semi-arid regions." *Indian Journal of Agroforestry* 21: 95-99. https://www.researchgate.net/publication/335220173_Integration_of_fruit_trees

_in_agroforestry_for_sustainability_and_profitability_of_farming_systems_in_arid_and_semi-arid_regions

Kumar, Sudhir, Rajendra Prasad, Veeresh Kumar, and Alok K. Krishna. 2021. "Organic source on productivity of pomegranate–lemongrass-based agroforestry system in central India." *Agroforestry Systems* 95(4): 615-624. doi:org/10.1007/s10457-021-00605-x.

Lai, R. 2004. "Soil Carbon Sequestration in Natural and Managed Tropical Forest Ecosystems." *Journal of Sustainable Forestry* 21: 1–30. doi:10.1300/j091v21n01_01.

Lai, Rattan. 1989. "Conservation Tillage for Sustainable Agriculture: Tropics Versus Temperate Environments." *Advances in Agronomy* 85–197. doi:10.1016/s0065-2113(08)60524-6.

Lenka, N. K., P. R. Choudhury, S. Sudhishri, A. Dass, and U. S. Patnaik. 2012. "Soil aggregation, carbon build up and root zone soil moisture in degraded sloping lands under selected agroforestry-based rehabilitation systems in eastern India." *Agriculture, ecosystems & environment* 150: 54-62. doi: 10.1016/j.agee.2012.01.003.

Lundgren, Björn, and John B. Raintree. 1983. *Sustained agroforestry*. Nairobi: ICRAF.

Manna, M. C., P. K. Ghosh, and C. L. Acharya. 2003. "Sustainable Crop Production Through Management of Soil Organic Carbon in Semiarid and Tropical India." *Journal of Sustainable Agriculture* 21: 85–114. doi:10.1300/j064v21n03_07.

Mathur, R. B., and M. P. Shiva. 1996. *Management of Minor Forest Produce for Sustainability*,573. New Delhi: Oxford & IBH Publ.

Mekonnen, Mulatie, Tigist Worku, Birru Yitaferu, Artemi Cerdà, and Saskia Keesstra. 2021. "Economics of Agroforestry Land Use System, Upper Blue Nile Basin, Northwest Ethiopia." *Agroforestry Systems*. doi:10.1007/s10457-021-00612-y.

Ministry of Environment, Forest and Climate Change (MoEF&CC), Govt. of India. 2018. Strategy for Increasing Green Cover Outside Recorded Forest. New Delhi.

Mohapatra S. C., Sasmita Behera, Prasannajit Mishra, Bibhuti Bhusan Behera, (2019). "Mango based Agri-horticultural System." In *Successful Agroforestry Models for Different Agro-Ecological Regions in India*. Edited by Handa, A. K., Inder Dev, R. H. Rizvi, Naresh Kumar, Asha Ram, Dheeraj Kumar, Anil Kumar, S. Bhaskar, S. K. Dhyani and Javed Rizvi, 113 – 116. Jhansi, Uttar Pradesh: Central Agroforestry Research Institute (CAFRI).

Montagnini, F., and P. K. R. Nair. 2004. "Carbon Sequestration: An Underexploited Environmental Benefit of Agroforestry Systems." *Agroforestry Systems* 61: 281–95. doi:10.1023/b: agfo.0000029005.92691.79.

Mukangango, Marguerite, Jean Nduwamungu, Francois Xavier Naramabuye, Gert Nyberg, and A. Sigrun Dahlin. 2020. "Biomass Production and Nutrient Content of Three Agroforestry Tree Species Growing on an Acid Anthropic Ferralsol under Recurrent Harvesting at Different Cutting Heights." *Agroforestry Systems* 94: 857–67. doi:10.1007/s10457-019-00455-8.

Murmu, Suren, Pratap Kumar Dhara, and Debjit Roy. 2016. "Agroforestry: A Sustainable Agro-technique for Red and Laterite Zone of West Bengal." *Journal of Agroecology and Natural Resource Management*. 3:280-283. https://www.krishisanskriti.org/vol_image/16Dec2016081209155%20%20%20%20%20S%20%20Murmu%20%20%20%20%20%20%20%20%20280-283.pdf.

Mutanal S. M., S. J. Patil, H. Y. Patil, S. L. Madiwalar. 2019. "Agroforestry Model: Teak based Agrisilvicultural System". In *Successful Agroforestry Models for Different Agro-Ecological Regions in India*, edited by Handa, A. K., Inder Dev, R. H. Rizvi, Naresh Kumar, Asha Ram, Dheeraj Kumar, Anil Kumar, S. Bhaskar, S. K. Dhyani and Javed Rizvi, 44 – 47. Jhansi, Uttar Pradesh: Central Agroforestry Research Institute (CAFRI).

Nair, P. K. Ramachandran. 1993. *An introduction to agroforestry*. Dordrecht, The Netherlands: Kluwer Academic Publishers.

Narain, Pratap, K. D. Sharma, D. C. Joshi, L. N. Harsh, and J. C. Tiwari. 2000. *Glimpses of Agroforestry and Soil Conservation in India*, 1 – 17. Jodhpur: ICAR - Central Arid Zone Research Institute.

Naugraiya, M. N., and Puri Sunil. 2001. "Performance of multipurpose tree species under agroforestry systems on Entisols of Chattisgarh plains." *Range Management and Agroforestry* 22: 164-172. https://www.cabdirect.org/cabdirect/abstract/20023061834.

Naugraiya, M. N., and S. S. Shaw. 2014. "Gmelina arborea as high-density plantation to reclaim the red lateritic wasteland lands of Chhattisgarh." *Journal of Tree Sciences* 33: 34-42. http://www.ists.in/articleFiles/270814043219JTS_VI.pdf.

Neha, B. S. Bhople, and Sandeep Sharma. 2020. "Seasonal Variation of Rhizospheric Soil Properties under Different Land Use Systems at Lower Shivalik Foothills of Punjab, India." *Agroforestry Systems* 94: 1959–1976. doi:10.1007/s10457-020-00512-7.

Newaj R., Asha Ram (2019). "Aonla based Agri-horticultural System." In *Successful Agroforestry Models for Different Agro-Ecological Regions in India.* Edited by Handa, A. K., Inder Dev, R. H. Rizvi, Naresh Kumar, Asha Ram, Dheeraj Kumar, Anil Kumar, S. Bhaskar, S. K. Dhyani and Javed Rizvi, 31 – 36. Jhansi, Uttar Pradesh: Central Agroforestry Research Institute (CAFRI).

Newaj, Ram, and R. K. Tewari., Ajit and R. S. Yadav. 2006. *Aonla based agroforestry system: A promising system for rainfed areas*,1 – 10. NRCAF, Jhansi (India).

Noumi, V. Noiha, L. Zapfack, M. R. Hamadou, V. Awe Djongmo, N. Witanou, B. Nyeck, J. D. Ngossomo, R. B. Tabue Mbobda, and P. M. Mapongmetsem. 2018. "Floristic diversity, carbon storage and ecological services of eucalyptus agrosystems in Cameroon." *Agroforestry systems* 92: 239-250. doi:10.1007/s10457-017-0130-5.

Panda, S., P. K. Dhara, S. Sarkar, and N. C. Das. 2018. "Agroforestry: A Multipurpose Multi – Storeyed Renewable Plant Treasure with Medicinal Values Giving Multi – Returns for Livelihood and Benefits Over Climate Change." In *Climate Change and Agroforestry: Adaptation, Mitigation and Livelihood Security*, Edited by C. B. Pandey, Mahesh K. Gaur, and R. K. Goyal, 627–637. New Delhi: New India Publishing Agency.

Pandey, D. 2000. "Trees outside forests: India." *Food and Agriculture Organization.* http://www.fao.org/3/y2328e/y2328e18.htm#TopOfPage.

Pandey, D. 2008. "Trees Outside the Forest (TOF) Resources in India." *International Forestry Review* 10: 125–33. doi:10.1505/ifor.10.2.125.

Pandey, Deep Narayan. 2002. "Carbon sequestration in agroforestry systems." *Climate policy* 2 (4): 367-377. doi:10.1016/S1469-3062(02)00025-6.

Pandey, Deep Narayan. 2007. "Multifunctional agroforestry systems in India." *Current science* 92: 455-463. Accessed June 2 https://www.jstor.org/stable/24097558?seq=1.

Partey, Samuel T., Naresh V. Thevathasan, Robert B. Zougmoré, and Richard F. Preziosi. 2016. "Improving Maize Production through Nitrogen Supply from Ten Rarely-Used Organic Resources in Ghana." *Agroforestry Systems* 92:375–387. doi:10.1007/s10457-016-0035-8.

Parthiban K T. (2019). "Melia dubia based Agroforestry Systems: Mini Clonal Technology for Mass Multiplication." In *Successful Agroforestry Models for Different Agro-Ecological Regions in India.* Edited by Handa, A. K., Inder Dev, R. H. Rizvi, Naresh Kumar, Asha Ram, Dheeraj Kumar, Anil Kumar, S. Bhaskar, S. K. Dhyani and Javed Rizvi, 71 – 84. Jhansi, Uttar Pradesh: Central Agroforestry Research Institute (CAFRI).

Pathak, P. S., H. M. Pateria, and K. R. Solanki. 2000. "Agroforestry systems in India: a diagnosis and design approach." *NRCAF (ICAR), New Delhi* 1 - 223.

Pérez-Flores, Julián, Alejandra Arias Pérez, Yesenia Primo Suárez, Vinicio Calderón Bolaina, and Asunción López Quiroga. 2017. "Leaf Litter and Its Nutrient Contribution in the Cacao Agroforestry System." *Agroforestry Systems* 92:365–374. doi:10.1007/s10457-017-0096-3.

Prasad, R., and S. K. Dhyani. 2010. "Review of agroforestry contributions and problems in arid ecosystem for livelihood support in India." *Journal of Soil and Water Conservation*, India 9 (4): 277-287.

Qaiser, K. N., V. Dutt., A. H. Mughal., J. A. Mugloo. 2016. "Apple based Horti-agri-pasture system under temperate conditions of Kashmir valley." In *Agroforestry technologies for different agro-climatic zones of the country*, edited by O. P. Chaturvedi., A. K. Sikka, A. K. Handa, and C. K. Bajpai, 1-177. All India Coordinated Research project on Agroforestry, Jhansi: ICAR-Central Agroforestry Research Institute.

Raddy, K. Sammi., and Arun K. Shanker. 2018. "Climate Change and Agriculture: Global and Local Perspectives." In *Agroforestry Opportunities for Enhancing Resilience to Climate Change in Rainfed Areas*, edited by G. Rajeshwar Rao, M. Prabhakar, G. Venkatesh, I. Srinivas, and K. Sammi Raddy, 1–11. Hyderabad: ICAR - Central Research Institute for Dryland Agriculture.

Rallings, Anna M., Sean M. Smukler, and Kent Mullinix. 2020. "A Novel Methodology to Characterize and Quantify Regional Farmscape Non-Production Perennial Vegetation Carbon Storage and Potential for Loss in Southwest British Columbia." *Agroforestry Systems* 94: 1947–58. doi:10.1007/s10457-020-00514-5.

Ramachela, K. "Edaphic and Climatic Factors and the Nutrient Dynamics in Uapaca Kirkiana Dominated Miombo Woodland Agroforestry Ecosystems. 2018." Agroforestry Systems 93: 1501 -1511. doi:10.1007/s10457-018-0256-0.

Ramos, Helen Monique Nascimento, Steel Silva Vasconcelos, Osvaldo Ryohei Kato, and Débora Cristina Castellani. 2018. "Above-and belowground carbon stocks of two organic, agroforestry-based oil palm production systems in eastern Amazonia." *Agroforestry systems* 92: 221-237. doi:10.1007/s10457-017-0131-4. 221.

Rao, K. P. C., Louis V. Verchot, and Jan Laarman.2007. "Adaptation to climate change through sustainable management and development of agroforestry systems." *Journal of SAT agricultural research* 4(1): 1-30. http://oar.icrisat.org/id/eprint/2561.

Rao, M. R., M. P. Singh, and R. Day. 2000. "Insect pest problems in tropical agroforestry systems: Contributory factors and strategies for management." *Agroforestry Systems* 50 (3): 243-277. doi:10.1023/A:1006421701772.

Raveendra, S. A., Sarath P. Nissanka, Deepakrishna Somasundaram, Anjana J. Atapattu, and Sylvanus Mensah. 2021. "Coconut-Gliricidia Mixed Cropping Systems Improve Soil Nutrients in Dry and Wet Regions of Sri Lanka." *Agroforestry Systems* 95: 307–19. doi.10.1007/s10457-020-00587-2.

Rocha, Delphy, T. K. Kunhamu, B. Mohan Kumar, A. V. Santhoshkumar, and V. Jamaludheen. 2021. "Density manipulation and pruning on fine root production and decomposition dynamics in a mature Acacia mangium Willd. stand in Kerala, India." *Agroforestry Systems* 1-15. doi:10.1007/s10457-021-00601-1.

Roy, G. B., S. K. Majhi., R. K. Ghosh., S. Panda. 1989. "Study of soil moisture status in different planting methods of Akashmoni (*Acacia auriculiformis* A. Cunn.) trees in an agroforestry system of cultivation on laterite soil of west Bengal." *Ind J Landscape Systems and Ecological Studies* 12(2):31 – 34.

Rudramurthy, H. V., E. T. Puttaiah, and B. R. Gurumurthy. 2008. "Water retention characteristics of red soils under different land use ecosystems in southern transitional zone of Karnataka." *Indian Journal of Agricultural Research* 42:79-85. https://www.researchgate.net/publication/268011465_Water_retention_characteristics_of_red_soils_under_different_land_use_ecosystems_in_southern_transitional_zone_of_Karnataka.

Saha, Debasish, S. S. Kukal, and S. Sharma. 2011. "Landuse Impacts on SOC Fractions and Aggregate Stability in Typic Ustochrepts of Northwest India." *Plant and Soil* 339(1): 457–470. doi:10.1007/s11104-010-0602-0.

SAHA, SAJAL, P. K. DHARA, and P. K. PATRA. 2014. "Agri-Horti-Silviculture–An Alternative Land Use System for Rainfed Lands in Red and Laterite Zone of West Bengal." *Natural Resource Management Opportunities and Technological Options*,1: 223–231. https://www.researchgate.net/profile/Prasanta-Patra-2/publication/288828980_Agri-Horti-Silvi_culture_-an_alternative_land_use_system_for_rained_lands_in_Red_and_Laterite_Zone_of_West_Bengal/links/59de61df0f7e9bcfab23ff6a/Agri-Horti-Silvi-culture-an-alternative-land-use-system-for-rained-lands-in-Red-and-Laterite-Zone-of-West-Bengal.pdf.

Sanchez, P. A. 1995. "Science in Agroforestry." *Agroforestry: Science, Policy and Practice*, 5–55. doi:10.1007/978-94-017-0681-0_1.

Sanna, Federico, Giovanni Antonio Re, Giovanna Piluzza, Giuseppe Campesi, and Leonardo Sulas. 2018. "Forage Yield, Nutritive Value and N-Fixation Ability of Legume Based Swards Are Affected by Light Intensity in a Mediterranean Agroforestry System." *Agroforestry Systems* 93: 2151–61. doi:10.1007/s10457-018-0331-6.

Sarto, Marcos V. M., Wander L. B. Borges, Jaqueline R. W. Sarto, Charles W. Rice, and Ciro A. Rosolem. 2020. "Deep soil carbon stock, origin, and root interaction in a tropical integrated crop–livestock system." *Agroforestry Systems* 94: 1865-1877. doi:10.1007/s10457-020-00505-6.

Sehgal, J. 1998. "Red and Lateritic Soils: An Overview." In *Managing Red and Lateritic Soils for Sustainable Agriculture,* edited by J. Sehgal, W. E. Blum, and K. S. Gajbhiya, 3–9. Netherlands: Rotterdam.

Sharma, K. L. 2018. "Improving Soil Health and Crop Productivity amidst Climate Change through Conservation Agriculture." In *Agroforestry Opportunities for Enhancing Resilience to Climate Change in*

Rainfed Areas, edited by G. Rajeshwar Rao, M. Prabhakar, G. Venkatesh, I. Srinivas, and K. Sammi Reddy, 112–121. Hyderabad: ICAR - Central Research Institute for Dryland Agriculture.

Sharma, Prashant, M. K. Singh, Prabhat Tiwari, and Kamlesh Verma. 2017. "Agroforestry systems: Opportunities and challenges in India." *Journal of Pharmacognosy and Phytochemistry* 1(sp): 953-957. "https://www.researchgate.net/publication/337167646_ Agroforestry_systems_Opportunities_and_challenges_in_India.

Sharma, S. K. 2004. "Hortipasture based land use systems for enhancing productivity of degraded lands under rain fed and partially irrigated conditions." *Uganda Journal of Agricultural Sciences* 9: 320-325.

Shiva, M. P., and R. B. Mathur. 1996. "Management of minor forest produce for sustainability." Management of minor forest produce for sustainability. New Delhi: India.

Shukla, A. K., Sunil Kumar, and S. N. Ram. 2018. "Yield, Soil Health and Moisture Status in Bael (Aegle Marmelos)-Based Hortipasture System in Semi-Arid Ecosystem of India." *Agroforestry Systems* 93: 1275–83. doi:10.1007/s10457-018-0241-7.

Singh, Mahendra, K. B. Sridhar, Dhiraj Kumar, A. R. Uthappa, R. P. Dwivedi, Inder Dev, R. H. Rizvi, R. K. Tewari, and O. P. Chaturvedi. 2017. "Doubling farmers' income through agroforestry in north-western India: A policy perspective." *Indian Journal of Agroforestry* 19: 90-95. https://www.indianjournals.com/ijor.aspx?target=ijor:ijaf&volume=19&issue=2&article=016.

Singh, R. P. (1996). "Alternate land use system for sustaining development." *Range Management Agroforestry* 17(2):155–178.

Singh, R. P. 1996. "Alternate land use system for sustainable development." *Rang Management and Agroforestry*. 17 (2): 155 – 177.

Sobola, O. O., D. C. Amadi, and G. Y. Jamala. 2015. "The role of agroforestry in environmental sustainability." *IOSR Journal of Agriculture and Veterinary Science* 8: 20-25. https://www.researchgate.net/publication/320710948_The_Role_of_Agroforestry_in_Environmental_Sustainability.

Soil Science Society of America. 1996. "Glossary of Soil Science Terms." Soil Science Society of America, Madison, WI.

Soil Survey Staff USA. 1999. "Soil Taxonomy: Basic System of Soil Classification for Making and Interpreting Soil Surveys." USDA- Natural Resource Conservation Service, Washington, DC.

Talbot, W. A. 1902. "The Trees, Shrubs and Woody Climbers of the Bombay Presidency." Nature 67: 148. doi:10.1038/067148a0.

Tewari, S. K. 2008. "Farm Forestry in Agroforestry." Agro-forestry Project Dept. of Genetics and Plant Breeding, G. B. Pant University of Agriculture and Technology, Pantnagar.

Tewari, V. P. 2009. "Trees Outside Forests: its assessment, role and importance in Arid and Urban areas." Paper presented at the 2nd GAForN International Symposium on 'Multipurpose Forestry: Managing and Enhancing Ecosystem Services and Production Functions of Forests, Woodlands and Trees Outside Forests,' Dehradun (India). doi:10.13140/2.1.4741.8562.

Van Den Berge, Sanne, Pieter Vangansbeke, Lander Baeten, Thomas Vanneste, Fien Vos, and Kris Verheyen. 2021. "Soil Carbon of Hedgerows and 'Ghost' Hedgerows." *Agroforestry Systems* 1 - 17 doi:10.1007/s10457-021-00634-6.

Verchot, Louis V., Meine Van Noordwijk, Serigne Kandji, Tom Tomich, Chin Ong, Alain Albrecht, Jens Mackensen, Cynthia Bantilan, K. V. Anupama, and Cheryl Palm. 2007. "Climate Change: Linking Adaptation and Mitigation through Agroforestry." *Mitigation and Adaptation Strategies for Global Change* 12: 901–18. doi:10.1007/s11027-007-9105-6.

World Bank. 2004. *Sustaining Forest: A Development Strategy*. Washington, DC: World Bank. doi:10.1596/0-8213-5755-7.

Yengwe, Jones, Mesfin Tsegaye Gebremikael, David Buchan, Obed Lungu, and Stefaan De Neve. 2017. "Effects of Faidherbia Albida Canopy and Leaf Litter on Soil Microbial Communities and Nitrogen Mineralization in Selected Zambian Soils." *Agroforestry Systems* 92:349–363. doi:10.1007/s10457-016-0063-4.

Young, A. 1989. *Agroforestry for Soil Conservation*. CAB International, Wallingford: UK.

Zaro, Geovanna C., Paulo H. Caramori, George M. Yada Junior, Carlos R. Sanquetta, Armando Androcioli Filho, Amanda L. Nunes, Cássio E. Prete, and Paul Voroney. 2019. "Carbon Sequestration in an

Agroforestry System of Coffee with Rubber Trees Compared to Open-Grown Coffee in Southern Brazil." *Agroforestry Systems* 94: 799–809. doi:10.1007/s10457-019-00450-z.807.

Zhang, W., W. L. Xu, P. Ahanbieke, B. J. Wang, X. D. Hao, Y. Zhu, and L. H. Li. 2018. "Competition for 15N-Labeled Nitrogen in a Jujube Tree (*Zizyphus Jujuba* Mill.)/Wheat (*Triticum Aestivum* L.) Agroforestry System in North-western China." *Agroforestry Systems* 93: 2097–2110. doi:10.1007/s10457-018-0332-5.

Chapter 19

Sustainable Technology for Replenishing Ecological Soil Health

Rahul Chopra[1,*], Ajeet Singh[2] and Vishnu D. Rajput[3]

[1]Agriculture University, Kota, Rajasthan, India

[2]Department of Soil Science and Agricultural Chemistry,
Sri Karan Narendra Agriculture University, Jobner, Jaipur, Rajasthan, India

[3]Academy of Biology and Biotechnology,
Southern Federal University, Rostov-on-Don, Russia

Abstract

Soil health is a prerequisite for long-term, sustainable agricultural productivity. Achieving food security, healthy ecosystems and climate change mitigation benefits requires understanding the value and impacts of soil health. A healthy soil has excellent physical, chemical and biological properties. Healthy soil has excellent soil tilth, good soil drainage and enriched with organic matter content, beneficial microorganisms and essential plant nutrients. Good soil health is the key factor for sustainable crop production. Healthy soils sustain plant and animal productivity, maintain environmental quality along with water, and enhance plant and animal health. But intensive and conventional agricultural practices degrade the soil health worldwide. In intensive agriculture, poor management decreases physico-chemical and biological properties of soil and exhausts the essential plant nutrients from soil. Therefore to improve and replenish soil health sustainably, it is essential to incorporate sustainable technologies in soil health management system. These technologies are friendly to farmers and the environment. Soil management technology is important for soil health, crop productivity, environmental sustainability, and human health. Sustainable technologies are holistic in nature to replenish the soil health. To achieve sustainability in food security, with sustainable soil management practices will be the challenging. It can be accomplished through proper nutrient management and appropriate soil conservation practices. Thus, this chapter emphasizes the use sustainable technologies to replenish soil health as well as crop production.

Keywords: soil health, principles, sustainable technologies

---

* Corresponding Author's Email: r.chopra051988@gmail.com.

In: Ecosystem Services
Editor: Hanuman Singh Jatav
ISBN: 978-1-68507-614-6

1. Introduction

Healthy soil is the best supporting pillar for productive, profitable, and environmentally sounds agricultural systems. A good quality soil defined as the capacity of a specific kind of soil to function, within the naturally managed ecosystem boundaries, to sustain plant and animal productivity, maintain and enhance water and air quality as well as support human health and habitation (Seifu and Elias, 2018; Karlen et al., 1997). A healthy and good quality soil support plant growth, nutrient cycling, biological control of diseases and pests as well as regulation of water and air supply. All these functions are influenced by the interrelated physico-chemical and biological properties of soil. Healthy soil is the key component for sustainable crop production. According to (FAO, 2015) 95% of our food is directly or indirectly produced on our soils. Healthy soils are the foundation of the food system. Our soils are the base and medium for agriculture where all food-producing plants are grown. Healthy soils produce healthy crops that in turn nourish people and animals. Indeed, soil quality is directly linked to food quality and quantity. Soils supply the essential nutrients, water, oxygen and root support that our food-producing plants need to grow and flourish. They are also serving as a buffer to protect delicate plant roots from drastic fluctuations in temperature (FAO, 2015).

Intensive agriculture practices along with intensive use of modern crop varieties has resulted in deterioration of soil quality and health (Pretty and Bharucha, 2014; Jahir and Satter, 2010). Intensive cultivation practices, indiscriminate use of agro-chemicals, reduced soil organic carbon is the key reason for poor soil health. Poor soil health has many ill effects such as low crop productivity, low soil fertility and high environmental risk.

The climate change globally has adverse effect on land degradation and losses in biodiversity due to low soil quality and fertility. Improved management practices in the soils, which help to maintaining sustainable soil quality (Lal, 2015b). Healthy and quality soils are essential for productive and resilient agricultural systems (CGIAR, 2017).

When a holistic and system based approach to be applied in soil management, productivity of soil increases (Lal, 2015). Efficient management of soil, water and nutrients improves soil quality, soil fertility, land productivity and ultimately food security in sustainable manner. Proper management of soil, water and nutrient also ensure the conservation of natural resources and reduces the GHG emissions (Lal, 2003; Kassam et al., 2009; Palm et al., 2014).

This is possible only by understanding how the soil function to support plant growth and regulate soil physical, chemical, biological as well as environmental quality are affected by different soil management practices. To maintain soil quality in sustainable manner it is necessary to design a soil management system that improves and maintains soil health over time.

2. 4 'M' Principles of Soil Health Management (USDA-NRCS, 2019)

Good soil health is the prerequisite for sustainable crop production. There are 4 principles which are backbone of soil health management (Figure 1). These principles are described as follows:

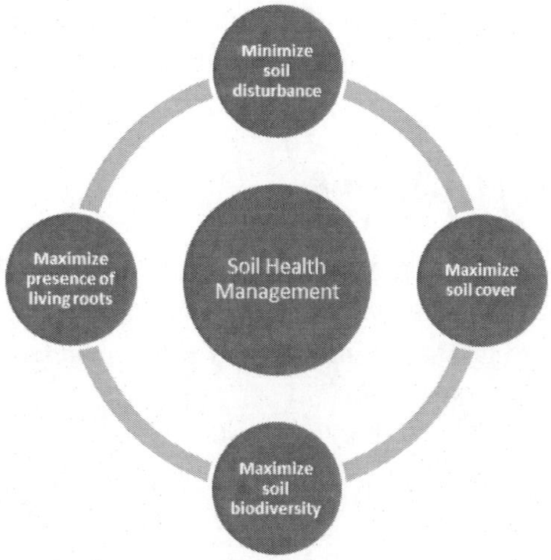

Figure 1. 4 'M' Principles of Soil Health Management.

2.1. Minimize Soil Disturbance

There are 3 key factors namely physical, chemical and biological, which disturb the soil health. Physical disturbances occur due to the conventional tillage practices; chemical disturbances induced due to indiscriminate use of fertilizer and pesticide, and soil disturbed biologically due to over-grazing of animals that can lead to compaction and reduction in perennial root systems, introduction of invasive species and use of monocultures can cause biological imbalances which all affects soil functions.

2.2. Maximize Soil Cover

There are two main forms of soil cover: 1) living plant canopy such as a growing crop, cover crop, or grassland; and 2) mulch, either as dead plant material such as crop residues, prunings from trees and shrubs, thatch in grasslands and compost, bark chips as an amendment. Maximum soil cover helps to mitigate the problem of soil erosion, nutrient leaching, weed infestation, insect-pest. It also enhances the carbon sequestration. So soil cover is important to manage soil health sustainably.

2.3. Maximize Soil Biodiversity

Biodiversity is the variation of life forms within a given ecosystem or field. The different life forms include all of the plants, animals and microorganisms, and their secretions. For good soil health, biodiversity can be increased through a variety of approaches including: plant diversity through the use of diversified crop rotations, cover crop mixes, diversity through the proper integration of grazing animals (e.g., livestock) into the system and includes animals living within the soils or microbial diversity, as well as direct additions with biological amendments. All four soil health management principles contribute to biodiversity.

2.4. Maximize Presence of Living Roots

Maximizing the presence of living roots in the soil can be accomplished through eliminating fallow, diverse crop rotation, inclusion of cover crops, and or through dedicated grasslands. Mixing up which plants are grown during the year or over the course of multiple years may help to break disease and pest cycles.

3. Healthy Soils Are Able to:

- Store carbon and nutrients in soil organic matter
- Cycle nutrients effectively
- Provide good aeration to promote root growth
- Improve farm resiliency and profitability
- Improve yield stability
- Decrease erosion and runoff
- Improve water storage and plant available water while protecting water quality
- Be resilient to, heavy rainfall events, temperature extremes and drought
- Decrease pest and disease problems.

4. Sustainable Soil Management

"Management of soil in sustainable manner possible if the supporting, provisioning, regulating, and cultural services provided by soil are maintained or enhanced without significantly hampering either the soil functions that enable those services or biodiversity. The balance between the supporting and provisioning services for plant production and the regulating services the soil provides for water quality and availability and for atmospheric greenhouse gas composition is a particular concern" (FAO, 2017).

5. Sustainable Technologies to Replenish Soil Health

Good soil health is the pillar of sustainable agriculture. According to FAO healthy soils are the basis of healthy food production. So it is essential to restore soil properties and maintain soil health sustainable manner. There are various technologies which improve soil health sustainably are described as follows:

5.1. Reduce Soil Erosion

Water and wind erosion are the most significant threat to world soils and the ecosystem. Soil erosion by water and wind causes the loss of surface soil layers, soil horizons, subsoil, as well as reduced water quality and sedimentation. Soil erosion is triggered by anthropogenic activities of human, reduced plant cover, deforestation, conventional tillage practices and other field operations. Soil erosion adversely hinders the growth of plants, crop yields, quality of water, and recreation. It is a key cause of degradation of soils as it occurs naturally on all lands (Bai et al., 2010; Ding et al., 2015).

Soil erosion should be reduced through implementation of appropriate measures such as reduce tillage, mulching, cover cropping, decreased use of herbicide, mulching, adopt crop rotation, strip cropping, afforestation, establish wind breaks and shelter belts helps to replenish soil health (FAO, 2017).

5.2. Improve Soil Organic Matter Content

Soil organic matter is the lifeblood of fertile and productive soil. It plays a key role to maintain various soil functions as well as helps to prevent soil degradation. It improves soil physical, chemical and biological properties such as soil structure, aggregation, water retention, buffering capacity, the cycling and storage of plant nutrients and soil biodiversity respectively. SOM increases soil fertility by providing cation exchange sites and acting as reserve of plant nutrients, especially nitrogen (N), phosphorus (P), and sulfur (S), along with micronutrients, which are slowly released upon SOM mineralization. Soil organic matter also acts as a major sink and source of soil carbon (C) (Jangir et al., 2019).

A loss of soil organic carbon (SOC) from soil due to in-appropriate use of land, mis-management of soil and faulty cropping practices can cause a decline in soil quality and health and enhances the emissions of carbon into the atmosphere.

It is essential to improve soil organic matter to replenish soil quality. Organic matter content can be increased through practices such as: crop residues management, organic farming, integrated soil fertility management, integrated pest management, use of organic manures, cover crops, reduced tillage practices, crop rotation and mulching. Adaptation of these practices enhances soil organic matter (FAO, 2017).

5.3. Balance Nutrition

Indiscriminate and imbalance use of chemical fertilizers decrease soil fertility and productivity. There are many consequences of excess use of chemical fertilizers such as loss of fertilizers especially nitrogen and phosphorus from agricultural fields through many ways, emission of GHGs to the atmosphere, leaching of nitrate nitrogen into the drinking water with potential human health effects, and crop failure. Poor fertilization hampers physical, chemical and biological properties of soil. (Singh et al., 2015; 2017). Deficiency of essential nutrients reduces the development of plants, yields as well as quality of crop. To maintain the good soil health and quality, plant nutrition should be based on crop requirement, local soil and environmental conditions. Balanced use of fertilizer is very important to enhance soil fertility and nutrient use efficiency of the applied plant nutrients for maintaining soil productivity. Right source, right quantity, appropriate time and correct placement are four important pillars (Figure 2) of the balanced nutrient management to increase soil fertility, NUE and reduce excess use of fertilizers as well as check the emissions of green house gases in the atmosphere (Johnston and Bruulsema, 2014; Withers et al., 2018). Sufficient and balanced nutrient supply improve crop productivity, boost nutrient use efficiency, enhance carbon sequestration, reduce plant protection measures, decrease application of soil amendments and mineral fertilizers and reduce pollution (FAO, 2017; Singh et al., 2018).

Figure 2. 4R's Principle of Efficient Nutrient Management.

5.4. Reduce Soil Salinity and Alkalinity

Soil salinity and alkalinity reduces the physico-chemical properties of soil. Salt affected soil (Soil salinity and alkalinity) develop due to high evapotranspiration rates, inland sea water intrusion, accumulation of soluble salts and exchangeable sodium, use of poor quality irrigation water, poor drainage and improper irrigation processes (Minhas and Sharma, 2003) . A high salt concentration in soil solution reduces soil quality and crop yields. To prevent the salt accumulation in soil, mulching is excellent which reduce evaporation losses; Water use efficiency of irrigation water should be increased through improved methods of conveyance, distribution, and field application. Proper management of irrigation water should ensure adequate supply of water for plant growth and efficient drainage to overcome the problems of salt accumulation; Irrigation water quality should be tested and monitored; appropriate drainage systems should be installed in surface and sub-surface soil layer to control rising groundwater tables and soil salinity. Amelioration of saline and alkali soils can be done by using a various management practices such as scraping, leaching of soluble salts, grow salt tolerant crop varieties, use organic manures, domestication of native wild halophytes for use in agro-pastoral systems, use chemical and organic amendments (FAO, 2017).

5.5. Reduce Soil Contaminations

Heavy metals are considered as one of the major sources of soil degradation. Heavy metal decrease soil quality by various metals especially Cu, Ni, Cd, Zn, Cr, and Pb (Hinojosa et al., 2004). The negative consequence of heavy metals on soil biological and biochemical properties are well recognized. The soil properties such as clay contents, organic matter and pH have major influences on the extent of the effects of metals on biological and biochemical properties (Speira et al., 1999). High concentration of metal adversely affects soil microbial properties e.g., respiration rate, enzyme activity, which appears to be very useful indicators of soil pollutions. Agricultural inputs (agrochemicals), use of by-products, atmospheric deposition, flood and irrigation water, accidental spills, poor management of urban waste and wastewater these are the sources from which heavy metal can enter into the soil. Accumulation and contamination of heavy metals occur if the rate of deposition higher than the rate of removal from the soil system. There are many negative impacts of heavy metals such as plant toxicities and subsequent productivity declines. Therefore prevent soil contamination is the best way to maintain healthy soils as well as food (FAO, 2017).

5.6. Improve Acid Soil

Acid soil reduces microbial and macro faunal diversity. It impairs soil structure which makes it more sensitive to erosion. There are less availability of nutrients in the soil, larger impact of toxic elements to plants, and consequences to soil biological functions (such as nitrogen fixation) (Hollier et al, 2005). To reduce soil acidity it is required to monitor soil acidity and reducing the upper surface and sub-surface soil acidity by using adequate amount of amendments (such as lime, gypsum and clean ash); Balanced and efficient use of fertilizer and organic amendment; grow acid tolerant crops and minimum or appropriate use of acidic fertilizer (FAO, 2017).

5.7. Boost Soil Biodiversity

Soil biodiversity (including macro and micro flora and fauna) is the backbone of good soil health. More than 25% of the earth's total biodiversity found in soil. Soil biodiversity supports life on land and water, regulate climatic conditions, helps in nutrient cycling and retention, food production as well as reduce pollution (Bach et al., 2020). Soil biodiversity helps to regulate the dynamics of soil organic matter, increase soil carbon sequestration, nutrient cycling and GHGs, modifying soil physical properties, enhancing the nutrient use efficiency and enhancing plant health (FAO, 2015), but poor cropping practice exert impact on soil microbes, including their activities and their biodiversity. Soil biodiversity is under threat by land degradation, climate change, pollution, urbanization, and over-use and misuse of fertilizers (Montanarella et al., 2015). Deforestation, reduction in the plant species with different rooting system, reduction in soil organic matter content and indiscriminate use of agro-chemicals affects the soil environment, decreases the number and species of soil organisms and maximizes soil degradation. The quality and health of soils largely influences the sustainable agricultural production as well as environmental quality. It is necessary to improve soil biodiversity for maintaining soil quality and nutritional security. Good agricultural practices such as organic farming, zero-tillage operation, crop rotations and conservation agriculture practices, can sustainably increase farm productivity without degrading the soil and water resources as well as nurturing soil biodiversity (FAO, 2015).

5.8. Reduce Soil Compaction

Soil compaction is the important physical properties of soil which affects the soil health. Soil compaction is increased due to intensive agriculture, imposed stresses by machinery and mechanization of farm operations. Soil compaction increases bulk density as well as decreases the soil aeration, water drainage and infiltration as well as generating higher runoff. Due to high mechanical impedance compacted soil limits root growth, poor seed germination, reduce soil biodiversity and causing surface soil crusting (Badalıkova and Hruby, 2006; Radford et al., 2000; Greenland, 1977; Schwen et al., 2011; Chan et al., 2006). To prevent soil compaction it is essential to reduce inappropriate or excessive tillage, minimize vehicular traffic on bare soils, reduce the number and frequency of field operations, develop controlled traffic systems, avoid use of heavy machinery, agriculture/forestry operations should be done only when the soil moisture content is suitable; adopt suitable cropping systems and maintain adequate amount of soil organic matter to improve and stabilize soil structure (FAO, 2017; Singh et al., 2015).

5.9. Soil Water Management

Soil water management is a holistic approach to enhance soil quality and productivity of crop; it can play an important role in addressing the challenges for sustainable agriculture. There are 3 important quality of sustainably managed soil first is rapid water infiltration, second is storage of plant available water at optimal level and efficient drainage facility. However, when these conditions are not met, water logging and deficit water problems arise. In water logging conditions plant roots of many plants not established properly, thereby reducing yields, and can cause contaminants the soil with heavy metals (such as arsenic and methyl mercury) to become mobile in the soil. On the other hand, deficiency of water occurs in areas where water is lost by evaporation, surface runoff and percolation, result crop failure. To avoid these problems soil water management practices should be required. To overcome the problem of water logging drainage systems should be installed in surface and sub-surface soil layer to control increasing groundwater; improve water use efficiency by various methods. In dryland cropping systems, several measures should be implemented to enhance water-use efficiency such as mulching, preparation of water harvesting structures which reduces the runoff and evaporative losses from the soil surface; and ensure adequate water availability at each stage of crop growth. Adaptation of these measures improves soil health sustainably as well as crop productivity (FAO, 2017).

5.10. Minimum Soil Disturbance

Intensive soil tillage is one of the main causes for poor soil health because it degrades the soil in many ways. Conventional tillage is the most detrimental practice to decrease soil health. Deep tillage is highly destructive because it reduces the soil microbial population which significantly contributes to maintain and improve soil health. Due to modern intensive tillage operations the SOM (soil organic matter) reduces rapidly.

Therefore, to maintain soil health adopts reduced tillage or conservation tillage methods to improve the soil organic matter content and increase the population soil microorganisms (Darapuneni et al., 2019; Idowu et al., 2019). Zero tillage practices, can have important economic and environmental benefits such as reduced field operations, reduced erosion, retention of nutrient-rich top soil, improved water capture and retention (Delgado et al., 2019; Hobbs et al., 2007; Mosquera et al., 2019). It can be concluded that by adopting reduced tillage practices, crop growers can remain profitable in terms of yield and at the same time conserve and improve the soil health of their fields.

5.11. Crop Rotation

Crop rotation is a method of farming in which farmers grow different crops sequentially on the same field. There are several benefits crop rotation such as improve soil health, optimize nutrients in the soil, increase soil organic matter content, enhance microbial population, diversity and reduces weed, pest and disease infestation (Vukicevich et al., 2016, Berendsen et al., 2012; Govaerts et al., 2009). The greater the diversity of crops, the greater the diversity of soil micro-biomass, resulting in better soil health.

5.12. Cover Cropping

Cover crops are plants that are grown in between main crop which reduces the soil erosion, runoff, nutrient losses, pest and weed infestation, increases nutrient scavenging and cycling support beneficial soil fauna, increase N fixation by symbiotic bacteria in legume and side by side improve soil organic matter content by adding root exudates and biomass (Singer et al, 2007; Amundson et al., 2015; Blanco-Canqui et al., 2015; Chen & Weil, 2011; Dabney et al., 2010; Dabney et al., 2001; Delgado et al., 2007; Schipanski et al., 2014; Snapp et al., 2005; Valkama et al., 2015).Therefore, the growing of cover crop to increase soil C content and organic matter, which also enhances C sequestration while reducing net greenhouse gas emissions from crop lands and agroecosystems (Delgado and Gantzer 2015; Lal 2015). This means that cover cropping is important tool to improve soil health.

5.13. Diversified Farming Systems

Adaptation of diversified farming into production system will allow buildup a healthy soil. Diversification of production system can be done in many ways such as use of mixed varieties, mixed cropping, integrating livestock into crop production, cover cropping, crop rotation, fallowing fields, hedgerows, buffer strips, and many other practices (Kremen et al., 2012). Adaptation of such practices improves nutrient cycling, soil water management, pest control, habitats for pollinators and support agricultural biodiversity.

5.14. Application of Soil Organic Amendments

Plants and animals residues materials are using as organic amendments. These materials such as manure, compost, biochar, and similar materials when added to the soil, improves soil conditions such as organic carbon and stimulate soil biodiversity (Dolan et al., 2006; Devêvre and Horwáth 2000). Research studies have shown that continues and regular use of organic amendments to the soil can improve soil fertility as well as soil physical, chemical and biological properties (Diacono and Montemurro, 2011; Bouajila and Sanaa, 2011).

5.15. Integrate Livestock into Cropping Systems

Inclusion of livestock into the cropping system, beneficial for growers by increasing the resource utilization efficiency and reducing the need for agrochemicals such as synthetic fertilizers and pesticides through winter grazing. It helps to improve soil health (Ghimire et al., 2013). Integration of livestock in cropping system increases the growth of microbial communities that increase soil organic matter accumulation which helps to improve soil health. Grazing by animals elevate the decomposition of plant residues, thus increasing the availability of nutrients. However, overgrazing causes excessive trampling, aggregate breakdown and hoof-induced compaction of the soil surface which reduces the soil health. Therefore, an optimal balance is required in livestock-integrated systems for improving soil health.

5.16. Reduce Use of Agrochemicals

In modern agriculture system farmer keep faith on an extensive use of agrochemicals such as pesticides, fungicides and herbicides to enhance crop productivity by controlling harmful pests, pathogens, and weeds. However use of agrochemicals in farming systems may adversely influence soil microorganisms that are mainly involved in nutrient cycling processes and other essential nutrient biotransformation. These agrochemicals reduce biochemical reaction and activities of soil enzymes that are key indicators of soil microbiological health (Mandal et al., 2020). To maintain the soil as well as environmental health for sustainable production it is necessary avoid or minimize the use of agrochemicals.

Conclusion

Soil health plays a prominent role in agricultural production systems. Healthy soil is more resilient against fluctuations in growing conditions. Good soil health will ensure sustainable productivity, enhance farmers' incomes, and promote food security. But whole world facing the problem of poor soil health. To overcome this problem various sustainable technologies are very helpful. These technologies are farmers and environmental friendly to replenish soil health. Adoptions of these soil health practices are maintain soil fertility and crop productivity sustainably.

References

Amundson, R., A. A. Berhe, J. W. Hopmans, C. Olson, A. E. Sztein, and D. L. Sparks. (2015). Soil and human security in the 21st century. *Science* 348 (6235):1261071-1-6. doi:10.1126/science.1261071.

Bach, E. M., Ramirez, K. S., Fraser, T. D. and. Wall, D. H. (2020). Soil Biodiversity Integrates Solutions for a Sustainable Future. *Sustainability* 2020, 12, 2662, doi:10.3390/su12072662.

Badalıkova B and Hruby J (2006). Influence of minimum soil tillage on development of soil structure. In: Soil management for sustainability. *Advances in Geoecology* 38 430-35.

Bai, Z. G., Wu, Y. J., Dent, D. L., Zhang, G. L., Dijkshoorn, J. A., van Engelen, V. W. P. and van Lynden, G. W. J. (2010). *Land degradation and improvement in China 2. Accounting for soils, terrain and land use change*. ISRIC Report 2010-05.

Berendsen, R. L., Pieterse, C. M. J. and Bakker, P. A. H. M. (2012). The rhizosphere microbiome and plant health. *Trends in Plant Science*, 17, 478–486.

Blanco-Canqui, H., Shaver, T. M., Lindquist, J. L., Shapiro, C. A., Elmore, R. W., Francis, C. A., and Hergert, G. W. (2015). Cover crops and ecosystem services: Insights from studies in temperate soils. *Agronomy Journal*, 107, 2449–2474. http://doi.org/10.2134/ agronj15.0086.

Bouajila, K and Sanaa, M. (2011). Effects of organic amendments on soil physico-chemical and biological properties, *J. Mater. Environ. Sci.* 2 (S1) 485-490.

CGIAR, (2017) Research Program on Water, Land and Ecosystems (WLE). Healthy soils for productive and resilient agricultural landscapes. Colombo, Sri Lanka: International Water Management Institute (IWMI). *CGIAR Research Program on Water, Land and Ecosystems* (WLE). 12p. (WLEtowardss Sustainable Intensification: Insights and Solutions Brief 2). doi: 10.5337/2017.211.

Chan KY, Oates A, Swan AD, Hayes RC, Dear BS and Peoples MB (2006). Agronomic consequences of tractor wheel compaction on a clay soil. *Soil and Tillage Research* 89 13-21.

Chen, G., and Weil, R. R. (2011). Root growth and yield of maize as affected by soil compaction and cover crops. *Soil and Tillage Research*, 117, 17–27. http://doi.org/10. 1016/j.still.2011.08.001.

Dabney, S. M., Delgado, J. A., Meisinger, J. J., Schomberg, H. H., Liebig, M. A., Kaspar, T., and Reeves, W. (2010). Using cover crops and cropping systems for nitrogen management. In J. A. Delgado & R. F. Follett (Eds.) *Advances in Nitrogen Management for Water Quality* (231–282). Ankeny, IA: Soil andWater Conservation Society.

Dabney, S., Delgado, J., and Reeves, D. (2001). Using winter cover crops to improve soil and water quality. *Communications in Soil Science and Plant Analysis*, 32, 1221–1250. http://doi.org/10.1081/CSS-100104110.

Darapuneni, M. K., Idowu, O. J., Lauriault, L. M., Dodla, S. K., Pavuluri, K., Ale, S., Grover, K. and Angadi., S. V. (2019). Tillage and nitrogen rate effects on corn production and residual soil characteristics. *Agronomy Journal*, 111, 1524–1532.

Davis, B., Walker, N., Ball, D. and Fitter, A. *Impacts of acid soils in Victoria: a report*. Rutherglen, Vic. ISBN 1741062462. OCLC 1034691965.

Delgado, J. A., and Gantzer. C. J. (2015). The 4Rs for cover crops and other advances in cover crop management for environmental quality. *Journal of Soil & Water Conservation* 70:142A–145A.

Delgado, J. A., Dillon, M. A., Sparks, R. T., & Essah, S. Y. (2007). A decade of advances in cover crops. *Journal of Soil and Water Conservation*, 62, 110A–117A.

Delgado, J., Barrera Mosquera, V., Escudero López, L., Cartagena Ayala, Y., Alwang, J., Stehouwer, R., Valverde, F. (2019). Conservation agriculture increases profits in an Andean region of South America. *Agrosystems, Geosciences & Environment* 2, 180050.

Devêvre, O. C., and Horwáth, W. R. (2000). Decomposition of rice straw and microbial carbon use efficiency under different soil temperatures and moistures. *Soil Biol.* Biochem. 32: 1773–1785.

Diacono, M., and Montemurro, F. (2011). Long-term effects of organic amendments on soil fertility. In E. Lichtfouse, M. Hamelin, M. Navarrete, and P. Debaeke (Eds.), *Sustainable Agriculture Volume 2* (pp. 761–786). Dordrecht, The Netherlands: Springer.

Ding, L., Chen, K. L., Cheng, S. G., and Wang, X. (2015). Water ecological carrying capacity of urban lakes in the context of rapid urbanization: A case study of East Lake in Wuhan. *Physics and Chemistry of the Earth, Parts A/B/C, 89–90*, 104–113.

Dolan, M. S., Clapp, C. E., Allmaras, R. R., Baker, J. M., and Molina, J. A. E. (2006). Soil organic carbon and nitrogen in a Minnesota soil as related to tillage, residue and nitrogen management. *Soil Till. Res.* 89:221-231.

FAO (2015). *Healthy soils are the basis for healthy food production.*

FAO (2017). *Voluntary Guidelines for Sustainable Soil Management Food and Agriculture Organization of the United Nations Rome*, Italy.

Ghimire, R., Norton, J. B., Norton, U., Ritten, J. P., Stahl, P. D. and Krall, J. M. (2013). Long-term farming systems research in the central High Plains. *Renewable Agriculture and Food Systems*, 28, 183–193.

Govaerts, B., Sayre, K. D., Goudeseune, B., De Corte, P., Lichter, K., Dendooven, L., and Deckers, J. (2009). Conservation agriculture as a sustainable option for the central Mexican highlands. *Soil Till. Res.* 103:222-230.

Greenland, D. J. (1977). Soil damage by intensive arable cultivation: temporary or permanent? *Philosophical Transactions of the Royal Society B: Biological Sciences* 281 193-208.

Hinojosa M. B., Carreira J. A., Ruız R. G., and Dick R. P. (2004). Soil moisture pre-treatment effects on enzyme activities as indicators of heavy metalcontaminated and reclaimed soils. *Soil Biology & Biochemistry*, 36, 1559–1568.

Hobbs, P. R., Sayre, K. and Gupta, R. (2007). The role of conservation agriculture in sustainable agriculture. Philosophical Transactions of the Royal Society B: Biological Sciences, 363, 543–555. http://doi.org/10.1098/rstb.2007.2169.

Hollier, Carole, Reid, Michael. (2005). "*Acid Soils.*" ISSN 1329-8062.

Idowu, O. J., Sultana, S., Darapuneni, M., Beck, L. and Steiner, R. (2019). Short-term conservation tillage effects on corn silage yield and soil quality in an irrigated, arid agroecosystem. *Agronomy*, 9, 455.

Jahir, M., Satter, M.A. (2010). Agricultural Research Priority: Vision- 2030 and beyond-land and soil resour ce management. Final Report, March 2010. *Bangladesh Agricultural Research Council*, Farmgate, Dhaka.

Jangir, C. K., Kumar, S. and Meena, R. S. (2019). Significance of soil organic matter to soil quality and evaluation of sustainability. *Sustainable Agriculture* pp. 357-381 Publisher: Scientific Publishers.

Johnstona, A. M. and T. W. Bruulsemab 2014. 4R Nutrient Stewardship for Improved Nutrient Use Efficiency. *Procedia Eng.* 83, 365- 370.

Karlen, D. L., Mausbach, M. J., Doran, J. W., Cline, R. G., Harris, R. F., and Schuman, G. E. (1997). Soil quality: A concept, definition, and framework for evaluation. *Soil Sci. Soc. Am. J.* 61:4-10.

Kassam, A., Friedrich, T., Shaxson, F., and Pretty, J. 2009. The spread of conservation agriculture: justification, sustainability and uptake. *Int. J. Agric. Sustain.* 7: 292–320.

Kremen, C., Iles, A. and Bacon, C. (2012). Diversified farming systems: An agroecological, systems-based alternative to modern industrial agriculture. *Ecology and Society*, 17, 44.

Lal, R. (2015). Sequestering carbon and increasing productivity by conservation agriculture. *J. Soil Water Conserv.* 70: 56A-62A.

Lal, R. (2015). Soil carbon sequestration and aggregation by cover cropping. *Journal of Soil & Water Conservation* 70:329–39.

Lal, R. 2003. Global potential of soil C sequestration to mitigate the greenhouse effect. *Crit. Rev. Plant Sci.* 22: 151–184.

Mandal, A., Sarkar, B, Mandal, S., Vithanage, M., Patra, A. and Manna, M. C. (2020). Impact of agrochemicals on soil health. *Agrochemicals Detection, Treatment and Remediation Pesticides and Chemical Fertilizers.* Pages 161-187.

Minhas, P. S. and Sharma, O. P. (2003) Management of Soil Salinity and Alkalinity Problems in India, *Journal of Crop Production*, 7:1-2, 181-230, doi: 10.1300/J144v07n01_07.

Montanarella, L., Pennock, D. J., McKenzie, N. J., Badraoui, M., Chude, V., Baptista, I., Mamo, T., Yemefack, M., Singh Aulakh, M., Yagi, K. (2015). World's soils are under threat. *SOIL Discuss.* 2, 1263–1272. [CrossRef].

Mosquera, V., Delgado, J., Alwang, J., López, L., Ayala, Y., Andrade, J., and D'Adamo, R. (2019). Conservation agriculture increases yields and economic returns of potato, forage, and grain systems of the Andes. *Agronomy Journal*, 111, 2747–2753. http://doi.org/ 10.2134/agronj2019.04.0280.

Palm, C., Blanco-Canqui, H., DeClerck, F., Gatere, L., and Grace, P. 2014. Conservation agriculture and ecosystem services: an overview. *Agric. Ecosyst. Environ*. 187: 87–105.

Pretty, J. and Bharucha, Z. P. (2014). Sustainable intensification in agricultural systems. *Annals of botany*, *114*(8), 1571–1596. https://doi.org/10.1093/aob/mcu205.

Radford BJ, Bridge BJ, Davis RJ, McGarry D, Pillai UP, Rickman JF, Walsh PA and Yule DF (2000). Changes in the properties of Vertisol and responses of wheat after compaction with harvester traffic. *Soil and Tillage Research* 54 155-70.

Schwen A, Ramirez GH, Smith EJL, Sinton SM, Carrick S, Clothier BE, Buchan GD and Loiskandl W (2011). Hydraulic properties and the water-conducting porosity as affected by subsurface compaction using tension infiltrometers. *Soil Science Society of America Journal* 75 822-31.

Seifu, W. and Elias, E. (2018). Soil quality attributes and their role in sustainable agriculture: a review. *International Journal of Plant & Soil Science* 26(3): 1-26.

Singer, J. W., Nusser, S. M. and Alf, C. J. (2007). Are cover crops being used in the US corn belt? *J. Soil Water Conserv*. 62:353–358.

Singh SR, et al., (2015) Impact of balanced fertilization on nutrient acquisition, fibre yield of jute and soil quality in New Gangetic alluvial soils of India. *Applied Soil Ecology* 92, 24–34.

Singh SR, et al., (2017) Identification of minimum data set under balanced fertilization for sustainable rice production and maintaining soil quality in alluvial soils of Eastern India. *Communications in Soil Science and Plant Analysis* 48, 2170–2192.

Singh, J., Salaria, A. and Kaul, A. (2015). Impact of soil compaction on soil physical properties and root growth: a review. *International Journal of Food, Agriculture and Veterinary Sciences*, Vol. 5 (1) pp. 23-32.

Singh, S. R., Kundu, D. K., Dey, P., Singh, P. and Mahapatra, B. S. (2018). Effect of balanced fertilizers on soil quality and lentil yield in Gangetic alluvial soils of India. *The Journal of Agricultural Science* 1–16. https://doi.org/10.1017/S0021859618000254.

Snapp, S., Swinton, S., Labarta, R., Mutch, D., Black, J., Leep, R. and O'Neil, K. (2005). Evaluating cover crops for benefits, costs, and performance within cropping system niches. *Agronomy Journal*, 97, 322–332.

Speira T. W., Kettlesb H. A., Percivalc H. J. and Parshotam A. (1999). Is soil acidification the cause of biochemical responses when soils are amended with heavy metal salts? *Soil Biology and Biochemistry*, 31, 1953-1961.

USDA-NRCS. (2019). *Soil health*. Retrieved from https://www.nrcs.usda.gov/wps/portal/nrcs/main/soils/health/.

Valkama, E., Lemola, R., Känkänen, H., and Turtola, E. (2015). Metaanalysis of the effects of undersown catch crops on nitrogen leaching loss and grain yields in the Nordic countries. *Agriculture, Ecosystems & Environment*, 203, 93–101.

Vukicevich, E., Lowery, T., Bowen, P., Úrbez-Torres, J. R. and Hart. M. (2016). Cover crops to increase soil microbial diversity and mitigate decline in perennial agriculture. A review. *Agronomy for Sustainable Development*, 36, 48.

Withers, P. J. A., Rodrigues, M., Soltangheisi, A., Carvalho, T. S. de and Guilherme, L. R. G. (2018). Transitions to sustainable management of phosphorus in Brazilian agriculture. *Sci. Rep*., 8:2537.

Chapter 20

Emergence of Nanotechnology in Efficient Nutrient Management in Soil

Rakesh Kumar[1,*], Satdev[1], Shruti Kumari[1], Mahesh Jajoria[2], Nintu Mandal[1], Arun Kumar Jha[1] and Rahul Raj[1]

[1]Department of Soil Science & Agricultural Chemistry, Bihar Agricultural University, Bhagalpur, Bihar, India

[2]Department of Soil Science & Agricultural Chemistry, SKNAU, Jobner, Rajasthan, India

Abstract

For addressing the increased challenges of agricultural production and food security, significant technological advancements and innovations have been made in recent years in the field of sustainable agriculture. Where Imbalanced and inappropriate synthetic chemical fertilizer application adversely affected soil health and limits the capacity of the soil to produce with effective nutrient management. Nutrient use efficiency for major commercial fertilizers is very low and attains the values as 30–35%, 18–20%, 35–40% & 1-3% for N, P, K &micronutrients respectively. Major causes of low nutrient use efficiency are losses due to leaching, denitrification, fixation, and low solubility or rapid oxidation to unavailable forms. Also, the larger size of foliar-applied fertilizers limits the efficient use of nutrients. Low nutrient efficient use of fertilizers leads to higher cost of cultivation, eliminates natural resources of chemical fertilizers and creates environmental pollution issues. Hence, an urgent demand raised for the development of nutrient efficient, low-cost fertilizers and techniques for efficient nutrient management. One of the most revolutionary technologies came to emergence as "nanotechnology" which can be defined as the technology of using novel nano-sized material (size in between 1 to 100 nm), devices and systems having some novel properties like thermal, optical, electromagnetic, and morphological. Nanotechnology in efficient nutrient management serves in the forms of novel nano fertilizers, nano clay polymers, nano coated biofertilizers, and nano biosensors. Nano fertilizers can enhance the nutrient use efficiency by 3 times and also provide tolerance against different biotic and abiotic factors. Irrespective of the type of crop nano technology can be used, nano technology that can enhance the use of bio sources and is ecologically safe in nature, can influence carbon uptake along with improving soil stability and aggregation. Nano fertilizers are required to be applied in very small doses and have lower per unit nutrient value cost. Since these nano clay polymer-based fertilizers contain nutrients, growth promoters encapsulated in nano scale polymers, they will also have slow and targeted nutrient delivery and release. In present days innovations of sensing technology are booming. When it comes to determining a parameter from the soil or plants causing disturbance in the agricultural field there are methods that give accurate results

In: Ecosystem Services
Editor: Hanuman Singh Jatav
ISBN: 978-1-68507-614-6

with live pictures and monitors changes in chemical or physical properties of soil. Advances in nano science and technology have a remarkable impression on nano sensing and nano biosensors due to high sensitivity, faster response time by using functionalized nanorods, nanobelts, or nanowires. And thus, they can help in better detection of nutrient requirements and can alarm the farmer for the best suitable times of application to enhance the nutrient use efficiency. Nano-biofertilizer is a step ahead in the innovative field of nanotechnology in agriculture, toward the joint formulation of biofertilizer and nano fertilizer into nano-biofertilizer, a more nutrient effective alternative for eco sustainable agriculture. A lot more has to come from nanotechnology in the field of nutrient management through nano genetic engineering tools for developing nutrient efficient crop genotypes. Thus, nanotechnology can become the most used technology for sustainable fertility management in near future with more farmer and ecosystem-friendly options.

Keywords: nanotechnology, nutrient management, nano bio sensors, nutrient use efficiency

1. Introduction

Sustainability of food production is a most concerned factor for India and the world even after attaining surplus food production for feeding everyone with technological advancement viz. green revolution. But still, it doesn't seem able to feed the flourishing population effectively in the upcoming future due to static crop production. Agriculture with being the backbone of developing countries with having 60% or more individuals relying on the agricultural sector for their livelihood (Brock et al. 2011) is most important sector. Sustainable agriculture is addressing countless challenges like climate change, the existence of nutrient deficiency, declining land holding size, increasing cost of production, and leading to economic losses to the farming communities and causing malnutrition in poor populations. Some of the evilest issues among them are the high cost of chemical fertilizers, lower use efficiency (N.U.E.) of conventional fertilizers, mismanagement of soil nutrient resources, and fertility. On an average 90% of Indian soils are deficient in nitrogen and phosphorus status along more than 50% samples are found to be deficient in potassium. While, 39.9%, 12.9%, 6.0% and 4.3% samples are deficient in Zn, Fe, Mn, and Cu deficient respectively (Shukla and Tiwari 2014). Deficiency of more than one micro or macro nutrients along with lower organic carbon status, lower fertilizer efficiency is causing nutrient mining, fertility loss, stagnant crop productivity even after having high yielding improved varieties. Major factors contributing towards low nutrient use efficiency are loss of nutrients by fixation, leaching, oxidation, volatilization, and immobilization due to microbial actions. Nutrient use efficiency of plants has been reported in between 30-35%, 18-20%,35-40% & 1-3% for N, P, K & micronutrients respectively, with the conventional fertilizers (Subramanian et al. 2015). The remaining parts of these applied fertilizers are wasted in the environment and contribute to large economical and nutrient resource loss and environmental pollution in the forms of groundwater pollution, eutrophication, etc. Nutrient use efficiency (NUE) is a critical term in crop production system evaluation. Plant growth and development are dependent on all-important plant nutrients. Fertilizer usage is inversely proportional to yield, with further fertilizer use resulting in soil, water, and environmental emissions. The most important factor to consider is the response of applied fertilizers and their efficiency in achieving maximum production. Surface runoff, leaching, volatilization, denitrification, and fixation all contribute to the low nutrient usage performance of applied fertilizers. While the highest nutrient usage efficiency exists at the lower sections of the yield response curve, where fertilizer inputs are lowest, the efficacy of fertilizers in growing crop yields and maximizing farmer profitability should not be sacrificed solely for efficiency's sake. It's essential to strike a balance between nutrient efficiency and crop productivity. Hence, to enhance the N.U.E. and to increase crop production and profitability from the present static level, an advanced technological emergence was much needed. In need of hour, nanotechnology emerged in human life as most advanced technology in sector of medicine, pharma, engineering, and energy as the most promising interdisciplinary science and technology. Recently, the advent of nanotechnology in the field of nano fertilizers made a revolutionary change in this direction as nano fertilizers were found to be more nutrient efficient than conventional fertilizers because of its' very minute size and easier to plant uptake (Shang et al. 2019). Nanotechnology emerged as a new concept, idea, and tool for increasing nutrient use and plant-soil health management. Although, agricultural nanotechnology is in its budding stage with more theoretical knowledge only, but it has the potential to revolute entire agriculture and fertilizer production market through nano scale fertilizers, nano polymers, nano biofertilizers, and nano sensors by the development of controlled target based nutrient delivery systems and fertilizers. Also, smart nutrient management techniques can be developed to assess soil nutrient status, soil moisture, pH, soil temp. and plant nutrient deficiency identification through nano science. Although immense care has to be taken while adopting and exploiting the potential of nano fertilizers because of environmental safety and ethical issues. Some limitations and adverse effects of NFs have also been reported (Dimkpa et al. 2013). However, these can be overcome by proper standardization of NF dose and selection of befitting Nano material for the test crops. Therefore, detailed information on nanotechnology in agriculture is much needed and is explained in further sections of this chapter.

2. Nanotechnology in Agriculture

Terminology "nanotechnology" was coined by Norio Taniguchi in 1974, where he described dimensional validities of nano materials (Taniguchi 1996). Word nanotechnology has been derived from the Greek word "Nano" means "dwarf" and "Technology" means "science of craft (visualization, production or manipulation of size & shape)". Nanotechnology is the design, production, and use of structures through control of size and shape of materials at nano meter scale. The original idea and concept of nanoscience and nanotechnology were given by "Nobel Laureate physicist Richard Feynman" (Father of nanotechnology) in his talk "There's Plenty of Room at the Bottom" during a meeting of the American physical society in December1959.

European Food Safety Authority (2011) explained nano material as "A materials having following criteria".

- Consists of particles with one or more external dimensions in the size range of 1-100 nm, for more than 1% of their size distribution
- Has internal or surface structure in one or more dimensions in the size range of 1-100 nm

- Has a specific surface area to volume ratio greater than 60 m^2/cm^3, excluding particles with a size lower than 1 nm.

Some other authors also referred to nanoparticles as colloidal systems with particle sizes in a range of 10 and 1000 nm (Nakache et al. 1999). Nano materials has amazing unique properties like a smaller size, high surface area, high reactivity along with optical, electrical, thermostability, and catalytic properties due to quantum size effect (Ghormade et al., 2011). Therefore, nanomaterials are getting immense limelight due to wide application in agricultural science too as nano fertilizers, nano gene transporters, food preservatives, water treatment, heavy metal remediation, and pesticides (Weir et al., 2012; Mukherjee et al. 2016; Rajput et al. 2017; Servin et al. 2017). There are two major approaches of nano particle synthesis which are: 1. Top-down as high-pressure ball milling, Arc discharge, laser ablation, sonication, and 2. Bottom-up: chemical vapour deposition, homogenous nucleation, sol gel synthesis, hydrothermal synthesis (Shah et al. 2014).

3. Reasons of Low N.U.E. with Conventional Fertilizers

Nutrient use efficiency with conventional fertilizers is found to be low based on the following grounds.

3.1. Solubility and Dispersion of Fertilizers

For nutrient management practices in crops, water-soluble fertilizers are extensively used that also mainly in sulfate and nitrate forms. But these type of micronutrient fertilizers carries a very lower amount of nutrient per unit volume when compared to their insoluble oxide forms. Also, the larger size of these conventional fertilizers limits the uptake and efficient use of conventional micronutrient fertilizers.

3.2. Loss in Environment

Commercially available fertilizer sources of primary nutrients are lost in environment through leaching (nitrates), fixation (phosphate and potassium), volatilisation (ammonium), and microbial losses (denitrification) and limit the efficient use nutrients and also reduces the uptake ratio for nutrients.

3.3. Uncontrolled Release

Conventional micronutrient fertilizers have no specific release mechanism and can cause toxicity of nutrients to plants specifically when the range of deficiency and toxicity remains narrow. This not only causes low NUE, economical loss but also results in environmental pollution and ecological imbalance.

3.4. Release Period and Pattern

Plant nutrients are very essential throughout the entire life cycle of the plant. But nutrient deficiency symptoms occur generally at the last stages of crops. Excess release of nutrients at the time of application leads to losses in forms of leaching, fixation and volatilization, while later stages suffer from nutrient deficiency due to uneven release patterns.

4. Advantages with Nano-Fertilizers

Cui et al. (2011) explained the advantages of nano fertilizers over conventional fertilizers.

4.1. High Solubility of Metal oxide-based Micronutrient Fertilizers

Under nanotechnological fabrication of oxide materials into smaller size, high surface area and shapes it can be converted into soluble or dispersible oxides of metallic micronutrients like Zn, Mn, Cu, and Fe. And can enhance the bioavailability of these nutrients to plants into the soil.

4.2. Easy to Uptake

Nano fertilizer may increase the N.U.E and nutrient uptake ratio due to its nano size particle which can easily enter the root hairs and cells of leaf cuticles when applied via soil and foliar applications, respectively.

4.3. Controlled Release and Target Based Delivery

Nano encapsulated fertilizers have ability to release nutrients slowly and evenly for the long period which can deliver the nutrients in more précised way for the whole period of growth for a crop. Surface coatings to fertilizers with nanomaterials can bind the materials and can reduce the susceptibility of fertilizers to leaching losses due to controlled release (Brady and Weil 1999). Which can increase the N.U.E. as well as eco safety of fertilizers due to reduced losses into the environment.

4.4. Nanotechnology in Efficient Nutrient Management

In the present nanotechnological world use of engineered nano particles (EgNPs) assumed greater importance in day-to-day life due to their increasing reliance in beauty products, medical science, energy, mechanical engineering, and the agriculture sector. EgNPs are deliberately manufactured and designed particles having very peculiar properties related to their size, shape, surface area, and chemical reactivity. These properties of EgNPs can be demonstrated in solution, aerosol, colloidal suspension or crystalline powder forms. EgNPs can be further classified in four kinds according to USEPA: Metal/Metal oxide nano particles and carbon-based nano particles. Among these EgNPs specifically, metal oxide nanoparticles are getting wider attentiveness due to their use in agricultural science as nano fertilizers. A larger number of researches are ongoing to evaluate the ecotoxicological effect of these nano particles. Rajput et al. (2018) impressively explicated the consequences of nanoparticles in agriculture due to their imbalanced use.

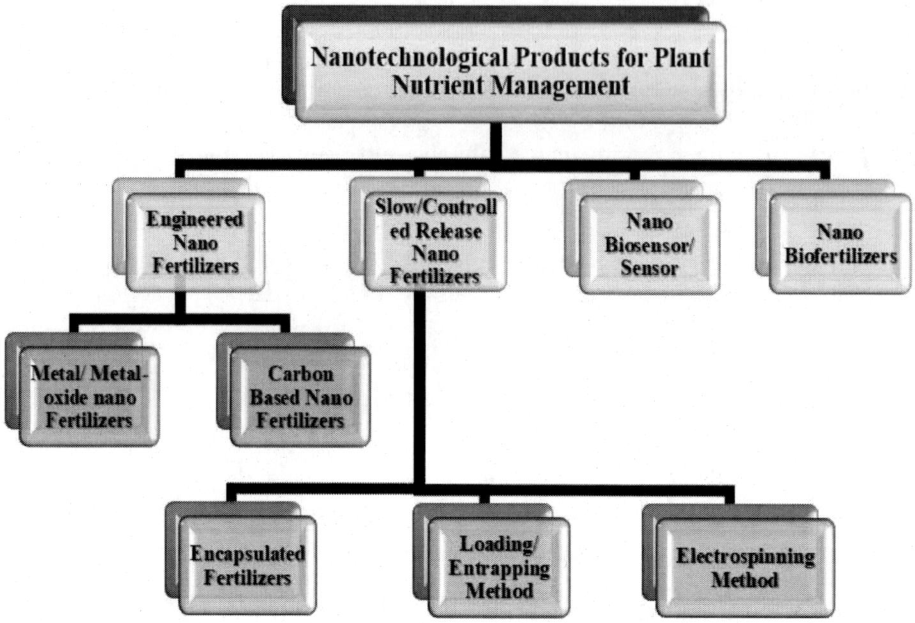

Figure 1. Nanotechnological Products for Plant Nutrient Management.

5. Types Nanotechnological Products

5.1. Engineered Nano Fertilizers

5.1.1. Metal/Metal Oxide Nano Fertilizers

Converting insoluble forms of high nutrient-containing chemicals like metal oxides/metals into soluble forms is one of the most key benefits. Nano structures can be achieved through reducing their size or changing their shape by chemical, physio-chemical or biological ways. The use of EgNPs thus can increase the N.U.E. due to higher solubility and increased bioavailability in soil. Metal/metal-oxide based nano fertilizers can be used in plant nutrient management in different ways explained below.

5.1.1.1. Metal Oxide Nanoparticles on Seed Germination

Various types of metal/metal oxide nano fertilizers are being used as seed treatment or seed priming for improved crop production, which has shown a vital degree of positive or negative effects. Zn, Cu, Fe, Ti, Al, Au, and Ag Metal or metal oxides are mostly used for seed treatment. Tha use of nano TiO_2 can increase the seed germination, seedling vigour (Sani 2012), can promote radicle and plumule growth (Mahmoodzadeh et al. 2013), and can result in more dry matter, higher photosynthesis, increased chlorophyll 'a' content (Zheng et al. 2005) due to increased seed stress resistance and higher capsule penetration for water and oxygen intake. A large number of researches confined to ZnO nano particles as seed germinated has shown the similar type of results in different legume crops viz. green gram, Bengal gram and groundnut. The functioning of ZnO nanoparticles is much more associated with structure and dose of application. But seed treatment with nano Fe oxide has shown negative effects on seedling parameters except very low doses up to 100 ppm. Satdev et al. (2020) reported the higher cob yield in maize crops with seed treatment of zinc oxide nanomaterials at 1000 ppm when compared to control and bulk ZnO. Sundaria et al. (2019) reported an increase of 64% in shoot length and 46% in root length in IITR26, WL711 genotypes of wheat with seed priming at a concentration of 200 ppm with nano iron oxide particles as a result of triggered iron uptake in the plant. Silicon as a beneficial element for various physiological activities and metabolic reactions in plant cells can increase nitrate reductase enzymatic activity in *Glycine max* when applied in nano form with TiO_2 (Lu et al. 2002). It can also enhance the capabilities of water adsorption, and nutrient uptake along with encouraged antioxidant and accelerated seed germination. It has been reported that seed treatment with nano SiO_2 of nanoscale SiO_2 at 400 ppm concentration alone can improve seed germination, root-shoot length, and percentage water content in maize crop (Rad et al. 2014). Application of seed prechilling along with SiO_2 nano fertilizers seed treatment can break the seed dormancy for wheatgrass. So, it can been be understood that lower doses of nanoparticles with seed treatment can improve the crop productivity and nutrient use while higher doses show some negative effects due to higher reactivity of nano particles. The use of biocompatible materials as a coating on nano particles can achieve the protection from the phytotoxic impact of nano materials.

5.1.1.2. Nano Metal Oxides as Foliar Application

Other than the application of nano particles as seed treatment/priming and soil applications a lot of work has been done in direction of penetration and uptake of nutrients through foliar application. Application of metal oxide nano fertilizers as a foliar spray can be a potential method for improving crop production and nutrient use efficiency due to its smaller size and high penetration ability. The foliar spray method can be applied during certain unfavorable soil and weather situations. Specially under high pH or calcareous soils, application of nano sized oxides of Fe, Zn, Mn, Cu as a foliar spray can be promising due to rapid oxidation of available nutrient to unavailable forms in arable lands. Foliar application can be done during unfavorable soil and weather conditions. In addition to this, Foliar application promotes the direct stimulated entry of fertilizers into the plant shoot system, thus reducing the wastage of applied nutrients. Hence, foliar application of EgNPs results in higher nutrient use efficiency (NUE) and gives a rapid counter-response to the growth of crops. EgNPs are more reactive in nature and can enter through the leaf, stem, and cuticle, ensuring controlled release behaviour of nano fertilizers and targeted delivery. Foliar spray of nano metal oxides of iron (Fe_2O_3) resulted in an increase in chlorophyll content in groundnut leaves. Also, it increased antioxidant enzymatic activities (especially catalase) and the yield of crop (Liu et al. 2005). Similarly, Ghafari and Razmjoo (2013) found that foliar application of 0.2% solution of nano-iron oxide more responsive on yield improvement than that of other iron fertilizers viz. iron sulphate and iron chelated fertilizers.

5.1.2. Carbon Based Nano Materials

Carbon is the most important non-metallic element for life on earth and the history of this element is far older than human civilization. Carbon materials like coal, activated, and inactivated charcoals are being used by humans for various purposes including water purification and discoloration through the adsorption principle. In the past few years carbon-based nano-materials have received huge attention as a powerful resource for fertilizers, optical tools, biological contamination remediation, molecular engineering, pharmaceuticals, medical science, and drug delivery carriers. A broad range of carbon-based nanomaterials has been prepared viz. Zero dimensional fullerenes, three-dimensional carbon nanotubes (single and multiwalled). Fullerenes were 1st carbon-based nanomaterials produced artificially, followed by multi-walled CNTs, and Single walled CNTs. Recently these materials have been introduced into fertilizer sector to improve crop productivity. Application of fullerenes as seed treatment has shown a 112 to 128% increase in fruit yield of bitter melon as well as total biomass production and Phyto medical contents (cucurbitacin, lycopene, charatins & insulin content) (Kole et al. 2013). The experimental findings also showed positive or negative types of effects and Thus it was noted that carbon-based nanomaterials have diverse effects on different crops based on their reaction due to the presence of different reaction mechanisms. But these effects can also be altered in presence of functionalization of these materials. Several functionalized carbon-based nanomaterials are being used as fertilizer sources to improve crop production and nutrient use efficiency.

5.2. Slow/Controlled Release Nano Fertilizers (SRNFs)

Efficient nutrient management involves one of the most famous principle of Four "R"s. Which are Right time, Right dose, Right source, and Right method of fertilizer application. Fertilizers carry an important role in improving crop production, especially high nutrient-containing water-soluble fertilizers which have an efficiency limiting property of quicker release, resulting in environmental losses due to leaching, volatilization, and immobilization. On an average 40 to 70% N, 80 to 90% P, 50 to 70% K, and 97 to 99% micronutrients applied through conventional fertilizer sources are lost in the environment and lead to heavy economical loss and pollution of soil and water. To mitigate these problems, slow/control released fertilizers (SRFs) production from nanotechnology has gained immense applause over the whole agriculture sector. It can supply nutrients with slower and more controlled release making it available during a longer period of plant growth and improving nutrient use efficiency (Junejo et al. 2011). Amount and speed of release mainly rely on the chemicals used to coat the fertilizers. Trenkel (1997) defined 3 criteria for slow/control release fertilizer: 1. 15% of total nutrient content should be delivered within 24 hours, 2. Not more than 75% should be delivered before 28 days, 3. Minimum 75% nutrient should be released with a static rate between 40 to 360 days of application. Presently nanotechnology as an advanced tool for fertilizer production has developed these "Intelligent target delivery-based fertilizers" which can be manufactured using coating/loading, encapsulation/entrapping of nano sized nutrient particles, or using nano materials for manufacturing core shells, hollow nano tubes and electrospinning methods.

5.2.1. Encapsulated Nano Fertilizers

Control or CRFs are an attractive option for water-soluble fertilizers because they may decline the dose and time interval between fertilizer applications by gradually delivering nutrients present in fertilizers to meet the plant's nutritional requisition. Not long ago, Coating fertilisers with synthesized or naturally derived polymers have gained popularity in the slow release of Agro-chemicals. Wu et al. (2008) produced such product which has shown slower release and water holding capability, as well as its nontoxicity in soil and ecological safety, which could be particularly beneficial in the crop production sector. Corradini et al. (2010) studied the application of NPK nutrient content through sources like urea, calcium-phosphate, and MOP into plants. Controllable incorporation of methacrylic acid (MAA) nanoparticles into chitosan. The coating's permeability decreased the Fertiliser solubilization rate, allowing for slower, sustainable delivery of sulphur coated fertilisers. Fernando et al. (2010) investigated the rate of phosphate release by coating calcium phosphate pallets with wax-nano calcium mixtures, discovered that the wax's increasing nano calcium carbonate content. Because of their higher surface size, larger adsorption capacity, ease of modifying surface, and colloidal properties, clay minerals are now used to encapsulate agricultural chemicals such as fertilizer, PGPs, insecticides, and fungicides. The word "clay" comes from the minerals that are natural and comparatively inexpensive portions of soil that are used in a nutrient reservoir and a nutrient niche. The layer type nano size clay minerals are having higher surface area, which are used in improving the interaction of surfaces with fertilizer nutrients. Man et al. (2004) prepared a controlled release urea-based fertilizer by intercalating urea into the montmorillonite layers. Furthermore, this type of clay minerals was used for hybridization with polymers to create nano clay polymer composites (NCPCs). Reduced P-fixation and improved P availability were reported in studies with pillared clay-modified phosphate fertilizers (Basak et al. 2012). Nano or sub nano material coating-cementing may have been capable to control the burst of nutrients from capsules of nano coated fertilizers (Liu et al. 2006). A patent on nano-composite made up of nitrogen, phosphorus, sugars like mannose, and AA (amino acid) has been found effective in increasing nutrient uptake and utilization by grain crops (Jinghua 2004).

5.2.2. Loading/Entrapping Method

Serious limitations with the formulation of control release nano fertilizers by coating or encapsulation method like problems of a sudden release of nutrients from nano coated water-soluble fertilizers or issues related to release slower than expected. Thus, a unique system of nano material synthesis was developed by loading nano particles inside hollow tubes made up of silica, clays, or metal oxides. Advancement with this method was capacity to load more quantity of nutrients, and better release pattern than capsulation method. Perforated zeolite (Clinoptilolite) clay minerals were used as a carrier of nutrients most popularly. Thus, also named as zeolite-based nano fertilizers. Zeolite minerals are cationic aluminosilicates of group I-a or II-a with hollow, porous 3D lattice structures. They have very high C.E.C. and being used as cheap source of cation exchange for different uses (Breck 1974). These highly negatively charged natural minerals are used to trap potassium or other cationic nutrients for slow prolonged release. But for entrapping of negatively charged nutrients like P, S, or B these minerals need to be surface charge modified, which can be done through adding compounds like hexa-decyl tri methyl ammonium bromide (Preetha et al. 2014). Zeolite minerals has been found effective in improving N.U.E. due to its use as carrier for nutrient elements like N, P and K. Ammonium-loaded zeolite fertilizers have shown remarkable improvement in yield of

Lolium multiflorum and also the N use efficiency, because of the slow nitrogen release pattern of fertilizer. In an experiment of nano sized nitrogen entrapped zeolite fertilizer formulation, it was found that a significantly higher amount of nitrogen (up to 28%) can be loaded and a prolonged release pattern of approx. 48 days was achieved in comparison to Urea (Manikandan and Subramanian 2014). Similarly, hydroxyapatite nano fertilizer after modification with urea was entrapped into porous *G. sepium* soft-wood and shown a more even, longer release pattern of sixty days against uneven and shorter release pattern of Conventional N fertilizer for thirty days. Another way of production of slow-release nano fertilizers is the manufacturing of nano sized hollow tubes or cores, which can be used as the carrier of delivery agents for nutrients, drugs, insecticides, or biotoxins for use in the agriculture sector. Different sized, shaped or diameter shells or tubes may be produced using carbonates of Ca or Mn and can be used for loading or water-soluble nano fertilizers for control release and highly nutrient efficient fertilizers. Yuvraj and Subramanian (2015) manufactured nano zinc enriched fertilizer with entrapping soluble $ZnSO_4$ into hollow Mn core shells. It was found that this Mn core shell-based Zn fertilizer provided Zn to the soil up to 696 hours when compared to 408 days in the case of conventional $ZnSO_4$ and increased N.U.E of Zn in rice crop by decreasing nutrient loss into the environment.

5.2.3. Electrospinning Method

Electrospinning is a novel technology which uses electrostatic force to form ultra-minute sized fibres of size in the range of some micro meter to nano meter and can be used with any polymer compounds or composites. The major principle behind this simple and easy carried out procedure is development of two different liquid compounds from one due to side wise electrospinning and the creation of complex ultrafine nano sized fibres with unique shape, structure and properties of nano materials. Several kinds of researches have been done in this direction and obtained various patents in this sector of controlled release nano materials production during the past few years. But most of them are related to the use of these chemicals in the medical science and pharmacy sector. A few years back Castro-Enriquez et al. (2012) manufactured slow release-based urea fertilizer using electrospinning of wheat gluten compound. It was reported that urea release from the product was rapid for initial 10 minutes followed by controlled release up to 300 minutes of application where it attained an equilibrium and released approx. 98 percent of total urea present in the compound. Coaxial electrospinning was used to synthesize a unique fertilizer containing poly hydroxy butyrate and poly lytic acid as shell and core of nano fibres by Kampeerapappun and Phanomkate (2013). The slower rate of shell solution resulted in decreased rate of fertilizer release due to low solubilization of nano fibre produced through electrospinning. The core shell-type structure of nano-fibre fertilizers made it able to hold the nutrient content up to 30 days of application. Also, both the polymer compounds were found to be ecologically safer for use due to their easier biodegradability.

5.3. Nano Sensor/Nano Bio Sensor

As soon as the introduction of nano science and technology in devices related to diagnosis, nano or nano-bio-sensors got immense applause over different sectors of science including agriculture. Nano-sensors are devices that have a very minute size in the range of nano scale (1-100 nm) and have specific qualities like nano materials. While, nano-bio-sensors are diagnostic devices with nanosized biosensing facilities having fixed structures containing protein, lipids, nucleic acids, or complete cells with a sensing system that can detect the biological responses and transmit them in form of information or data. A large quantity of studies has been carried out in this direction with the use of these types of sensors in the detection and treatment of tumor cells. These nano-bio-sensors are equally efficient in sensing programs in the agriculture sector as a sensor for nutrient status and deficiency, diseases, pests, soil water and pH status to improve crop production and sustainability. Nano-bio-sensors are going to be one of the most important components of the precision farming technique. Nano-bio-sensors in farms can be incorporated with a GPS system and can be used for live monitoring of crops for soil-plant constraints. Bio-sensors can be established in crop fields wirelessly and can be used to make maximum output from efficient use of farm inputs (Scott and Chen 2003). This type of sensing is useful in the best suitable time of cropping, crop harvest, irrigation scheduling, fertigation, protective measures against pests, and other intercultural operations needed to administrate environment safety and crop physiological conditions. For identification of deficiency of elements, nano-bio-sensors can be incorporated with nano sized particles, which will be useful in the helping for application of protective or corrective practices. Such types of practices are being widely followed in developed countries like Australia & U.S.A. Some of the orchards of grapes in California have established wireless technology with the help of one information technology corporate named "Accenture". The higher establishment cost of this system can be overshadowed by the very high cost of finest wines produced from grapefruits coming out of these orchards. Kumar (2011) explained that how nano-sensors can be used for deriving precise data of plant growth, climate, and soil such as SMC, fertility status, nutrient level, pests. Which may be helpful for producers to reach out to the maximum efficiency of input use and productivity of crops. Recently use of bio-sensors in the determination of cationic micronutrient (especially Zn) deficiency & toxic levels. A Zn-specified bio-sensor Pseudomonas putida X4 (pczcR3GFP), was manufactured by inserting "EGFP (Enhanced-Green-Fluorescent-Protein)" gene with external promotor "CZC-R3" in the genome of *Pseudomonas putida.* This modified strain of bacteria has identified almost the maximum (90%) Zn present in the extract obtained from the soil in which Zn was applied. Researchers have given the conclusion that such bio-sensor can be an efficient option for the determination of zinc toxicity in soil-ecosystem (Liu et al. 2012). Such type of research has been also reported with the use of genes in *E. coli* strains HB-101 and MG-1655 in bio-sensors named as "puCd-607" and "pZNTlux" for Zn assessment in soil and on-field trails (Maderova and Paton 2013).

Further nano-bio-sensor based on aptamers appeared as an important technique for investigating the production and interactions of metabolic compounds released by root cells in the rhizosphere and to study the reactions of control/SRN Fertilizers in the soil system. Introduction of aptamer nano devices layered poly electrolyte polymers can be used as a potent tool for use in target-based deliveries, slow-release nutrients, medicines, and pest protecting chemicals (Sultan and DeRosa 2011, Zhang et al. 2013). An efficient nano/nano-bio sensor can be helpful in detecting and treating deficiencies of plant nutrients by application of nutrients as per crop requisition throughout the plant life cycle. Monreal et al. (2015) manufactured a smart nano fertilizer delivery system for zinc, iron, manganese, and copper, where the principle mechanism of fertilizer delivery is identification and reading of particular sensing by some nano-bio-sensor present in a polymer's coating which can bind Zinc, iron, manganese and copper nutrients sources like nano fertilizer of soluble conventional fertilizers.

5.4. Nano Bio Fertilizers

Nano-biofertilizer is one advanced technique in the agricultural science and fertilizer sector, toward the combined formulation of bio-fertilizer and NFs into nano-bio-fertilizer, an efficient alternative for ecologically sustainable agricultural practices. Although there is no such definition has been described till present time but still, we can understand it by the explanation given by Thirugnanasambandhan (2019) as "*Nano-biofertilizer is a hybrid combination of nano and biofertilizer developed by the formulation of biofertilizer to nano size (1–100 nm) with the help of certain nanomaterial (preferably biosynthesized) coating*".

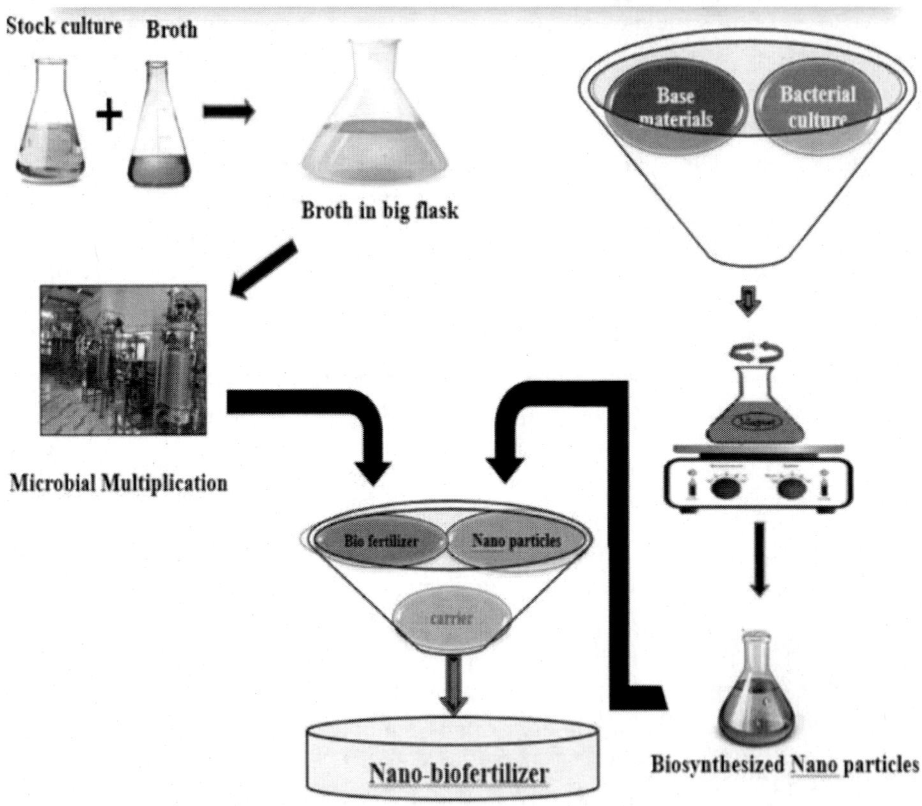

Figure 2. Graphical presentation for production prototype for Nano Biofertilizers.

"Nano-biofertilizer" have several benefits including longer shelf life, high-temperature resistance, large-scale production within very short time, less dose required, environmentally safer, and regeneratable fertilizer application properties. Due to surface coating and encapsulation of bio-fertilizer by NPs, the stability of biofertilizers increases. Improved surface area, nano-size, and high activity of nanocoated biofertilizer can permit for increased interaction with nutrient supply. Nano biofertilizers can be produced by the coating cultured microbial inoculants with biosynthesized nano materials using crosslinking agents/carriers like amino acids or chelating complexes (Figure 2). Biosynthesis of nanomaterial can be a top-down approach using microbes to break down the bulky materials into nanosized particles.

6. Uptake Mechanism of Nano Fertilizers

Generally, all the essential and functional nutrients can be absorbed by the plants with ease, but metal/metal oxide-based nano fertilizers in water suspension may be absorbed/rejected basis on plant selectivity. EgNPs may approach plants through direct application by foliar or soil application, accidental contamination of fertilizers or chemicals, and sometimes through atmospheric precipitations. These things can have both positive or negative effects on plant nutrition, cropping system, and food web. With the advent of nano fertilizers in agriculture, it was necessary to initiate the study on nanoparticle uptake, transport, bio-accumulation, and transformation to different bioproducts in the shoot as well as root and seed parts of plant. Studies revealed that nano particles have capabilities to invade and enter the cell membrane of root epidermic and can reach to cortex, xylem cells of the root vesicular system of crop. Minute-sized nano particles are subjected to penetrate into cell walls because of the perforated (average size 3 to 8 nm diameter) polysaccharide network of cell membranes. These porous polysaccharide networks behave as the natural screen for nanoparticles. However, nanoparticles above this size can't pass the cell wall and remain aggregated at the cell surface. Again, Navarro et al. (2008) reported that nano sized particles can stimulate the development of new larger sized pores on interaction with cell walls and these large pores allow entry of other bigger nano particles into cells. Carbon-based nano-materials like fullerenes, CNTs, graphene oxides are applied by foliar spray, seed treatment, and direct soil application. Multiwalled CNTs has capacities to develop pores in the seed coat of tomato and thus facilitate water uptake resulting in better seed germination and seedling growth (Khodakovskaya et al. 2009). When working on uptake of carbon-based nano materials through plant root cells it was reported that 5-50 nm sized carbon-coated nanomagnetic fertilizers were absorbed and transported in the plant vesicular system with a broad range of efficiency in different plant species. Other than that, CNTs can remain aggregated at surface of root epidermic or it may get translocated to plant reproductive shoot portion. Aggregation or deposition of CNTs was reported over root cell surface of different crops like tomato, onion, cabbage, lettuce, carrot, cucumber, rice, and wheat. Organic compounds solubilized in CNTs may help in the absorption and translocation of nanomaterials into plant shoot due to capillary action. These carbon-based nanomaterials are also genetically carried into upcoming generations of plants. Rico et al. (2011) reported the accumulation of C-70 nano material in various stages and parts of rice plant when grown for 2 continuous generations under the experiment of C-70 nano material insertion in first-generation, which got transmitted to a subsequent generation. HR-TEM analysis exhibited accumulation of larger multi-walled CNTs (size above 200 nm) in subcellular parts and minute-sized (size 30 to 100 nm) in vacuoles, nucleus, and plastids (Serag et al. 2013). But single-walled CNTs remain deposited on the outer surface of cucumber seedling roots (Lou et al. 2011) and fullerene nanoparticles were found aggregated into xylem cells of stem and leaf (Santos et al. 2013). Because particular fullerene nanomaterials entered into root cells due to osmatic, capillary force, and pore space of plasmodesmata or other symplastic pathways. It was also reported that fullerene nanoparticles smaller than cell wall pore diameter may enter the cell wall and get accumulated in between cell wall and plasma membrane or may occur at the cell wall of nearby epidermic cells (Gao et at. 2011, Kole et al. 2013). No proper report or research is available on the use of carbon-based nano materials through foliar application. Giraldo et al. (2014) reported the capability of single-walled CNTs to transmit through stomata and penetration into the chloroplast and even into thylakoids of spinach leaves on foliar application. Transportation of metal or metal oxide nanoparticles from root to foliage of plants has been studied vigorously through various research on the use of nano oxides of Ti, Zn, Fe, Ag, and Au in foliar, soil as well as seed treatment methods. Leaf epidermis contains cells covered with water proofing compounds like cutins and fatty acids and bears stomata with pore width 3-12 micro-meter and length 10-30 micro-meter on expansion due to turgor pressure during gaseous in-out. These stomatal pores provide a vital facility for penetration of metal or metal oxide nano materials through the foliar spray and provide more success than seed treatment and soil application. Some researchers reported that nano fertilizers have capabilities to perforate into seeds without interfering germination and growth of plant seedlings. However, nano fertilizers of pure metals may get absorbed by plant root cells and are capable of getting transported to shoot portions depending on physiochemical properties of nanoparticles, plant species, and growth media.

Conclusion

Crops are always thirsty for water and hungry for nutrients leading to starvation and consequently the nutritional starvation of the whole human population. So, Nutrients are not only essential for the life cycle completion of plants but also for the nutritional stability of humans. Under inherent nutrient deficiency of soils crop production, profitability and sustainability have been lowered. The application of nutrients from external sources such as fertilizers played an important part in improving crop productivity and sustainability. But Surface runoff, leaching, volatilization, denitrification, and fixation all contribute to the low nutrient usage performance of conventional fertilizers due to less solubility of these fertilizers, low nutrient availability, uncontrolled and rapid release pattern. There are opportunities and tools available to accomplish the task of improving the efficiency of applied nutrients through nanotechnology. Because of the high surface area and high reactivity of EgNPs, these can activate plant and microbial activities, resulting in greater nutrient use efficiency. Nanoparticles may act as effective enzymes and polysaccharide that triggers plant and microbial metabolism. Nanotechnology through the formulation of smart and control release nano fertilizers during various stages and various parts of the plant life cycle may work as a problem solver for the agriculture sector. EgNPs and carbon-based nano fertilizers can penetrate into root and leaf cells much easier than conventional fertilizers and may stimulate the metabolic functionalization of cells. As an advancement nano bio-sensor gives a better hand for choosing the best suitable time for planting, irrigation, fertigation, pesticidal spray, and other intercultural operations. It can also provide the information and signals on soil physicochemical properties, nutrient content, deficiency symptoms by plants to assist in taking proper and timely measures to safeguard against crop losses. Nano biofertilizers may help in improving the soil microbial health and nutritional status. Nanotechnological products like nano clays, polymers, NCPCs, hydrogels can remediate heavy metal contamination and may increase the water-holding ability of soils. Although, some of the ecotoxicological challenges exist with nanotechnological products but, after proper systemic and quantitative researches on potential health issues, environmental hazards, and safe disposal of nano materials, nanotechnology may become the most improved technology through application of nano fertilizers for efficient nutrient management and agricultural sustainability.

References

Basak B.B., Pal S., Datta S.C. 2012. Use of modified clays for retention and supply of water and nutrients. *Curr Sci.*102: 1272-1279.

Brady N.R. and Weil R.R. 1999. *The nature and properties of soils*. 415-473. Prentice Hall, Upper Saddle River.

Breck D.W. 1974. *Zeolite Molecular Sieves*. Wiley, New York.

Brock D.A., Douglas T.E., Queller D.C. and Strassmann J.E. 2011. *Primitive agriculture in a social amoeba.* Nature 469: 393-396.

Castro-Enríquez D.D., Rodríguez-Félix F., Ramírez-Wong B., Torres-Chávez P.I., Castillo-Ortega M.M., Rodríguez-Félix D.E., Armenta-Villegas L. and Ledesma-Osuna A.I. 2012. Preparation, characterization and release of urea from wheat gluten electrospun membranes. *Materials*. 5: 2903-2916.

Corradini E., de Moura M.R. and Mattoso L.H.C. 2010. A preliminary study of the incorporation of NPK fertilizer into chitosan nanoparticles. *Express Polym Lett*. 4: 509-515.

Cui H., Jiang J. and Liu Q. 2011. On plant nutrition smart delivery systems and precision fertilization. *Acta Metall Sin*.17: 494-499.

Dimkpa C.O., McLean J.E., Martineau N., Britt D.W., Haverkamp R. and Anderson A.J. 2013. Silver nanoparticles disrupt wheat (Triticum aestivum L.) growth in a sand matrix. *Environmental Science and Technology*. 47: 1082-1090.

Fernando W.J.N., Ahmad N.N.R. and Ahmad A.L. 2010. Release rates of phosphate from coated fertilizer through nano CaCO3-wax mixtures. *Book of abstracts international conference on environmental engineering and applications*. 34-37. ICEEA 2010, Singapore.

Gao J., Wang Y. and Folta K.M. 2011. Polyhydroxy fullerenes (fullerols or fullerenols): beneficial effects on growth and lifespan in diverse biological models. *PLoS ONE*. 6(5): Article ID e19976.

Ghafari H. and Razmjoo J. 2013. Effect of foliar application of nano-iron oxidase, iron chelate and iron sulphate rates on yield and quality of wheat. *Int J Agron Plant Prod*. 4: 2997-3003.

Ghormade V., Deshpande M.V. and Paknikar K.M. 2011. Perspectives for nano-biotechnology enabled protection and nutrition of plants. *Biotechnol Adv*. 29: 792-803.

Giraldo J.P., Landry M.P., Faltermeier S.M., Mc Nicholas T.P., Iverson N.M., Boghossian A.A., Reuel N.F., Hilmer A.J., Sen F. and Brew J.A. 2014. Plant nanobionics approach to augment photosynthesis and biochemical sensing. *Nat Mater*. 13: 400-408.

Jinghua G. 2004. Synchrotron radiation, soft X-ray spectroscopy and nano-materials. *J Nanotechnol*. 1: 193-225.

Junejo N., Khanif M.Y., Hanfi M.M., Yunus W.M.Z. and Dharejo K.A. 2011. Role of inhibitors and biodegradable material in mitigation of nitrogen losses from fertilized lands. *Afr J Biotechnol*.10: 3504-3514.

Kampeerapappun P. and Phanomkate N. 2013. Slowrelease fertilizer from core-shell electrospunfibers. *Chiang Mai J Sci*. 40(4): 775-782.

Khodakovskaya M., Dervishi E., Mahmood M., Xu Y., Li Z., Watanabe F. and Biris A. 2009. Carbon nanotubes are able to penetrate plant seed coat and dramatically affect seed germination and plant growth. *ACS Nano*. 3: 3221-3227.

Kole C., Kole P. and Randunu K.M. 2013. Nanobiotechnology can boost crop production and quality: first evidence from increased plant biomass, fruit yield and phytomedicine content in bitter melon (*Momordica charantia*). *Biotechnol* 13:37.

Kole C., Kole P. and Randunu K.M. 2013. Nanobiotechnology can boost crop production and quality: first evidence from increased plant biomass, fruit yield and phytomedicine content in bitter melon (*Momordica charantia*). *BMC Biotechnol*. 13:37.

Kumar P. 2011. Nanotechnology in agriculture. *Financ Agric*. 43: 8-10.

Liu P., Huang Q. and Chen W. 2012. Construction and application of a zinc-specific biosensor for assessing the immobilization and bioavailability of zinc in different soils. *Environ Pollut.* 164: 66-72.

Liu X., Feng Z., Zhang S., Zhang J., Xiao Q. and Wang Y. 2006. Preparation and testing of cementing nano-subnano composites of slow or controlled release of fertilizers. *Sci Agric Sin.* 39: 1598-1604.

Liu X.M., Zhang F.D., Zhang S.Q., He X.S., Fang R., Feng Z. and Wang Y. 2005. Effects of nano-ferric oxide on the growth and nutrients absorption of peanut. *Plant Nutr Fert Sci.* 11: 14-18

Lou J.C., Jung M.J., Yang H.W., Han J.Y. and Huang W.H. 2011. Removal of dissolved organic matter (DOM) from raw water by single-walled carbon nanotubes (SWCNTs). *J Environ Sci Health.* Part A46: 1357-1365.

Lu C.M., Zhang C.Y.W. and Tao M.X. 2002. Research of the effect of nanometer on germination and growth enhancement of *Glycine max L.* and its mechanism. *Soybean Sci.* 21: 168-172.

Maderova L. and Paton G.I. 2013. Deployment of microbial sensors to assess zinc bioavailability and toxicity in soils. *Soil Biol Biochem.* 66: 222-228.

Mahmoodzadeh H., Nabavi M. and Kashefi H. 2013. Effect of nanoscale titanium dioxide particles on the germination and growth of canola (*Brassica napus*). *J Ornamental Hortic Plants.* 3: 25-32.

Man P., Chan Y.K., Dong H.L., Choong L.C., Jyung C. and SungReol L. 2004. Intercalation of magnesium urea complex into swelling clay. *J Phys Chem Solids.* 65: 409-412.

Manikandan A. and Subramanian K.S. 2014. Fabrication and characterization of nanoporouszeolite based N fertilizer. *Afr J Agric Res.* 9: 276-284.

Monreal C.M., DeRosa M., Mallubhotla S.C., Bindraban P.S. and Dimkpa C. 2015. *The application of nanotechnology for micronutrients in soil-plant systems.* pp 44. VFRC (Virtual Fertilizer Research Center) Report 2015/3, Washington, DC. 44.

Mukherjee A., Majumdar S., Servin A.D., Pagano L., Dhankher O.P. and White J.C. 2016. Carbon nanomaterials in agriculture: a critical review. *Frontier in Plant Science.* 7: 172.

Nakache E., Poulain N., Candau F., Orecchioni A.M. and Irache J.M. 1999. *Biopolymer and polymer nanoparticles and their biomedical applications.* In *Handbook of nanostructured materials and nanotechnology* Edited by Nalwa H.S. 577-63 Academic, New York.

Navarro E., Piccapietra F., Wagner B., Marconi F., Kaegi R., Odzak N., Siqq L. and Behra R. 2008. Toxicity of silver nanoparticles to *Chlamydomonas reinhardtii. Environ Sci Technol.* 42: 8959.

Preetha P.S., Subramanian K.S. and Rahale C.S. 2014. Characterization of slow release of sulphur nutrient-a zeolite based nano-fertilizer. *Int J Dev Res.* 4: 229-233.

Rad J.S, Karimi J., Mohsenzadeh S., Rad M.S. and Moradghol J. 2014. Evaluationg SiO2 nanoparticles effects on developmental characteristic and photosynthetic pigment contents of Zea mays. *Bull Envt Pharm Life Sci.* 3:194-201.

Rajput V.D., Minkina T., Fedorenko A., Tsitsuashvili V., Mandzhieva S., Sushkova S. and Azarov A. 2018. *Metal oxide nanoparticles: applications and effects on soil ecosystems.* In Soil Contamination. Edited by Jesper E. Lund. 81-106. Nova Science Publishers, Inc.

Rajput V.D., Minkina T., Sushkova S., Fedorenko A., Tsitsuashvili V., Mandzhieva S., and Azarov A. 2017. Effect of nanoparticles on crops and soil microbial communities. *J. Soils Sediments.* 1-9. doi 10.1007/s11368-017-1793-2.

Rico C.M., Majumdar S., Duarte-Gardea M., Peralta-Videa J.R. and Gardea-Torresdey J.L. 2011. Interaction of nanoparticles with edible plants and their possible implications in the food chain. *J Agric Food Chem.* 59: 3485-3498.

Sani B. 2012. Energy efficiency management (EEM) by use of nanoparticles in maize seedling growth. *Eur J Exp Biol.* 2(5): 1918-1921.

Santos S.M.A., Dinis A.M., Rodrigues D.M.F., Peixoto F., Videira R.A. and Jurado A.S. 2013. Studies on the toxicity of an aqueous suspension of C60 nanoparticles using a bacterium (gen. Bacillus) and an aquatic plant (Lemnagibba) as in vitro model systems. *AquatToxicol.* 142: 347-354.

Satdev, Zinzala V.J., Chavda B.N. and Saini L.K. 2020. Effect of nano ZnO on growth and yield of sweet corn under South Gujarat condition.*International Journal of Chemical Studies.* 8(1): 2020-2023.

Scott N. and Chen H. 2003. *Nanoscale science and engineering for agriculture and food systems.* A report submitted to cooperative state research, education and extension service, USDA, National Planning Workshop, Washington, DC.

Serag M.F., Kaji N., Habuchi S., Bianco A. and Baba Y. 2013. Nanobiotechnology meets plant cell biology: carbon nanotubes as organelle targeting nanocarriers. *RSC Adv 3.*

Servin A.D., De la Torre-Roche R., Castillo-Michel H., Pagano L., Hawthorne J., Musante C., Pignatello J., Uchimiya M. and White J.C. 2017. Exposure of agricultural crops to nanoparticle CeO2 in biochar amended soil. *Plant Physiol. Biochem.* 110: 147-15.

Shah V., Collins D., Walker V.K. and Shah, S. 2014. The impact of engineered cobalt, iron, nickel and silver nanoparticles on soil bacterial diversity under field conditions. *Environ. Res. Lett.* 9(2). 024001.

Shang Y., Hasan M.K., Ahammed G.J., Li M., Yin H. and Zhou J. 2019. Applications of nanotechnology in plant growth and crop protection: a review. *Molecules*. Jul. 24, 2558.

Shukla A.K. and Tiwari P.K. 2014. Micro- and Secondary Nutrients and Pollutant Elements ResearchinIndia.*CoordinatorReport-AICRPonMicro-and Secondary Nutrients and Pollutant Elements in Soils and Plants.* IISS, Bhopal. 1-146.

Subramanian K.S., Manikandan A.,Thirunavukkarasu M. andRahale C.S. 2015.*Nano-fertilizers for balanced crop nutrition*. In*Nanotechnologies inFood and Agriculture.* Edited by Rai M., Ribeiro C., Mattoso L. and Duran N.69-80.Springer. Cham, Switzerland.

Sultan Y. and DeRosa M.C. 2011. Target binding influences permeability in aptamer-polyelectrolyte microcapsules. *Small.* 7: 1219-1226.

Sundaria N., Singh M., Upreti P., Chauhan R.P., Jaiswal J.P. and Kumar A. 2019. Seed Priming with Iron Oxide Nanoparticles Triggers Iron Acquisition and Biofortification in Wheat (Triticum aestivum L.) Grains. *Journal of Plant Growth Regulation*. 38: 122-131.

Taniguchi N. 1996. *Nanotechnology: integrated processing systems for ultra-precision and ultrafine products.* Oxford University Press, USA.

Thirugnanasambandan T. 2019. *Advances and trends in nano-biofertilizers.* pp 59. SSRN. https://doi.org/10.2139/ssrn.3306998

Trenkel M.E. 1997.*Controlled-release and stabilized fertilizers in agriculture*. InternationalFertilizer Industry Association, Paris.

Weir A., Westerhoff P., Fabricius L., Hristovski K. and Von Goetz N. 2012. Titanium dioxide nanoparticles in food and personal care products. *Environ. Sci. Technol.* 46: 2242-2250.

Wu L., Liu M. and Liang R. 2008. Preparation and properties of a double-coated slow-release NPK compound fertilizer with superabsorbent and water-retention. *Bioresour Technol.* 99(3): 547-554.

Yuvaraj M. and Subramanian K.S. 2015. Controlled-release fertilizer of zinc encapsulated by a manganese hollow core shell. *Soil Sci Plant Nutr*. 61: 319-326.

Zhang Y., Chan H.F. and Leong K.W. 2013. Advanced materials and processing for drug delivery: the past and the future. *Adv Drug Deliv Rev*. 65: 104-120.

Zheng L., Hong F., Lu S. and Liu C. 2005. Effect of nano TiO2 on strength of naturally aged seeds and growth of spinach. *Biol Trace Elem Res*. 104: 83-91.

Biographical Sketch

Rakesh Kumar

Bihar Agricultural University, Bhagalpur, Bihar, India.

Education: BSc Ag (Hons), Masters of Science (Agriculture) with specialization in Soil Science & Agricultural Chemistry

Business Address: Department of SSAC, Bihar Agricultural University, Bhagalpur. Bihar (India) PIN- 813210

Honors: 2nd Position as team leader in national level "BAU A-Hackathon 2021" for Title "Nano-biofertilizer technology: Future tool of research for agriculture".

Publications from the Last 3 Years: 2 Research papers, 3 book chapters, 5 general articles, 3 seminar abstracts and 1 poster presentation at International Seminar.

Chapter 21

Current State of Knowledge in Diagnosis and Mitigation of Micronutrients Deficiency in Crop Production from an Indian Prospective

Satish Kumar Singh[1], Astha Pandey[1], Ayush Bahuguna[1], Kiran Kumar Mohapatra[2], Abhik Patra[1], Eetela Sathyanarayana[4], Hanuman Singh Jatav[4,*], Surendra Singh Jatav[1] and Vishnu D. Rajput[6]

[1]Department of Soil Science and Agricultural Chemistry,
Institute of Agricultural Sciences, Banaras Hindu University, Varanasi, Uttar Pradesh
[2]Department of Soil Science and Agricultural Chemistry, College of Agriculture,
Odisha University of Agriculture and Technology, Bhubaneswar, Odisha, India
[3]Sri Karan Narendra Agriculture University, Jobner, Jaipur, Rajasthan, India
[4]Southern Federal University, Rostov-on-Don, Russia
[5]Agricultural College-Palem, Nagarkurnool, PJTSAU, Hyderabad, Telangana, India
[6]Southern Federal University, Rostov-on-Don, Russia

Abstract

Indian soils are fairly satisfactory with respect to total micronutrient content. But in spite of the relatively high total contents, micronutrient deficiencies have been frequently reported in many crops due to low levels of available micronutrients in soils. Based on the critical limits followed in different states of India, the status of the micronutrients deficiencies was assessed in different soils under the leadership of ICAR All India Coordinated Research Project on Micro- and Secondary Nutrients and Pollutant Elements in Soils and Plants (AICRP-MSPE). In addition to single micronutrient deficiencies, multi-micronutrient deficiencies have emerged in different areas of the country over the years, posing a threat to the sustainability of agriculture. Various diagnostic procedures for micronutrients deficiency in soil and plants have been briefly discussed in this paper. In total 1421 field experiments were conducted on specific crops to determine the critical nutrient concentration of different micronutrients. Based on field experiments and crop response to micronutrients, generalized transition zones were worked out for different nutrients across the soil types. Various factors affecting micronutrients availability in soil and deficiency symptoms of micronutrients have been discussed along with their mitigation strategies. Hence this paper represents the current status of micronutrients in Indian soil with emerging solutions for micronutrient deficiency.

---

[*] Corresponding Author's Email: hanumaniasbhu@gmail.com.

In: Ecosystem Services
Editor: Hanuman Singh Jatav
ISBN: 978-1-68507-614-6

Keywords: biofortification, critical limits, customized fertilizer, micronutrients, micronutrient-efficient cultivars

1. Introduction

Agriculture and agricultural products link soil to the well-being of animals and humans. For plant growth and development, soils are the primary source of nutrients. Plants take nutrients from the soil in varied proportions depending on their needs. Plants require more macronutrients, whereas micronutrients (also known as trace elements) are required in lesser amounts, yet, all kinds of nutrients play equally significant roles in plant metabolism. "If plants are thought of as biological machines, their bodies are made up of macro-elements, their working parts are proteins and enzymes revolving around N atoms, and the 'MICRONUTRIENTS' provide the special lubricants required for a variety of energy transfer mechanisms within the plants," writes Stout (1962). Micronutrients are required in lesser amounts than macronutrients, yet they are just as important. Micronutrients are also known as minor or trace elements because their quantities in plant tissues are minor or trace in comparison to macronutrients (Mortvedt, 2000). The micronutrients required by the crops are zinc (Zn), copper (Cu), iron (Fe), manganese (Mn), boron (B), molybdenum (Mo), nickel (Ni) and chlorine (Cl). The sequence in which these micronutrients are accumulated by plants is as follows: Fe > Mn > Zn > B > Cu > Mo. Depending on the plant species and growth conditions, this order may change. Except for B and Cl, the necessary plant micronutrients are metals based on physicochemical qualities. Despite the fact that micronutrients are only required in trace amounts by field crops, their impact on crop productivity is just as essential as macronutrients (Table 1). The micronutrients, with the exception of B and Cl, are typically found in prosthetic groups that accelerate redox reactions by electron transfer, such as the primary transition elements Fe and Mn, and to a lesser extent Cu and Mo.

Enzymes are formed by micronutrients in their natural state. Micronutrients help the human body create enzymes, hormones, and other compounds necessary for appropriate growth and development. Micronutrients are necessary for energy production, haemoglobin synthesis, bone development, appropriate immunological function and body defense against oxidative damage. They aid in the synthesis and repair of muscle tissue during exercise and injury rehabilitation. Although agriculture's primary goal is to provide food, there is a significant gap in the development community between human nutrition and agricultural sciences. If human capital and capacity building are the goals, recent advances in our understanding of population nutritional demands imply that integrating agriculture and human nutrition should be a primary emphasis of future development initiatives. The creation of a more sophisticated concept of nutritional status, as well as a broader debate of relevant development initiatives, with a focus on agricultural and food-based solutions to malnutrition.

Table 1. Function and deficiency symptoms of micronutrients in plants

Nutrient	Function	Deficiency symptoms	Critical limit in soil (mg kg^{-1})
Boron	• Helps in: cell-wall formation and stability; • Cell division; • Sugar/energy transport to growing parts; • Pollination, seed set and integrity of fruits	• Terminal leave lose colour • Loss of colour starts at the base with eventual death of terminal bud • Deformed young leaves • Deformed fruits	0.5
Chlorine	• Regulates stomatal opening, turgor control and water loss and electrical charge balance for physiological functions in plants; restricts disease susceptibility	• Shows up as wilting, restricting and highly branched root with shoot tips; leaf mottling, tip wilting and cholosis	8.0
Copper	• Constitutes of oxidase enzymes; helps in: carbohydrate and N metabolism, lignin formation imparting thereby cell-wall strength and averting wilting	• Noticed as stunting of growing leaves, die-back of stems and twigs and light over-all yellowing of leaves giving a withered appearance	0.2
Iron	• Key nutrient for chlorophyll production and happening of photosynthesis, being component of many enzymes, Fe is associated with energy transfer, N reduction and fixation	• Chlorosis of young leaves is the key sign, which appears first on upper leaves in interveinal tissues followed by complete yellowing/ whitening and even browning and death	4.5
Manganese	• Necessary indirectly for creation of chlorophyll, photosynthesis, N metabolism, chloroplast production and is a cofactor in many physiological reactions.	• Manifested as interveinal cholorsis of younger leaves, brown necrotic spots and even premature drop of leaves	2.0
Molybdenum	• Indispensible for N fixation by legumes and is also involved in N and S metabolism of plants	• Resemble yellowing of older leaves, which spread to younger foliage also; development of irregular leaf blades known as whip tail	0.1
Nickel	• Being part of urease enzyme- helps converting urea to ammonim, which is plant usable form	• Symptoms seldom develop, nevertheless if occur these show up as necrosis of the leaf tips.	0.1
Zinc	• Important component of many enzymes essential for energy production, protein synthesis and growth regulation	• Show up as reduced length of internodes, smaller leaves, interveinal cholorosis and brown spots, delayed maturity	0.6

In India, the use of leaf/soil analysis to detect micronutrient disorders/toxicities, particularly at the early stages of hidden hunger, is limited. Micronutrient deficiency in country soils is reaching an alarming level. The copper deficit in plants could be caused by soils with high organic content, which favors Cu complexation with organic compounds. Micronutrient deficiencies (Zn, Fe, folic acid) during pregnancy have been linked to low birth weight and lowered immunity. The majority of micronutrients required for human and animal growth are either required or beneficial for crop plants. Boron is included as a useful trace element for humans to defend against post-menopausal women's loss of calcium and magnesium. Each essential element has a unique role in metabolism, either as a cofactor in a variety of enzymatic activities or as a necessary component of critical structural proteins that cannot be substituted in any way. Trace elements are necessary for dietary food in amounts < 100 mg kg^{-1}, however a lack of any micronutrient leads to poor health, sickness, morbidity, mortality, infertility and death.

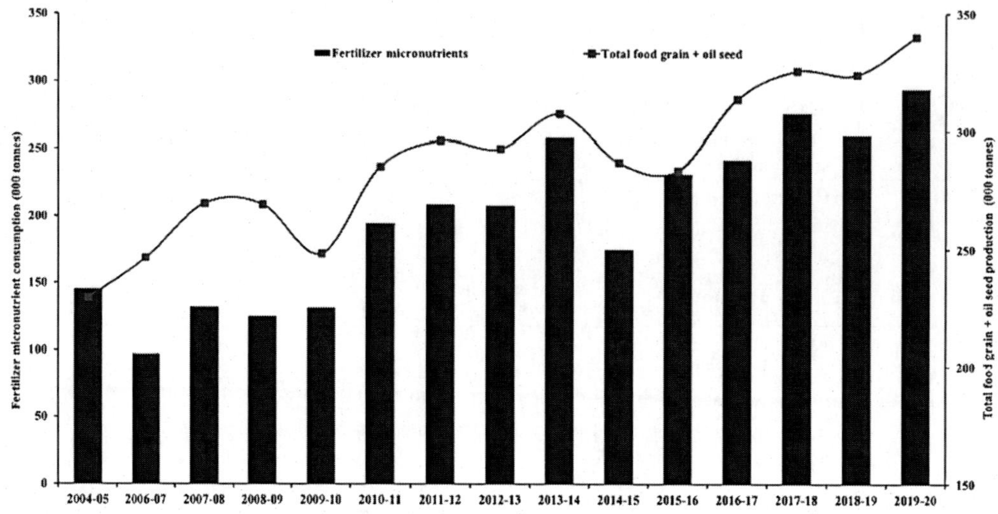

Figure 1. Micronutrient fertilizer consumption vis-a-vis food grain production in the country during last decade.

Over the years, the consumption of micronutrient fertilizer has increased in India. The increasing trend in consumption of micronutrients fertilizer and food grain production of the country reveals the importance of micronutrients in sustainable food production (Figure 1). The consumption of micronutrient fertilizer followed the order: zinc sulphate > iron sulphate > boric acid/borax > manganese sulphate > copper sulphate. Of the total Zn used, 70% goes to the field crops and the remaining 30% to vegetable and fruit crops, while the reverse is true for Mn, Fe and Cu. Of the total B application, about 60% are applied to vegetables and fruit crops and 40% to food and oilseed crops. Based on the thousands of experiments conducted at farmers' fields, the best micronutrient management practices have been developed for efficient management of nutrients to enhance system productivity (Behera et al., 2019). Micronutrient management must be done based on micronutrient demand and supply in the soil-plant system. Crops, soil types, deficit severity, source, technique, time, rates, and frequency of application all affect it. Micronutrient management best practices include:

- Make sure that poor crop development in a field or area of a field isn't due to anything else.
- Look for particular micronutrient deficiency symptoms in the damaged crop.

- Take soil and plant tissue samples from the affected areas separately for a thorough examination.
- Keep soil from deteriorating
- Use of organic manures on a regular basis (FYM)
- Use micronutrient fertilizers in addition to NPK fertilizers regularly.
- Use of chelated micronutrient
- Adaptation of customized fertilization
- Selection of crop varieties that can efficiently absorb micronutrients from soil solid phases
- Consider that other crop inputs such as pesticides, lime or manure can affect the availability of micronutrients in soil

2. Current Scenario of Micronutrients Deficiencyin India

Micronutrient content in soil is dependent on several factors such as geochemical composition (total micronutrient contents of the soil parent material); soil type *i.e.*, clay mineralogy, particle size distribution, soil horizon, soil age, soil formation processes; intrinsic soil properties like pH, redox potential (Eh), soluble salt concentration (EC), quality and quantity of soil organic matter and calcium carbonate content; inputs of trace elements supplied through atmospheric deposition, pesticides, manures, fertilizers, available content of macronutrients, micronutrient interactions and vegetation (Alloway 2008; Shukla et al., 2016). In addition to these factors, low levels of available micronutrients and their rampant deficiencies in Indian soils are attributed to the intensification of agriculture without proper replenishment of micronutrients through fertilization. Further, loss of micronutrients through leaching, liming of soils, scanty use of manures and accelerated use of macronutrient based chemical fertilizers without micronutrient additions have resulted in accelerated exhaustion of available micronutrients in the soils. Total soil micronutrient content is a complex function of parent material and pedogenic processes. Indian soils are fairly satisfactory with respect to total micronutrient content. But in spite of the relatively high total contents, micronutrient deficiencies have been frequently reported in many crops due to low levels of available micronutrients in soils (Behera and Shukla, 2014; Shukla and Tiwari, 2016).

2.1. Plant Available Micronutrient Deficiencies in Soils

Availability can be defined as the quantity of a soil nutrient that is accessible to plant roots during the period of the growing season. Since plant roots accumulate micronutrients directly from the soil solution, the total pool of soil micronutrients is not directly available. Based on the critical limits followed in different states of India, the status of the micronutrients deficiencies was assessed in different soils under the leadership of ICAR All India Coordinated Research Project on Micro- and Secondary Nutrients and Pollutant Elements in Soils and Plants (AICRP-MSPE) during 2011-2017. Analysis of more than 2.0 lakhs soil samples during 2011-2017 revealed that on an average, 36.5, 12.8, 4.2, 7.1 and 23.4% soils were deficient in Zn, Fe, Cu, Mn and B, respectively (Shukla et al., 2018; Shukla and Behera, 2019) (Table 2). Spatial distribution of Zn deficiency exhibited inter-state variations. More than 50% soils of the states such as Tamil Nadu (63.3%), Rajasthan (56.5%), Madhya Pradesh (57.1%) and Goa (55.3%) exhibited Zn deficiency whereas states like Arunachal Pradesh, Himachal Pradesh, Meghalaya, Mizoram, Nagaland, Tripura and Uttarakhand had Zn deficiency in less than 10% of soils. More than 20% soils had Fe deficiency in states like Rajasthan (34.4%), Gujarat (25.9%), Haryana (21.7%) and Maharashtra (23.1%) while the states like Uttar Pradesh, Telangana, Andhra Pradesh, Bihar, Goa, Tamil Nadu had a deficiency in 10-20% of soils. In general, Cu deficiency was <5% soils in any of the states, except Haryana (5.1%), Rajasthan (9.2%) and Tamil Nadu (12.0%). Higher Mn deficiency was reported in light textured rice growing soils (especially in rice-wheat systems) of the states like Rajasthan (28.3%), Punjab (26.2%), Goa (16.9%), Uttar Pradesh (15.8%) and Chhattisgarh (14.8%). The soils of the states like Bihar, Haryana, Himachal Pradesh and Telangana exhibited Mn deficiency to the tune of 5-10%. The states having acid soils like Jharkhand, Odisha, Karnataka, Jammu and Kashmir, Himachal Pradesh, Manipur, Meghalaya, Mizoram and West Bengal exhibited B deficiency in 35-60% soils (Shukla et al., 2014).

Table 2. State-wise per cent distribution of micronutrient deficiencies in India

State	Zn	Cu	Fe	Mn	B
Andhra Pradesh	22.92	17.24	1.33	1.63	4.08
Arunachal Pradesh	4.63	1.44	1.40	3.01	39.15
Assam	28.11	0.00	2.80	0.01	32.75
Bihar	45.25	12.00	3.19	8.77	39.39
Chhattisgarh	25.59	7.06	3.22	14.77	20.59
Goa	55.29	12.21	3.09	16.91	12.94
Gujarat	36.56	25.87	0.38	0.46	18.72
Haryana	15.42	21.72	5.13	6.16	3.27
Himachal Pradesh	8.62	0.51	1.43	6.68	27.02
Jammu & Kashmir	10.91	0.41	0.34	4.60	43.03
Jharkhand	17.47	0.06	0.78	0.26	60.00
Karnataka	30.70	7.68	2.28	0.13	36.79
Kerala	18.34	1.23	0.45	3.58	31.21
Madhya Pradesh	57.05	8.34	0.47	2.25	20.69
Maharashtra	38.60	23.12	0.14	3.02	37.17
Manipur	11.50	2.13	2.46	2.06	47.93
Meghalaya	3.84	1.33	1.03	2.95	37.17
Mizoram	1.96	0.49	0.98	1.22	47.93
Nagaland	4.62	2.00	0.53	3.05	32.76
Odisha	32.12	6.42	7.11	2.12	54.31
Punjab	19.24	13.04	4.67	26.20	18.99
Rajasthan	56.51	34.38	9.15	28.28	2.99
Tamil Nadu	63.30	12.62	12.01	7.37	20.65
Telangana	26.77	16.65	1.36	3.54	16.49
Tripura	5.51	1.57	2.36	0.00	23.62
Uttar Pradesh	27.27	15.56	2.84	15.82	20.61
Uttarakhand	9.59	1.36	1.51	4.82	13.44
West Bengal	14.42	0.03	1.76	0.98	37.05
India	36.50	12.8	4.20	7.10	23.4

Source: Shukla et al. (2018).

2.2. Periodic Changes in the Level of Micronutrient Deficiency

Monitoring the extent of micronutrient deficiencies on a time scale done in the AICRP-MSPE project showed that the Zn deficiency in the country declined from 48.8% in 1987-1999 to 36.5% in 2011-2017 (Figure 2). Extensive research and extension activities on micronutrients, especially on Zn by AICRP-MSPE helped in a linear increase in the consumption of zinc sulphate. Regular use of Zn fertilizers resulted in the Zn build-up in the soil, consequentially leading to a drop in Zn deficiency to 36.5% in 2017. Deficiencies of Fe and Mn increased slightly from 11.0 and 3.0% in 1967-1987 to 12.8 and 7.1%, respectively in 2011-2017.

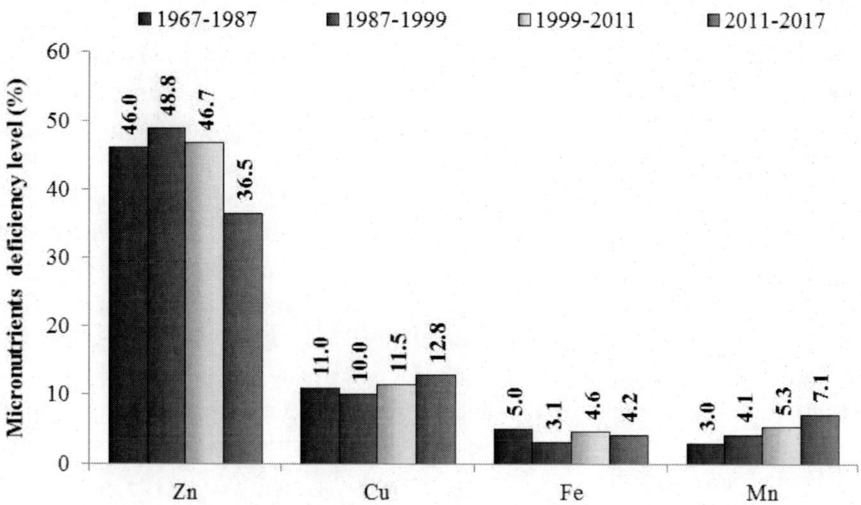

Figure 2. Periodic changes in the extent of micronutrient deficiencies in Indian soils on time scale (Shukla and Behera 2019).

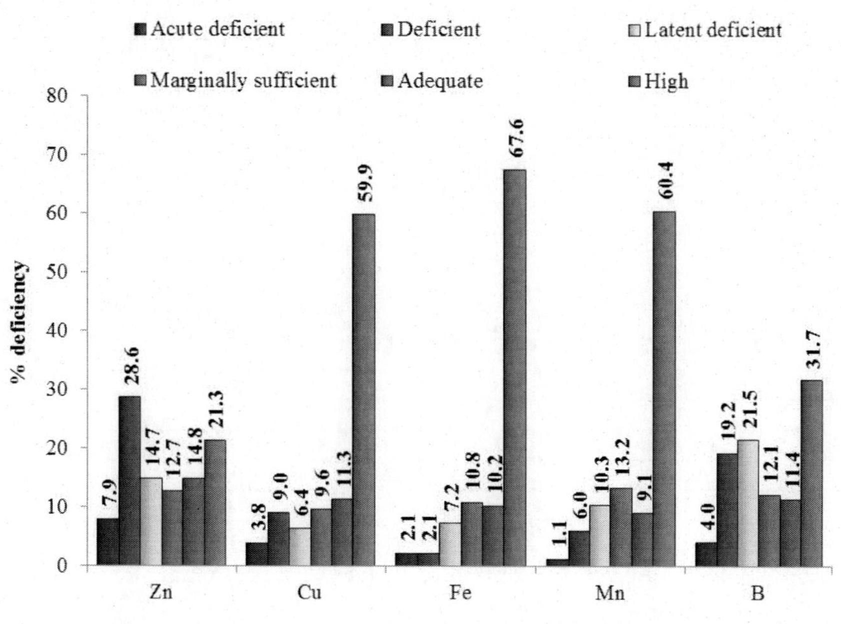

Figure 3. Extent of micronutrient deficiencies in Indian soils (Shukla et al. 2021).

An increase in Mn deficiency was due to >25% of the samples being Mn-deficient in the rice-wheat growing areas of Punjab, Haryana and western Uttar Pradesh. Deficiencies of Cu fluctuated between 3-5% over the years. Based on the newly devised critical limits, 7.9%, 28.6%, and 14.7% soil samples are acute deficient, deficient and latent deficient in Zn, respectively (Figure 3). Nearly 4.0%, 19.2%, and 21.5% soil samples are acute deficient, deficient and latent deficient in B, respectively. Deficiencies of Mn, Cu and Fe in different categories are witnessing rising trends in different parts of the country (Shukla et al., 2021a).

2.3. Multi-Micronutrient Deficiencies

In addition to single micronutrient deficiencies, multi-micronutrient deficiencies have emerged in different areas of the country over the years, posing a threat to the sustainability of agriculture (Table 3). The average per cent soil samples showing multi-micronutrients deficiencies followed the order Zn+B > Zn+Fe > Zn+Mn > Zn+Cu > Fe+B > Zn+Fe+B > Zn+Fe+Cu+Mn > Zn+Fe+Cu+Mn+B. The Zn+B deficiency varied from 0.6-20.3% spread over different states, with mean the value of 8.7% (Shukla et al., 2019; Shukla et al., 2021b). The prevalence of Zn+B deficiency was higher in states like Bihar, Karnataka, Odisha, and Tamil Nadu. On average, 5.8% of soils were deficient in Zn+Fe. Relatively higher per cent of soils of Gujarat, Karnataka, Maharashtra and Rajasthan were deficient in Zn+Fe. More than 20% of soils in 9 districts of Gujarat, 5 districts each of Madhya Pradesh and Maharashtra, and 7 districts of Rajasthan suffered from the twin deficiencies of Zn+Fe. Relatively more soils of Goa, Jammu and Kashmir, and Rajasthan were deficient in Zn+Mn, with a national average of 3.4%. The higher per cent of soils in Bihar and Karnataka were deficient in Zn+Fe+B. More than three micronutrient deficiencies like Zn+Fe+Cu+Mn and Zn+Fe+Cu+Mn+B were very less, recorded on only 0.30% and 0.10% of soils, respectively. Less than 5% of soils in 13 districts of Bihar, 7 districts each of Punjab and Uttar Pradesh, and 16 districts of Tamil Nadu were deficient in Zn+Fe+Cu+Mn; whereas, less than 5% of soils in 13 districts of Bihar, 4 districts of Punjab and 10 districts of Tamil Nadu were deficient in Zn+Fe+Cu+Mn+B (Majumdar and Prakash, 2018).

Table 3. Multi-micronutrient deficiencies (% soil samples deficient) in different states of India

State	Zn+B	Zn+Fe	Zn+Mn	Zn+Cu	Fe+B	Zn+ Fe+B	Zn+Fe+ Cu+Mn	Zn+Fe+Cu+ Mn+B
Andhra Pradesh	2.06	6.90	0.60	0.60	1.78	0.62	0.01	0.00
Arunachal Pradesh	2.41	0.00	0.50	0.30	0.00	0.00	0.00	0.00
Assam	7.88	0.00	0.00	0.90	0.14	0.00	0.00	0.00
Bihar	19.6	5.26	4.80	2.20	7.90	3.67	0.23	0.19
Chhattisgarh	8.87	3.04	0.90	0.90	0.87	0.63	0.02	0.00
Goa	6.32	2.21	10.30	2.10	0.59	0.29	0.00	0.00
Gujarat	7.29	11.7	0.30	0.30	4.35	1.94	0.01	0.00
Haryana	0.75	6.59	2.50	3.40	1.10	0.44	0.34	0.09
Himachal Pradesh	3.08	0.22	0.80	0.30	0.50	0.09	0.00	0.00
Jammu & Kashmir	0.57	0.54	7.00	0.20	0.00	0.00	0.07	0.00
Jharkhand	8.40	0.00	0.20	0.10	0.00	0.00	0.00	0.00
Karnataka	15.3	9.24	1.20	1.10	6.84	3.31	0.02	0.01

State	Zn+B	Zn+Fe	Zn+Mn	Zn+Cu	Fe+B	Zn+ Fe+B	Zn+Fe+ Cu+Mn	Zn+Fe+Cu+ Mn+B
Kerala	3.93	0.97	0.70	0.50	2.10	0.71	0.14	0.14
Madhya Pradesh	2.27	5.83	1.50	0.40	0.60	0.39	0.00	0.00
Maharashtra	5.19	9.82	1.80	0.10	5.23	1.48	0.01	0.00
Manipur	9.41	0.87	2.40	0.60	0.93	0.43	0.00	0.00
Meghalaya	1.31	0.07	0.30	0.10	0.62	0.07	0.00	0.00
Mizoram	1.11	0.00	0.00	0.20	0.22	0.00	0.00	0.00
Nagaland	2.19	0.40	0.30	0.10	0.60	0.30	0.00	0.00
Odisha	20.3	2.53	1.30	2.50	2.25	1.04	0.01	0.00
Punjab	2.83	4.64	7.00	2.70	1.82	0.68	0.48	0.15
Rajasthan	0.57	23.3	22.5	7.80	0.06	0.02	2.03	0.00
Tamil Nadu	13.1	7.71	5.30	9.10	2.00	1.38	0.66	0.10
Telangana	8.92	4.68	0.80	0.60	4.10	1.30	0.00	0.00
Tripura	0.56	0.00	0.00	0.00	0.00	0.00	0.00	0.00
Uttar Pradesh	5.52	5.25	4.80	1.20	2.75	0.97	0.32	0.19
Uttarakhand	1.26	0.44	1.10	0.70	0.49	0.09	0.09	0.00
West Bengal	4.17	0.00	0.40	0.50	0.02	0.00	0.00	0.00
India	8.70	5.80	3.40	2.80	2.70	1.20	0.30	0.10

Source: Shukla et al. (2021).

2.4. Current Status of Micronutrient Deficiencies in Indian Soils

Available contents of Zn, Fe, Mn, Cu and B in different soils of the country varied widely (Table 4) (Shukla and Behera, 2017). Available Zn content varied from traces to 45.4 mg kg^{-1} in different soil types. The mean values of available Zn varied from 0.40 mg kg^{-1} in desert-lithosolic soils to 2.4 mg kg^{-1} in glacier and eternal snow and sub-montane soils. The highest per cent of samples deficient in available Zn were recorded in desert soils-lithosolic (86.3%) followed by skeletal soils (63.2%), desert soil srhegosolic (54.0%), medium black soils (39.4%), mixed red and black soils (36.2%) and so on. The values of available Fe in different soils varied from 0.01 to 96 mg kg^{-1}. The highest mean value of 69.8 mg kg^{-1} was recorded in laterite soils whereas the lowest value of 7.0 mg kg^{-1} was recorded in desert soils-rhegosolic. The per cent samples deficient in available Fe followed the order: desert soils- lithosolic > desert soils-rhegosolic > grey brown soils > old alluvial soils > calcareous sierozemic soils > skeletal soils > medium black soils and so on. The concentration of available Mn in different soils ranged from 0.01 to 445 mg kg^{-1}. The lowest and highest mean values of available Mn concentration were recorded in desert soils-lithosolic (4.70 mg kg^{-1}) and deltaic alluvium (32.6 mg kg^{-1}) soils, respectively. Like available Zn and Fe, the highest per cent samples deficient in available Mn were obtained in desert soils-lithosolic (37.8%). Different soils had traced to 136 mg kg^{-1} of available Cu concentration. The mean values of available Cu varied from 0.40 to 3.90 mg kg^{-1}. The per cent sample deficient in available Cu was higher in desert soils followed by old alluvial soils, calcareous sierozemic soils and so on. The different soils varied widely with respect to available B concentrations. The available B concentration ranged from 0.01 to 170 mg kg^{-1} with mean values spanning from 0.54 mg kg^{-1} (sub-montane soils) to 12.9 mg kg^{-1} (deltaic alluvium). Higher values of per cent samples deficient in available B were recorded in grey brown soils (46.3%) and were followed by sub-montane soils (33.7%), desert soils-rhegosolic (24.7%), calcareous alluvial soils (24.4%) and red sandy soils (22.8%). The skeletal soils, desert soils-lithosolic and old alluvial soils had a lower percentage of B-deficient samples.

Table 4. Percentage wise distribution of micronutrient deficiencies in different soils of India

State	Per cent soil samples deficient (PSD) in				
	Zn	**Cu**	**Fe**	**Mn**	**B**
Alluvial soils	29.0	13.4	3.70	2.20	11.3
Brown hills soils	13.3	5.30	4.10	2.20	11.4
Calcareous alluvial soils	29.2	12.4	3.00	1.30	24.4
Calcareous sierozemic soils	13.4	22.8	9.90	5.00	5.80
Deep black soils	32.4	12.5	18.8	0.50	20.7
Deltaic alluvium	28.8	5.20	1.90	0.70	20.1
Desert soils-lithosolic	86.3	51.7	37.8	18.5	2.40
Desert soils-rhegoslic	54.0	39.4	16.3	7.50	24.7
Glaciers and eternal snow	8.60	1.30	1.10	0.50	12.3
Grey brown soils	32.7	25.1	0.30	0.60	46.3
Laterite soils	24.8	6.50	2.20	4.80	19.5
Medium black soils	39.4	22.3	1.10	0.30	16.9
Mixed red and black soils	36.2	17.9	3.50	2.20	12.7
Old alluvial soils	30.8	25.0	7.70	7.40	5.00
Red and yellow soils	31.7	11.9	1.50	0.80	17.2
Red loamy soils	27.3	11.4	9.50	1.60	11.5
Red sandy soils	20.9	15.0	4.80	0.40	22.8
Skeletal soils	63.2	22.4	2.10	0.30	0.60
Sub-montane soils (Podzolic)	9.90	1.00	0.00	0.20	33.7
Tarai soils	21.1	10.8	1.70	2.10	6.60

Source: Shukla and Behera (2017).

2.5. Current Status of Micronutrient Deficiencies in Eastern Uttar Pradesh Soils

Two thousand three hundred thirty four surface soil samples were collected from Chandauli, Mirzapur, Sant Ravidas Nagar and Varanasi districts of eastern Uttar Pradesh during 2011-12 under GPS and GIS based soil fertility mapping project to evaluate the fertility status of surface soil. The DTPA extractable Zn, Fe, Mn and Cu reported the respective deficiency percentage of the soils were 46, 15, 1 and 1 for Varanasi and 30, 18, 8 and 7 for Mirzapur districts. The soils of Sant Ravidas Nagar and Chandauli districts were deficient in Zn (16 and 32%) and Fe (3 and 1%) only (Singh et al., 2016). Soils were also found to be highly deficient in available B, with mean contents of 0.55, 0.49, 0.66 and 0.62 mg kg^{-1} in Chandauli, Mirzapur, Sant Ravidas Nagar and Varanasi districts, respectively showing deficiency in 55, 61, 30 and 37 per cent soil samples (Singh et al., 2015). A high magnitude of B deficiency was noticed in soils of Vikas Khand Nawgarh (94%) followed by Rajgarh (85%) and Marihan (80%), representing the area of low pH Vindhyan soils. For DTPA extractable Ni, Mirzapur district was most prone to Ni deficiency (7%) and among the blocks of this district Sikhar, had the highest (31.67%) deficiency followed by Kon (27.78%) and Majhawa (24.32%). The potential threat of Ni deficiency was highest in Sant Ravidas Nagar (22%) though it had only 4% Ni deficit soil, followed by Mirzapur (19%) (Singh and Patra, 2020).

3. Diagnostic Procedure for Micronutrients Deficiency in Soil and Plants

3.1. Use of Various Extractants to Assess the Available Micronutrients

Several researchers have worked on the evaluation of extractants for different soils for assessing the bioavailability of micronutrients in soils, which include: (i) mineral acids, (ii) chelating agents, (iii) buffered salts, (iv) neutral salts and (v) other extractants. Diethylene triamine penta acetic acid (DTPA) soil test of Lindsay and Norvell (1978) is most widely used for extraction of plant available micronutrient cations (Zn, Fe, Mn and Cu) in different soil types across the world including India (Singh, 2008; Alloway, 2008). Other extractants like ethylene diamine tetra acetic acid (EDTA), hydrochloric acid (HCl), ammonium bicarbonate-DTPA (AB-DTPA), Mehlich 1 and Mehlich 3, etc. are also quite popular. The DTPA soil test was originally developed to identify near-neutral and calcareous soils with insufficient available Zn to support the maximum yield of crops (Lindsay and Norvell, 1978). Studies on acid soils of India revealed the usefulness of DTPA, AB-DTPA, Mehlich 1, Mehlich 3 and 0.1 M HCl for extraction Zn, Cu, Fe and Mn (Behera et al., 2012; Behera and Shukla, 2014). By and large, the validity of the DTPA soil test for zinc and copper was evident, but this soil test did not hold good for Mn in some cases and Fe in general. In limited studies, the DTPA soil test was found to be more efficient than $Sr(NO_3)_2$ and $Ca(NO_3)_2$ for assessing nickel deficiency in the soil. Hot water extraction of B is commonly used for extraction of plant available B in soils of India despite having the problem for colorimetric estimation of B due to organic matter and turbidity from suspended fine clay particles obtained during boiling of soil in water. Other extractants like 0.05 M mannitol-0.01 M $CaCl_2$, NH_4OAc, dilute phosphoric acid, tartaric acid and salicylic acid are also being currently used for the assessment of available B in different soils. Studies conducted under AICRP-MSPE demonstrated that hot $CaCl_2$ was the most suitable for determining available B in acid soils of India, followed by hot water, salicylic acid, and ammonium acetate. However, salicylic acid appeared to be useful for routine soil analysis for available B, particularly when a large number of samples are to be analyzed (Datta et al., 2018). In case of Mo, Grigg's reagent has been used. The tediousness of this method is reflected in the meager number of soil samples analyzed so far across the country. There is an urgent need to develop a simple, rapid and reproducible soil test method for analyzing the minute quantities of available Mo in the soil. Commonly used extractants like DTPA and EDTA interfere with the estimation of micronutrient cations by ICP-MS (inductively coupled plasma mass spectrometer) in the soil extracts. It appears that the use of short sequential fractionation schemes for micronutrient cations may be a practical way of solid phase speciation. Considerable progress has been made in assessing the fraction of micronutrients, except Mo, in the soil for providing insight into their forms, availability and association with different soil constituents.

3.2. Critical Limits of Micronutrients in Different Crops

Critical level/limit (mostly deficiency) of a nutrient has been defined as the level in soil or plant tissues, below which there is a probability to get a response to application of that particular nutrient. Cate and Nelson procedure (graphical or statistical) has been used for determining the critical levels of micronutrients in soils and the same has been used for delineating the micronutrient-deficiencies. Critical levels of deficiency (CLD) of DTPA extractable Zn, Cu, Fe and Mn; and hot water extractable B has been developed for the soils of different states. As depicted in Table 5, CLDs varied with micronutrients, crops, soils, and the extractants used (Shukla and Tiwari, 2016). Earlier critical levels of different micronutrients determined under pot culture experiments revealed only two levels of concentration for each nutrient i.e., adequacy and deficiency (Takkar et al., 1989). Recently, critical levels were redefined by conducting on-farm trials in the cluster of villages (Shukla and Tiwari, 2016). In total 1421 field experiments were conducted on specific crops to determine the critical nutrient concentration of different micronutrients. Based on field experiments and crop response to micronutrients, generalized transition zones were worked out for different nutrients across the soil types. The first 3 categories indicate the level of deficiency (acute deficient, deficient, latent deficient) and the next 3 levels specify adequacy (marginally sufficient, adequate and high) (Table 6) (Shukla and Behera, 2017a; Shukla et al., 2018b).

Table 5. Use of various extractants to assess the available micronutrients

Micronutrients	**Extractants use**	**Critical level of deficiency**	**References**
Single micronutrient extractent			
Zinc	0.005 M DTPA + 0.1M TEA* + 0.01M $CaCl_2$ (pH 7.3)	0.5-1.0	Datta et al. (2018)
	0.1 N HCl	1-5	
	1 N NH4OAc (pH 4.6)	0.2-0.5	
	0.05 N HCl	1.0	
Copper	0.005 M DTPA + 0.1 M TEA + 0.01 M $CaCl_2$ (pH 7.3)	0.2-0.5	
	0.1 N HCl	1-3	
	1N NH4OAc (pH 4.8)	0.2	
Iron	0.005 M DTPA + 0.1 M TEA + 0.01 M $CaCl_2$ (pH 7.3)	2.5-5.8	
	1 N NH_4OAc (pH 4.8)	2	
Manganese	0.005 M DTPA + 0.1 M TEA + 0.01 M $CaCl_2$ (pH 7.3)	2-4	
	1 N NH_4OAc (pH 7.0)	3-4	
	3 N $NH_4H_2PO_4$ + 0.1 N H3PO4	15-20	
Nickel	0.005 M DTPA + 0.1 M TEA + 0.01 M $CaCl_2$ (pH 7.3)	0.17	
Boron	Boiling hot water	0.5-1.0	
	0.01 M $CaCl_2$ + 0.05 M mannitol	0.25	
	0.1 M salicylic acid	0.45	
Molybdenum	0.175 M Ammonium oxalate (pH 3.3)	0.05-0.2	
Multi micronutrient extractent			
Cu, Fe, Mn, Zn	0.72 N NaOAc + 0.52 N CH_3COOH		Morgan (1941)
B, Cu, Fe, Mn, Zn	0.073 M NaOAc + 0.52 N CH_3COOH + 0.001 M DTPA		Wolf (1982)
Mn, Zn	0.05 N HCl + 0.025 N H_2SO_4		Mehlich (1954)
Mn, Zn	0.2 N CH_3COOH + 0.015 N NH4F + 0.2 N NH_4Cl + 0.012 N HCl		Mehlich (1978)
Cu, Mn, Zn	0.2 *N* CH_3COOH + 0.25 *N* $NH4NO_3$ + 0.013 *N* HNO_3 + 0.015 N NH4F + 0.001 M EDTA		
Fe, Mn, Zn	0.005 M DTPA + 1 M NH_4HCO_3 (pH 7.6)		Soltanpour and Schwab (1977)

Table 6. Forecasting of critical limit of micronutrients in different soil

Micronutrients	Location	Soil type/ order	Plants	Extractent use	Critical limit	References
Zn	Ranga Reddy, Telangana	Alluvial soil	Rice	DTPA	0.7	Shukla and Tiwari, 2016
	Pusa, Bihar	Calcareous soil	Wheat	DTPA	0.8	Shukla and Tiwari, 2016
Cu	Coimbatore, Tamil Nadu	Sandy clay loam	Maize	DTPA	0.9	Shukla and Tiwari, 2016
Cu	Pantnagar, Uttarakhand	Silty clay loam	Maize	DTPA	0.75	Shukla and Tiwari, 2016
Cu	Jabalpur, Madhya Pradesh	Vertisol, clayey soil	Soybean	DTPA	0.70	Shukla and Tiwari, 2016
Cu	Erode, Tamil Nadu	Red and lateritic soil	Onion	DTPA	0.62	Shukla and Tiwari, 2016
Cu	Ludhiana, Punjab	Sandy loam	Rice	DTPA	0.30	Shukla and Tiwari, 2016
Cu	Ludhiana, Punjab	Loamy sand	Wheat	DTPA	0.40	Shukla and Tiwari, 2016
B	Jorhat, Assam	Acid Soils	Mustard	hot water extractable	0.55	Shukla and Tiwari, 2016
B	Hisar, Haryana	Calcareous soils	Mustard	hot water extractable	0.80	Shukla and Tiwari, 2016
B	Hisar, Haryana	Calcareous soils	Mustard	salicylic acid	0.61	Shukla and Tiwari, 2016
Ni	Varanasi, Uttar Pradesh	Alluvial soil	Barley	DTPA	0.22	Kumar et al. (2018)
Ni	New Delhi	Alluvial soil	Soybean	DTPA	0.18	Barman (2012)
Ni	New Delhi	Alluvial soil	Wheat	DTPA	0.15	Barman (2012)

3.3. Forecasting of the Critical Limit of Micronutrients in Different Soil

The critical limit of micronutrients in the soil is defined as the minimum soil test value associated with maximum crop yield. It is that the concentration below which deficiency occurs and it designates the lower end of the sufficiency range. The critical soil test value is the one that separates a group of soils that give a significant yield response to fertilizers from that of soils that don't respond. The critical limit in plant refers to a level at or below which plant either develops deficiency symptoms or causes a reduction in crop yields as compared to optimum yields. Critical limit is classified into 2 types upper critical limits (UCL) – toxicity after this and lower critical limits (LCL) – deficiency below this. The purpose of forecasting critical limits will help for revalidation of existing nutrient fertility ratings. Critical limits will help for the standardization and development of universally acceptable extractants for available soil nutrients. Different approaches of critical limits two different approaches were introduced by Cate and Nelson: Graphical method (1965) - Scattered diagram technique Statistical method (1971) - R2 value.

Due to the diversified nature of soils, it is not possible to establish a fixed value of the critical limit for the available nutrient in different soils due to the changed scenarios by intensive cropping with high yielding varieties. The statistically calculated critical level of soil Zn (0.83 mg kg^{-1}) for rice determined by DTPA extraction method was the same as that of the graphical method while the critical level values of HCl (1.8) and NH_4OAc (0.40 mg kg^1) extractable Zn varied considerably between graphical and statistical methods and thus it indicated that DTPA was better extractant for assessing available zinc status of calcareous soils (Rahman et al., 2007). Sanjeev and Raina (2008) established the critical range of 16-20 ppm DTPA extractable Zn for apple using the Cate-Nelson graphical model in Himachal Pradesh. Murthy et al. (2009) revealed that the critical level of DTPA extractable Zn of 0.325 mg kg^{-1} for castor in Alfisols grown in Ranga Reddy, Nalgonda, districts of Andhra Prades. The critical concentration of soil available B and plant tissues B was 0.35 and 12.0 mg kg^{-1} respectively below which appreciable responses to B application were observed in rice grown in alluvial soils of West Bengal (Debnath and Ghosh, 2012). Mahata et al. (2013) concluded that the critical limit of DTPA-Zn in soil and 3rd leaf of rice plants was 0.82 and 28.5 mg kg^{-1}, respectively. Meena et al. (2013) concluded that the critical limit of iron is 4.67 mg kg^{-1} for soils of sub-humid southern Zone (IV-b) of Rajasthan. The critical limit of iron in wheat plant is 43.52 mg kg^{-1}. Mahendran et al., (2016) reported that the critical limit of boron was found to be 0.39 mg kg^{-1} in soil and 42.7 mg kg^{-1} in groundnut plant of Madurai district of Tamil Nadu.

4. Factors Affecting Micronutrients Availability in Soil

Micronutrient cations (iron, manganese, zinc, copper, and nickel) and anion (B and Mo) are each influenced in a characteristic way by the soil environment Figure 4.

- *Soil pH:* The micronutrient cations are most soluble and available under acid conditions. In fact, under these conditions, the soil solution concentrations or activities of one or more of these elements (most commonly manganese) may become sufficiently high as to be toxic to many plants. If the pH increases, the ionic forms of the micronutrient cations change first to the hydroxyl ions and, finally, to the insoluble hydroxides or oxides of the elements. The principle is the same at low pH values, the solubility of micronutrient cations is high, and as the pH is raised, their solubility and availability to plants decrease. Acid soil conditions are most conducive for Mo, B deficiency. Similarly, high pH soils such as sodic soils show Zn, Fe, Mn and Cu deficiency.
- *Soil type:* Boron is available in more amounts in the coarse fractions, but fixed in fine textured soils. The soils whose parent material originally contains low amounts of micronutrients are most likely to show micronutrient deficiency under leaching and intensive cropping. Montmorillonite adsorbs more Zn and Cu.

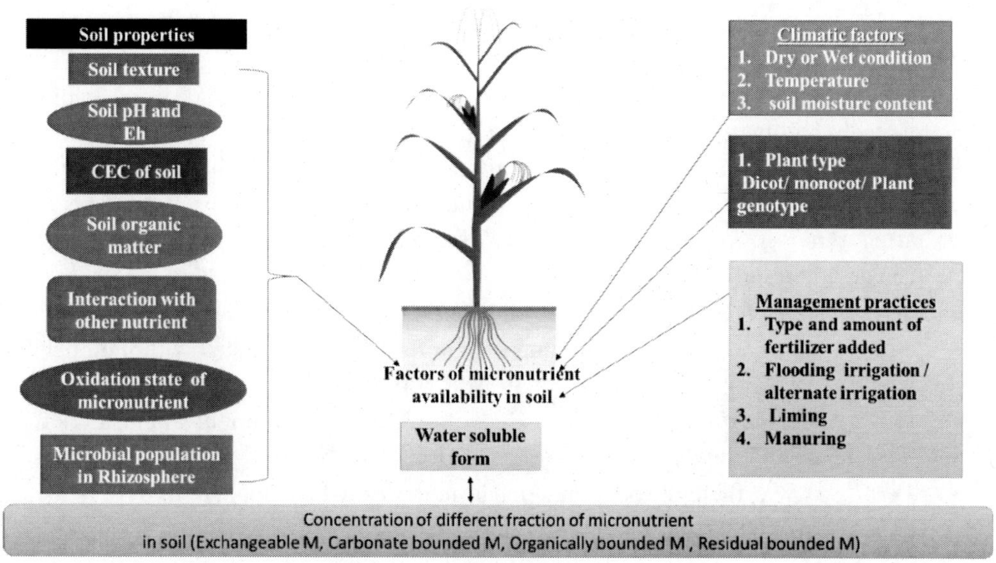

Figure 4. Factors affecting micronutrients availability in soil.

- *Interaction of soil reaction and aeration:* The interaction of soil acidity and aeration in determining micronutrient availability is of great practical importance. Iron, manganese, and copper are generally more available under conditions of restricted drainage or in flooded soils. Rice paddies established on red iron-rich upland soils after terracing and flooding provide an example of how dramatically reduction reactions can increase in the solubility of iron—and therefore its mobility and possible availability and even toxicity to plants. The influence of aeration can be quite complex in the case of zinc. Well-oxidized calcareous soils combine high soil pH levels with good aeration. These characteristics commonly cause plant deficiencies of available iron, zinc, or manganese even though adequate total quantities of these trace elements are present in the soil. Waterlogged conditions reduce the availability of Zn and Cu and increase the availability of Fe and Mn.
- *Other inorganic reactions:* Micronutrient cations interact with silicate clays in two ways. First, they may be involved in cation exchange reactions much like those of calcium or aluminum. Second, they may be tightly bound or fixed to certain silicate clays, especially the 2:1 type. Zinc, manganese, and iron ions sometimes occur in the crystal structure of these clays. Depending on conditions, they may be released from the clays or fixed by them in a manner similar to that by which potassium is fixed. The application of large quantities of phosphate fertilizers can adversely affect the uptake of both iron and zinc.
- *Organic matter:* Organic matter, organic residues, and manure applications affect the immediate and potential availability of micronutrient cations. Certain water-soluble organic compounds form organometallic complexes that enhance micronutrient availability. Other organic compounds react with micronutrient cations to form water insoluble complexes that protect the nutrients from interactions with mineral particles that can bind them in even more insoluble forms. Deficiencies of copper and, to a lesser extent, manganese are often found on poorly drained soils high in organic matter (e.g.,

Histosols and other wetland soils). Zinc is also retained by organic matter, but deficiencies stemming from this retention are not common.

- *Plant type:* There are marked differences among plant species and strains with regard to their ability to obtain micronutrients from low-solubility sources in soils. One example is seen in the sensitivity of different plant varieties to iron deficiency in soils with high pH. Certain plants can apparently solubilize iron in the immediate vicinity around their roots. HYV require more nutrients depleting the nutrients from the soil causing their deficiencies.
- *Over liming:* Over liming in acid soil often leads to deficiencies of iron, manganese, zinc, copper, and sometimes boron. Such deficiencies associated with high pH occur naturally in many of the calcareous soils of arid regions.
- *Imbalanced nutrient application:* High doses of macronutrient application induce micronutrient deficiencies.
- *Fertilizers:* Fertilizers containing macronutrients when applied to soil induce the deficiency of micronutrients. Basic nitrate salts reduce iron availability.

5. Critical Limits of Micronutrients in Different Crops

5.1. Physiological and Biochemical Studies in Plants

Table 7. Forecasting of critical limit of micronutrients in different soil

Micronutrients	Crop	Critical concentration (mg kg-1)
Zn	cereals	15
Zn	Millets	15-20
Zn	Legumes	7-20
Zn	Vegetable french bean	36
Zn	oilseed	12-25
B	Cereals	4-10
B	Millets	7-15
B	Legumes	3-15
B	Vegetables	3-5
B	oilseed	5-10
Mn	Cereals	25
Mn	Millets	10
Mn	Legumes	10-35
Mn	Vegetables	30-40
Mn	oilseed	5-18
Cu	Cereals	2-4
Cu	Millets	2-3.5
Cu	Legumes	4-8
Cu	Vegetables	2-6
Cu	oilseed	2-10
Ni	Barley	2.40
Ni	Soybean	0.22
Ni	Wheat	0.37

Sustained intensive investigations have helped in identifying useful physiological (dry matter partitioning, flowering/fruiting pattern) and biochemical (enzymes) indicators in resolving micro- and secondary nutrient deficiencies in crops including cereals, vegetables, fruits and ornamental plants. Under Zn deficiency stress, activities of catalase and carbonic anhydrase enzymes decreased and those of peroxidase and superoxide dismutase increased in the reproductive parts and immature seeds of mustard (Shukla and Tiwari, 2016) (Table 7). Net photosynthesis in plants was reduced due to the disturbance in the activity of carbonic anhydrase (CA), as CA limits the CO_2 fixation.

5.2. Deficiency Symptoms of Micronutrients in Plants

- *Zn:* Zinc deficiencies mainly occur when soil pH is high, high organic matter in soil, calcareous soils with high bicarbonate content, intensively cropped soils. Symptoms are common on younger or middle aged leaves. Brown to dusty brown spots on younger leaves in red soils, yellowing of leaves/midrib bleaching. Zinc deficiency has also been associated with high bicarbonate content, a Mg: Ca ratio in soils>1, intensive cropping, use of high yielding cultivars, and irrigation with alkaline water. Young mature or old leaves depending on plant show yellowing, usually interveinal, rosetting, and dwarfing of leaves, broad white band on either side of a corn leaf.
- *Fe:* Deficiency of iron causes chlorosis between the veins of leaves and the deficiency symptom show first in the young leaves of plants. It does not appear to be translocate from older tissues to the tip meristem and as a result, growth ceases. The main important deficiency symptoms are interveinal yellowing and chlorosis of emerging leaves, less dry matter production, reduced sugar metabolism enzymes, plants becoming stunted with narrow leaves, the entire leaf being white or yellow if severe. Roots grow profuse root hairs, bottle-brush effect.
- *B:* Deficiency occurs under moister stress and dry condition which cause reduced plant height. Plants fail to produce panicles if they are affected by B deficiency at the panicle formation stage. The tips of emerging leaves are white and rolled. Growing points of shoots and roots die. Reddish young leaves. Malformed buds, necrosis of internal tissues in fleshy stems, tubers, seeds. Stubby, bushy root system.
- *Mn:* Manganese deficiency is very common in upland, degraded soil high in Fe content, accumulation of H_2S, acid sandy or acid sulphate soil, excessive liming in acid soil etc. Symptoms include Young leaves, interveinal yellowing of leaves with smallest leaf veins remaining green to produce a chequered' effect tissue near veins remains green or mottled appearance on dicots. The plant may seem to grow away from the problem so that younger leaves may appear to be unaffected (Das, 2014). Brown spots may appear on leaf surfaces, and severely affected leaves turn brown and dead tissue in later stages. No lateral roots.
- *Cu:* Deficiency symptoms Main important deficiency symptoms of copper are chlorotic leaves, bluish green leaves, new leaves that don't unroll and leaf tips giving needle like appearance, reduced tillering, less pollen viability. Dieback of stems and twigs, yellowing of leaves, stunted growth and pale green leaves that wither easily.

Young leaves yellow, rolled or dead leaf tips, stunted leaves and plants. Drooping leaves on trees.

- *Mo:* Deficiency symptoms of Mo May mimic N deficiency (older leaves become chlorotic). Necrotic spots are seen at leaf margins because of NO3 accumulation. Other symptoms of Mo deficiency include pale leaves that may be scorched, cupped, or rolled narrow, whip-like leaves in brassicas, whitish color and leaf tip death in grasses (Das and Avasthe, 2015). Leaves may also appear thick or brittle, and will eventually wither, leaving only the midrib.
- *Ni:* Deficiency symptoms of Ni Newest leave chlorotic, small "mouse ear" leaves on pinnate-leafed trees, death of meristems.

5.3. Diagnosis Techniques to Examine the Micronutrients Deficiency in Plants

Micronutrient constraints for sustainable agricultural productivity or soil health due to their deficiencies have been diagnosed through (a) characteristic visual symptoms, (b) plant analysis, (Shukla et al., 2018a). Plant or tissue analysis has been found to be more useful in predicting micronutrient nutritional disorders in horticultural crops, especially fruit trees. Techniques for diagnosing micronutrient deficiencies are described below.

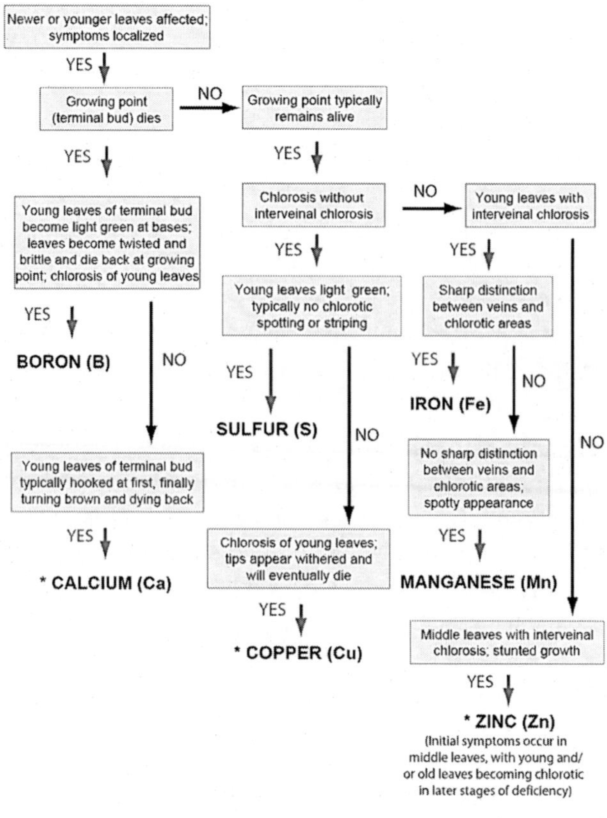

Figure 5. Visual diagnosis of micronutrient deficiency symptoms in crops.

Figure 6. Visual diagnosis of micronutrient deficiency symptoms in crops.

5.4. Visual Diagnosis in Plants

Diagnostic techniques of visual deficiency symptoms (nutritional disorder) of micronutrients on specific plant parts, i.e., on leaves, stem and root. Using refined sand culture techniques, typical micronutrient deficiency symptoms in cereals, pulses, oilseed and horticultural crops have been characterized and catalogued in Figure 5. The same has been published in the form of research bulletins/books (Shukla et al., 2012; Shukla et al., 2016). Visual symptoms of micronutrient deficiencies on various crops under field conditions have also been recorded, categorized and published. Some examples of the recently developed visual micronutrient deficiency symptoms are given in Figure 6.

6. Mitigation Strategy for Micronutrients Deficiency

Deficiencies in soils can be addressed, and the micronutrient balance can be restored, employing a variety of items available on the market that assist in restoring soil balance and enhancing micronutrient content. Different techniques which are employed to alleviate the micronutrient deficiencies are listed below:

- Straight micronutrient
- Chelated micronutrient
- Fortified and Coated fertilizer
- Customized fertilizer
- Fritted or slow-release fertilizer
- Nano-micronutrient fertilizers
- Bio release smart fertilizer
- Use of organic matrix
- Biofortification
- Fertilizer based on biological material.
- Water soluble fertilizer

6.1. Type of Micronutrient Fertilizer

6.1.1. Straight Micronutrient

They are most typically found as sulphate salts of minerals such as zinc sulphate and iron sulphate, which are water soluble. They come in crystalline and granular forms. They can be disseminated or banded into the soil, foliar sprayed and even applied by irrigation (fertigation). Straight fertilizers can be used in conjunction with primary fertilizers or agrochemicals to reduce application costs. However, there is a chance that nutrients will be separated, resulting in an uneven supply of nutrients to the plant. Sulphates are widely used because of their relative efficacy, low cost, widespread availability, and ease of handling. Micronutrient oxides are another source of micronutrients that are less expensive per unit of metal than sulphates but are water insoluble. As a result, when applied in granular form, the immediate effect of oxide may be minimal. To maximize soil contact for optimal uptake, finely milled oxide should be distributed uniformly and thoroughly into the soil. Oxysulpahtes are the oxides of copper, manganese and zinc, which are partially acidulated with sulphuric acid. The solubility of oxysulphates in water increases with increasing acidulation. There is detail information is given in Table 8.

Table 8. List of various micronutrients fertilizer products available in Indian market

Type of micronutrients	Trade name of micronutrients	Constituent (%nutrients)	Manufacturer
Straight	Anmol	52% (Ammonium molybdate)	Agri Search (I) Pvt. Ltd
Straight	Zealbor	20% (Boron)	Agri Search (I) Pvt. Ltd
Chelated	Lohadrip	6% (FeEDDHA)	Agri Search (I) Pvt. Ltd
Chelated	Amoga	12% (ZnEDTA)	Agriparie-Commerce Pvt., Ahmedabad
Straight	Active Boro	10.5%(B)	Amruth organic fertilizers, Karnataka
Straight	Uttam Zinc	33% (Zinc sulphate monohydrate)	Chambal fertilizer
Chelated	Volgamin-12	11% (FeEDTA)	Greencare Biosciences, Hyderabad
Micronutrient mixture	Mahadhan Combi	Cu @ 1%, Fe @ 2.5%, Mn @ 1%, Zn @ 3%, B @ 0.5%.	Mahadhan, Maharashtra
Straight	Copsomin25	24% (Cu) as copper sulphate	Greencare Biosciences, Hyderabad
Micronutrient mixture	Phasal Chakra	Zn@5%, Mn@2%, Cu@1% and B@1%	Pk Navratana Agritech, Pvt. Ltd
Micronutrient mixture	Balwan micro fertilizer	Zn@10%, Mn@2%, Cu@1%, Fe@5% and B@1%	Pk Navratana Agritech, Pvt. Ltd
Nano fertilizer	Geolife NanoZn	12%	Geolife Agritech India Pvt.Ltd
Nano fertilizer	Geolife Nano Combi	Zn+Mn+Cu (16.8 +3.8+3.8%)	Geolife Agritech India Pvt. Ltd.
Nano fertilizer	Geolife NanoFe	12%	Geolife Agritech India Pvt. Ltd
Nano fertilizer	Geolife NanoMn	26.6%	Geolife Agritech India Pvt. Ltd
Nano fertilizer	NanoBor	20%	Alert Biotech, Nasik
Customized	-	N (8%), P (16%), K (24%), S (6%), Zn (0.5%) and B (0.15%)	Tata Consultancy Limited
Customized	Grasim	N (14%), P (27%), K (10%), Zn(4%) and B(0.5%)	Indo-Gulf fertilizer

6.1.2. Chelated Micronutrient

They are a class of products in which a micronutrient is bonded to an organic ligand to boost plant availability. A chelate is an organic molecule in which two or more atoms can form a ring around a single metal atom. A ligand is a name for an organic molecule. These organic compounds may be synthesized synthetically, synthesized by the roots and released into the soil, or present in soil organic matter (SOM). The cations are shielded from interactions with inorganic soil components and so made available for plant uptake in this complex state. Chelating isn't possible for all nutrients. Others, such as iron, zinc, copper, manganese, calcium, and magnesium, cannot be chelated. Chelates are made from a variety of organic compounds (chelating agents). The most popular synthetic chelating agent is EDTA, which is used to chelate both soil and foliar nutrients. DTPA is primarily utilized in chelates for alkaline soils. Although it is more effective than EDTA, it is frequently more costly. The most efficient iron fertilizers on high pH soils are iron chelates produced with HEDTA and EDDHA, but they are also the most expensive (Shukla and Behera, 2017). Compounds like gluconate and citrate, which are naturally secreted from roots, are another cost-effective option for employing chelates. Chelated versions of micronutrient elements (e.g., Zn-EDTA) are known to be very effective because they protect the nutrient until it is used by the plant and prevent loss owing to soil fixation. Chelated micronutrients are more effective per unit of micronutrients, but they are also more expensive, limiting their use to high-value crops such as horticulture crops.

6.1.3. Mechanism of Chelation in Soil

Cells, in general, take up metals such as Zn and Mn from the soil solution only if they are present as free ions and not in the form of metal-organic complexes (Hassler et al., 2004), according to the free ion activity model (FIAM) and the biotic ligand model (BLM). The Fe chelate reductase enzyme (FCR) performs Fe uptake in all higher plants except gramineous (Strategy-I plants), which activity requires the availability of Fe in soil solution as a Fe chelate for further reduction and uptake on the root surface (Römheld and Marschner, 1986). As a result, distinct biochemical processes are involved in micronutrient uptake methods, which differ when considering soil versus hydroponic solutions. Figure 7 depicts a schematic representation of these operations. Chemical reactions limit the chelate mixtures' ability to stay in the nutrient/soil solution once they've been applied to the system.

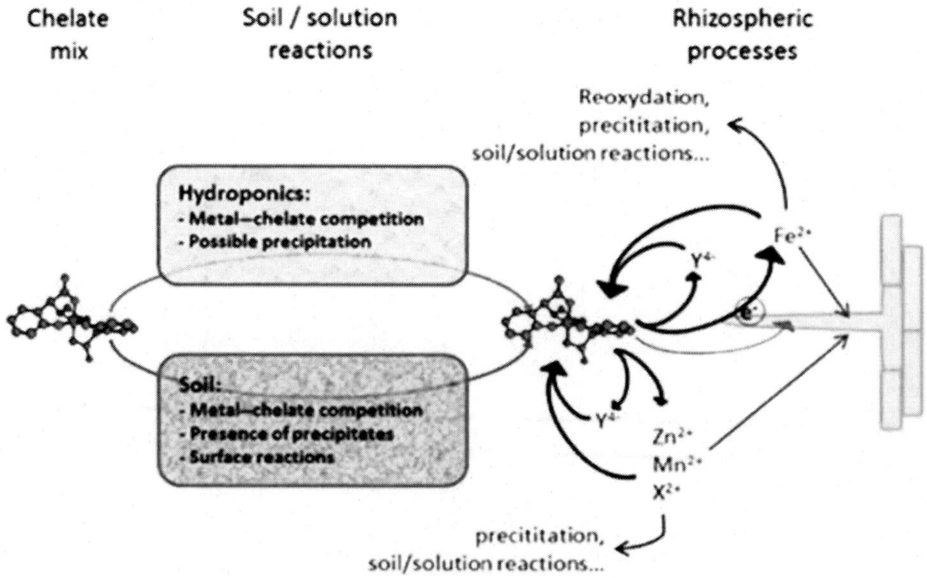

Figure 7. Chelation mechanism of micronutrients fertilizer in the soil.

6.1.4. Fortified and Coated Fertilizer

They're made by adding one or more micronutrients to ordinary fertilizers. Micronutrients can be mixed with NPK fertilizers by coating a micronutrient layer on granular primary fertilizers, adding micronutrients during manufacturing, or coating granular primary fertilizers with micronutrients. The use of fortified fertilizers allows for the uniform distribution of micronutrient fertilizers in tiny amounts, saving money on the cost of additional soil placement, broadcasting, or spraying operations. If direct micronutrient fertilizers are physically combined with primary fertilizers and applied concurrently, they may segregate, resulting in an unequal distribution of nutrients to plants. As a result, coating a primary fertilizer with a micronutrient element can aid in ensuring uniform nutrient supply to plants without the need for a second application process. There are 26 different fortified fertilizer formulas permitted, with zincated urea (2% Zn) and boronated SSP (1.2% B) being the most common.

6.1.5. Customized Fertilizer

They are multi-nutrient carriers that contain macro, secondary, and/or micronutrients from both inorganic and organic sources and are developed for certain crops and regions. Micronutrient mixtures are a step behind customized fertilizers. Fertilizers manufactured through fusion-blend granulation contain primary, secondary, and micronutrients. These are made through a systematic granulation process with strict quality controls, meeting the nutritional needs of the crop, and are unique to the site, soil, and stage, as validated by a scientific crop model, and designed by an accredited fertilizer manufacturing/marketing company. They also have water soluble specialty as a customized combination products (Rakshit et al., 2012). For crops such as rice, wheat, oil palm, sugarcane, chili, and potato, the Government of India has declared 34 tailored fertilizers for about 100 districts in the states of Andhra Pradesh, Telangana, Maharashtra, Uttar Pradesh, Uttarakhand, Tamil Nadu, and Karnataka. These fertilizers are thought to be the greatest alternative for correcting site-specific multi-nutrient deficits in soils and achieving optimum crop output through better nutrient usage efficiency (Table 9).

6.1.6. Fritted or Slow-Release Fertilizer

Frits are made by fusing one or more powdered micronutrient sources with silicates in a furnace at around 1350°C, resulting in a vitrified homogenous substance that is ground into a powder. Because of the gradual weathering of silicates, which may be controlled by the fineness of grinding and the inclusion of calcium, sodium, or potassium into a silicate matrix, it aids in the slow release of micronutrients. These products were intended to be used on coarse-textured soils in high-rainfall locations to prevent leaching. Frits' slow-release feature ensures a constant supply of the micronutrient source with no waste. Frits are more suited to maintenance applications than to alleviate the deficiency. Boron glass frits, boron phosphate, colemanite, and ulexite are some examples of fritted fertilizers.

6.1.7. Nano Fertilizers

The usage of Nano clay polymer composite (NCPC) to provide micronutrients slowly and steadily as per a plant's needs can be used to improve the efficiency of micronutrient fertilizers (Table 9). To boost the soil's health and vigor, nano-micronutrient formulations are sprayed on plants or supplied to the soil for root intake. However, research shows that going beyond traditional fertilization can have a positive influence (Abu-Elsaad and Abdel Hameed, 2019).

Table 9. Indicative list of beneficial effects of micronutrients fertilizer application in various agro-climatic zones of the India

Micronutrients fertilizer	Type of micronutrients fertilizers	Method of application	Crops	Amount	Benefits	Reference
Zinc sulphate heptahydrate	Straight	Soil and foliar	Rice, sesamum, wheat and maize	25 kg ha-1(soil) and 0.5% as foliar spray + 0.25% lime	Increase in no of panicle length, grain yield and no of spikes in wheat and corn yield	Tandon. (2013)
Ferrous sulphate hepthahydrate	Straight	Foliar	Sugarcane, pineapple, groundnut and rice	0.5-1%	Increase in sugar yield and no of pods in groundnut	Tandon. (2013)
Multi-micronutrient mix (Zn+B+Fe+Mn+Cu)	Multi-micronutrient	Pre plant soil	Okra, potato and cabbage	20 kg ha-1	Significant increase in yield	Patel et al. (2009)
Sodium molybdate	Straight	Soil and foliar	Soybean and chickpea	0.5 to 1 kg ha-1 and 0.1% as foliar spray	Increase in seed yield	Masood and Venkatesh (2009)
Manganese sulphate monohydrate	Straight	Seed priming, Soil and foliage	Wheat	0.01-0.10 mol L-1, 10-40 kg hm-2 and 0.5-1%	Increase in grain content and yield	Ullah et al. (2018)
Boric acid	Straight	Seed priming and foliage	Wheat and rice	0.1% (wheat), 0.001%-0.010% (rice)	Increase in yield and grain boron content	Nadeem et al. (2019)
Customized (N 20: P 10 : K 10 : S 5: Mg 2 : Zn 0.5 : B 0.3 : Fe 0.2)	Customized	Basal	Pomegranate	1000 kg ha-1	Increase in number of male and female flowers, yield of "A" grade fruits, yield of total marketable fruits, fruits size, aril percentage and juice percentage.	Goel et al. (2011)
Zinc oxide nano particle	Nanofertilizer	Seed treatment	Cowpea and soybean	250,750 and500ppm	Increase in seedling length, germination % and vigour index	Srinivasan et al. (2017)
Copper oxide nanoparticle	Nanofertilizer	Seed priming	Rice and maize	25-1000ppm	Increase in seedling length and germination%	Yang et al. (2015)
Zinc oxide nano particle	Nanofertilizer	Foliar	Foxtail millet	0.00026% (2.6 kg ha-1)	Increase in plant height, seed head length and oil content	Kolencik et al. (2019)
Micromax (6.00% Ca, 3.00% Mg, 12.00% S, 0.10% B, 1.00% Cu, 17.00% Fe, 2.50% Mn, 0.05% Mo and Zn 1.0%)	Control release fertilizer	Soil	Chinese Fan (Livistona chinensis) and queen (Syagrus romanzoffiana)	1.4 kg ha-1	Increase in basal diameter, leaf length and leaf width	Soti et al. (2015)

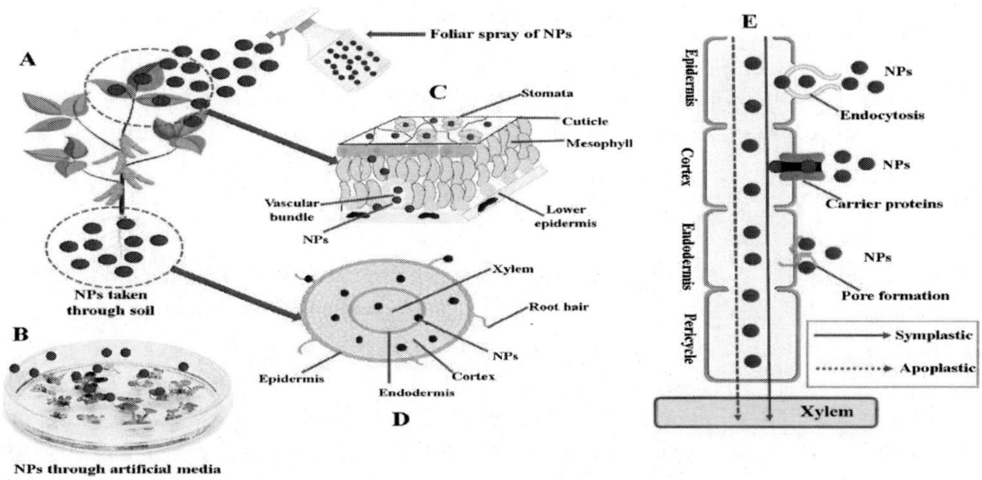

Figure 8. Schematic illustration of the different routes of uptake, entry and translocation of nanoparticles (NPs) into plant cells and organs (Anjum et al., 2019).

Duhan et al. (2017) reported that most microelements have been used to examine the efficacy of utilizing nano-fertilizers. Fe deficiency, for example, is a typical problem in plants that grow in high pH and calcareous soils. This difficulty could be solved by using nanoparticle technology to apply Fe nutrients to the leaves. The dosages are in the range of 0.01% to 0.04% because the contents of the iron nano-oxide solution are low. Laware and Raskar. (2014) reported that the usage of ZnO nanoparticles can impact flowering time, for example, onion flowering time is reduced by 12–14 days. Similarly, Hafeez et al. (2015) reported that the wheat growth and yield could be boosted by copper nanoparticles. Chlorophyll content, leaf area, and other metrics were all altered by their presence. In Figure 8, schematic illustration of the different routes of uptake, entry and translocation of nanoparticles (NPs) into plant cells and organs (Anjum et al., 2019).

6.1.8. Bio Release Smart Fertilizer

Bio-release smart fertilizer is a novel type of slow-release fertilizer that consists of a polyphosphate, one micronutrient from the group of B, Mo, or Cl, and one micronutrient from the group of Zn, Fe, Cu, and Mn. Organic acids generated by plants and microorganisms, not soil moisture, control the release of nutrients. These compounds are characterized by the fact that they are a solid, free-flowing, non-hygroscopic powder fertilizer that is insoluble in water but soluble in 2% citric acid, supplying nutrients for plant uptake. The fact that it is water insoluble avoids losses from leaching and conversion to non-available forms. Plants can absorb bio-release micronutrients at a lower dose and with a higher nutrient content throughout the crop cycle. When compared to standard micronutrient fertilizers, the approach claims to reduce dosage by 10% to 25% (Shukla and Behera, 2017).

6.1.9. Organic Matrix

The organic matrix allows nutrients to be released slowly. It is not necessary to cover the product because it is homogeneous. The product has an organic matrix with both positive and negative charges that hold cations such as ammonium, potassium, zinc, and iron, as well as anions such as sulphate and phosphorous. Nutrient binding to the organic matrix reduces the rate at which nutrients react with the soil. The soil microorganisms then begin to break down the matrix, gradually releasing nutrients for plant uptake. This method offers a cost-effective way to deliver key micronutrients in a base nutrient package while also improving soil tilth and water retention capacity. Entrapment of micronutrients in the organic matrix is one such example.

6.1.10. Biofortification

Biofortification is a nutrient enrichment technique in which nutrients is concentrated or made more accessible during plant growth. Because staple meals are very inexpensive and accessible to the general public, the biofortification of staple crops is critical for adequate micronutrient content (Riaz et al., 2020). Different strategies for the biofortification in the crops include agronomy, genetic, breeding and omics biofortification techniques are available. In Figure 9, the biofortification technique for micronutrients is shown and is given by Arpita et al. (2018).

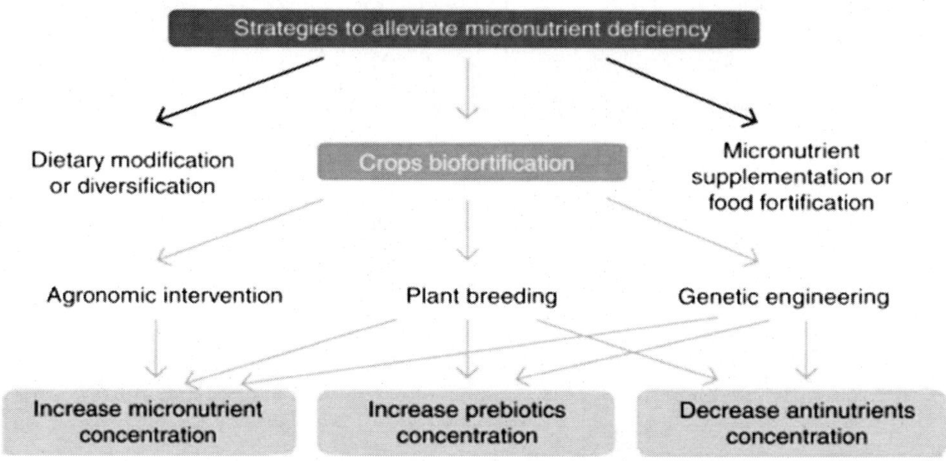

Figure 9. Biofortification techniques for micronutrient (Arpita et al., 2018).

6.1.11. Agronomic Biofortification

The deliberate application of mineral fertilizers to enhance the concentration of a target mineral in edible sections of crops in order to improve dietary intake of the target mineral is known as agronomic biofortification (Table 10). In resource-poor nations, agronomy biofortification is the most effective and accessible strategy for combating micronutrient deficiency (Ray et al., 2016). For the successful incorporation of micronutrients in crops, agronomic biofortification uses two approaches. The first is soil micronutrient application, while the second is foliar exertion. When these strategies are used together, however, they can produce better results (Riaz et al., 2020).

6.1.12. Conventional Plant Breeding

Traditional breeding is the most suited method of biofortification since it reduces the costs of transgenic instruments and agronomic methods (Table 11). For a realistic selection of amplified nutritional composition, genotypic heterogeneity among the crop of concern is required. The parent lines with the highest levels of minerals and vitamins are chosen and crossed with recipient lines in this process. This technique is repeated until the required results in the form of nutrient-concentrated plants are obtained. If there is not enough disparity and desired features are lacking, this can be overcome by crossing lines with close relatives. The characteristic is gradually introduced into commercial varieties/cultivars during this process (Garg et al., 2018).

Table 10. Effect of micronutrients application on grain yield, grain Zn/Fe/Mn concentration of different groups of cultivars at different locations

Crops	Efficient cultivar				Inefficient cultivar			
	Zinc (Zn)							
	Grain Yield ($t\ ha^{-1}$)		Grain Zn(ppm)		Grain Yield ($t\ ha^{-1}$)		Grain Zn(ppm)	
	-Zn	+Zn	-Zn	+Zn	-Zn	+Zn	-Zn	+Zn
1-IISS, Bhopal								
A Pigeon pea	1.41	1.54	32.6	43.8	1.06	1.41	35.1	48.2
B. Wheat	3.72	3.87	41.0	47.8	2.85	3.37	43.0	56.3
2-ANGRAU, Hyderabad								
A. Rice(dehusked)	5.98	6.18	11.0	16.7	5.36	7.92	9.5	16.9
B. Maize	5.04	6.13	24.2	27.4	4.39	6.59	23.7	29.5
3-GBPUAT, Pantnagar								
A. Rice(dehusked)	3.94	5.92	13.1	26.8	3.94	5.92	13.1	26.8
B. Wheat	3.71	3.95	20.3	43.1	3.26	4.23	15.1	43.8
	Iron (Fe)							
	Grain Yield ($t\ ha^{-1}$)		Grain Fe(ppm)		Grain Yield ($t\ ha^{-1}$)		Grain Fe(ppm)	
	-Fe	+Fe	-Fe	+Fe	-Fe	+Fe	-Fe	+Fe
4-AAU, Anand								
A Pigeon pea	2.50	2.42	34.1	36.0	2.27	2.55	33.7	38.5
B. Chickpea	3.15	3.27	59.0	62.8	2.36	2.91	56.0	67.5
5-RAU, Pusa								
A. Rice(dehusked)	5.04	15.6	21.4	81.7	2.99	4.06	13.8	24.3
B. Maize	5.19	5.55	46.8	66.2	5.22	6.22	41.33	63.2
	Manganese (Mn)							
	Grain Yield ($t\ ha^{-1}$)		Grain Fe(ppm)		Grain Yield ($t\ ha^{-1}$)		Grain Fe(ppm)	
	-Mn	+Mn	-Mn	+Mn	-Mn	+Mn	-Mn	+Mn
6-PAU, Ludhiana								
A. Rice	6.71	6.85	41.4	53.0	4.83	5.51	31.6	44.2
B. Wheat	5.02	5.45	25.0	33.1	4.24	5.20	19.9	30.2

6.1.13. Genetic Engineering

To cope with micronutrient deficiency, the genetic engineering method modifies one or more genes. Molecular approaches are used to relocate a specific genetic characteristic In comparison to traditional plant breeding, genetic modification provides two advantages. To begin with, it takes less time to produce a crop that consistently displays the desired feature (e.g., nutritional content). Second, it permits the transmission of a certain gene.in genetic modification from donor to recipient organism so that it expresses a particular trait (Ye et al., 2000).

6.1.14. Omics Technology

Modern improvements in omics analysis and analytical methodologies have resulted in gradual molecular alterations in research focusing on micronutrient biofortification (Winkel et al., 2012; Harris et al., 2014; Visioli et al., 2015). New perspectives in "omics technologies" (e.g., transcriptomics, genomics, proteomics, nutrigenomics, and metabolomics) are emerging, and cutting-edge analytical instruments such as chemical and elemental mapping by micro focused X-ray fluorescence and X-ray absorption near-edge structure spectroscopy are now available to reveal further speciation of micronutrients in association with the specific nutrient (Méplan, 2011; Winkel et al., 2012; Visioli et al., 2015).

Table 11. Genetically and agronomically Zn-efficient cultivars of different crops

IISS, Bhopal				ANGRAU, Hyderabad				GBPUAT, Pantnagar			
Pigeon pea		Wheat		Rice		Maize		Rice		Wheat	
Genetically efficient	Agronomically efficient	Genetically efficient	Agronomically efficient	Genetically efficient	Agronomically efficient	Genetically efficient	Agronomically efficient	Genetically efficient	Agronomically efficient	Genetically efficient	Agronomically efficient
C11	Hisar H02-60	GW-322	C-306	JGL11470	JGL-11727	Super9681	Harsha	Pant Dhan-18	Jaya	UP2554	UP 262
ICPL 87119	Hisar Manak	JW-3211	JW-17	MTU-7029	JGL-11118	30Y92	Laxmi 4950	Pant Sugandh	Pant Dhan-19	K9107	PBW590
GT-101	Hisar Paaras	HW-147	HW-2004	Erramallelu	MTU-1001	DHM-111	NK-6240	Kala namak	Pant Sankar-1	UP2628	VL804
T15-15	SKNP 05-05	HI-8627	AKW-4627	WGL-32100	SUREKHA	DHM-117	Ashwini	Pant Dhan-16	Basmati 370	PBW- 502	WH147
BSMR 853	GAUT 93 -17			BPT 5204	NLR33892			Pusa Sugandh 4	Type 3	UP256	WH1021
Virsa Arhar1	PKV Trombay									PBW 502	UP2572

Fe efficient cultivar								Mn efficient cultivar			
AAU, Anand				RAU, Pusa				PAU, Ludhiana			
Pigeon pea		Wheat		Rice		Maize		Rice		Wheat	
Genetically efficient	Agronomically efficient	Genetically efficient	Agronomically efficient	Genetically efficient	Agronomically efficient	Genetically efficient	Agronomically efficient	Genetically efficient	Agronomically efficient	Genetically efficient	Agronomically efficient
DT-23	BSMR853	GJG-506	ICCC-4	Birsa-mati	Boro3	Dewaki	Shakti-man-3	PAU 201	Pusa 44	BW 8989	PDW 291
AAUT 2007-4	BP-1-96	GG-1	GAG 839	RAU 759	Jeera-bati	Hemant	Shakti-man-4	3047	3138	BW 9149	PDW 314
BDN-2	C-11	GAG-838	GJG-305	Sanwal Basmati	Rajen-dra Kasturi	Pop-64	Lakshmi	PR 116	3140	PBW 550	BW 9-022
PKV Trombay	AAUT 2007-10			Swarna Sub-1	Rajendra Subhashini	Rajendra HM1	CM-400	3131	3141	PBW 636	BW-9178

"Omics" methods, combined with recent advances in analytical procedures and approaches, offer new perspectives on understanding the mechanisms of micronutrient biofortification, as well as identifying and possibly developing novel micronutrient transporters to improve plant micronutrient uptake and accumulation.

6.1.15. Fertilizer Based on Biological Material

Biological materials with a natural tendency to bind micronutrient ions could be a potential alternative to routinely used fertilizers. Functional groups (hydroxyl, phosphate, or amino groups) have an affinity for metal cations containing microelements, producing metal complexes, which has been discovered in the development of slow-release fertilizers (Arslanoglu, 2019). As a result, biosorption of micronutrient cations on bio-based materials provides an intriguing technique for producing control release fertilizer. Micronutrients would be delivered *via* a biodegradable carrier in these fertilizers. Biomass-based fertilizers have better bioavailability and a slower release of nutrients into the soil. This sort of bio-based fertilizer is completely biodegradable, reducing the environmental impact of fertilizers. When comparing the efficiency of biosorption fertilizers to conventional fertilizers, it was discovered that biosorption fertilizers are easily absorbed by plants and have a steady release of nutrients. In comparison to plants fed with traditional reference fertilizers (inorganic salts and chelate), the crop yielded by zinc-fertilized plants was higher (Tuhy et al., 2014). Immobilization of biosorbents in a polymer matrix, such as bio-based hydrogels, is one example. In Figure 10, a schematic representation of microelement release from hydrogels containing immobilized biomass is shown (Mikula et al., 2020).

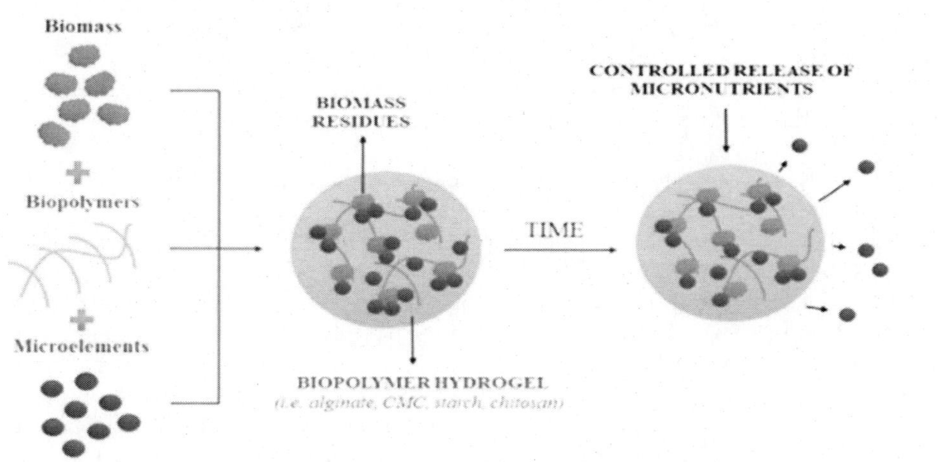

Figure 10. A schematic representation of microelement release from hydrogels containing immobilized biomass is shown (Katarzyna et al., 2019).

6.1.16. Water Soluble Fertilizer

Water-soluble fertilizers dissolve easily in water and are delivered to the crop in the irrigation water. One advantage of this method of fertilization is that gardeners may readily modify nutrient concentrations as a crop's needs change during the growing season. They increase the uptake of nutrients, promote root and shoot development being acidic aids in the removal of salts from drippers in drip irrigation systems. They can be used for a variety of crops as well as can be applied at any stage. Some of the examples of it are WSF-19-19-19 contains Zn, Fe Mn and B and WSF-13-40-13 enriched with micronutrients.

Conclusion

Micronutrients have been systematically extracted from limiting soil resources for a long time without external addition of these minerals through respective fertilizer sources, leaving Indian soils deficient in fertility. Identification of specific micronutrient-deficient and/or toxic areas under major soil types for various cropping systems, as well as the preparation of detailed geo-referenced digitized maps, is helpful in achieving higher sustainable agricultural productivity, nutritional quality, and improved soil and human health. Micronutrient insufficiency and crop response to micronutrient treatment are both variables. With varying soil crop circumstances, the extent of micronutrient deficiency and the responsiveness of crops to micronutrient application differed. To alleviate micronutrient deficit in soils and crops, increase crop yield, and create micronutrient-rich agricultural products, optimum micronutrient management strategies must be adopted.

References

Abu-Elsaad, N. I., and Abdel hameed, R. E. (2019) Copper ferrite nanoparticles as nutritive supplement for cucumber plants grown under hydroponic system. *Journal of Plant Nutrition 42*, 1645-1659.

Alloway, B. J. (2008) *Zinc in soils and crop nutrition*. International Zinc Association, Brussels. International Fertilizer Industry Association, Paris.

Anjum, S., Anjum, I., Hano, C., and Kousar, S. (2019) Advances in nanomaterials as novel elicitors of pharmacologically active plant specialized metabolites: Current status and future outlooks. *RSC Advances 9*, 40404-40423.

Arslanoglu, H. (2019) Adsorption of micronutrient metal ion onto struvite to prepare slow release multielement fertilizer: copper (II) doped-struvite. *Chemosphere* 217, 393-401.

Barman, M., and Datta, S. P. (2018) Assessing phytotoxic limits of nickel in intensively cultivated alluvial soils. *Journal of Environmental Biology*, 39, 358-364.

Behera, S. K. and Shukla, A. K. (2014) Total and extractable manganese and iron in some cultivated acid soils of India: status, distribution and relationship with some soil properties. *Pedosphere* 24, 196-208.

Behera, S. K., Shukla, A. K. and Singh, M. V. (2012) Distribution variability of total and extractable copper in cultivated acid soils and their relationship with some soil properties. *Agrochemical* 4, 28-41.

Behera, S. K., Singh, M. V., Singh, K. N. and Todwal, S. (2011) Distribution variability of total and extractable zinc in cultivated acid soils of India and their relationship with some selected soil properties. *Geoderma* 162, 242-250.

Behera, T. K., and Singh, S. (2019) Advances in vegetable breeding for nutraceuticals and quality traits. *Indian Journal of Genetics* 79, 216-226.

Brown, P. H., Welch, R. M., and Cary, E. E. (1987) Nickel: A micronutrient essential for higher plants. *Plant Physiology* 85, 801-803.

Cate, R. B. and Nelson, L. A. (1965) A rapid method for correlation of soil test analysis with plant response data. International soil testing series technical Bulletin No. I North Caroline. State University, *Agricultural Experiment Statistics*, Releigh (USA). pp. 135-136.

Das, A., Laha, S., Mandal, S., Pal, S., and Siddiqui, M. W. (2018) Preharvest biofortification of horticultural crops. In *Preharvest Modulation of Postharvest Fruit and Vegetable Quality*, 381-434).

Das, S. K. (2014) Role of Micronutrient in Rice Cultivation and Management Strategy in Organic Agriculture—A Reappraisal. *Agricultural Sciences* 5, 765-769.

Das, S. K. and Avasthe, R. K. (2015) Biochar as carbon negative in carbon credit under changing climate. *Current Science* 109, 1223.

Datta, S. P. and Meena, M. C. (2015) Micronutrients in soils and their management. In Soil Science: An Introduction (R. K. Rattan, J. C. Katyal, B. S. Dwivedi, A. K. Sarkar, Tapas Bhattacharyya, J. C. Tarafdar and S. S. Kukal, Eds.), pp. 623-648. *Indian Society of Soil Science*, New Delhi.

Datta, S. P., Meena, M. C., Barman, M., Golui, D., Mishra, R. and Shukla, A. K. (2018) Soil test for micronutrients: current status and future thrust. *Indian Journal of Fertilisers* 14, 32-51.

Debnath, P. and Ghosh, S. K. (2012) Critical limit of available boron for rice in alluvial zone soils of west Bengal. *Indian Journal Agricultural Research* 46, 275– 280.

Duhan, J. S., Kumar, R., Kumar, N., Kaur, P., Nehra, K., and Duhan, S. (2017) Nanotechnology: The new perspective in precision agriculture. *Biotechnology Reports* 15, 11-23.

Eskew, D. L., Welch, R. M., and Cary, E. E. (1983) Nickel: an essential micronutrient for legumes and possibly all higher plants. *Science* 222, 621-623.

Garg, M., Sharma, N., Sharma, S., Kapoor, P., Kumar, A., Chunduri, V., and Arora, P. (2018) Biofortified crops generated by breeding, agronomy, and transgenic approaches are improving lives of millions of people around the world. *Frontiers in Nutrition* 5, 12.

Graham, R. D. (2008) Micronutrient deficiencies in crops and their global significance. In Micronutrient deficiencies in global crop production. Springer, Dordrecht. pp. 41-61.

Hafeez, A., Razzaq, A., Mahmood, T., Jhanzab, H. M. (2015) Potential of copper nanoparticles to increase growth and yield of wheat. *J Nanosci Adv Tech* 1 (1).

Harris, J., Schneberg, K. A., and Pilon-Smits, E. A. (2014) Sulfur–selenium–molybdenum interactions distinguish selenium hyperaccumulator Stanleya pinnata from non-hyperaccumulator *Brassica juncea* (Brassicaceae). *Planta 239*, 479-491.

Kumar, O, Singh, S. K., Singh, A. P., Yadav, S. N., Latare, A. M. and Kumar, Maneesh (2018) Assessing a suitable extractant and critical limits of nickel in soil and plant for predicting the response of barley (Hordeum vulgare L.) to nickel grown in Inceptisols. *Communications in Soil Science and Plant Analysis* 49, 2602-2613.

Laware, S. L., and Raskar, S. (2014) Influence of zinc oxide nanoparticles on growth, flowering and seed productivity in onion. *International Journal of Current Microbiology Science 3*(7), 874-881.

Lindsay, W. L. and Norvell, W. A. (1978) Development of a DTPA soil test for zinc, iron, manganese and copper. *Soil Science Society of America Journal* 42, 421–428.

Mahata, M. K., Debnath, P. and Ghosh, S. K. (2013) Estimation of Critical limit of Zinc for Rice in Terai Soils of West Bengal. *Journal of the Indian Society of Soil Science* 61, 153-157.

Mahendran, P. P., Velmurugan, R. and Balasubramaniam, P. (2016) Identifying critical limit in soil and plant. *International Journal of Current Microbiology and Applied Science* 6, 241-249.

Majumdar, S., and Prakash, N. B. (2018) Prospects of customized fertilizers in Indian agriculture. *Current Science* 115, 242-248.

Meena R. S., Mathur A. K. and Sharma S. K. (2013) Determination of critical limit of iron for wheat in soils of subhumid southern zone (IV-B) of Rajasthan. *Green Farming* 3, 298- 302.

Mehlich, A. (1978) New extractant for soil test evaluation of phosphorus, potassium, magnesium, calcium, sodium, manganese and zinc. *Communications in Soil Science and Plant Analysis* 9, 477-492.

Mehlich, A. (1984) Mehlich 3 soil test extractant: A modification of Mehlich 2 extractant. *Communications in Soil Science and Plant Analysis* 15, 1409-1416.

Méplan, C. (2011) Trace elements and ageing, a genomic perspective using selenium as an example. *Journal of Trace Elements in Medicine and Biology* 25, S11-S16.

Mikula, K., Izydorczyk, G., Skrzypczak, D., Mironiuk, M., Moustakas, K., Witek-Krowiak, A., and Chojnacka, K. (2020) Controlled release micronutrient fertilizers for precision agriculture–A review. *Science of the Total Environment* 712, 136365.

Morgan, M. F. (1941) Chemical Diagnosis by the Universal Soil Testing System. Conneticut Agricultural Experiment Station (New Haven) Bulletin 450.

Mortvedt, J. J. (2000) Bioavailability of micronutrients. *Handbook of soil science.*

Murthy, I. Y. L. N., Padmavathi, P. and Padmaiah, M. (2009) Critical level of DTPA-Zn for castor (*Ricinus communis* L.) in Alfisols. *Agropedology* 19, 139-142.

Rahman, M. A., Jahiruddin, M. and Islam, M. R. (2007) Critical Limit of Zinc for Rice in Calcareous Soils. *Journal of Agricultural and Rural Development* 5, 43-47.

Rakshit, R., Rakshit, A., and Das, A. (2012) Customized fertilizers: Marker in fertilizer revolution. *International journal of agriculture, environment and biotechnology 5*, 67-75.

Ray, P., Datta, S. P., Rakshit, R., and Golui, D. (2016). Agronomic biofortification of food crops with zinc and iron for ameliorating their deficiencies in humans: Constraints and possibilities. *Indian Journal of Fertilisers 12*, 28-35.

Riaz, U., Aziz, H., Anum, W., Mehdi, S. M., Murtaza, G., and Jamil, M. (2020). Biofortification Technologies Used in Agriculture in Relation to Micronutrients. In *Plant Micronutrients*. 225-239.

Sanjeev K. Chaudhary and Raina, J. N. (2008) Zinc transformation of and its critical limits in apple orchards of Himachal Pradesh. *Journal of the Indian society of soil science* 56, 430-435.

Sharma, C. P. (Ed.) (1996) Deficiency Symptoms and Critical Concentration of Micronutrients in Crop Plants. Bulletin, Botany Dept, Lucknow University, Lucknow. Pp. 110.

Sharma, P. C. (2006) *Plant Micronutrients*. CRC Press, USA.

Shukla, A K., Behera, S. K., Pakhare, A. and Chaudhari, S. K. (2018) Micronutrients in soils, plants, animals and humans. *Indian Journal of Fertilisers* 14, 30-54.

Shukla, A. K., and Behera, S. K. (2017) Micronutrients research in India: Retrospect and prospects. In *Preprint, FAI Annual Seminar*.

Shukla, A. K. and Behera, S. K. (2012) Micronutrient fertilizers for higher productivity. *Indian Journal of Fertilisers* 8, 100-117.

Shukla, A. K. and Tiwari, P. K. (2016) Micro- and Secondary Nutrients and Pollutant Elements Research in India - Coordinator Report. AICRP on Micro and Secondary Nutrients and Pollutant Elements in Soils and Plants (2014-16), ICAR-IISS, Bhopal.

Shukla, A. K., and Behera, S. K. (2017) Micronutrients research in India: Retrospect and prospects. FAI Annual Seminar.

Shukla, A. K., and Behra, S. K. (2019) All India coordinated research project on micro-and secondary nutrients and pollutant elements in soils and plants: Research achievements and future thrusts. *Indian Journal of Fertilisers* 15, 522-543.

Shukla, A. K., and Tiwari, P. K. (2016) Micro and secondary nutrients and pollutant elements research in India. Coordinators Report-AICRP on micro-and secondary nutrients and pollutant elements in soils and plants. ICAR-IISS, Bhopal, 1-196.

Shukla, A. K., Behera, S. K., and Singh, G. (2021b) Micronutrient fertilizers in Indian agriculture–product profile, availability, forecast and agronomic effectiveness. *Indian Journal of Fertilisers* 17, 348-360.

Shukla, A. K., Behera, S. K., Lenka, N K., Tiwari, P. K., Prakash, C., Malik, R. S., Sinha, N. K., Singh, V. K., Patra, A. K., and Chaudhary, S. K. (2016) Spatial variability of soil micronutrients in the intensively cultivated Trans-Gangetic Plains of India. *Soil and Tillage Research* 163, 282-289.

Shukla, A. K., Behera, S. K., Pakhre, A., and Chaudhari, S. K. (2018) Micronutrients in soils, plants, animals and humans. *Indian Journal of Fertilisers* 14, 30-54.

Shukla, A. K., Behera, S. K., Prakash, C., Patra, A. K., Rao, C. S., Chaudhari, S. K., Das, S., Singh, A. K., and Green, A. (2021a) Assessing multi-micronutrients deficiency in agricultural soils of India. *Sustainability* 13, 9136.

Shukla, A. K., Behera, S. K., Satyanarayana, T., and Majumdar, K. (2019) Importance of micronutrients in Indian agriculture. *Better Crops-S Asia* 11, 6-10.

Shukla, A. K., Malik, R. S., Tiwari, P. K., Prakash, C., Behera, S. K., Yadav, H., and Narwal, R. P. (2015) Status of micronutrient deficiencies in soils of Haryana. *Indian Journal of Fertilisers* 11, 16-27.

Shukla, A. K., Tiwari, P. K., and Prakash, C. (2014) Micronutrients deficiencies vis-a-vis food and nutritional security of India. *Indian Journal of Fertilisers* 10, 94-112.

Singh, M. V. (2008) Micronutrient deficiencies in Indian soils and crops. In *Micronutrient Deficiencies in Global Crop Production* (B. J. Alloway, Ed.), pp. 93-125. Springer, Dordrecht.

Singh, M. V. (2009) Micronutrient nutritional problems in soils of India and Improvement for human and animal health. *Indian Journal of Fertilisers* 5,11-16.

Singh, S. K., and Patra, A. (2020) Status of soil available nickel (Ni) in different districts of eastern Uttar Pradesh, India. *Indian Journal of Agricultural Sciences* 90, 2209-2216.

Singh, S. K., Dey, P., Sharma, P. K., Singh, Y. V., Latare, A. M., Singh, C. M., Kumar, D., Kumar, O., Yadav, S. N., and Varma, S. S. (2016) Primary and cationic micronutrient status of soils in few districts of eastern Uttar Pradesh. *Journal of the Indian Society of Soil Science* 64, 319-332.

Singh, S. K., Dey, P., Singh, S., Sharma, P. K., Singh, Y. V., Latare, A. M., Singh, C. M., Kumar, D., Kumar, O., Yadav, S. N., and Verma, S. S. (2015) Emergence of boron and sulphur deficiency in soils of Chandauli, Mirzapur, Sant Ravidas Nagar and Varanasi districts of Eastern Uttar Pradesh. *Journal of the Indian Society of Soil Science* 63, 200-208.

Soltanpour, P. N. and Schwab, A. P. (1977) A new soil test for simultaneous extraction of macroand micronutrients in alkaline soils. *Communications in Soil Science and Plant Analysis* 8,195–207.

Stout, P. R. (1962) Micronutrient Research, Introduction to Micronutrient Elements. *Journal of Agricultural and Food Chemistry* 10, 170-170.

Tuhy, Ł., Samoraj, M., Michalak, I., and Chojnacka, K. (2014) The application of biosorption for production of micronutrient fertilizers based on waste biomass. *Applied biochemistry and biotechnology 174*, 1376-1392.

Visioli, G., D'Egidio, S., and Sanangelantoni, A. M. (2015) The bacterial rhizobiome of hyperaccumulators: future perspectives based on omics analysis and advanced microscopy. *Frontiers in plant science* 5, 752.

Winkel, L. H., Johnson, C. A., Lenz, M., Grundl, T., Leupin, O. X., Amini, M., and Charlet, L. (2012) Environmental selenium research: from microscopic processes to global understanding.

Wolf, B. (1982) An improved universal extracting solution and its use for diagnosing soil fertility. *Communications in Soil Science and Plant Analysis* 13, 1005-1033.

Ye, X., Al-Babili, S., Klöti, A., Zhang, J., Lucca, P., Beyer, P., and Potrykus, I. (2000) Engineering the provitamin A (β-carotene) biosynthetic pathway into (carotenoid-free) rice endosperm. *Science 287*, 303-305.

About the Editors

Dr. Hanuman Singh Jatav is an Assistant Professor in the Department of Soil Science and Agricultural Chemistry, Sri Karan Narendra Agriculture University, Jobner Jaipur (Rajasthan), India. He has more than 4 years of Teaching and 7 years of Research experience in the field of Sewage Sludge, Biochar, Micronutrients, Heavy Metals, Soil Fertility Management including MSc (Ag.) and PhD. He has completed B.Sc. (Hons.) Agriculture from Maharana Pratap University of Agriculture and Technology, Udaipur-Rajasthan, India. Dr. Jatav has completed his M.Sc. (Ag.) and PhD in Soil Science and Agricultural Chemistry from the Institute of Agricultural Sciences, Banaras Hindu University, Varanasi-U.P. India. At present, Dr. Jatav is an emerging person in the field of research transforming himself to an excellent scientific level by enhancing research capacity by promoting environmental sustainability. He has published more than 90 paper(s) including Research papers, Review papers, Books, Book chapters and Technical papers from various National and International reputed Journals and Publishers. He is serving as an Academic Editor and member of Subject expertise in Springer, Taylor and Francis, Elsevier, Frontiers, CARJ, JEB, ICAR Journals and Science Domain for the last five years. He is serving as an active member of more than 12 various national and international scientific societies. Dr. Jatav has been awarded a Certificate of Gratitude as an Ambassador by IUCN: International Union for Conservation of Nature, Gland, Switzerland (https://www.iucn.org/) in the World Conservation Congress during Global Youth Summit. He has also been awarded the best paper and poster presentation for his work. He has also attended more than 30 conferences; besides this, he has also completed more than 25 training programs pertaining to his research. He has been awarded as Outstanding Assistant Professor of the Sri Karan Narendra Agriculture University, Jobner Jaipur (Rajasthan) India. He may be contacted at hanumaniasbhu@gmail.com; hsjatav.soils@sknau.ac.in

Dr. Vishnu D Rajput earned degrees with good academic records in Agriculture and Biological Sciences: BSc Ag, subsequently MSc Ag from CS Azad University of Agriculture & Technology, India, PhD from Chinese Academy of Sciences, China, and Post-doc from Southern Federal University, Russia. Presently, he is working as a Leading Researcher (Assoc. Prof.) at Southern Federal University, Russia. His ongoing research is based on soil contaminations, i.e., PAHs, Benzo[a]pyrene, potentially toxic elements and metallic nanoparticles. Dr Rajput's group is one of the first who is investigating the bioaccumulation, bio/geo-transformations, uptake, translocation and toxic effects of metallic nanoparticles on plant physiology, morphology, anatomy, ultrastructure of cellular and subcellular organelles, cytomorphomatric modifications and DNA damage. He comprehensively detailed the state of research in environmental science in regard to how contaminants interact with plants, soil, microbial community and the larger environment. He has published (total 135 scientific publications in English) 67 peer-reviewed highly rated full-length articles, 5 books, 20 chapters (Scopus indexed), conference articles with h-Index 13, i10-index 19, and 620 citations (Google Scholar). He has also published several popular Hindi articles in popular Indian magazines, and his research was published by Russian and Indian newspapers as well. He is an internationally recognized reviewer and received an outstanding reviewing certificate by Elsevier, serving as a Guest Editor of Plants Journal. He received "Certificate for Appreciation 2019" "Certificate of Honor 2020" by Southern Federal University, Russia for outstanding contribution in academic, creative research and publication activities. He also received prestigious status "Highly Qualified Specialist" by SFedU and Russian Ministry of Internal Affairs. He attended more than 25 international conferences organized by university/institutes and delivered several presentations in plenary sections. He has more than 13 years of research experience in agriculture and allied sectors. In India, he worked at various Indian Council of Agricultural Research (ICAR) institutions, including IIPR Kanpur, CPB-HAU Hissar, ICAR-CSSRI Karnal.

Index

A

B

C

D

N

O

P

R

S

T

V

W